MODERN ANTENNAS

Modern Antennas

2nd edition

by

S. DRABOWITCH

Ecole Supérieure d'Electricité,
France

A. PAPIERNIK

University of Nice-Sophia Antipolis,
France

H.D. GRIFFITHS

University College London, U.K.

J. ENCINAS[†]

Institut Supérieur d'Electronique de Paris,
France

and

B.L. SMITH

Alcatel, Paris, France

 Springer

A C.I.P. Catalogue record for this book is available from the Library of Congress.

ISBN-10 1-4020-3216-1 (HB)
ISBN-13 978-1-4020-3216-5 (HB)
ISBN-10 0-387-26231-8 (e-book)
ISBN-13 978-0-387-26231-4 (e-book)

Published by Springer,
P.O. Box 17, 3300 AA Dordrecht, The Netherlands.

www.springeronline.com

Printed on acid-free paper

Contents

List of contributors

Serge Drabowitch
Docteur Ingénieur. Former Professor at the Ecole Supérieure d'Electricité. Scientific Advisor Thales Group; Senior Member of the IEEE and the French SEE.

Albert Papiernik
Professor at the University of Nice-Sophia Antipolis, France.

Hugh Griffiths
Professor, and Head of Department of Electronic and Electrical Engineering, University College London; Fellow of the IEE, Fellow of the IEEE and Fellow of the Royal Academy of Engineering.

Jean Encinas
(now deceased)
Institut Supérieure d'Electronique de Paris, France.

Bradford L. Smith
Senior Intellectual Patent Attorney and Engineer, Corporate Research Centre, Alcatel in Paris, France; Senior Member of the IEEE and the French SEE.

Foreword

Why does the world need yet another book on antennas ?

For the reason which made the success of the first edition of 'Modern Antennas' (published in 1997): it has served well both graduate-level students (with exercises at the end of each chapter), and practicing engineers by combining a solid theoretical treatment with a practical development allowing the serious reader to undertake antenna design from first principles.

In this second edition, coverage of the signal processing aspects, which will be key for the antenna systems of tomorrow, is amplified and the balance between theory and practical design techniques is maintained. This reflects, as well as the European character of the book, the origins and experience of its eminent contributors from both Academia and Industry:

Albert Papiernik, as Professor at the University of Nice - Sophia Antipolis, France, has led one of the most prestigious printed antenna laboratories in Europe and made major theoretical and practical contributions in this field.

Hugh Griffiths, Professor, and Head of the Department of Electronic and Electrical Engineering at University College London, has been a pioneer in the field of signal processing antennas and radar systems.

Jean Encinas (now deceased) was a specialist of adaptive antennas and circuits at the Institut Supérieur d'Electronique de Paris.

Serge Drabowitch, presently Engineering Advisor for the Thales Group, former Head of the Surface Radar Antenna Group of Thomson CSF and Professor at the Ecole Supérieure d'Electricité, has made major contributions to radar, in particular in the application of signal theory to antennas.

Bradford L. Smith, Senior Patent Attorney for the Alcatel Group, has made fundamental contributions and authored books in the fields of antennas and microwave components.

This new edition maintains the same principle and takes into account the progress in the design of classical antennas and the recent developments of antennas for personal and mobile communications as well as the evolution towards active, integrated and signal processing antennas.

Throughout this book, the presentation of the material is simple and understandable and the reader is guided progressively through the stages of the design process. Many examples of practical designs applied to real engineering situations are presented and discussed.

After a very readable introduction on the history of electromagnetism and antennas, the first four chapters cover fundamentals in electromagnetism, radiation and antennas in the transmit and receive modes.

Simple antennas and printed antennas are presented in the two subsequent chapters.

The next chapters cover in details large aperture antennas, primary feeds, reflectors and direct radiating arrays.

Chapters 12, 13 and 14, which cover in depth polarimetry, antennas and signal theory, and signal processing antennas, constitute distinctive features of this book.

Finally the last chapter covers the essentials of antenna measurements.

This new edition features a number of major additions and changes to the first edition, including in particular:

> A completely new chapter on electromagnetism and antennas - a historical perspective;
>
> A new section in chapter 4 on antenna RCS;
>
> New material in chapter 6 on low profile, wideband or multiband antennas for mobile communications and short-range applications;
>
> New material in chapter 7 on Fresnel zone plate antennas;
>
> Additions in chapter 11 on circular and spherical arrays;
>
> A new section in chapter 11 on MEMS technology;
>
> A new section in chapter 11 on retrodirective and self-phasing arrays;
>
> New results in chapter 14 on superresolution and the MUSIC algorithm;
>
> Suppression from chapter 14 of somewhat outdated material on analogue adaptive arrays;
>
> A new section in chapter 14 on smart antennas for communications (including MIMOs) and for radar applications;
>
> New material in chapter 15 on measurements of mobile handset antennas;
>
> Many additional references and website addresses;
>
> Several recent pictures of recent antenna systems.

Distinctive features of the book include chapters on polarimetry (chapter 12), on antennas and signal theory (chapter 13), and on signal processing antennas (chapter 14).

The chapter on polarimetry is both complete and very readable: after presenting very clearly all the relevant definitions for fully polarised, partially polarised and unpolarised waves, it addresses the representation of radar targets and their response to different incident polarisations. Application of dual polarisation to communications and the design of polarisers and of polarisation separators are also discussed in this chapter.

The chapter on antennas and signal theory is very unique. By exploiting the analogy between an antenna radiation pattern and its illumination law on the one hand, and a time varying signal and its frequency spectrum on the other, fundamental properties and limitations of aperture antennas can be derived. The antenna can be viewed as a spatial filter. Shannon's sampling theorem applies to radiation patterns of finite antennas and this brings about a better understanding and significant savings in their analysis, their synthesis and in near field and far field measurements techniques.

The chapter on signal processing antennas is very complete, nearly a book in itself. It covers mostly radar applications but also includes a section on MIMOs currently under study for mobile and personal telecommunication systems. It starts with an excellent presentation on synthetic antennas, including those for Synthetic Aperture Radar systems, and continues with imaging of coherent and of incoherent sources. It covers high resolution imaging techniques and various super resolution methods, including the MUSIC algorithm. Finally, it concludes with a study of adaptive antennas which spatially filter out interfering signals while maximizing the wanted signal strength.

With a mix of electromagnetics, integrated solid state technologies and signal processing, the field of antennas is becoming more and more multidisciplinary.

This new edition of *Modern Antennas* reflects this multidisciplinary evolution and comes at the right time both for the student in modern antenna theory and for the antenna system designer of tomorrow.

Antoine Roederer
Head – Electromagnetics and Space Environment, European Space Agency

Acknowledgements

In a work of this kind it is impossible to acknowledge individually all the people who have contributed, either directly or indirectly. However, we would like to mention by name certain colleagues and friends whose knowledge and understanding of the subject has been a particular inspiration: Professor Elie Roubine, Professor Jean-Charles Bolomey, Patrice Brachat and Professor Alex Cullen.

We thank our numerous colleagues and students, past and present, from the Antennas Laboratory of THOMSON-CSF (now THALES), from the Laboratoire d'Electronique Antennes et Télécommunications of the University of Nice-Sophia Antipolis, from CNET - France Télécom, from Roke Manor Research, and from the Department of Electronic and Electrical Engineering at University College London. Many have helped by reading early versions of the manuscript and seeking out errors, or suggesting better ways of expressing ourselves.

Special thanks are due to several people for helping obtain material, photographs and illustrations, and for obtaining permission for their publication, both in this second edition and the first. In particular, Dr Paul Brennan and Erik De Witte of University College London; Dr Wolfram Titze, formerly of University College London; Prof. David Olver, Prof. Peter Clarricoats and Prof. Clive Parini of Queen Mary, University of London; Prof. Vince Fusco of Queen's University Belfast; Dr Abbas Baghai of Aerial Facilities Ltd; Georges Bienvenu of Thales Underwater Systems; David Hall and Jane Ward of EADS-Astrium; Antoine Roederer, John Reddy and Kees van't Klooster of the European Space Agency; Gabriel Rebeiz of the University of Michigan and Dr Jeff DeNatale of the Rockwell Scientic Company; Dr Roni Eiges of RAFAEL; Darren Seymour-Russell formerly of the Defence Research Agency; David Gentle of the National Physical Laboratory; and John Holloway, Prof. David Money and Dr Peter Tait of AMS. We particularly thank Prof. Yiannis Vardaxoglou of Loughborough University, and Schmidt and Partner Engineering AG, for providing material on the measurement of mobile handset antennas.

Finally we acknowledge the contribution of our friend Jean Encinas, who died suddenly while the first edition of this book was being written.

This book is dedicated to our wives and our families, who have put up with the time that we have spent on the project with great forbearance, and to our students.

Electromagnetism and antennas – a historical perspective

Michael Faraday (1791–1867)

INTRODUCTION

The techniques of antennas have their roots in the science of electromagnetic radiation. It is therefore of interest to explain briefly the principal steps in the development of the subject.

The history of electromagnetism crosses many frontiers. It is a human story which bears the mark of individuals who have each made their contributions, with the diversity of their genius. The various portraits which we give in these pages provide an illustration.

It is necessary to go back to the 17th century to find the origins of the subject in the context of optics. It was in effect the measurement of the velocity of light by the Danish astronomer Roemer in 1676 which marked the start of the subject. Roemer had noticed an advance or a delay of 8 minutes in the eclipses of a satellite of Jupiter, according to whether the Earth was in conjunction or in opposition to this planet with respect to the Sun. He deduced from this a good approximation to the velocity of light, for which the value of 3×10^8 m/s is generally adopted.

Later, the Frenchman Augustin Fresnel (1788–1827) developed the wave theory of light, in which the velocity of light c appears as the product of the frequency f with the wavelength λ. Recognizing that he considered light as a vibration in a hypothetical medium, the *ether*, it can be seen that he paved the way for the future discoveries of Maxwell.

A first link between electricity and light was found by two scientists who are relatively little known: the Germans Weber and Kohlrausch. In 1857 they had had the idea of expressing the same physical phenomenon – the intensity of electric current – successively in two systems of units, electric and magnetic. They were surprise to notice the ratio of the values found was equal to the velocity of light. But they were not sure how to interpret the result.

It was left to the Scottish physicist James Clerk Maxwell (1831–1879) to find the key to this enigma. The Englishman Michael Faraday (1791–1867) had had the idea of illustrating the concept of an electric or magnetic *field* by *lines of force*. In his paper *On Faraday's Lines of Force* in 1855, and particularly *On Physical Lines of Force* in 1862, Maxwell likened them to cords or strings in tension, and mutually repelling each other. This remark led him to the idea of the *field* existing in a sort of elastic medium which he identified as the *ether*. Maxwell then determined the equivalent mechanical constants by means of the system of differential equations which bear his name, and he studied the form of the propagation of a disturbance in this medium – and it was the same as the velocity of light !

The mark of genius of Maxwell was his intellectual audacity when he identified light as an electromagnetic oscillation; with this numerical equivalence he established the nature of light. It was a grandiose synthesis of two sciences which had hitherto been separate: optics and electricity.

In spite of its beauty, Maxwell's work was still not complete. It was not till three years later in Denmark that Lorenz gave to his theory a very elegant form by introducing the concept of *retarded potentials*, which better allowed him to express the mechanisms of radiation and propagation. The classical teaching of electromagnetism relies upon these concepts.

In fact Maxwell's formulation was not in the elegant form of the four vector equations which we now use. This was due to the rather eccentric Englishman, Oliver Heaviside (1850–1925), who also made great contributions in the subjects of transmission line theory and operational calculus. Heaviside also postulated (in 1902) the existence of a reflecting layer in the atmosphere that would reflect high-frequency radio waves and allow long-distance propagation.

It is interesting to see how the consequences of Maxwell's equations have given birth to two great physical theories of the twentieth century – namely Relativity and Quantum Theory.

The theories of Maxwell were based on mechanical models of the field propagating in a hypothetical medium, the ether, which constituted 'absolute' space. The interferometer experiments of the American, Michelson in 1881 had shown the impossibility of stating an *absolute* position of the Earth in space. Thus optics and mechanics could no longer explain a uniform translational movement. This is what is known as the Special Theory of Relativity.

In 1905 Albert Einstein (1879–1955) postulated that Relativity is a universal principle, which applies to all natural phenomena and renders the concept of the ether obsolete. He showed that in order to conform to this principle, Maxwell's equations and those of kinematics should abandon classical mechanics and obey the law of composition of velocities defined by the Dutchman Lorentz (1853–1928). This stated that a velocity greater than that of light is impossible, and introduced the concept of relative time. Another important consequence of this was a simplification and a better understanding of the phenomenon of induction in moving bodies.

In 1892 Lorentz had put forward the hypothesis in which Maxwell's equations, established at a *macroscopic* scale, are equally valid at the *microscopic* scale of elementary particles. After some success (the Zeeman effect), this extension encountered a difficulty in explaining the photoelectric effect and the laws of black body radiation. The hypotheses of Max Planck (1856–1947) and of Einstein on the quantum nature of light (photons) resulted in Quantum Theory and, later, in the Wave Mechanical theory due to Louis De Broglie (1892–1987).

We turn now to the history of antennas. The German Heinrich Hertz (1857–1894) was a theoretician, but he is particularly celebrated for having proved experimentally the existence of electromagnetic waves. It was he who constructed, in 1887, the first antennas. Hertz's spark gap was a dipole of about 1 metre in length at the terminals of which it was necessary to generate a spark by means of a Ruhmkorff induction coil. The dipole radiated a decaying sinusoidal wave of around a metre in wavelength. This was detected by means of a resonator: a circular loop of size according to the radiated frequency. A spark appeared at the terminals of the resonator up to a distance of a few tens of metres.

With this rudimentary equipment, Hertz was able to reproduce the fundamental phenomena of optics: reflection, refraction, diffraction and polarization. These results encouraged other researchers, and numerous laboratories succeeded in improving the results that he had obtained.

The reception of electromagnetic waves improved rapidly after 1890 thanks to the *coherer* developed by the Frenchman Edouard Branly (1844–1940). This was in effect a semiconductor in a circuit comprising a battery and a galvanometer or bell. The high-frequency signal caused the coherer to conduct, so completing the circuit and ringing the bell !

The year 1895 saw a significant development in Russia with the antenna of Popov (1859–1906). This was a long metallic wire connected to a coherer. On 24 March 1896 he succeeded in transmitting a telegraphic signal over a distance of 250 m. At the same time in Italy, Guglielmo Marconi (1874–1937) had been conducting similar experiments. He developed them further in England, and succeeded in 1901 in the first transatlantic radio communication, from Cornwall to Newfoundland. The antenna used by Marconi in 1895 was the prototype of the classical antenna: a vertical quarter-wavelength metallic wire.

In London, John Ambrose Fleming, who had designed and built some of the equipment used for Marconi's experiments, was searching for an improved detector, and in 1904 invented and patented the diode valve.

From these beginnings at the start of the twentieth century one can say that the techniques of antennas left the purely scientific domain and became a subject in their own right. Their history developed along three principal lines:

- the great increase in the number of practical applications;

- the progressive development of a certain empirical *savoir-faire*;

- the development of theoretical models and new methods of calculation.

Among the first applications was maritime communications. It is known, for example, that in 1912 the *S.S. Titanic* was warned by radio telegraphy of the presence of the icebergs which caused her sinking, and by the same means her wireless operator was able to summon help. After the First World War the number of applications increased greatly, especially for broadcast applications, as well as for intercontinental communications. These links at first used very long wavelengths to take advantage of diffraction around the Earth, then on the other hand frequencies in the short wave (MHz) band once the phenomenon of ionospheric reflection had been discovered.

The first radar was developed by the German, Christian Hülsmeyer, and patented in 1904. The importance of this invention was not immediately recognized, and it was not till the 1930s and 1940s that further advances were made, in several countries simultaneously.

All these applications produced a wealth of new types of antennas. They led progressively to a practical understanding which developed at first independently from the underlying theory. They created the fundamental concepts which now form a classical part of the subject: *input impedance, radiation resistance, gain, directivity, sidelobes, bandwidth, polarization*, and so on. A body of knowledge and set of analytical techniques therefore became established around the middle of the twentieth century.

At this time antennas were usually masts or towers or systems of metallic wires. At a theoretical level the problem of an antenna was to determine the distribution of current on its surface, and hence to determine the input impedance, as well as the distribution in space of the radiated field, and its directional properties. At a rigorous level this problem, notably that of the currents, leads to a complex integral equation. The methods for solving these require great knowledge and skill, and it was not until the beginning of the Second World War that satisfactory results were obtained for the case of the dipole (the theories due to Hallén, Schelkunoff,etc.).

From the Second World War to the present, the evolution has been marked by the great diversity of applications: satellite communications, radionavigation, radar, radioastronomy, and by the development of microwave antennas, which form the principal subject of this book. The tremendous progress in methods of numerical analysis and in the speed of computers will allow us to respond to this evolution by putting in place new methods of calculation and new mathematical models.

It is thus that in order to study the radiation from a microwave antenna, the classical methods mentioned above, starting from the current distribution on the surface of the antenna is often not the most appropriate, since the antenna is often of complex form.

It is then preferable to consider a small closed surface of simple form enclosing the antenna. In most cases the useful part of this surface reduces to a plane (§3.2.2). It is often possible to determine approximately the field on this surface by theoretical or experimental means (near field measurement §15.3.4). The radiated field is deduced from this directly either by the Kirchhoff integral (§2.2), or by spectral analysis (§2.1). These two methods are also related via the Fourier Transform (§2.2.3).

We finish this introduction by emphasizing one of the most attractive and appealing aspects of the work of an engineer: *invention* and *innovation*. When looking at a modern antenna such as a *corrugated horn* (§8) or a complex *signal processing antenna array* (§14), it can be understood that for such innovations, practice and theory are not by themselves sufficient; it also needs these mysterious specifically-human qualities – *intuition* and *creativity*.

FURTHER READING

1. Bussey, G., *Marconi's Atlantic Leap*, Marconi Communications, 2000.
2. Hamilton, J., *Faraday – The Life*, Harper Collins, 2002.
3. Latham, C. and Stobbs, A., *Pioneers of Radar*, Sutton Publishing, 1999.
4. Nahin, P., *Oliver Heaviside - Sage in Solitude*, IEEE Press, 1988; republished by Johns Hopkins University, 2001.
5. O'Hara, J.G. and Pricha, W., *Hertz and the Maxwellians*, Peter Peregrinus, Stevenage, 1987.
6. Sarkar, T.P. (ed), *A History of Wireless*, Wiley, 2005.
7. Swords, S.S., *Technical History of the Beginnings of Radar*, Peter Peregrinus, Stevenage, 1986.
8. Tolstoy, I., *James Clerk Maxwell*, Canongate Publishing, Edinburgh, 1981.

1

Fundamentals of electromagnetism

James Clerk Maxwell (1831−1879)

1.1 MAXWELL'S EQUATIONS[1]

1.1.1 Maxwell's equations in an arbitrary medium

Electromagnetic fields and sources

Electromagnetism is the science that describes the macroscopic interactions between electric charges, which may be either stationary or moving. This description is

---

[1] Maxwell, J.C., *A Treatise on Electricity and Magnetism*, Clarendon Press, Oxford, 1873 (Dover, New York, 1954).

carried out by means of four vector quantities which make up the electromagnetic field:

- the electric field **E**
- the electric displacement **D**
- the magnetic field **H**
- the magnetic flux density **B**

This field is created by stationary charges or by moving charges, known as field sources and described by

- the current density **J**
- the charge density ρ

related by the charge-conservation equation

$$\text{div } \mathbf{J} + \frac{\partial \rho}{\partial t} = 0$$

In the time domain the electromagnetic field and the sources are functions of time and position of the observation point M, such that $\mathbf{r} = \mathbf{OM}$. They are related by Maxwell's equations

$$\text{curl } \mathbf{E} = -\frac{\partial \mathbf{B}}{\partial t} \qquad\qquad \text{curl } \mathbf{H} = \mathbf{J} + \frac{\partial \mathbf{D}}{\partial t}$$

$$\text{div } \mathbf{D} = \rho \qquad\qquad\qquad \text{div } \mathbf{B} = 0$$

Complex notation

We restrict ourselves to a purely sinusoidal variation with time of these quantities (frequency f, angular frequency $\omega = 2\pi f$) and we use complex notation. Each physical quantity is represented by its complex envelope, or phasor.

Physical quantity: $\qquad\qquad A(r,t) = A_0(r)\cos\left[\omega t + \varphi(r)\right]$

Associated complex envelope: $A(r) = A_0(r)\exp\left[j\varphi(r)\right]$

From a knowledge of the complex quantity A(r), we can obtain the physical quantity $A(r,t) = \text{Re}\left[A(r)\exp(j\omega t)\right]$.

The complex notation transforms derivatives with respect to time into a multiplication by $j\omega$. The complex notation is applied to ρ, and to the components of **J**, **E** and **H**.

$$\rho(r,t) = \mathrm{Re}\left[\rho(r)\exp(j\omega t)\right] \qquad \mathbf{J}(r,t) = \mathrm{Re}\left[\mathbf{J}(r)\exp(j\omega t)\right]$$

$$\mathbf{E}(r,t) = \mathrm{Re}\left[\mathbf{E}(r)\exp(j\omega t)\right] \qquad \mathbf{H}(r,t) = \mathrm{Re}\left[\mathbf{H}(r)\exp(j\omega t)\right]$$

This allows us to simplify the charge-conservation equation

$$\mathrm{div}\ \mathbf{J} + j\omega\rho = 0 \tag{1.1}$$

and to write Maxwell's equations in the form

$$\mathrm{curl}\ \mathbf{E} = -j\omega\ \mathbf{B} \tag{1.2a}$$

$$\mathrm{curl}\ \mathbf{H} = \mathbf{J} + j\omega\ \mathbf{D} \tag{1.2b}$$

$$\mathrm{div}\ \mathbf{D} = \rho \tag{1.2c}$$

$$\mathrm{div}\ \mathbf{B} = 0 \tag{1.2d}$$

Continuity conditions

Let us consider a surface separating two media (denoted 1 and 2) and a point M on which there exist a surface current density $\mathbf{J}_s$ and a surface charge density ρ_s. Let us compare the various components of the electromagnetic field at points very close to M, but located on either side of the boundary. Some components are continuous, while others are not. The continuity equations, deduced from Maxwell's equations, relate the discontinuities to the surface charge and current densities

$$\mathbf{E}_{t2} - \mathbf{E}_{t1} = 0 \qquad\qquad \mathbf{H}_{t2} - \mathbf{H}_{t1} = \mathbf{J}_s \times \mathbf{n}$$

$$\mathbf{D}_{n2} - \mathbf{D}_{n1} = \rho_s \qquad\qquad \mathbf{B}_{n2} - \mathbf{B}_{n1} = 0 \tag{1.3}$$

n is the unit vector normal to the boundary, directed from region 1 towards region 2. The subscripts t and n denote respectively the tangential and the normal components.

Electromagnetic potentials

Maxwell's equations (1.2d) and (1.2a) allow us to introduce a vector potential **A** and a scalar potential Φ, related to the electromagnetic field by the following relations

$$\mathbf{B} = \text{curl } \mathbf{A} \qquad (1.4a)$$

$$\mathbf{E} = -\text{grad } \Phi - j\omega\mathbf{A} \qquad (1.4b)$$

These relations do not completely determine the potentials from the fields. We may impose an additional condition (see §1.1.2 - Lorentz's condition - equation (1.10)) to determine them completely.

1.1.2 Linear media

In numerous types of medium, linear relationships between fields (or between fields and sources) allow a simplification of Maxwell's equations. We shall examine linear dielectric media, linear magnetic media, and conducting media.

Linear dielectric media

These are media for which a linear relationship exists between **D** and **E**

$$\mathbf{D} = \varepsilon\mathbf{E} \qquad (1.5)$$

ε is the dielectric constant or absolute permittivity of the medium. In the case of a vacuum (and in practice, in air), this permittivity is denoted ε_0. The relative permittivity of a medium is defined by $\varepsilon_r = \varepsilon/\varepsilon_0$

Let us limit ourselves to the case of an isotropic medium for which ε is a scalar, thus excluding linear anisotropic dielectrics for which ε is a tensor. The permittivity is real for *perfect* linear dielectrics in which there is no heat loss by dielectric hysteresis. It is complex for *imperfect* linear dielectrics in which there is heat loss by dielectric hysteresis. For such dielectrics, **D** shows a phase delay δ with respect to **E**, termed the dielectric loss angle. Permittivity can be expressed in the following form (with ε', ε'' and δ being positive)

$$\varepsilon = \varepsilon' - j\varepsilon'' = |\varepsilon|\exp(-j\delta) \qquad (1.6)$$

The quantity $\tan\delta = \varepsilon''/\varepsilon'$ is known as the *loss tangent*. For good dielectrics it is of the order of 10^{-3} radians at microwave frequencies.

Linear magnetic media

Linear magnetic media are those for which there is a linear relation between **B** and **H**

$$\mathbf{B} = \mu \mathbf{H} \qquad (1.7)$$

μ is the absolute permeability of the medium, equal to μ_0 in the case of free space (and in practice, in air). The relative permeability of the medium is defined by $\mu_r = \mu/\mu_0$. We only consider the case of an isotropic medium for which μ is a scalar, thus excluding linear anisotropic magnetic media for which μ is a tensor. The magnetic permeability μ is real for *perfect* linear magnetic media which do not present heat losses by magnetic hysteresis. It is complex for *imperfect* linear magnetic media which do present heat losses by magnetic hysteresis. For such media, **B** shows a phase delay with respect to **H** and the permeability can be expressed in the following form (with μ' and μ'' being positive)

$$\mu = \mu' - j\mu''$$

Except in media with specific magnetic properties (ferrites for example), the permeability of media used at microwave frequencies is very close to $\mu_0 = 4\pi \times 10^{-7}$ H/m .

Maxwell's equations in a linear medium

In a medium with both dielectric and magnetic properties, Maxwell's equations become

$$\text{curl } \mathbf{E} = -j\omega\mu \, \mathbf{H} \qquad (1.8a)$$

$$\text{curl } \mathbf{H} = \mathbf{J} + j\omega\varepsilon \, \mathbf{E} \qquad (1.8b)$$

$$\text{div } \mathbf{E} = \frac{\rho}{\varepsilon} \qquad (1.8c)$$

$$\text{div } \mathbf{H} = 0 \qquad (1.8d)$$

Propagation equations (see exercise 1.1)

In a linear medium (ε,μ) free of charges and currents, elimination of **H** (or **E**) in Maxwell's equations yields the following equations, in which $k^2 = \omega^2 \varepsilon\mu$

$$\nabla^2 \mathbf{E} + k^2 \mathbf{E} = 0 \qquad (1.9a)$$

$$\nabla^2 \mathbf{H} + k^2 \mathbf{H} = 0 \qquad (1.9b)$$

These equations are called the Helmholtz equations. If sources are present, charges and currents appear in the second term, in a relatively complicated form. It is then more convenient to introduce the potentials **A** and Φ and to impose the Lorentz condition

$$\text{div } \mathbf{A} + j\omega\varepsilon\mu\Phi = 0 \tag{1.10}$$

We obtain under these conditions the propagation equations in terms of potentials, in which the second terms are relatively simple

$$\nabla^2 \mathbf{A} + k^2 \mathbf{A} = -\mu \mathbf{J} \tag{1.11a}$$

$$\nabla^2 \Phi + k^2 \Phi = -\frac{\rho}{\varepsilon} \tag{1.11b}$$

Taking into account the Lorentz condition, the **E** and **H** fields can be expressed in terms of the vector potential **A** alone (the scalar potential is not necessary)

$$\mathbf{E} = -\frac{j}{\varepsilon\mu\omega}\text{grad}\left(\text{div }\mathbf{A}\right) - j\omega\mathbf{A} \tag{1.12a}$$

$$\mathbf{H} = \frac{1}{\mu}\text{curl }\mathbf{A} \tag{1.12b}$$

1.1.3 Conducting media

In a conducting medium the electric field **E** creates a current density **J** related to **E** by Ohm's law

$$\mathbf{J} = \sigma\mathbf{E} \tag{1.13}$$

The conductivity σ is purely real at microwave frequencies. It is very large (some 10^7 S m^{-1} for metals such as copper, silver or gold, which are considered as excellent conductors.

Perfect conductor

A perfect conductor has by definition an infinite conductivity σ. Taking into account the finite nature of the current density, this implies that the electric field **E** should be zero in a perfect conductor. According to Maxwell's equations (1.8), **H**, **J** and ρ must be zero too. There are no currents or charges inside a perfect conductor, only on its surface. The conditions of continuity allow us then to determine the **E** and **H** fields in the vicinity of a point M on the surface of a perfect conductor, in a dielectric of permittivity ε, in terms of current surface densities $\mathbf{J}_s$ and charges ρ_s in the vicinity of the point M (Fig. 1.1).

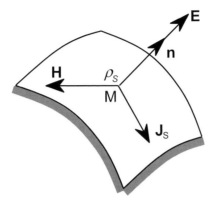

Fig. 1.1 Electromagnetic field in the vicinity of a perfect conductor.

Thus we obtain the following properties for a perfect conductor:

- the electromagnetic field is zero inside the conductor.

- there are neither currents nor volume charges.

- the electric field is normal to the surface. It is related to the surface charge density by

$$\mathbf{E} = \frac{\rho_s}{\varepsilon} \mathbf{n} \qquad (1.14)$$

- the magnetic field is tangential to the surface. It is related to the surface current by

$$\mathbf{H} = \mathbf{J}_S \times \mathbf{n} \qquad (1.15)$$

Conducting and linear dielectric media

In a medium characterized by a conductivity σ and a complex permittivity $\varepsilon = \varepsilon' - j\varepsilon''$, the second of Maxwell's equations takes the form

$$\operatorname{curl} \mathbf{H} = (\sigma + j\omega\varepsilon)\mathbf{E} = \left[(\sigma + \omega\varepsilon'') + j\omega\varepsilon'\right]\mathbf{E}$$

from which it appears that a fictitious conductivity $\omega\varepsilon''$, related to the dielectric hysteresis, is added to the ohmic conductivity σ. This relation shows that div $\mathbf{E} = 0$ and therefore that the volume charge density ρ is zero in the conductor. In other words negative charges are at each point exactly compensated by positive charges.

In a 'good' conductor, defined by the inequality $\sigma \gg \omega|\varepsilon|$, the second of Maxwell's equations takes the form

$$\text{curl } \mathbf{H} = \sigma \mathbf{E} \tag{1.16}$$

1.1.4 Reciprocity theorem

This theorem, which establishes a relation between two solutions of Maxwell's equations, will be used in Chapter 4 to relate the properties of a particular antenna on transmission and on reception.

Consider two distinct solutions of Maxwell's equations. The first $(\mathbf{E}_1, \mathbf{H}_1)$ is due to the current $\mathbf{J}_1$ and the second $(\mathbf{E}_2, \mathbf{H}_2)$ is due to the current $\mathbf{J}_2$. Using Maxwell's equations (1.8a, 1.8b) and the identity div $(\mathbf{a} \times \mathbf{b}) = \mathbf{b}.\,\text{curl } \mathbf{a} - \mathbf{a}.\,\text{curl } \mathbf{b}$ we obtain

$$\text{div} \left(\mathbf{E}_1 \times \mathbf{H}_2 - \mathbf{E}_2 \times \mathbf{H}_1 \right) = \mathbf{E}_2.\mathbf{J}_1 - \mathbf{E}_1.\mathbf{J}_2$$

Integrating over a volume V enclosed by the surface S (Fig. 1.2)

$$\iint_S \left(\mathbf{E}_1 \times \mathbf{H}_2 - \mathbf{E}_2 \times \mathbf{H}_1 \right).\mathbf{n}\,dS = \iiint_V \left(\mathbf{E}_2.\mathbf{J}_1 - \mathbf{E}_1.\mathbf{J}_2 \right) dV \tag{1.17}$$

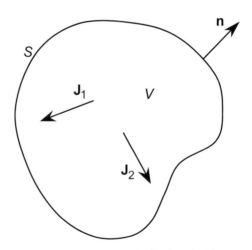

Fig. 1.2 Reciprocity theorem.

We shall demonstrate that the two terms of these two relations are zero when $\mathbf{J}_1$ and $\mathbf{J}_2$, assumed to be zero at infinity, are enclosed by the surface S. In effect, it is easy to verify that the first term is zero when a surface impedance relation of the type $\mathbf{E}_{\text{tangential}} = -Z\mathbf{n} \times \mathbf{H}$ is imposed between $\mathbf{E}$ and $\mathbf{H}$ on the surface S. This is the case[2] if S is a sphere S_∞ of infinite radius since the electromagnetic field at infinity has the character of a plane wave (see §1.3.1), with

$$Z = \eta = \sqrt{\frac{\mu}{\varepsilon}}$$

The second term of equation (1.17) is therefore zero when S is a sphere of infinite radius. It then follows that for any surface S enclosing the sources, we have

$$\iint_S \left(\mathbf{E}_1 \times \mathbf{H}_2 - \mathbf{E}_2 \times \mathbf{H}_1 \right).\mathbf{n}\,dS = \iiint_V \left(\mathbf{E}_2.\mathbf{J}_1 - \mathbf{E}_1.\mathbf{J}_2 \right) dV = 0 \qquad (1.18)$$

This property constitutes the reciprocity theorem, which may be written as follows:

$$\iiint_V \mathbf{E}_1.\mathbf{J}_2\,dV = \iiint_V \mathbf{E}_2.\mathbf{J}_1\,dV \qquad (1.19)$$

1.2 POWER AND ENERGY

1.2.1 Power volume densities

Let us consider an element of volume dV, around a point M. At this point the current density is $\mathbf{J}$ and the electric field is $\mathbf{E}$. The mean power transferred by the field to the charges in the volume dV is denoted dP. The work of the electromagnetic force applied to the mobile charges in the volume dV leads to the expression for the mean power density transferred by the field to the current.

$$\frac{dP}{dV} = \frac{1}{2}\text{Re}(\mathbf{E}.\mathbf{J}^*) \qquad (1.20)$$

In the case of a conductor, the power transferred by the current is supplied to the medium in the form of heat. This formula, associated with Ohm's law $\mathbf{J} = \sigma \mathbf{E}$,

---

[2] This also the case if S encloses a perfect conductor ($Z = 0$) or a good conductor (see §1.3.2 - skin effect).

allows us to calculate the mean power dP_J dissipated by the Joule effect in an element of volume dV

$$\frac{dP_J}{dV} = \frac{1}{2}\sigma\mathbf{E}.\mathbf{E}* \tag{1.21}$$

In the case of an imperfect dielectric, we can obtain the mean power dP_D dissipated by dielectric hysteresis in the volume dV by replacing the ohmic conductivity σ by the fictitious conductivity $\omega\varepsilon''$

$$\frac{dP_D}{dV} = \frac{1}{2}\omega\varepsilon''\mathbf{E}.\mathbf{E}* \tag{1.22}$$

The mean power dP_M dissipated by magnetic hysteresis is given by an analogous formula

$$\frac{dP_M}{dV} = \frac{1}{2}\omega\mu''\mathbf{H}.\mathbf{H}* \tag{1.23}$$

1.2.2 Energy volume densities

An element of volume dV contains the mean electric energy dW_E and the mean magnetic energy dW_M. Let us introduce

- the mean density of electric energy: $u_E = \dfrac{dW_E}{dV}$,

- the mean density of magnetic energy: $u_M = \dfrac{dW_M}{dV}$,

- the mean density of electromagnetic energy: $u = u_E + u_M$.

In a perfect linear medium (ε and μ real), these densities are given by the following expressions:

$$u_E = \frac{1}{4}\varepsilon\mathbf{E}.\mathbf{E}*, \quad u_M = \frac{1}{4}\mu\mathbf{H}.\mathbf{H}*, \quad u = \frac{1}{4}\varepsilon\mathbf{E}.\mathbf{E}* + \frac{1}{4}\mu\mathbf{H}.\mathbf{H}* \tag{1.24}$$

In an imperfect linear medium (ε and μ complex), the density expressions are more complicated

$$u_E = \frac{1}{4}\frac{\partial(\omega\varepsilon')}{\partial\omega}\mathbf{E}.\mathbf{E}* \qquad u_M = \frac{1}{4}\frac{\partial(\omega\mu')}{\partial\omega}\mathbf{H}.\mathbf{H}*$$

We shall consider the case where ε' and μ' are in practice independent of frequency.

We find again the expressions (1.22) in which we should introduce the real parts ε' and μ' of the permittivity and of the permeability.

$$u_E = \frac{1}{4}\varepsilon'\mathbf{E}.\mathbf{E}*, \quad u_M = \frac{1}{4}\mu'\mathbf{H}.\mathbf{H}*, \quad u = \frac{1}{4}\varepsilon'\mathbf{E}.\mathbf{E}* + \frac{1}{4}\mu'\mathbf{H}.\mathbf{H}* \qquad (1.25)$$

1.2.3 Poynting vector and power

The Poynting vector, defined by $\frac{1}{2}\mathbf{E}\times\mathbf{H}*$, allows us to calculate the power transported by the electromagnetic field. From the first two of Maxwell's equations, we can find the Poynting identity

$$-\frac{1}{2}\operatorname{div}\mathbf{E}\times\mathbf{H}* = \frac{1}{2}\mathbf{E}.\mathbf{J}* + \frac{1}{2}\omega\varepsilon''\mathbf{E}.\mathbf{E}* + \frac{1}{2}\omega\mu''\mathbf{H}.\mathbf{H}* + 2j\omega(u_M - u_E)$$

which, by integration over a volume V enclosed by the surface S and whose normal $\mathbf{n}$ is oriented towards the interior, leads to Poynting's theorem

$$-\frac{1}{2}\iint_S (\mathbf{E}\times\mathbf{H}*).\mathbf{n}dS$$

$$= \iiint_V \left(\frac{1}{2}\mathbf{E}.\mathbf{J}* + \frac{1}{2}\omega\varepsilon''\mathbf{E}.\mathbf{E}* + \frac{1}{2}\omega\mu''\mathbf{H}.\mathbf{H}*\right)dV + 2j\omega(W_M - W_E)$$

- The first term $-\frac{1}{2}\iint_S (\mathbf{E}\times\mathbf{H}*).\mathbf{n}dS$ is the flux of the Poynting vector penetrating inside the closed surface S,
- W_M and W_E are the magnetic and the electric energies stored in the volume V.

From equations (1.20) to (1.23)

$$P = \operatorname{Re}\left[\iiint_V \left(\frac{1}{2}\mathbf{E}.\mathbf{J}* + \frac{1}{2}\omega\varepsilon''\mathbf{E}.\mathbf{E}* + \frac{1}{2}\omega\mu''\mathbf{H}.\mathbf{H}*\right)dV\right]$$

is the total power supplied by the electromagnetic field to the volume V. This power appears in the form of an increase of the kinetic energy of the mobile particles or in the form of heat energy dissipated by the Joule heating, dielectric hysteresis or magnetic hysteresis. The power crossing the surface S, in the direction of the normal **n**, is therefore equal to the real part of the Poynting vector flux across this surface

$$P = \frac{1}{2}\text{Re}\left[\iint_S (\mathbf{E} \times \mathbf{H}^*).\mathbf{n} \; dS \right] \tag{1.26}$$

This relation may be interpreted by introducing the transmitted power dP through an element of surface dS to which the normal is **n**

$$\frac{dP}{dS} = \text{Re}\left[\left(\frac{1}{2}\mathbf{E} \times \mathbf{H}^* \right).\mathbf{n} \right] \tag{1.27}$$

1.3 PLANE WAVES IN LINEAR MEDIA

1.3.1 Plane waves in an isotropic linear medium

Definition of a plane wave

Let us introduce a direction **u,** termed the direction of propagation, and planes normal to **u**, denoted wavefronts. A quantity propagates as plane waves when it takes the same value at any point on a planar wavefront. In the case of electromagnetic plane waves, this is the electromagnetic field (**E, H**) which is the same at each point of a wavefront. In particular, the phase is the same at each point of the wavefront of a plane wave. Rigorously speaking it is impossible to realize a plane wave experimentally. The plane wave is an abstract concept, but is useful for the following reasons:
- the equations which govern the propagation of plane waves are very simple;
- any radiated electromagnetic field can be considered as a superposition of plane waves (cf Chapter 2);
- when propagation is not according to plane waves, the equiphase surfaces (for which the phase of the field is the same) are not plane. But in the neighbourhood of a point on an equiphase surface the field most usually has the form of a plane wave. This is particularly so for the spherical waves considered at the end of this paragraph.

The most important properties of plane waves are covered below. Also, more details and complementary properties are provided in Chapter 13, on polarimetry.

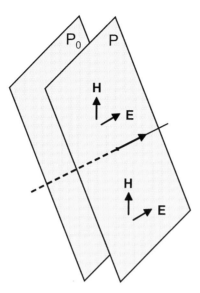

Fig. 1.3 Structure of a plane wave.

Propagation constant, wavenumber, phase velocity, refractive index, attenuation

We shall denote by (α, β, γ) the Cartesian coordinates of a unit vector in the direction of propagation

$$\mathbf{u} = \alpha\,\mathbf{u}_x + \beta\,\mathbf{u}_y + \gamma\,\mathbf{u}_z \quad \text{with} \quad \alpha^2 + \beta^2 + \gamma^2 = 1$$

Each field component at a point M (referred to by $\mathbf{r} = \mathbf{OM} = x\mathbf{u}_x + y\mathbf{u}_y + z\mathbf{u}_z$), contains the factor $\exp(-jk\mathbf{u}.\mathbf{r}) = \exp\left[-jk\left(\alpha\,x + \beta\,y + \gamma\,z\right)\right]$, where k is called the wavenumber.

In a lossless medium, k is purely real. The phase velocity v_ϕ, refractive index n and wavelength λ take the form

$$v_\phi = \frac{\omega}{k} \qquad n = \frac{c}{v_\phi} \qquad \lambda = \frac{2\pi}{k}$$

In a lossy medium, k is complex ($k = k' - jk''$). Each field component contains the factor $\exp(-k''\mathbf{u}.\mathbf{r})\exp(-jk'\mathbf{u}.\mathbf{r})$. We deduce:

- the complex refractive index of the medium $n = n' - jn'' = \dfrac{c}{\omega}(k' - jk'')$;

- the phase velocity $v_\phi = \dfrac{\omega}{k'}$;

- the wavelength $\lambda = \dfrac{2\pi}{k'}$;

- the penetration depth $\varDelta = \dfrac{1}{k''}$.

New composite materials known as *metamaterials*, having a negative permittivity and permeability have recently been conceived[3] and realized[4,5]. They possess unusual properties: negative refractive index, phase and group velocities of opposite sign (backward waves).

Table 1.1 gives, in terms of the characteristics of the medium $(\varepsilon, \mu, \sigma)$, the expressions for k and the quantities related to it: phase velocity v_ϕ, wavelength λ and penetration depth $\varDelta$. As noted in §1.3.2, for good conductors $\varDelta$ is very small and it is known as the *skin depth*.

Structure, wave impedance

The electromagnetic field $(\mathbf{E}, \mathbf{H})$ and the direction $\mathbf{u}$ are related by the equations

$$\mathbf{H} = \frac{1}{Z}\mathbf{u}\times\mathbf{E} \qquad \mathbf{E} = Z\,\mathbf{H}\times\mathbf{u} \tag{1.28}$$

which shows that the fields are transverse and that their relative amplitude depends on the impedance of the medium Z. The expression for the impedance in terms of the characteristics $(\varepsilon, \mu, \sigma)$ of the medium is given in Table 1.1.

---

[3] Veselago, V.G., 'The electrodynamics of substances with simultaneous negative values of ε and μ', *Sov. Phys. Uspekhi.*, Vol.10, No.4, pp509-514, Jan/Feb. 1968.

[4] Ziolkowski, R.W., 'Design, fabrication and testing of double negative metamaterials', *IEEE Trans. Antennas & Propagation*, Vol.51, No.7, pp1516-1529, July 2003.

[5] Grbic, A. and Eleftheriades, G.V., 'Experimental verification of backward-wave radiation from a negative refractive index material', *J. Appl. Phys.*, Vol.92, No.10, 15 Nov. 2002.

Medium	$\varepsilon, \mu, \sigma, n$	$k = n\omega/c, v_\phi, \Delta$	Z
Free space (or air)	ε_0 μ_0 $\sigma = 0$ $n = 1$	$k_0 = \omega\sqrt{\varepsilon_0\mu_0} = \dfrac{\omega}{c}$ $v_\phi = c = \dfrac{1}{\sqrt{\varepsilon_0\mu_0}}$	$\sqrt{\dfrac{\mu_0}{\varepsilon_0}} = 120\pi$
Perfect dielectric	$\varepsilon = \varepsilon_0\varepsilon_r$ positive $\mu = \mu_0\mu_r$ positive $\sigma = 0$ $n = \sqrt{\varepsilon_r\mu_r}$	$k = \omega\sqrt{\varepsilon\mu} = \dfrac{\omega}{v_\phi}$ $v_\phi = \dfrac{1}{\sqrt{\varepsilon\mu}} = \dfrac{c}{n}$	$\sqrt{\dfrac{\mu}{\varepsilon}}$
Imperfect dielectric non conducting	$\varepsilon = \varepsilon_0\varepsilon_r = \varepsilon' - j\varepsilon''$ $= |\varepsilon|\exp(-j\delta)$ $\mu = \mu_0\mu_r$ positive $n = n' + jn'' = \sqrt{\varepsilon_r\mu_r}$	$k' = \omega\sqrt{|\varepsilon|\mu}\cos\delta = \dfrac{2\pi}{\lambda}$ $v_\phi = \dfrac{1}{\cos\delta\sqrt{|\varepsilon|\mu}}$ $k'' = \omega\sqrt{|\varepsilon|\mu}\sin\delta = \dfrac{1}{\Delta}$	$\sqrt{\dfrac{\mu}{|\varepsilon|}}\exp(j\delta/2)$
Good conductor	ε and μ positive $\sigma \gg \omega|\varepsilon|$	$k' = k'' = \dfrac{1}{\Delta} = \dfrac{2\pi}{\lambda}$ $\Delta = \sqrt{\dfrac{2}{\omega\sigma\mu}}$	$\dfrac{1}{\sigma\Delta} = \sqrt{\dfrac{\omega\mu}{2\sigma}}$
Meta-material	$\varepsilon = \varepsilon_0\varepsilon_r$ negative $\mu = \mu_0\mu_r$ negative $\sigma = 0$ $n = -\sqrt{\varepsilon_r\mu_r}$ negative	$k = -\omega\sqrt{\varepsilon\mu} = \dfrac{\omega}{v_\phi}$ $v_\phi = \dfrac{c}{n} = -\dfrac{1}{\sqrt{\varepsilon\mu}}$	$-\sqrt{\dfrac{\mu}{\varepsilon}}$

Table 1.1 Refractive index, wavenumber, phase velocity, wavelength, penetration depth and impedance in different media.

Polarization

Polarization describes the movement of the tip of the electric field vector. In the general case, this polarization is elliptical. It is said to be *linearly polarized* when the electric field remains parallel to a constant direction, known as the polarization direction. If the tip of the electric field traces a circle the wave is said to be *circularly polarized*. In the case of circular or elliptical polarization, the sense of the rotation is defined. This is the *left* sense of rotation (trigonometric) or *right* (clockwise) of the tip of the field, seen by an observer (located at the point O, for instance) who sees the wave move away (Fig. 1.4).

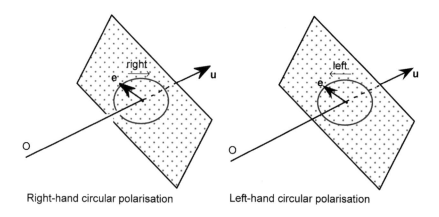

Right-hand circular polarisation   Left-hand circular polarisation

Fig. 1.4 Right-handed and left-handed circular polarization.

Any plane wave $(\mathbf{E}, \mathbf{H})$ may be expressed as a decomposition into two linearly polarized waves $(\mathbf{E}_1, \mathbf{H}_1)$ and $(\mathbf{E}_2, \mathbf{H}_2)$ according to two orthogonal directions $\mathbf{u}_1$ and $\mathbf{u}_2$ of the wavefront (Fig. 1.5). The components of these electric fields are characterized by:

• the amplitude $|E_1|$ and phase ϕ_1 of the electric field component $\mathbf{E}_1$ along $\mathbf{u}_1$;

• the amplitude $|E_2|$ and phase ϕ_2 of the electric field component $\mathbf{E}_2$ along $\mathbf{u}_2$.

The character of the polarization depends on the phase difference $\phi_2 - \phi_1$ between the two waves. It is linear when $\phi_2 - \phi_1 = 0$ and when $\phi_2 - \phi_1 = \pm\pi$. It is circular when $|E_1| = |E_2|$ and $\phi_2 - \phi_1 = \pm\pi/2$. A more complete study of polarization is presented in Chapter 12.

Power transported by a plane wave in a lossless medium

A lossless medium (ε, μ) is characterized by a real impedance $\eta = \sqrt{\mu/\varepsilon}$. The Poynting vector $\frac{1}{2}\mathbf{E} \times \mathbf{H}^*$ allows the calculation of the power dP which crosses a surface element dS on a wavefront (Fig. 1.5)

$$\frac{dP}{dS} = \frac{\mathbf{E}.\mathbf{E}^*}{2\eta} = \frac{\left|E_1\right|^2 + \left|E_2\right|^2}{2\eta} \qquad (1.29)$$

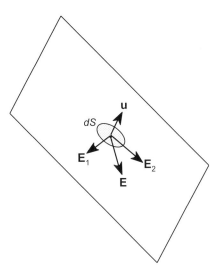

Fig. 1.5 Decomposition of a plane wave into two linearly polarized waves according to two orthogonal directions of the wavefront.

In the case of an arbitrarily polarized wave, where the two linear components have amplitudes $\left|E_1\right|$ and $\left|E_2\right|$, we obtain

$$\frac{dP}{dS} = \frac{\left|E_1\right|^2 + \left|E_2\right|^2}{2\eta} \qquad (1.30)$$

In the case of a linearly polarized wave of amplitude $\left|E_0\right|$, we have

$$\frac{dP}{dS} = \frac{\left|E_0\right|^2}{2\eta} \qquad (1.31)$$

The wave impedance η has the value $\eta_0 = 377\Omega$ in the case of free space and of air.

Spherical waves in a lossless medium

A wave from a point source is said to be 'spherical' (Fig. 1.6). We indicate below some properties of spherical waves at large distance from the source ($r \gg \lambda$).

- Locally the character is that of a plane wave. Thus, equations (1.28) are satisfied, with $Z = \eta$

- The amplitude decreases as $1/r$ and its phase varies as $-kr$. Thus, the factor $\exp(-jkr)/r$ is present in all the field components.

The electromagnetic field is therefore expressed in the form

$$\mathbf{E}(r, \mathbf{u}) = \frac{\exp(-jkr)}{kr}\,\mathbf{e}(\mathbf{u}) \qquad \mathbf{H} = \frac{1}{\eta}\,\mathbf{u} \times \mathbf{E} \qquad (1.32)$$

where $\mathbf{e}(\mathbf{u})$ is a vector which depends solely on the distribution of currents in the source and the direction $\mathbf{u}$ of observation. This expression will be used in section 3.1.1 to describe the radiation at a large distance from an antenna.

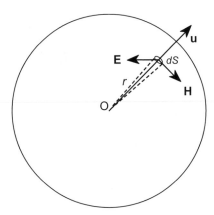

Fig. 1.6 Geometry of a spherical wave.

1.3.2 Skin effect

At microwave frequencies, the depth of penetration is very small (for example, $\Delta \approx 10^{-6}$ m in copper at 1 GHz). The study of the refraction of a plane wave (exercise 1.2) at the interface between a dielectric medium and a good conductor allows us to describe the main property of the skin effect:

- the magnetic field **H** is tangential in the vicinity of a good conductor;

- in addition to the normal component **E**$_n$, the electric field has a tangential component

$$\mathbf{E}_t = Z_S \mathbf{n} \times \mathbf{H} \qquad (1.33)$$

$Z_S = R_S (1 + j)$ is the surface impedance of the conductor,

$R_S = \dfrac{1}{\sigma \Delta}$ is the surface resistance,

$\Delta = \sqrt{\dfrac{2}{\omega \sigma \mu}}$ is the depth of penetration or skin depth,

n is the unit vector normal to the interface (Fig. 1.7), directed into the dielectric.

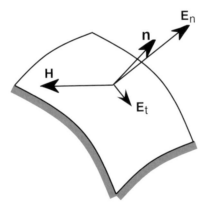

Fig. 1.7 Skin effect. Field in the vicinity of the surface of a good conductor.

- the Joule effect causes a heat dissipation in the immediate vicinity of the dielectric-conductor interface. The heat power dissipated per unit area is

$$\frac{dP}{dS} = \frac{1}{2} R_S \mathbf{H}.\mathbf{H}^* \qquad (1.34)$$

where **H** is the magnetic tangential field at the dielectric-conductor interface.

FURTHER READING

1. Balanis, C.A., *Advanced Engineering Electromagnetics*, Wiley, New York, 1989.
2. Benson, F.A. and Benson, T.M., *Fields, Waves and Transmission Lines*, Chapman & Hall, London, 1991.
3. Clarke, R.H. and Brown, J., *Diffraction Theory and Antennas*, Ellis Horwood, 1980.
4. Collin, R.E., *Foundations for Microwave Engineering*, McGraw-Hill, New York, 1966.
5. Collin, R.E., *Field Theory of Guided Waves*, IEEE Press, 1991.
6. Franklin, J., *Classical Electromagnetism*, Pearson / Addison-Wesley, 2005.
7. Harrington, R.F., *Time-Harmonic Electromagnetic Fields*, McGraw-Hill, New York, 1961.
8. Jordan, E.C. and Balmain, K.G., *Electromagnetic Waves and Radiating Systems*, Prentice-Hall, New Jersey, 1968.
9. Kong, J.A., *Electromagnetic Wave Theory*, Wiley, 1986
10. Kraus, J.D. and Carver, K.R., *Electromagnetics*, McGraw-Hill, New York, 1973.
11. Ramo, S., Whinnery, J.R. and van Duzer, T., *Fields and Waves in Communication Electronics*, Wiley, New York, 1984.
12. Stratton, J.A., *Electromagnetic Theory*, McGraw-Hill, 1941.
13. Ulaby, F., *Fundamentals of Applied Electromagnetics* (media edition), Prentice Hall, 2004.
14. Van Bladel, J., *Electromagnetic Field*, Hemisphere Publishing Corporation, New York, 1985.
15. Wait, J.R., *Introduction to Antennas and Propagation*, Peter Peregrinus, Stevenage, 1966.

EXERCISES

1.1 *Field propagation equations in a linear medium*

Q1 Show that in a linear medium ($\mathbf{D} = \varepsilon\mathbf{E}$, $\mathbf{B} = \mu\mathbf{H}$) free of charges ($\rho = 0$) and currents ($\mathbf{J} = 0$), the following equations, known as Helmholtz equations, hold:

$$\nabla^2\mathbf{E} + \frac{\omega^2}{v^2}\mathbf{E} = 0 \qquad \nabla^2\mathbf{H} + \frac{\omega^2}{v^2}\mathbf{H} = 0$$

where $v = \dfrac{1}{\sqrt{\varepsilon\mu}}$ and $\nabla^2\mathbf{a} = \operatorname{grad}(\operatorname{div}\mathbf{a}) - \operatorname{curl}(\operatorname{curl}\mathbf{a})$

Q2 What is the form of these equations if the medium includes charges ($\rho \neq 0$) and currents ($\mathbf{J} \neq 0$) ?

1.2 *Electromagnetic potentials*

Q1 Starting from Maxwell's equations in the form for sinusoidal signals, show that it is possible to introduce a vector potential $\mathbf{A}$ and a scalar potential Φ related to the fields by:

$$\mathbf{B} = \operatorname{curl}\mathbf{A} \qquad \mathbf{E} = -\operatorname{grad}\Phi - j\omega\mathbf{A} \qquad (1)$$

Show that there are an infinite number of ways of determining these potentials.

Q2 Consider a linear medium ($\mathbf{D} = \varepsilon\mathbf{E}$, $\mathbf{B} = \mu\mathbf{H}$) which includes a current density $\mathbf{J}$. Show that $\mathbf{A}$ satisfies:

$$\nabla^2\mathbf{A} + \omega^2\varepsilon\mu\mathbf{A} = -\mu\mathbf{J} + \operatorname{grad}(\operatorname{div}\mathbf{A} + j\omega\varepsilon\mu\Phi) \qquad (2)$$

Q3 We can make use of the fact that $\mathbf{A}$ and Φ are undetermined by imposing the Lorentz relationship:

$$\operatorname{div}\mathbf{A} + j\omega\varepsilon\mu\Phi = 0$$

What do relations (1) and (2) become ?

1.3 *Propagation in the direction Oz*

The electromagnetic field in a waveguide in direction Oz in a medium characterized by (ε, μ) is given by:

$$\mathbf{E}(x,y,z) = \mathbf{E}(x,y)\exp(-j\beta z) \qquad \mathbf{H}(x,y,z) = \mathbf{H}(x,y)\exp(-j\beta z)$$

Q1   Show that the transversal components $\mathbf{E}_{ot}$ and $\mathbf{H}_{ot}$ can be obtained from the longitudinal components $\mathbf{E}_{oz}$ and $\mathbf{H}_{oz}$ by the relations (where $k^2 = \omega^2 \varepsilon \mu$)

$$\left(k^2 - \beta^2\right)\mathbf{E}_t = j\omega\mu\,\mathbf{u}_z \times \operatorname{grad} H_z - j\beta\operatorname{grad} E_z$$
$$\left(k^2 - \beta^2\right)\mathbf{H}_t = -j\omega\mu\,\mathbf{u}_z \times \operatorname{grad} E_z - j\beta\operatorname{grad} H_z$$

Q2   Show that E_z and H_z obey the Helmholtz equation:

$$\nabla^2 E_z + \left(k^2 - \beta^2\right)E_z = 0 \qquad \nabla^2 H_z + \left(k^2 - \beta^2\right)H_z = 0$$

Q3   What happens when $k = \beta$?

1.4 *Polarization*

We study the evolution of the electric field $\mathbf{E}$ in a plane wave propagating in the direction Oz.

Q1   We wish to determine the locus described by the tip M of the vector $\mathbf{E}$ observed at a point O. Show that this locus is of the second degree (conic).

Q2   In which cases does this locus reduce to a straight line (rectilinear polarization)? What is its direction with respect to Ox ?

Q3   In which case is this locus a circle (circular polarization) ? What is its radius and sense (for an observer in the region $z < 0$ looking in the plane xOy) ?

Q4   Show that in the general case M describes an ellipse (elliptical polarization whose sense is given by the sign of ϕ).

Q5   What is the nature of the polarization for values of $\alpha = E_y/E_x$

(a) $\alpha = \sqrt{3}$ (b) $\alpha = +j$ (c) $\alpha = -j$ (d) $\alpha = (1+j)/\sqrt{2}$

1.5 *Volume density of charge*

In an ohmic conductor ($\mathbf{J} = \sigma\mathbf{E}$) whose properties can be treated as linear ($\mathbf{D} = \varepsilon\mathbf{E}$), we wish to study the evolution of the volume density of charge ρ.

Q1 Using Maxwell's equations (and the equation of conservation of charge), show that $\rho(x, y, z, t)$ obeys a partial differential equation of the form

$$\frac{\partial \rho}{\partial t} + \frac{1}{\tau}\rho = 0$$

where τ is a constant which characterizes the conductor, and is a function of ε and the conductivity σ.

Q2 At the instant $t = 0$, the charge density at the point $M(x, y, z)$ is ρ_0. What is the expression for $\rho(x, y, z, t)$? Show that ρ decreases and tends to zero more rapidly when τ is small.

Q3 Calculate τ for copper ($\varepsilon \approx \varepsilon_0$ and $\sigma = 0.6 \times 10^8 \ \Omega^{-1}\text{m}^{-1}$). Compare τ to the period T_1 for a television signal of frequency $f_1 = 500$ MHz. Make the same comparison between τ and the period T_2 for visible light for which $f_2 = 6 \times 10^{14}$ Hz. What conclusions do you draw ?

1.6 *Skin effect*

Consider a conducting medium (ε_0, μ_0 and $\sigma = 5.9 \times 10^7 \ \Omega^{-1}\text{m}^{-1}$) and a second medium ($\varepsilon$, μ) separated by a plane surface xOy which contains neither charges nor surface currents. The conductor occupies the region $z > 0$. An electromagnetic wave, polarized along Ox and of angular frequency $\omega = 2\pi f$, propagates along Oz in the dielectric and penetrates into the conductor. We denote the amplitude of the electric field in $z = 0^+$ by E_0.

Q1 Verify that $\sigma \gg \omega\varepsilon_0$ and write down Maxwell's equations for the conductor, taking into account this inequality.

Q2 Show that E_x obeys the equation $\dfrac{d^2 E_x}{dz^2} - \dfrac{2j}{\delta^2}E_x = 0$ in the conductor. Give the expression for δ, which is known as the skin depth. Calculate its numerical value.

Q3 Integrate the preceding equation, and hence deduce the volume current density J_x and the magnetic field H_y.

Q4 Show that in the metal and its immediate neighbourhood we have the following relation:

$$\mathbf{E} = R_S (1 + j)\, \mathbf{H} \times \mathbf{u}_z \qquad where \qquad R_S = \frac{1}{\sigma \delta}$$

Q5 Evaluate the Joule losses P_J per unit area. Show that

$$P_J = \frac{1}{2} R_S |H_0|^2$$

where H_0 is the tangential magnetic field in $z = 0^+$.

Q6 Determine the numerical values of σ and R_s for frequencies of 10 MHz, 1 GHz and 100 GHz.

2

Radiation

Jean Baptiste Joseph Fourier (1768–1830)

We have chosen to present the theoretical foundations of electromagnetic radiation by means of the plane wave spectrum method, which makes use of the two-dimensional Fourier transform. This powerful method leads to relatively simple expressions for the far-field radiation from plane apertures and from elementary dipoles. These expressions will be used in Chapters 3, 5 and 6 to study antennas in transmission, antennas of simple geometry and printed antennas.

There is, however, a method which is older than the plane wave spectrum approach, based on the use of electromagnetic potentials. This recalls the expression for the retarded vector potential $\mathbf{A}(M)$ created at a point M by a distribution of currents $\mathbf{J}(M')$ distributed within a volume V

$$\mathbf{A}(M) = \iiint_V \mathbf{J}(M') \frac{\exp(-jk\,MM')}{MM'} dV$$

By this formula the effect of an element of current $\mathbf{J}(M')dV$ located at a point M' appears at an observation point M with a phase delay $kMM' = 2\pi MM'/\lambda$, corresponding to the propagation time between the points M and M'. The vector potential and the electromagnetic field are thus calculated from the source currents.

The reader will find in references [1,5 and 7] a detailed description of this method.

2.1 PLANE WAVE SPECTRUM

The electromagnetic field radiated by currents and apertures may be written in a very simple way, as the superposition of plane waves. To do this we introduce the spectral domain, deduced from the spatial domain by 2-D Fourier transformation with respect to the x and y coordinates.

2.1.1 Spectral domain

2-D Fourier transform

To any arbitrary function G(x, y) of the spatial coordinates (x, y), we associate a function G(α, β) by means of the Fourier transform:

$$G(\alpha,\beta) = \frac{1}{\lambda^2} \int_{-\infty}^{\infty} \int_{-\infty}^{\infty} G(x,y) \, \exp\left[jk(\alpha x + \beta y)\right] dx dy \qquad (2.1)$$

$$G(x,y) = \int_{-\infty}^{\infty} \int_{-\infty}^{\infty} G(\alpha,\beta) \exp\left[-jk(\alpha x + \beta y)\right] d\alpha d\beta \qquad (2.2)$$

- α and β are the variables associated with x and y;
- $k = \dfrac{2\pi}{\lambda} = \dfrac{\omega}{v}$ is the wave number in the medium (ε, μ);
- (x, y) represents the spatial domain and (α, β) the spectral domain.

Setting normalized variables $\eta = x/\lambda$ and $\xi = y/\lambda$, equation (2.1) can be written

$$G(\lambda\eta, \lambda\xi) = \int_{-\infty}^{\infty} \int_{-\infty}^{\infty} G(\alpha,\beta) \exp\left[-2\pi j(\alpha\eta + \beta\xi)\right] d\eta \, d\xi$$

which shows that $G(\alpha,\beta)$ is the inverse Fourier transform of G($\lambda\eta$, $\lambda\xi$).

Properties

Derivatives with respect to x (or y) correspond to multiplying by $-jk\alpha$ (or by $-jk\beta$).

Spatial domain (x, y)	Spectral domain (α, β)
$\dfrac{\partial G(x, y)}{\partial x}$	$-jk\alpha G(\alpha, \beta)$
$\dfrac{\partial G(x, y)}{\partial y}$	$-jk\beta G(\alpha, \beta)$

- Parseval's theorem, applied to functions $F(x, y)$ and $G(x, y)$ takes the form:

$$\int_{-\infty}^{\infty}\int_{-\infty}^{\infty} F(\alpha,\beta)G^{*}(\alpha,\beta)\,d\alpha\,d\beta = \frac{1}{\lambda^2}\int_{-\infty}^{\infty}\int_{-\infty}^{\infty} F(x,y)G^{*}(x,y)\,dx\,dy \qquad (2.3)$$

Examples:

- $G = 1$ inside the rectangle of sides a, b and $G = 0$ outside;

$$G(\alpha, \beta) = \frac{ab}{\lambda^2}\frac{\sin(k\alpha a/2)}{k\alpha a/2}\frac{\sin(k\beta b/2)}{k\beta b/2} \qquad (2.4a)$$

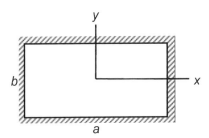

Fig. 2.1 Rectangle of sides a, b.

- $G(\alpha, \beta)$ is a Dirac delta function located at (α', β').

$$G(\alpha,\beta) = \delta(\alpha-\alpha')\delta(\beta-\beta')$$
$$G(x,y) = \exp\left[-jk(\alpha'x+\beta'y)\right]$$

(2.4b)

The electromagnetic field in the spectral domain

In the spatial domain, the field is described by the vectors **E** and **H** or by the six Cartesian coordinates (E_x, E_y, E_z, H_x, H_y, H_z) which are functions of (*x, y, z*). In the spectral domain the electromagnetic field is described by the vectors **E** and **H** or by the six Cartesian coordinates (E_α, E_β, E_z, H_α, H_β, H_z) which are functions of (α,β,z).

The Fourier transform allows us to pass from the spatial domain to the spectral domain, and vice versa:

Passing from the spatial domain into the spectral domain

$$\mathbf{E}(\alpha,\beta,z) = \frac{1}{\lambda^2}\int_{-\infty}^{\infty}\int_{-\infty}^{\infty}\mathbf{E}(x,y,z)\exp\left[jk(\alpha x+\beta y)\right]dxdy$$

(2.5)

$$\mathbf{H}(\alpha,\beta,z) = \frac{1}{\lambda^2}\int_{-\infty}^{\infty}\int_{-\infty}^{\infty}\mathbf{H}(x,y,z)\exp\left[jk(\alpha x+\beta y)\right]dxdy$$

(2.6)

Passing from the spectral domain to the spatial domain

$$\mathbf{E}(x,y,z) = \int_{-\infty}^{\infty}\int_{-\infty}^{\infty}\mathbf{E}(\alpha,\beta,z)\exp\left[-jk(\alpha x+\beta y)\right]d\alpha d\beta$$

(2.7)

$$\mathbf{H}(x,y,z) = \int_{-\infty}^{\infty}\int_{-\infty}^{\infty}\mathbf{H}(\alpha,\beta,z)\exp\left[-jk(\alpha x+\beta y)\right]d\alpha d\beta$$

(2.8)

Similarly, we can introduce in the spectral domain:

- the charge density ρ and current density *J*;

- the vector potential *A* and the scalar potential Φ.

2.1.2 Electromagnetic field in a semi-infinite space with no sources

The semi-infinite space $z > 0$ is assumed to be free of charges and currents (this does not mean that sources and surface currents are not present in the plane $z = 0$). In this region the Cartesian components of the electromagnetic field obey Helmholtz's equation

$$\nabla^2 F + k^2 F = 0 \qquad (2.9)$$

where $F(x, y, z)$ is one of the components of **E** or **H**.

Component expressions

In the spectral domain, setting $\gamma^2 = 1 - \alpha^2 - \beta^2$, Helmholtz's equation can be written

$$\frac{\partial^2 F}{\partial z^2} + k^2 \gamma^2 F = 0 \qquad (2.10)$$

where $F(\alpha, \beta, z)$ is any one of the components of the field, expressed in the spectral domain.

The general solution of this equation is

$$F(\alpha, \beta, z) = A(\alpha, \beta) \exp(-jk\gamma z) + B(\alpha, \beta) \exp(jk\gamma z) \qquad (2.11)$$

where $A(\alpha, \beta)$ and $B(\alpha, \beta)$ are determined by the boundary conditions of the region under consideration ($z = 0^+$ and $z = \infty$). The relation $\gamma^2 = 1 - \alpha^2 - \beta^2$ defines γ, (though not its sign). Taking into account the form of the solution (2.11), and conserving the generality of this solution, we introduce the following more specific determination

$$\gamma = \sqrt{1 - \alpha^2 - \beta^2} \quad \text{when } \alpha^2 + \beta^2 < 1$$

$$\qquad (2.12)$$

$$\gamma = -j\sqrt{\alpha^2 + \beta^2 - 1} \quad \text{when } \alpha^2 + \beta^2 \geq 1$$

When $\alpha^2 + \beta^2 < 1$ the term $B(\alpha, \beta) \exp(jkz)$ is interpreted as an incoming wave from $z = \infty$. When $\alpha^2 + \beta^2 \geq 1$, the term $B(\alpha, \beta) \exp(jkz)$ increases exponentially with z, or remains constant if $\alpha^2 + \beta^2 = 1$. If the region $z > 0$ is free of sources this term should be zero. Under these conditions the solution is

$$F(\alpha, \beta, z) = A(\alpha, \beta) \exp(-jk\gamma z) \qquad (2.13)$$

When $\alpha^2 + \beta^2 < 1$, this reduces to only conserving the incoming waves (those which originate from the plane $z = 0$). When $\alpha^2 + \beta^2 < 1$, $F(\alpha, \beta, z)$ decreases exponentially with z. Equation (2.13) allows us to easily interpret the term $A(\alpha, \beta)$ which is the value of $F(\alpha, \beta, z)$ at $z = 0^+$.

$$A(\alpha,\beta) = F\left(\alpha,\beta,0^+\right)$$

From which the solution is

$$F\left(\alpha,\beta,z\right) = F\left(\alpha,\beta,0^+\right)\exp\left(-jk\gamma z\right) \qquad (2.14)$$

and returning to the spatial domain (x, y, z)

$$F(x,y,z) = \int_{-\infty}^{\infty}\int_{-\infty}^{\infty} F\left(\alpha,\beta,0^+\right)\exp\left[-jk\left(\alpha x + \beta y + \gamma z\right)\right] d\alpha d\beta \quad (2.15)$$

where

$$F\left(\alpha,\beta,0^+\right) = \frac{1}{\lambda^2}\int_{-\infty}^{\infty}\int_{-\infty}^{\infty} F\left(x,y,0^+\right)\exp\left[jk\left(\alpha x + \beta y\right)\right] dxdy$$

This relation shows that any component is entirely determined by the value that the components take in the vicinity of the xy plane ($z = 0^+$). Since it is valid for each component, this property may be applied to both the **E** and **H** fields.

Electric field

$$\mathbf{E} = \int_{-\infty}^{\infty}\int_{-\infty}^{\infty} \mathbf{E}\left(\alpha,\beta,0^+\right)\exp\left[-jk\left(\alpha x + \beta y + \gamma z\right)\right] d\alpha d\beta \qquad (2.16)$$

where

$$\mathbf{E}\left(\alpha,\beta,0^+\right) = \frac{1}{\lambda^2}\int_{-\infty}^{\infty}\int_{-\infty}^{\infty} \mathbf{E}\left(x,y,0^+\right)\exp\left[jk\left(\alpha x + \beta y\right)\right] dxdy$$

Magnetic field

$$\mathbf{H} = \int_{-\infty}^{\infty}\int_{-\infty}^{\infty} \mathbf{H}\left(\alpha,\beta,0^+\right)\exp\left[-jk\left(\alpha x + \beta y + \gamma z\right)\right] d\alpha d\beta \qquad (2.17)$$

where

$$\mathbf{H}\left(\alpha,\beta,0^+\right) = \frac{1}{\lambda^2}\int_{-\infty}^{\infty}\int_{-\infty}^{\infty} \mathbf{H}\left(x,y,0^+\right)\exp\left[jk\left(\alpha x + \beta y\right)\right] dxdy$$

Interpretation: angular plane-wave spectrum

Equations (2.15), (2.16) and (2.17) can then be easily interpreted when $\alpha^2 + \beta^2 < 1$, and therefore when $\gamma > 0$. The field appears as a superposition of plane waves (see Fig. 2.2) that propagate in the direction of the unit vector **u** whose components are (α, β, γ). When $\alpha^2 + \beta^2 > 1$ (and therefore for $\gamma = -j\sqrt{\alpha^2 + \beta^2 - 1}$), the field appears as a superposition of evanescent waves along the direction Oz, localized in the vicinity of the *xy* plane. Fig. 2.3 gives, according to the values of α and β, the propagating or evanescent character of the waves making up the plane wave spectrum. Inside the circle $\alpha^2 + \beta^2 = 1$ the waves propagate in the direction defined by the vector (α, β, γ). Outside the circle the waves are localized in the vicinity of the plane $z = 0$ and propagate in a direction parallel to it (surface waves).

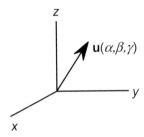

Fig. 2.2 Plane wave spectrum: direction of propagation of an elementary wave.

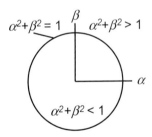

Fig. 2.3 Propagating waves (inside the circle) and surface waves (outside the circle).

Field component relations in the spectral domain

Taking into account the general expression (2.13) for the components, differentiation with respect to *z* corresponds to a multiplication by $-jk$ in the spectral domain. In a medium (ε, μ), free of sources, Maxwell's equations lead to the following relations, valid in the region $z > 0$.

- Expressions for E_z, H_x, H_y, H_z as function of E_x and of E_y

$$E_z = -\frac{1}{\gamma}\left(\alpha E_x + \beta E_y\right) \tag{2.18a}$$

$$H_x = -\frac{1}{\eta\gamma}\left[\alpha\beta E_x + \left(1-\alpha^2\right)E_y\right] \tag{2.18b}$$

$$H_y = \frac{1}{\eta\gamma}\left[\left(1-\beta^2\right)E_x + \alpha\beta E_y\right] \tag{2.18c}$$

$$H_z = \frac{1}{\eta}\left(-\beta E_x + \alpha E_y\right) \tag{2.18d}$$

- Expressions for E_x, E_y, E_z, H_z as a function of H_x and of H_y

$$E_x = \frac{\eta}{\gamma}\left[\alpha\beta H_x + \left(1-\alpha^2\right)H_y\right] \tag{2.19a}$$

$$E_y = \frac{\eta}{\gamma}\left[\left(1-\beta^2\right)H_x + \alpha\beta H_y\right] \tag{2.19b}$$

$$E_z = -\eta\left(-\beta H_x + \alpha H_y\right) \tag{2.19c}$$

$$H_z = -\frac{1}{\gamma}\left(\alpha H_x + \beta H_y\right) \tag{2.19d}$$

Thus, from a knowledge of the two components (E_x, E_y) or (H_x, H_y) we can obtain all the other components.

The electric (or magnetic) field can be expressed from the tangential electric (or magnetic) field in the vicinity of the xy plane.

Equation (2.16) suggests that the calculation of **E** at any point requires a knowledge of three components: $E_x(\alpha, \beta, 0^+)$, $E_y(\alpha, \beta, 0^+)$, $E_z(\alpha, \beta, 0^+)$ of the field $E(\alpha, \beta, 0^+)$ in the vicinity of the xy plane. However, equation (2.18a) shows that $E_x(\alpha, \beta, 0^+)$ and $E_y(\alpha, \beta, 0^+)$ are sufficient. Under these conditions the electric field **E** at any point can be expressed as a function only of the tangential component in the vicinity of the *xy* plane. Considering the tangential fields in the vicinity of the *xy* plane, we have

$$\mathbf{E}_{0t} = E_x\left(x, y, 0^+\right)\mathbf{u}_x + E_y\left(x, y, 0^+\right)\mathbf{u}_y$$
$$\mathbf{H}_{0t} = H_x\left(x, y, 0^+\right)\mathbf{u}_x + H_y\left(x, y, 0^+\right)\mathbf{u}_y$$

Equations (2.16), (2.17), (2.18) and (2.19) show that a knowledge of one of these tangential fields ($\mathbf{E}_{0t}$ or $\mathbf{H}_{0t}$) is sufficient to completely determine the electromagnetic field ($\mathbf{E}$, $\mathbf{H}$).

2.1.3 The far field

We are aiming at obtaining the electromagnetic field at a point M, in the direction of the unit vector $\mathbf{u}$ of components (α, β, γ) when the distance $r =$ OM is very large compared with the wavelength

$$kr = 2\pi\frac{r}{\lambda} \to \infty$$

The electromagnetic field is determined from equations (2.16) and (2.17), rewritten in terms of the variables (α', β') to describe the spectral domain.

$$\mathbf{E} = \int_{-\infty}^{\infty}\int_{-\infty}^{\infty} E\left(\alpha', \beta', 0^+\right)\exp\left[-jk\left(\alpha'x + \beta'y + \gamma'z\right)\right] d\alpha' d\beta' \quad (2.20)$$

$$\mathbf{H} = \int_{-\infty}^{\infty}\int_{-\infty}^{\infty} H\left(\alpha', \beta', 0^+\right)\exp\left[-jk\left(\alpha'x + \beta'y + \gamma'z\right)\right] d\alpha' d\beta' \quad (2.21)$$

where

$$\gamma'^2 = 1 - \alpha'^2 - \beta'^2$$

The observation direction $\mathbf{u}(\alpha, \beta, \gamma)$ and the direction of propagation $\mathbf{u}'(\alpha', \beta', \gamma')$ of the elementary wave of the plane wave spectrum are represented in Fig. 2.4.

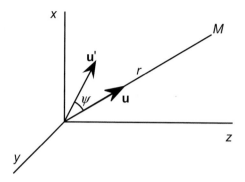

Fig. 2.4 Observation direction **u** and direction of propagation **u′** of the elementary wave of the plane wave spectrum.

By introducing the angle ψ between the vectors **u** and **u′**, the preceding equations become

$$\mathbf{E} = \int_{-\infty}^{\infty} \int_{-\infty}^{\infty} \boldsymbol{E}\left(\alpha',\beta',0^+\right)\exp\left(-jkr\cos\psi\right) d\alpha' d\beta'$$

$$\mathbf{H} = \int_{-\infty}^{\infty} \int_{-\infty}^{\infty} \boldsymbol{H}\left(\alpha',\beta',0^+\right)\exp\left(-jkr\cos\psi\right) d\alpha' d\beta'$$

where

$$\cos\psi = \alpha\alpha' + \beta\beta' + \gamma\gamma'$$

Field calculation by the stationary phase method

When r is very large, the phase $kr\cos\psi$ varies very rapidly with ψ. The terms being summed have phases which take all possible values and their contribution is in general zero. Nevertheless, an important exception exists in the vicinity of the value $\psi = 0$. The phase $kr\cos\psi$ is then stationary ($\cos\psi$ is close to 1). This means that the preceding integrals may be evaluated by considering only the plane waves whose propagation direction **u′** is close to that of the observation direction **u**. This process of integral calculations is known as the 'method of stationary phase'. Its validity increases as r becomes larger.

Calculation of integrals with rapidly-varying argument by the stationary phase method: The calculation of the integral

$$I = \iint_{D} f\left(u,v\right)\exp\left[jKg\left(u,v\right)\right]dudv \tag{2.22}$$

where $f(u, v)$ and $g(u, v)$ are real and continuous functions and K is a real, positive and very large number, is possible when $g(u, v)$ is stationary in the vicinity of the point $g(u_0, v_0)$ pertaining to the integration domain D. The only significant contribution is that originating from the vicinity of this point, since at a large distance from it, the phase varies too rapidly.

The function g is expanded in a Taylor series in the vicinity of the point (u_0, v_0). The stationary conditions are $g'_u(u_0, v_0) = 0$ and $g'_v(u_0, v_0) = 0$. Setting $u = u_0 + \eta$ and $v = v_0 + \xi$, we obtain

$$g(u,v) = g(u_0,v_0) + \frac{1}{2}\left[\eta^2 g''_u(u_0,v_0) + 2\xi\eta g''_{uv}(u_0,v_0) + \xi^2 g''_v(u_0,v_0)\right]$$

When K is large, only the vicinity of $g(u_0,v_0)$ contributes to the value of the integral

$$I \approx f(u_0,v_0)\exp\left[jKg(u_0,v_0)\right] \int_{-\infty}^{\infty}\int_{-\infty}^{\infty}\exp\left[\frac{1}{2}jK\left(\eta^2 g''_u + 2\xi\eta g''_{uv} + \xi^2 g''_v\right)\right]d\eta d\xi$$

Using the following integral identities (which are only valid for $a > 0$):

$$\int_{-\infty}^{\infty}\cos\left(ax^2 + bx + c\right)dx = \sqrt{\frac{\pi}{a}}\cos\left(\frac{\pi}{4} + \frac{ac - b^2}{a}\right)$$

$$\int_{-\infty}^{\infty}\sin\left(ax^2 + bx + c\right)dx = \sqrt{\frac{\pi}{a}}\sin\left(\frac{\pi}{4} + \frac{ac - b^2}{a}\right)$$

we obtain, when g''_u and g''_v are positive and $g''_u g''_v > g''^2_{uv}$

$$I \approx \frac{2\pi j}{K\sqrt{g''_u g''_v - g''^2_{uv}}}f(u_0,v_0)\exp\left[jKg(u_0,v_0)\right] \qquad (2.23)$$

Application to the calculation of the far field

The electric field **E** at a point M with coordinates $x = \alpha r$, $y = \beta r$ and $z = \gamma r$ in the direction **u** (α, β, γ) is expressed by the integral (2.20)

$$\mathbf{E} = \int_{-\infty}^{\infty}\int_{-\infty}^{\infty}\mathbf{E}\left(\alpha',\beta',0^+\right)\exp\left[-jkr\left(\alpha'\alpha + \beta'\beta + \gamma'\gamma\right)\right]d\alpha'd\beta'$$

where

$$\gamma = \sqrt{1 - \alpha^2 - \beta^2} \text{ and } \gamma' = \sqrt{1 - \alpha'^2 - \beta'^2}$$

Setting $K = kr$, $\alpha' = u$, $\beta' = v$, $f = \mathbf{E}(\alpha', \beta', 0^+)$ and $g(\alpha', \beta') = -(\alpha\alpha' + \beta\beta' + \gamma\gamma')$, we obtain for the point $(\alpha' = \alpha, \beta' = \beta)$

$$g(\alpha' = \alpha, \ \beta' = \beta) = -1$$

$$\left(\frac{\partial g}{\partial \alpha'}\right)_{\substack{\alpha'=\alpha \\ \beta'=\beta}} = 0, \quad \left(\frac{\partial g}{\partial \beta'}\right)_{\substack{\alpha'=\alpha \\ \beta'=\beta}} = 0 \ \rightarrow \ g(\alpha', \beta') \text{ is stationary at } (\alpha' = \alpha, \ \beta' = \beta)$$

$$\left(\frac{\partial^2 g}{\partial \alpha'^2}\right)_{\substack{\alpha'=\alpha \\ \beta'=\beta}} = 1 + \frac{\alpha^2}{\gamma^2}, \quad \left(\frac{\partial^2 g}{\partial \beta'^2}\right)_{\substack{\alpha'=\alpha \\ \beta'=\beta}} = 1 + \frac{\beta^2}{\gamma^2}, \quad \left(\frac{\partial^2 g}{\partial \alpha' \partial \beta'}\right)_{\substack{\alpha'=\alpha \\ \beta'=\beta}} = \frac{\alpha\beta}{\gamma^2}$$

From which the radiated field in the direction $\mathbf{u}(\alpha, \beta, \gamma)$ at a distance r very large compared with the wavelength is

$$\mathbf{E} = 2\pi j \frac{\exp(-jkr)}{kr} \gamma \, E\left(\alpha, \beta, 0^+\right) \tag{2.24}$$

Since $\gamma = \mathbf{u}.\mathbf{u}_z$ and using equations (2.18) and (2.19) which relate the field components in the frequency domain, as well as introducing the tangential component $\mathbf{E}_{0t}(\alpha, \beta, 0^+)$ of the Fourier transform of the electric field at 0^+, we obtain

$$\mathbf{E} = 2\pi j \frac{\exp(-jkr)}{kr} \mathbf{u} \times (\mathbf{E}_{0t} \times \mathbf{u}_z), \quad \mathbf{H} = \frac{1}{\eta} \mathbf{u} \times \mathbf{E} \tag{2.25}$$

Therefore, the Fourier transform $\mathbf{E}_{0t}(\alpha, \beta, 0^+)$ of the tangential electric field in the vicinity of the xy plane determines the far electromagnetic field $(\mathbf{E}, \mathbf{H})$ in the direction $\mathbf{u} \ (\alpha, \beta, \gamma)$.

Equally, the far magnetic field is expressed as a function of the Fourier transform $\mathbf{H}_{0t} \ (\alpha, \beta, 0^+)$ of the magnetic field in the plane $z = 0^+$.

$$\mathbf{H} = 2\pi j \frac{\exp(-jkr)}{kr} \gamma \, H\left(\alpha, \beta, 0^+\right) \tag{2.26}$$

Similarly, we obtain

$$\mathbf{H} = 2\pi j \frac{\exp(-jkr)}{kr} \mathbf{u} \times (\mathbf{H}_{0t} \times \mathbf{u}_z), \quad \mathbf{E} = \eta \mathbf{H} \times \mathbf{u} \tag{2.27}$$

Application to the radiation from an elementary dipole (Hertzian dipole)

An electric dipole is composed of an element of current I, of length ℓ assumed to be infinitely small, in the direction of the unit vector $\mathbf{u}_\mathrm{D}$. The current density describing an electric dipole at a point M of Cartesian coordinates (x', y', z') is

$$\mathbf{J} = I\ell\,\delta(x-x')\,\delta(y-y')\,\delta(z-z')\,\mathbf{u}_\mathrm{D} \tag{2.28}$$

If the dipole is located at a point (x', y') of the xy plane in the direction $\mathbf{u}_\mathrm{D}$ in this plane, it may be described by a surface current density

$$\mathbf{J}_\mathrm{S}(x,y) = I\ell\,\delta(x-x')\,\delta(y-y')\,\mathbf{u}_\mathrm{D} \tag{2.29}$$

An electric dipole located at the origin ($x' = 0$, $y' = 0$, $z' = 0$) in the direction Ox is described by the surface current density $\mathbf{J}_S = I\ell\,\delta(x)\,\delta(y)\,\mathbf{u}_x$. Taking into account the symmetry in the xy plane and the discontinuity of the magnetic field when crossing this plane, we easily obtain the tangential component of the magnetic field

$$\mathbf{H}_{0t}(x,y) = \mathbf{H}_t\left(x,y,0^+\right) = -\mathbf{H}_t\left(x,y,0^-\right) = -\frac{1}{2}I\ell\,\delta(x)\,\delta(y)\,\mathbf{u}_y \tag{2.30}$$

whose inverse 2-D Fourier transform is

$$H_{0t}(\alpha,\beta) = -\frac{1}{2\lambda^2}I\ell\,\mathbf{u}_y \tag{2.31}$$

Equation (2.27) leads to

$$\mathbf{H} = j\frac{k}{4\pi}I\ell\,\frac{\exp(-jkr)}{r}\,\mathbf{u}_x \times \mathbf{u} \tag{2.32}$$

Since the choice of Ox is arbitrary, the preceding result may be generalized for a dipole in any direction $\mathbf{u}_\mathrm{D}$.

$$\mathbf{H} = j\frac{k}{4\pi}I\ell\,\frac{\exp(-jkr)}{r}\,\mathbf{u}_\mathrm{D} \times \mathbf{u} \qquad \mathbf{E} = \eta\,\mathbf{H} \times \mathbf{u} \tag{2.33}$$

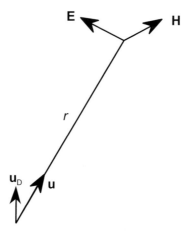

Fig. 2.5 Electromagnetic field radiated by an electric dipole.

2.2 KIRCHHOFF'S FORMULATION

The representation of the field by a spectrum of plane waves allows us to express the field at any point of a source-free domain from the tangential field which exists on a plane. Kirchhoff's formulation is more general: it gives the field from the tangential field which exists on any closed surface enclosing the sources.

2.2.1 Green's identity and Green's functions

We recall below some mathematical properties which are in common use in electromagnetism.

Green's identity

Two functions f and g are introduced, and a closed surface S enclosing a volume V, to which the outward normal is **n**. Green's identity

$$\iiint_V \left(f \nabla^2 g - g \nabla^2 f \right) dV = \iint_S \left(f \frac{\partial g}{\partial n} - g \frac{\partial f}{\partial n} \right) dS \qquad (2.34)$$

relates f, g, and their Laplacians $\nabla^2 f$ and $\nabla^2 g$ and the normal derivatives

$$\frac{\partial f}{\partial n} = \mathbf{n}.\text{grad}\, f\,, \quad \frac{\partial g}{\partial n} = \mathbf{n}.\text{grad}\, g$$

Green's function in an unbounded space

The Green's function in an unbounded space is

$$G(x,y,z|x_0,y_0,z_0) = \frac{1}{4\pi} \frac{\exp\left(-jk\sqrt{(x-x_0)^2+(y-y_0)^2+(z-z_0)^2}\right)}{\sqrt{(x-x_0)^2+(y-y_0)^2+(z-z_0)^2}} \tag{2.35}$$

This is the solution of the inhomogeneous Helmholtz's equation

$$\nabla^2 G + k^2 G = -\delta(x-x_0, y-y_0, z-z_0) \tag{2.36}$$

which is zero at infinity.

Odd Green's function in the half-space z > 0

This is the function

$$G_1(x,y,z|x_0,y_0,z_0) = G(x,y,z|x_0,y_0,z_0) - G(x,y,-z|x_0,y_0,z_0)a \tag{2.37}$$

It satisfies the inhomogeneous Helmholtz's equation in the region $z > 0$

$$\nabla^2 G_1 + k^2 G_1 = -\delta(x-x_0, y-y_0, z-z_0) \tag{2.38}$$

It is zero in the plane $z = 0$ and at infinity.

Even Green's function in the half-space z > 0

This is the function

$$G_2(x,y,z|x_0,y_0,z_0) = G(x,y,z|x_0,y_0,z_0) + G(x,y,-z|x_0,y_0,z_0) \tag{2.39}$$

This satisfies the inhomogeneous Helmholtz's equation in the region $z > 0$ and it is zero at infinity.

$$\nabla^2 G_2 + k^2 G_2 = -\delta(x-x_0, y-y_0, z-z_0)$$

Its normal derivative $\partial G_2/\partial n$ is zero in the plane $z = 0$.

2.2.2 Kirchhoff's integral formulation

By means of Green's function in an unbounded space

Consider a surface S enclosing all the sources (Fig. 2.6). We apply Green's identity to $G(x, y, z|x_0, y_0, z_0)$ and $E_x(x_0, y_0, z_0)$, then to G and E_y and finally to G and E_z, in the volume between infinity and the surface S, the integration variables being the coordinates (x_0, y_0, z_0) of a point M_0 of S. We obtain the electric field at a point $M(x, y, z)$ of the space

$$\mathbf{E}(M) = \iint_S \left(\frac{\partial \mathbf{E}_0}{\partial n} - \mathbf{E}_0 \frac{\partial G}{\partial n} \right) dS(M_0) \qquad (2.40)$$

where $\mathbf{E}_0$ is the electric field at the point M_0 on the surface S, and

$$\frac{\partial \mathbf{E}_0}{\partial n} = \frac{\partial \mathbf{E}_{0x}}{\partial n}.\mathbf{u}_x + \frac{\partial \mathbf{E}_{0y}}{\partial n}.\mathbf{u}_y + \frac{\partial \mathbf{E}_{0z}}{\partial n}.\mathbf{u}_z$$

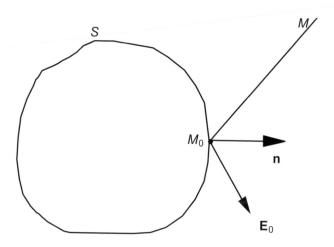

Fig. 2.6 General Kirchhoff's integral formulation.

By means of the odd Green's function G_1 in a half-space

We apply Green's identity inside the infinite hemisphere in the region $z > 0$

$$E(M) = -\iint_S E_0 \frac{\partial G_1}{\partial n} dS(M_0)$$

S is the xy plane, M_0 is a point on this plane and $\partial G_1/\partial n = \partial G_1/\partial z$. Also, we have

$$\left[\frac{\partial G_1}{\partial z}\right]_{z_0=0} = 2\left[\frac{\partial G}{\partial z}\right]_{z_0=0}$$

from which

$$E(M) = -\frac{1}{2\pi} \iint_S E_0 \frac{\partial}{\partial z}\left[\frac{\exp\left(-jk\sqrt{(x-x_0)^2+(y-y_0)^2+(z-z_0)^2}\right)}{\sqrt{(x-x_0)^2+(y-y_0)^2+(z-z_0)^2}}\right] dS(M_0)$$

$$(2.41a)$$

or, by introducing $r = \sqrt{x^2+y^2+z^2}$ in the form of the convolution product (denoted $\otimes$) of the variables x and y

$$E(M) = -\frac{1}{2\pi} E_0 \otimes \frac{\partial}{\partial z}\left[\frac{\exp(-jkr)}{r}\right] \qquad (2.41b)$$

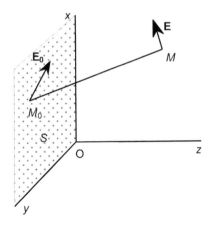

Fig. 2.7 Kirchhoff's integral formulation by means of half-space Green's functions.

By means of the even Green's function G_2 in a half-space

With the even Green's function G_2 we would have obtained

$$E(M) = \frac{1}{2\pi} \iint_S \left[\frac{\exp\left(-jk\sqrt{(x-x_0)^2 + (y-y_0)^2 + (z-z_0)^2}\right)}{\sqrt{(x-x_0)^2 + (y-y_0)^2 + (z-z_0)^2}} \right] \frac{\partial E_0}{\partial z} dS(M_0) \quad (2.42a)$$

or in the form of a convolution product

$$E(M) = \frac{1}{2\pi} \frac{\exp(-jkr)}{r} \otimes \frac{\partial E_0}{\partial z} \quad (2.42b)$$

2.2.3 Plane wave spectrum and Kirchhoff's formulation

The formulation in terms of the plane wave spectrum described by equation (2.16)

$$E = \int_{-\infty}^{\infty} \int_{-\infty}^{\infty} E(\alpha, \beta, 0^+) \exp\left[-jk(\alpha x + \beta y + \gamma z)\right] d\alpha d\beta$$

is equivalent to Kirchhoff's formulation (2.41b)

$$E(M) = -\frac{1}{2\pi} E_0 \otimes \frac{\partial}{\partial z} \left[\frac{\exp(-jkr)}{r} \right]$$

We can obtain one from the other by application of the convolution theorem. We can show (see exercise 2.2) that the 2-D Fourier transform of $\exp(-j\gamma z)$ is

$$-\frac{1}{2\pi} \frac{\partial}{\partial z} \left[\frac{\exp(-jkr)}{r} \right]$$

The evaluation of the field by plane waves (spectral method) or spherical waves (Kirchhoff's formulation) can be considered as a process of linear filtering, and introduces a first link to the ideas of antennas and signal theory (Chapter 13).

FURTHER READING

1. Balanis, C.A., *Advanced Engineering Electromagnetics*, Wiley, New York, 1989.
2. Benson, F.A. and Benson, T.M., *Fields, Waves and Transmission Lines*, Chapman & Hall, London, 1991.
3. Booker, H.G. and Clemmow, P.C., 'The concept of an angular spectrum of plane waves, and its relation to that of polar diagram and aperture distribution', *Proc. IEE,* Vol. 97, Pt. III, pp11–17, January 1950.
4. Clarke, R.H. and Brown, J., *Diffraction Theory and Antennas*, Ellis Horwood, 1980.
5. Collin, R.E., *Antennas and Radiowave Propagation*, McGraw-Hill, New York, 1985.
6. Collin, R.E. and Zucker, F.J. (eds), *Antenna Theory*, Parts 1 and 2, McGraw-Hill, New York, 1969.
7. Harrington, R.F., *Time-Harmonic Electromagnetic Fields*, IEEE Press Series on Electromagnetic Wave Theory, 2001.
8. Jordan, E.C. and Balmain, K.G., *Electromagnetic Waves and Radiating Systems*, Prentice-Hall, New Jersey, 1968.
9. Jull, E.V., *Aperture Antennas and Diffraction Theory*, Peter Peregrinus, Stevenage, 1981.
10. Kraus, J.D. and Carver, K.R., *Electromagnetics*, McGraw-Hill, New York, 1973.
11. Ramo, S., Whinnery, J.R. and van Duzer, T., *Fields and Waves in Communication Electronics*, Wiley, New York, 1984.

EXERCISES

2.1 2D Fourier transforms

The function F(x, y) is zero outside the rectangle (a, b). Inside the rectangle (Fig. 2.1) it is of the form

$$F(x, y) = f_1(x)f_2(y)$$

Q1 Show that $F(\alpha, \beta)$ can be put in the form

$$F(\alpha, \beta) = f_1(\alpha)f_2(\beta)$$

Q2 Determine $F(\alpha, \beta)$ in the following cases

(a) $f_1 = f_2 = 1$

(b) $f_1(x) = 1 - 2\dfrac{|x|}{a}, \quad f_2(y) = 1 - 2\dfrac{|y|}{b}$

(c) $f_1(x) = \cos \pi \dfrac{x}{a}, \quad f_2(y) = \cos \pi \dfrac{y}{b}$

(d) $f_1(x) = \cos^2 \pi \dfrac{x}{a}, \quad f_2(y) = \cos^2 \pi \dfrac{y}{b}$

(e) $f_1(x) = \dfrac{1}{3}\left(1 + 2\cos^2 \pi \dfrac{x}{a}\right), \quad f_2(y) = \dfrac{1}{3}\left(1 + 2\cos^2 \pi \dfrac{y}{b}\right)$

Q3 Give the shape of F(x, 0) and of $F(\alpha, 0)$ in each case.

Q4 Evaluate the widths of the principal maxima and the level of the first sidelobe.

2.2 2D Fourier transform of exp(−jkr)/r

The free space Green's function

$$G(r) = \frac{1}{4\pi}\frac{\exp(-jkr)}{r} \quad (\text{where } r = \sqrt{x^2 + y^2 + z^2})$$

is the unique solution of the equation $\nabla^2 G + k^2 G = -\delta(x)\delta(y)\delta(z)$, which is zero when $r = 0$. You are asked to determine its inverse Fourier transform by finding this solution directly in the spectral domain.

Q1   Show that $G(\alpha, \beta, z)$ satisfies $\dfrac{\partial^2 G}{\partial z^2} + k^2 \gamma^2 G = -\dfrac{1}{\lambda^2}\delta(z)$, where $\gamma^2 = 1 - \alpha^2 - \beta^2$.

Q2 Hence show that $G(\alpha, \beta, z) = -\dfrac{j}{4\pi\lambda\gamma}\exp\left(-jk\gamma|z|\right)$

Q3 Show that the 2D Fourier transform of $\exp\left(-jk\gamma|z|\right)$ is $-\dfrac{1}{2\pi}\dfrac{\partial}{\partial z}\left[\dfrac{\exp(-jkr)}{r}\right]$,

for $z > 0$.

3
Antennas in transmission

Heinrich Hertz (1857–1894)

In this Chapter we are concerned with the radiation at great distances from antennas. We will introduce the concepts of directivity, gain and input impedance. We will make use of the results of Chapter 2 to determine the properties of elementary dipoles and aperture antennas.

3.1 FAR FIELD RADIATION

3.1.1 Vector characteristic of the radiation from the antenna

We suppose that the antenna is of finite dimensions. The point O is a reference point of the antenna. We consider the radiated field at a point M in the direction **u**, at a distance r which is very large compared with the wavelength λ (Fig. 3.1). The electromagnetic field has the form of a plane wave

$$\mathbf{H} = \frac{1}{\eta}\mathbf{u}\times\mathbf{E} \quad \text{and} \quad \mathbf{E} = \eta\mathbf{H}\times\mathbf{u} \tag{3.1}$$

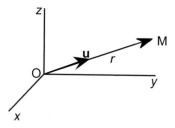

Fig. 3.1 Reference coordinate system.

It propagates with velocity $v = 1/\sqrt{\varepsilon\mu} = \omega/k$, with a propagation term $\exp(-jkr)$ and an amplitude which varies as $1/r$ so that energy is conserved. The electric field then has the general form

$$\mathbf{E}(r,\mathbf{u}) = \frac{\exp(-jkr)}{kr}\mathbf{e}(\mathbf{u}) \tag{3.2}$$

where the term **e(u)** is the vector function which represents the radiation pattern of the antenna. This term contains both the directional properties (amplitude and phase of the radiation in terms of the transmission direction) and the polarization (orientation of the radiated electric field vector). The radiated field is generally expressed in spherical coordinates (r, θ, ϕ) and equation (3.2) takes the form

$$\mathbf{E}(r,\theta,\varphi) = \frac{\exp(-jkr)}{kr}\mathbf{e}(\theta,\varphi) \tag{3.3}$$

3.1.2 Translation theorem

An antenna with reference point located at O creates an electric field at a point M situated at a distance r in the direction $\mathbf{u}$ (Fig. 3.2). Let us translate the antenna by a vector $\mathbf{d}$ to a new point O', such that the distance OO' is small in comparison with the distance $r = OM$. When it is at O', the antenna creates an electric field $\mathbf{E}'$ at point M in the same direction $\mathbf{u}$ (Fig. 3.2), at the distance $r' = O'M = r - \mathbf{u}.\mathbf{d}$.

$$\mathbf{E}' = \frac{\exp(-jkr')}{kr'}\mathbf{e}(\mathbf{u}) = \frac{\exp(-jkr)}{k(r+\mathbf{u}.\mathbf{d})}\exp(jk\mathbf{u}.\mathbf{d})\mathbf{e}(\mathbf{u}) \qquad (3.4)$$

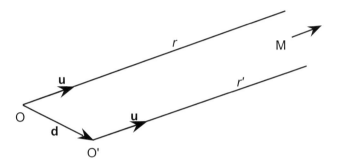

Fig. 3.2 Translation theorem.

Since the shift is small ($\mathbf{u}.\mathbf{d} \ll r$), we can neglect $\mathbf{u}.\mathbf{d}$ in comparison with r and write

$$\mathbf{E}' = \mathbf{E}\exp(jk\mathbf{u}.\mathbf{d}) \qquad (3.5)$$

This is the translation theorem (or shift theorem): when an antenna is translated by a vector $\mathbf{d}$, its radiated field in the direction $\mathbf{u}$ at distance r is simply multiplied by the factor $\exp(jk\mathbf{u}.\mathbf{d})$.

3.1.3 Application: radiation produced by an arbitrary current

From the magnetic field produced by an electric dipole we can calculate the electromagnetic field created by an arbitrary current flowing in a wire by the following steps

- The electromagnetic field radiated in the direction $\mathbf{u}$ by a dipole $I\ell\mathbf{u}_D$ located at point O is

$$\mathbf{H} = \frac{jk}{4\pi}\frac{\exp(-jkr)}{r}I\ell(\mathbf{u}_D \times \mathbf{u}), \qquad \mathbf{E} = \eta\mathbf{H}\times\mathbf{u} \tag{3.6}$$

- The electromagnetic field radiated in the direction $\mathbf{u}$ by an element of current $I\mathbf{dc}$ located at point O is

$$d\mathbf{H} = \frac{jk}{4\pi}\frac{\exp(-jkr)}{r}I(\mathbf{dc}\times\mathbf{u}), \qquad \mathbf{E} = \eta\mathbf{H}\times\mathbf{u} \tag{3.7}$$

- The electromagnetic field radiated in the direction $\mathbf{u}$ by an element of current $I\mathbf{dc}$ located at point M', defined by $\mathbf{r}' = OM'$ of a wire is

$$d\mathbf{H} = \frac{jk}{4\pi}\frac{\exp(-jkr)}{r}I\exp(jk\mathbf{u}.\mathbf{r}')(\mathbf{dc}\times\mathbf{u}), \qquad \mathbf{E} = \eta\mathbf{H}\times\mathbf{u} \tag{3.8}$$

- The electromagnetic field radiated in the direction $\mathbf{u}$ by a wire is

$$\mathbf{H} = \frac{-jk}{4\pi}\frac{\exp(-jkr)}{r}\mathbf{u}\times\int_C I(M')\exp(jk\mathbf{u}.\mathbf{r}')\,\mathbf{dc}, \qquad \mathbf{E} = \eta\mathbf{H}\times\mathbf{u} \tag{3.9}$$

- In the case of a surface distribution of current $\mathbf{J}_S(\mathbf{r}')$ over a surface S, we replace $I\mathbf{dc}$ by $\mathbf{J}_S dS$, the integral being taken over the surface S.

$$\mathbf{H} = \frac{-jk}{4\pi}\frac{\exp(-jkr)}{r}\mathbf{u}\times\iint_S \mathbf{J}_S(\mathbf{r}')\exp(jk\mathbf{u}.\mathbf{r}')\,dS, \qquad \mathbf{E} = \eta\mathbf{H}\times\mathbf{u} \tag{3.10}$$

- In the case of a volume distribution $\mathbf{J}(\mathbf{r}')$ in a volume V, we replace $I\mathbf{dc}$ by $\mathbf{J}\,dV$, the integral being taken over the volume V.

$$\mathbf{H} = \frac{-jk}{4\pi}\frac{\exp(-jkr)}{r}\mathbf{u}\times\iiint_V \mathbf{J}_S(\mathbf{r}')\exp(jk\mathbf{u}.\mathbf{r}')\,dV, \qquad \mathbf{E} = \eta\mathbf{H}\times\mathbf{u} \tag{3.11}$$

Application: calculation of the field radiated by a closed circuit of small dimensions; magnetic dipole

The circuit C is assumed to be closed, but not necessary planar (Fig. 3.3). A constant current I flows in it (the same at each point M' of the circuit). Its dimensions are small compared with the wavelength λ. Such a circuit is known as a magnetic dipole.

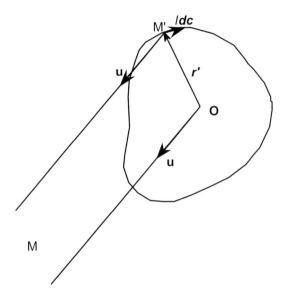

Fig. 3.3 Magnetic dipole.

The small dimensions allow us to replace $\exp(jk\mathbf{u}.\mathbf{r}')$ by its first order expansion

$$\exp(jk\mathbf{u}.\mathbf{r}') \approx 1 + jk\mathbf{u}.\mathbf{r}'$$

The electric field (equation (3.9)) may then be written

$$\mathbf{E} = \frac{jk\eta}{4\pi} \frac{\exp(-jkr)}{r} I\mathbf{u} \times \left[\mathbf{u} \times \int_C (1 + jk\mathbf{u}.\mathbf{r}') \, d\mathbf{c}\right] \qquad (3.13)$$

The integral is split into two

$$\int_C (1 + jk\mathbf{u}.\mathbf{r}') \, d\mathbf{c} = \int_C d\mathbf{c} + jk \int_C (\mathbf{u}.\mathbf{r}') \, d\mathbf{c} \qquad (3.14)$$

The first integral is zero since the circuit is closed. The second remains to be calculated

$$\int_C (\mathbf{u}.\mathbf{r}') \, d\mathbf{c} = \int_C (\alpha x' + \beta y' + \gamma z') \, d\mathbf{c} \qquad (3.15)$$

where (α, β, γ) are the components of **u** and (x', y', z') are the coordinates of the point M' of the circuit. This integral is calculated using the identity[1]

$$\iint_S \mathbf{n} \times \operatorname{grad} f \, dS = \int_S f \, d\mathbf{c} \tag{3.16}$$

Setting $f = (\alpha x' + \beta y' + \gamma z')$ and noting that $\mathbf{u} = \operatorname{grad}(\alpha x' + \beta y' + \gamma z')$, we obtain

$$\int_C (\mathbf{u}.\,\mathbf{OM}')d\mathbf{c} = \iint_S \mathbf{n} \times \mathbf{u} dS = \mathbf{S} \times \mathbf{u} \tag{3.17}$$

$\mathbf{S} = \int_S \mathbf{n} dS$ is called the surface-vector associated with the circuit C. The surface-vector is independent of the chosen surface S (as an exercise you should be able to show this). In the case of a plane surface whose area is S, we simply have $\mathbf{S} = S\mathbf{n}$.
Inserting this into the expression for the electromagnetic field, we obtain

$$\mathbf{E} = -\frac{k^2\eta}{4\pi}\frac{\exp(-jkr)}{r}\mathbf{u} \times I\mathbf{S}, \qquad \mathbf{H} = \frac{1}{\eta}\mathbf{u} \times \mathbf{E} \tag{3.18}$$

$\mathbf{M} = I\mathbf{S}$ is known as the magnetic torque of the circuit.

3.1.4 Radiated power

A wave from a (quasi) point source is said to be 'spherical'. Below we review some properties of spherical waves at a large distance ($r \gg \lambda$) from the source.

- Locally the character is that of a plane wave: $\mathbf{H} = \frac{1}{\eta}\mathbf{u} \times \mathbf{E}$

- The field amplitude decreases as $1/r$ and its phase varies as $-kr = -2\pi r/\lambda$.

- The field components include the factor $\exp(-jkr)/r$.

- The power dP crossing a surface element dS of the plane wave is (equation (1.29))

$$\frac{dP}{dS} = \frac{\mathbf{E}.\mathbf{E}^*}{2\eta}$$

---

[1] This identity is deduced from Stokes' theorem $\iint_S \mathbf{n}.\operatorname{curl}\mathbf{A} dS = \int_C \mathbf{A}.d\mathbf{c}$ setting $\mathbf{A} = f\mathbf{a}$, where **a** is a constant vector.

- In the case of an arbitrarily polarized wave, whose linear components have amplitudes $|E_1|$ and $|E_2|$

$$\frac{dP}{dS} = \frac{|E_1|^2 + |E_2|^2}{2\eta}$$

- In the case of a linearly polarized wave of amplitude E_0, this relation becomes

$$\frac{dP}{dS} = \frac{|E_0|^2}{2\eta}$$

dP/dS is the power flux density and η is the wave impedance ($\eta = 377\,\Omega$ in vacuum or in air).

The radiated power dP may also be expressed as a function of the solid angle $d\Omega = dS/r^2$ from the point O to the characteristic radiation vector $\mathbf{e}(\mathbf{u})$.

$$\frac{dP}{d\Omega} = \frac{1}{2\eta k^2}\, \mathbf{e}(\mathbf{u}) . \mathbf{e}*(\mathbf{u})$$

3.2 FIELD RADIATED FROM AN ANTENNA

3.2.1 Elementary dipoles

These are antennas made up of elementary currents.

Electric dipole

This is an element of rectilinear current of length ℓ which is very small compared with the wavelength λ, directed in a direction $\mathbf{u}_D$, and carrying a current of complex amplitude I.

The field radiated by an electric dipole in the direction $\mathbf{u}$ is

$$\mathbf{H} = -\frac{jk}{4\pi}\frac{\exp(-jkr)}{r} I\ell\, \mathbf{u} \times \mathbf{u}_D \qquad \mathbf{E} = \eta\, \mathbf{H} \times \mathbf{u} \qquad (3.19)$$

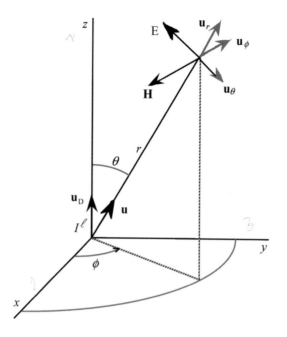

Fig. 3.4 Electric dipole.

- the electric field lies in the plane $(\mathbf{u}, \mathbf{u}_D)$;
- the magnetic field is normal to the plane $(\mathbf{u}, \mathbf{u}_D)$;
- the electromagnetic field is zero in the direction $\mathbf{u}_D$ of the dipole axis. It is maximum in the plane perpendicular to the dipole axis.

In spherical coordinates $\mathbf{u} \times \mathbf{u}_D = -\mathbf{u}_\phi \sin\theta$. The field comprises only components E_θ and H_ϕ, expressed as

$$E_\theta = \frac{jk}{4\pi}\eta\frac{\exp(-jkr)}{r}I\ell\sin\theta \tag{3.20a}$$

$$H_\phi = \frac{jk}{4\pi}\frac{\exp(-jkr)}{r}I\ell\sin\theta \tag{3.20b}$$

The flux density of the radiated power is

$$\frac{dP_{\mathrm{rad}}}{dS} = \frac{\mathbf{E}.\mathbf{E}^*}{2\eta} = \frac{E_\theta E_\theta^{\,*}}{2\eta} \tag{3.21a}$$

$$\frac{dP_{\text{rad}}}{dS} = \frac{\eta}{8\lambda^2}|I\ell|^2 \frac{\sin^2\theta}{r^2}$$ (3.21b)

The total radiated power P_{rad} is obtained by integrating over a sphere of radius r

$$P_{\text{rad}} = \frac{\pi}{3\lambda^2}\eta|I\ell|^2$$ (3.22)

Magnetic dipole

This is a single turn of wire of dimensions very small compared with λ, characterized by its surface vector **S**, by the complex amplitude I of the current it carries, and by its magnetic-torque vector **M** = I**S**. The simplest case is that of a plane single turn bounding a surface S, whose normal points in the direction $\mathbf{u}_D$. Then, we have **S** = $S\mathbf{u}_D$ and **M** = $IS\mathbf{u}_D$.

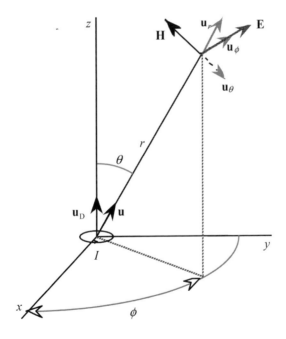

Fig. 3.5 Magnetic dipole.

The expression for the field radiated by the magnetic dipole is

$$E = -\frac{k^2}{4\pi}\eta\frac{\exp(-jkr)}{r}SI\ \mathbf{u}\times\mathbf{u}_D\ , \qquad H = \frac{1}{\eta}\mathbf{u}\times\mathbf{E} \qquad (3.23)$$

- the electric field is normal to the plane $(\mathbf{u}, \mathbf{u}_D)$
- the magnetic field lies in the plane $(\mathbf{u}, \mathbf{u}_D)$
- the electromagnetic field is zero in the direction $\mathbf{u}_D$ of the dipole axis. It is maximum in the plane perpendicular to the dipole axis (plane of the single turn).

In spherical coordinates $\mathbf{u}\times\mathbf{u}_D = -\mathbf{u}_\phi \sin\theta$. The field comprises only components E_ϕ and H_θ, expressed as

$$E_\phi = \frac{k^2}{4\pi}\eta\frac{\exp(-jkr)}{r}SI\sin\theta \qquad (3.24a)$$

$$H_\theta = -\frac{k^2}{4\pi}\frac{\exp(-jkr)}{r}SI\sin\theta \qquad (3.24b)$$

The flux density of radiated power is

$$\frac{dP_{rad}}{dS} = \frac{\mathbf{E}.\mathbf{E}^*}{2\eta} = \frac{E_\phi E_\phi{}^*}{2\eta} \qquad (3.25a)$$

$$\frac{dP_{rad}}{dS} = \frac{\pi^2\eta}{2\lambda^4}|SI|^2\frac{\sin^2\theta}{r^2} \qquad (3.25b)$$

The total radiated power P_{rad} is obtained by integrating over a sphere of radius r

$$P_{rad} = \frac{4\pi^3}{3\lambda^4}\eta|SI|^2 \qquad (3.26)$$

3.2.2 Plane-aperture radiation

General formula

Let us consider a plane 'aperture' S in the xy plane. The components $E_x(x,y,0)$ and $E_y(x,y,0)$ in the plane of the aperture are assumed to be zero outside this aperture.

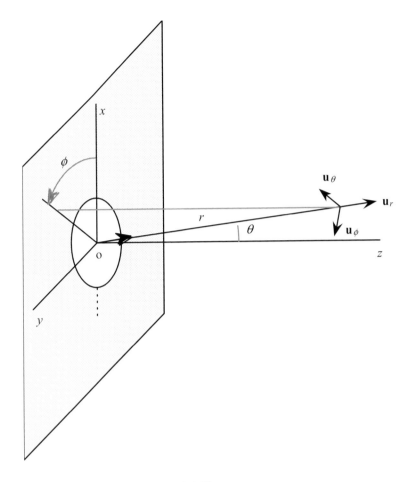

Fig. 3.6 Plane aperture.

We have demonstrated (Chapter 2, §2.1.3) that the field at large distance in the direction **u** can be calculated from the components $E_x(x,y,0)$ and $E_y(x,y,0)$ of the tangential electric field over the xy plane of the aperture. We limit ourselves to the case where one of the components (for instance E_y) is zero. The other one is denoted $E(x,y)$

$$E_x(x,y,0) = E(x,y)$$
$$E_y(x,y,0) = 0$$

The calculation of the radiated field introduces the inverse Fourier transform $E(\alpha, \beta)$ of the tangential electric field over the aperture

$$E(\alpha, \beta) = \frac{1}{\lambda^2} \iint E(x, y) \exp\left[jk(\alpha x + \beta y)\right] dxdy \qquad (3.27)$$

The integration domain is the xy plane, but we may limit it to the aperture S.

The theory of plane apertures gives the expression for the far-electric field at a distance r which is very large in comparison with λ and with the aperture dimensions. Using equations (2.25), where $E_{0t}(\alpha, \beta) = E(\alpha, \beta)\mathbf{u}_x$, we obtain

$$\mathbf{E} = -j\lambda \frac{\exp(-jkr)}{r} E(\alpha, \beta)\mathbf{u} \times \mathbf{u}_y \qquad (3.28)$$

and making use of the fact that $\mathbf{u}_\theta \cos \phi - \mathbf{u}_\phi \cos \theta \sin \phi = -\mathbf{u} \times \mathbf{u}_y$,

$$\mathbf{E} = j\lambda \frac{\exp(-jkr)}{r} E(\alpha, \beta)\left(\mathbf{u}_\theta \cos \phi - \mathbf{u}_\phi \cos \theta \sin \phi\right) \qquad (3.29)$$

We may compare the aperture radiation (equation 3.28) to the electromagnetic field of a magnetic dipole oriented according to Oy and characterized by SI.

$$\mathbf{E} = -\frac{k^2}{4\pi} \eta SI \frac{\exp(-jkr)}{r} \mathbf{u} \times \mathbf{u}_y$$

We see that the polarization of the electromagnetic field is the same.

Radiated power

The density of radiated power is

$$\frac{dP_{\text{rad}}}{dS} = \frac{\mathbf{E}.\mathbf{E}^*}{2\eta} = \frac{\lambda^2\left(1 - \beta^2\right)}{2\eta r^2}\left|E(\alpha, \beta)\right|^2 \qquad (3.30)$$

The total power radiated is obtained by integration (difficult) over a hemisphere of radius r. But in many cases the apertures are large (compared to λ) and their radiation is concentrated about the Oz axis (property of Fourier transform). Neglecting the second order term in θ we obtain in the case of apertures radiating in the Oz direction

$$\frac{dP_{rad}}{dS} = \frac{\lambda^2}{2\eta r^2}\left|E\left(\alpha,\beta\right)\right|^2 \tag{3.31}$$

$$P_{rad} = \frac{\lambda^2}{2\eta}\iint\left|E\left(\alpha,\beta\right)\right|^2 d\alpha\ d\beta \tag{3.32}$$

The integration domain, which in principle extends from $-\infty$ to $+\infty$ may be limited since $E(\alpha,\beta)$ is only non-zero in the vicinity of $(\alpha=0,\ \beta=0)$.

Uniformly illuminated rectangular aperture (see exercise 3.7 and §7.4.4)

$$E(x,y)=\begin{cases}\mathbf{E}_0\ \text{inside the aperture}\\0\ \text{outside the aperture}\end{cases}$$

$$E\left(\alpha,\beta\right)=\mathrm{E}_0\frac{ab}{\lambda^2}\frac{\sin\left(\pi\dfrac{\alpha a}{\lambda}\right)}{\pi\dfrac{\alpha a}{\lambda}}\frac{\sin\left(\pi\dfrac{\beta b}{\lambda}\right)}{\pi\dfrac{\beta b}{\lambda}} \tag{3.33}$$

in which $\alpha=\sin\theta\cos\phi,\beta=\sin\theta\sin\phi$

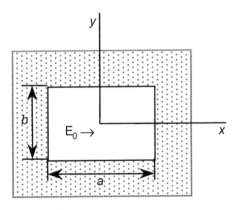

Fig. 3.7 Rectangular aperture.

Uniformly illuminated circular aperture (see exercise 3.8 and §7.4.4)

$$E(x, y) = \begin{cases} \mathbf{E}_0 \text{ inside the aperture} \\ 0 \text{ outside the aperture} \end{cases}$$

$$E = \mathbf{E}_0 \frac{\pi a^2}{\lambda^2} \frac{2\mathrm{J}_1(ka\sin\theta)}{ka\sin\theta} \qquad (3.34)$$

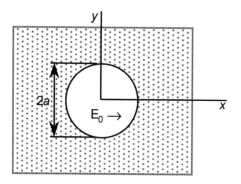

Fig. 3.8 Circular aperture.

3.3 DIRECTIVITY, GAIN, RADIATION PATTERN

3.3.1 Radiated power

The total power radiated by the antenna is obtained by integration using one of the following formulae:

- Integration over the sphere of centre O and radius r

$$P_{\text{rad}} = \frac{1}{2\eta} \iint_S \frac{\mathbf{E}.\mathbf{E}^*}{2\eta} dS \qquad (3.35a)$$

- In spherical coordinates $dS = r^2 \sin\theta\, d\theta\, d\varphi$

$$P_{rad} = \frac{1}{2\eta k^2} \int\limits_{\theta=0}^{\theta=\pi} \int\limits_{\varphi=0}^{\varphi=2\pi} \mathbf{e}.\mathbf{e}* \sin\theta \, d\theta \, d\varphi \qquad (3.35b)$$

- By using the components $\alpha = \sin\theta\cos\varphi, \beta = \sin\theta\sin\varphi, \gamma = \cos\theta$ of the unit vector $\mathbf{u}$ (the domain of integration is then the circle C defined by $\alpha^2 + \beta^2 = 1$)

$$P_{rad} = \frac{1}{2\eta k^2} \iint_C \mathbf{e}.\mathbf{e}* \frac{d\alpha \, d\beta}{\gamma} \qquad (3.35c)$$

- By using solid angles (the domain of integration, corresponding to all space, is then 4π steradians)

$$P_{rad} = \frac{1}{2\eta k^2} \iint_{(4\pi)} \mathbf{e}.\mathbf{e}* \, d\Omega \qquad (3.35d)$$

3.3.2 Directivity

The average flux density of radiated power is the value that the power density would have if the antenna radiated isotropically.

$$\left(\frac{dP_{rad}}{dS}\right)_{average} = \frac{P_{rad}}{4\pi r^2}$$

$$\left(\frac{dP_{rad}}{d\Omega}\right)_{average} = \frac{P_{rad}}{4\pi}$$

The directivity $D(\mathbf{u})$ describes the spatial distribution of the radiation as

$$D(\mathbf{u}) = \frac{\text{power flux density in direction } \mathbf{u}}{\text{average power flux density}} = \frac{\dfrac{dP_{rad}}{dS}}{\dfrac{P_{rad}}{4\pi r^2}} = \frac{\dfrac{dP_{rad}}{d\Omega}}{\dfrac{P_{rad}}{4\pi}} \qquad (3.36a)$$

Taking into account the expressions for dP/dS (§ 3.1.4) and those for the total radiated power (equations 3.35), the directivity is expressed as a function of the characteristic radiation vector radiation by

$$D(\mathbf{u}) = 4\pi r^2 \frac{\mathbf{E}.\mathbf{E}^*}{\displaystyle\iint_S \mathbf{E}.\mathbf{E}^* \, dS} = 4\pi \frac{\mathbf{e}.\mathbf{e}^*}{\displaystyle\iint_{(4\pi)} \mathbf{e}.\mathbf{e}^* \, d\Omega} \tag{3.36b}$$

Table 3.1 gives some examples (see also exercises 3.1 – 3.5). They show a general property of apertures, that the maximum theoretical directivity of a plane aperture of area S is

$$D_{\max} = 4\pi S / \lambda^2 \tag{3.37}$$

Maximum directivity may only be obtained with uniform illumination of the aperture (same amplitude and phase)[2].

3.3.3 Gain

The antenna feed line receives a power P_a from the transmitter (P_a is the difference between the incident and reflected powers). The radiated power P_{rad} is less than or equal to P_a due to possible ohmic losses in the antenna

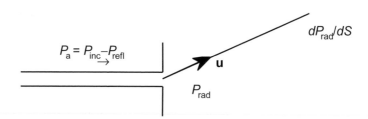

Fig. 3.9 Incident, reflected and radiated power from the antenna.

Gain (or absolute gain): The antenna gain $G(\mathbf{u})$ takes into account these losses, comparing the power density effectively radiated to the average density $P_a/4\pi r^2$ that would exist if all the power accepted by the antenna were radiated (assuming zero losses).

$$G(\mathbf{u}) = \frac{\text{effective power density in direction } \mathbf{u}}{\text{average density assuming zero losses}} = \frac{\dfrac{dP_{\text{rad}}}{dS}}{\dfrac{P_a}{4\pi r^2}} = \frac{\dfrac{2\eta\,\mathbf{E}.\mathbf{E}^*}{P_a}}{4\pi r^2} \tag{3.38}$$

---

[2] This property does not take into account the phenomenon of superdirectivity. This is treated in §13.4.

Antenna	Radiated power	Directivity	D_{max}
Electric dipole	$$\frac{dP_{rad}}{dS} = \frac{\eta}{8\lambda^2}|I\ell|^2 \frac{\sin^2\theta}{r^2}$$ $$P_{rad} = \frac{\pi}{3\lambda^2}\eta|I\ell|^2$$	$\frac{3}{2}\sin^2\theta$	$\frac{3}{2}$
Magnetic dipole	$$\frac{dP_{rad}}{dS} = \frac{\pi^2\eta}{2\lambda^4}|IS|^2 \frac{\sin^2\theta}{r^2}$$ $$P_{rad} = \frac{4\pi^3}{3\lambda^4}\eta|IS|^2$$	$\frac{3}{2}\sin^2\theta$	$\frac{3}{2}$
$\cos^\nu\theta$ power radiation law over the half space	$$A\frac{\cos^\nu\theta}{r^2}\ (\theta \le \pi/2)$$ $$P_{rad} = \frac{2\pi A}{\nu+1}$$	$2(\nu+1)\cos^\nu\theta$	$2(\nu+1)$
Plane aperture S of large dimensions uniformly illuminated	$$\frac{dP_{rad}}{dS} = \frac{\lambda^2}{2\eta r^2}|E|^2$$ $$P_{rad} = \frac{\lambda^2}{2\eta}\iint|E|^2 d\alpha d\beta$$	$\frac{2\pi\lambda^2}{\eta P_{rad}}|E|^2$	$\frac{4\pi S}{\lambda^2}$
Rectangle $(a,b)\gg\lambda$ uniformly illuminated	$$E = E_0\frac{ab}{\lambda^2}\frac{\sin(\pi\alpha a/\lambda)}{\pi\alpha a/\lambda}\frac{\sin(\pi\beta b/\lambda)}{\pi\beta b/\lambda}$$		$\frac{4\pi S}{\lambda^2}$
Circle of radius $a\gg\lambda$, uniformly illuminated	$$E = E_0\frac{\pi a^2}{\lambda^2}\frac{2J_1(ka\sin\theta)}{ka\sin\theta}$$		$\frac{4\pi S}{\lambda^2}$

Table 3.1 Radiated power and directivity of simple antennas.

Realized gain: the gain (or *absolute gain*) defined by eqn. (3.38) is in terms of the power P_a accepted by the antenna, i.e. the power which effectively penetrates into the antenna. This is less than or equal to the power P_{inc} on the line (or waveguide) feeding the antenna, because of the mismatch between the antenna and the line. Replacing in eqn. (3.38) the accepted power P_a by the incident power P_{inc} defines the *realized gain*. This is less than (or equal to in the case of a matched impedance) to the absolute gain. The reference[3] gives a precise definition of the various gains used in the domain of antennas.

Gain and directivity are related by the expression

$$\frac{G(\mathbf{u})}{D(\mathbf{u})} = \frac{P_{rad}}{P_a} \le 1 \tag{3.39}$$

A knowledge of the gain and of the power P_a entering the antenna allows the amplitude $E(r, \mathbf{u})$ of the field radiated by the antenna at a distance r and direction $\mathbf{u}$ to be determined

$$E(r,\mathbf{u}) = \frac{1}{r}\sqrt{\frac{\eta P_a G(\mathbf{u})}{2\pi}} = \frac{1}{r}\sqrt{60 P_a G(\mathbf{u})}$$

3.3.4 Radiation pattern

Definition

The tip of the vector $G(\mathbf{u})\mathbf{u}$, traced from the point O describes a surface known as the radiation pattern of the antenna (Fig. 3.10). This pattern is characterized by the angular width of the main lobe at its −3 dB points, and its sidelobe levels.

---

[3] IEEE Standard Definitions of Terms for Antennas; IEEE Std 145-1993; approved 18 March 1993, IEEE Standards Board.

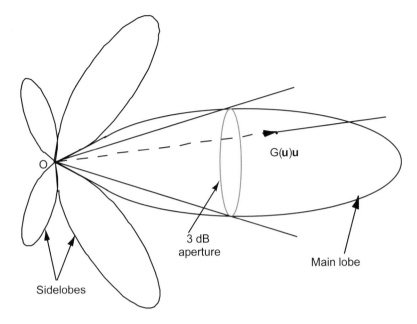

Fig. 3.10 Radiation pattern.

By the same principle, radiation patterns can also be plotted for the directivity, field amplitude (Fig. 3.11 and Fig. 3.12) and field phase.

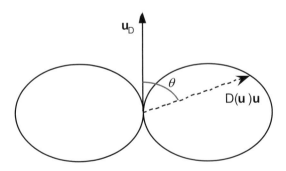

Fig. 3.11 Directivity radiation pattern of an elementary dipole.

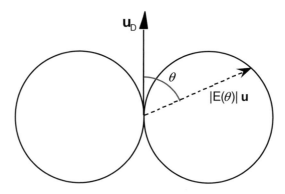

Fig. 3.12 Field amplitude radiation pattern of an elementary dipole.

3.3.5 Input impedance

At the end of the transmission line which feeds it, the antenna presents a load, through which flows a current I. This load impedance $Z_a = R_a + jX_a$ is called the *input impedance*.

- The input resistance R_a (which is related to the power P_a absorbed by the antenna) is defined by

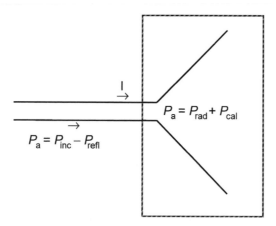

Fig. 3.13 Radiated power.

$$P_a = \frac{1}{2} R_a |I|^2 \tag{3.40}$$

P_a comprises two parts: the radiated power P_{rad} and the heat power P_{cal}, thus $P_a = P_{\text{rad}} + P_{\text{cal}}$ (Fig. 3.13). We then define the radiation resistance R_{rad} and the resistance R_{cal} describing heat losses by

$$P_{\text{rad}} = \frac{1}{2} R_{\text{rad}} |I|^2 \qquad P_{\text{cal}} = \frac{1}{2} R_{\text{cal}} |I|^2 \tag{3.41}$$

- The input reactance X_a is related to the electromagnetic energy stored in the vicinity of the antenna.

Radiation resistance of elementary dipoles

From equations (3.22), (3.26) and (3.41) we deduce

- the radiation resistance of an electric dipole

$$R_{\text{rad}} = \frac{2\pi}{3} \eta \left(\frac{\ell}{\lambda} \right)^2 = 800 \left(\frac{\ell}{\lambda} \right)^2 \quad (\Omega) \tag{3.42}$$

- the radiation resistance of a magnetic dipole

$$R_{\text{rad}} = \frac{8\pi^3}{3} \eta \left(\frac{S}{\lambda^2} \right)^2 = 31200 \left(\frac{S}{\lambda^2} \right)^2 \quad (\Omega) \tag{3.43}$$

FURTHER READING

1. Balanis, C.A., *Antenna Theory and Design*, John Wiley & Sons, New York, 2nd edition, 1997.
2. Collin, R.E., *Antennas and Radiowave Propagation*, McGraw-Hill, New York, 1985.
3. Collin, R.E. and Zucker, F.J. (eds), *Antenna Theory*, Parts 1 and 2, McGraw-Hill, New York, 1969.
4. Schelkunoff, S.A. and Friis, H.T., *Antenna Theory and Practice*, Wiley, New York, 1952.

EXERCISES

3.1 Realization of circular polarization

Suggest a way to radiate circular polarization in the direction Oz in each of the following configurations:

(a) using two electric dipoles at the origin O;
(b) using an electric dipole and a magnetic dipole at O.

Which solution do you think is better ?

3.2 Crossed dipoles

Two elementary electric dipoles are placed at O and oriented respectively along Ox and Oy. They are fed by currents of the same amplitude, and phases ϕ_1 and ϕ_2 respectively.

Determine the form of the polarization of the electromagnetic field radiated in the directions $\mathbf{u} = \mathbf{u}_z$ and $\mathbf{B} = \dfrac{1}{\sqrt{3}}[1, 1, 1]$ when:

(a) $\phi_1 = \phi_2 = 0$
(b) $\phi_1 - \phi_2 = \pi/2$

3.3 Two electric dipoles

Two identical electric dipoles, oriented along Oz, are placed on the Ox axis at a separation of $d = 2\lambda/3$ (the first is at $x = +\lambda/3$, the second at $x = -\lambda/3$). They are used on transmit, and fed with the same amplitude.

What should the phase shift $\phi = \phi_2 - \phi_1$ between the two dipoles be in order to obtain:

(a) a maximum in the direction Ox
(b) a null in the direction Ox
(c) a maximum in the direction bisecting xOz ?

3.4 Electric dipole and magnetic dipole

An elementary electric dipole $(I\ell)$, oriented along Ox, is used with a magnetic dipole along Oy and of moment M.

(a) Compare the amplitudes and directions of the electric fields radiated along the axis Oz by the two dipoles. What should the relation between M and $I\ell$ be in order that these fields should be equal ?

(b) Suppose that the preceding relation is satisfied. Determine the total electric field at an arbitrary point (r, θ, ϕ) by calculating the following:

3.5 *Antenna radiating into a half-space with a function $\cos^\nu\theta$*

An antenna radiates into the region $z > 0$ according to $\dfrac{dP}{dS} = \dfrac{A}{r^2}\cos^\nu\theta$.

Q1 Show that the total radiated power is $P_{rad} = \dfrac{2\pi A}{\nu+1}$

Q2 Show that the directivity is $D(\theta) = 2(\nu+1)\cos^\nu\theta$

3.6 *Maximum directivity of an aperture antenna*

Consider an aperture of area S in the plane xOy, whose dimensions are large compared with the wavelength λ. The tangential electric field in the aperture is oriented along Ox:

$$\mathbf{E}_t(x,y,0) = E(x,y)\mathbf{u}_x$$

where $E(x,y)$ is a real function.

Q1 Show that the radiated field will in general be concentrated around the axis Oz.

Q2 Show that the power density radiated at a distance r in the direction $\mathbf{u}$ defined by (α, β) is:

$$\frac{dP}{dS} = \frac{\lambda^2}{2\eta r^2}|E(\alpha,\beta)|^2$$

Q3 Using Parseval's theorem, show that the total radiated power is

$$P_{rad} = \frac{1}{2\eta}\iint_S |E(x,y)|^2 dS$$

Q4 Using Parseval's theorem (equation 2.3), show that the general expression for the directivity is:

$$D(\alpha,\beta) = \frac{4\pi}{\lambda^2} \frac{\left| \iint_S E(x,y)\exp[\,jk(\alpha x + \beta y]dxdy \right|^2}{\iint_S |E(x,y)|^2\,dxdy}$$

Q5 Using Schwartz's inequality $\left| \int_S f\,g\,dS \right|^2 \leq \int_S |f|^2\,dS.\int_S |g|^2\,dS$, deduce from this that

$$D(\alpha,\beta) \leq \frac{4\pi S}{\lambda^2}$$

Q6 Show that the maximum directivity $D_{max} = \dfrac{4\pi S}{\lambda^2}$ is obtained when the illumination function is uniform (E(x, y) = constant) and in direction ($\alpha = 0$, $\beta = 0$).

3.7 *Rectangular aperture*

A rectangular aperture of dimensions (*a*, *b*) much greater than the wavelength, and whose centre lies at the origin of the coordinate system, is uniformly illuminated. Show that its directivity is given by

$$D(\alpha,\beta) = \frac{4\pi A}{\lambda^2}\left[\frac{\sin(k\alpha a/2)}{k\alpha a/2}\right]^2\left[\frac{\sin(k\beta b/2)}{k\beta b/2}\right]^2$$

3.8 *Circular aperture*

Consider a circular aperture of radius *a* whose centre lies at the origin of the coordinate system. The tangential field $\mathbf{E}_t$ at a point P is oriented along O*x* and depends only on the distance $\rho = \mathrm{OP} = \sqrt{x^2 + y^2}$

$$\mathbf{E}_t = E(\rho)\mathbf{u}_x$$

Q1 Making use of the relations:

$$\int_0^{2\pi} \exp(ju\cos x)dx = 2\pi J_0(u) \quad \text{and} \quad \frac{d}{dx}\left[xJ_1(x)\right] = xJ_0(x)$$

show that the 2D inverse Fourier transform of E(ρ) is

$$E(s) = \frac{2\pi}{\lambda^2} \int\limits_0^\infty E(\rho) J_0(k\rho s) \rho d\rho \quad \text{where} \quad s = \sqrt{\alpha^2 + \beta^2}$$

Q2 Deduce from this the field radiated in direction **u** making an angle θ with Oz (note that $s = \sin\theta$).

3.9 *Uniformly illuminated circular aperture*

A circular aperture of radius $a \gg \lambda$, whose centre lies at the origin of the coordinate system, and is uniformly illuminated. Show that its directivity is

$$D(\theta) = \frac{4\pi^2 a^2}{\lambda^2} \left[\frac{2 J_1(ka\sin\theta)}{ka\sin\theta} \right]^2$$

4

Receiving antennas

Guglielmo Marconi (1874–1937)

4.1 ANTENNA RECIPROCITY THEOREM

4.1.1 Reciprocity theorem applied to a source-free closed surface

Consider a source-free closed surface and two solutions of Maxwell's equations ($\mathbf{E}_1$, $\mathbf{H}_1$) and ($\mathbf{E}_2$, $\mathbf{H}_2$), generated by sources outside the surface S. The reciprocity theorem (see equation (1.18)) is expressed as follows:

$$\iint_S \left(\mathbf{E}_1 \times \mathbf{H}_2 - \mathbf{E}_2 \times \mathbf{H}_1 \right).\mathbf{n}\,dS = 0 \qquad (4.1)$$

The fields $(\mathbf{E}_1, \mathbf{H}_1)$ and $(\mathbf{E}_2, \mathbf{H}_2)$ are respectively those associated with an antenna when it is transmitting and when it is receiving, and S is a closed surface as depicted in Fig. 4.1. The vector $\mathbf{n}$ is the inward normal to the surface.

Surface used for the application of the reciprocity theorem

The closed surface S is a sphere of large radius r, centred on the antenna and including a portion S_g of the feed waveguide located in the reference plane $z = 0$ and a surface enclosing the metal surroundings of the antenna (Fig. 4.1). The contribution of the metal surface of the antenna to the integral of equation (4.1) is zero if the metal is perfectly conducting and even if the metal is imperfectly conducting (skin effect). Only the contribution of the waveguide cross-section S_g and that of the sphere of radius r remains.

Field $(\mathbf{E}_1, \mathbf{H}_1)$ when the antenna is transmitting

Over an antenna feed waveguide: The antenna (Fig. 4.2) is connected to a waveguide (or transmission line) whose reference plane is the origin plane $z = 0$. Over the antenna feed waveguide, the electromagnetic field is of the form

$$\mathbf{E}_1 = A\mathbf{e}_g(x, y)\exp(-j\beta z), \quad \mathbf{H}_1 = \frac{1}{Z_g}\mathbf{u}_z \times \mathbf{E}_1 \qquad (4.2)$$

• $\mathbf{e}_g$ is the normalized transverse electric distribution in the straight section S_g of the waveguide

$$\iint_{S_g} |\mathbf{e}_g|^2 = 1 \qquad (4.3)$$

• Z_g is the impedance of the propagation mode of the waveguide.

• A is a complex number expressing the amplitude and phase of the field. It is related to the power P_T delivered to the antenna by the relation

$$P_T = \frac{|A|^2}{2Z_g} \qquad (4.4)$$

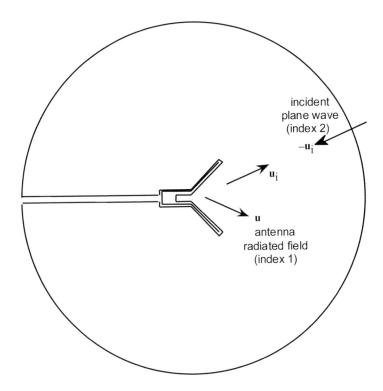

Fig. 4.1 Surface used for the application of the reciprocity theorem.

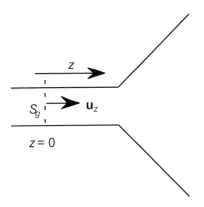

Fig. 4.2 Antenna fed by waveguide.

Over free space: In the direction **u** at a distance r, the antenna radiates an electromagnetic field

$$\mathbf{E}_1 = \frac{\exp(-jkr)}{kr}\mathbf{e}(\mathbf{u}), \quad \mathbf{H}_1 = \frac{1}{\eta}\mathbf{u} \times \mathbf{E}_1 \tag{4.5}$$

The vector function $\mathbf{e}(\mathbf{u})$ of the radiation (§3.1.1) contains all the information about the amplitude, phase and polarization of the field transmitted in the direction $\mathbf{u}$. The power flux density in the direction $\mathbf{u}$ is then deduced

$$\left[\frac{dP}{dS}\right]_{\text{trans}} = \frac{1}{2\eta k^2 r^2}\left|\mathbf{e}(\mathbf{u})\right|^2 \tag{4.6}$$

from which we obtain the gain

$$G(\mathbf{u}) = \frac{\left[\dfrac{dP}{dS}\right]_{\text{trans}}}{\dfrac{P_T}{4\pi r^2}} = \frac{\lambda^2 Z_g}{\pi\eta}\frac{\left|\mathbf{e}(\mathbf{u})\right|^2}{\left|A\right|^2} \tag{4.7}$$

Field $(\mathbf{E}_2,\mathbf{H}_2)$ when the antenna is receiving

An incoming plane wave travels toward the antenna from the direction $\mathbf{u}_i$. Its propagation direction is $-\mathbf{u}_i$. Its polarization is represented by the unit vector $\mathbf{p}$ normal to $\mathbf{u}_i$. Its amplitude and phase are determined by the complex number E_0. The field at any arbitrary point, whose vector position with respect to O is $r\mathbf{u}$, is

$$\mathbf{E}_2 = E_0\mathbf{p}\exp(jkr\mathbf{u}.\mathbf{u}_i), \quad \mathbf{H}_2 = -\frac{1}{\eta}\mathbf{u}_i \times \mathbf{E}_2 \tag{4.8}$$

The power flux density of this plane wave is

$$\left[\frac{dP}{dS}\right]_{\text{inc}} = \frac{\left|E_0\right|^2}{2\eta}$$

This power flux density is received by the antenna, where the wave creates a transversal electromagnetic field in the feed waveguide such that

$$\mathbf{E}_2 = B\mathbf{e}_g(x,y)\exp(j\beta z), \quad \mathbf{H}_2 = -\frac{1}{Z_g}\mathbf{u}_z \times \mathbf{E}_2 \tag{4.9}$$

where B is a complex number describing the amplitude and phase of the field. The power P_R transported in the waveguide towards the receiver is

$$P_R = \frac{|B|^2}{2Z_g} \qquad (4.10)$$

Calculation of the reciprocity integral

The contribution to the reciprocity integral (4.1) of the waveguide cross-section S_g is calculated by taking into account $\mathbf{n} = -\mathbf{u}_z$, and using equations (4.2), (4.3) and (4.9) we find

$$\iint_{S_g} (\mathbf{E}_1 \times \mathbf{H}_2 - \mathbf{E}_2 \times \mathbf{H}_1).\mathbf{n}\, dS = 2\frac{AB}{Z_g} \qquad (4.11)$$

The contribution of the integral over the sphere of radius r is evaluated by taking into account $\mathbf{n} = \mathbf{u}$, and using equations (4.5) and (4.8), we obtain

$$\iint_{\text{sphere}} (\mathbf{E}_1 \times \mathbf{H}_2 - \mathbf{E}_2 \times \mathbf{H}_1).\mathbf{n}\, dS$$

$$= \frac{\exp(-jkr)}{kr\eta} \iint_{\text{sphere}} \left[(\mathbf{p}.\mathbf{e})(\mathbf{u}.\mathbf{u}_i + 1) + (\mathbf{p}.\mathbf{u})(\mathbf{e}.\mathbf{u}_i) \right] \exp(jkr\mathbf{u}.\mathbf{u}_i)\, dS$$

When kr tends to infinity, this integral is evaluated by the method of stationary phase (§2.1.3)

$$\iint_{\text{sphere}} (\mathbf{E}_1 \times \mathbf{H}_2 - \mathbf{E}_2 \times \mathbf{H}_1).\mathbf{n}\, dS = \frac{j\lambda^2}{\pi\eta} \mathbf{p}.\mathbf{e}(\mathbf{u}) \qquad (4.12)$$

4.1.2 Relation between the field on transmit and the field on receive

Combining equations (4.11) and (4.12), we obtain a relationship between the fields on transmit and on receive

$$AB = \frac{-j\lambda^2}{2\pi} \frac{Z_g}{\eta} E_0\, \mathbf{p}.\mathbf{e}(\mathbf{u})$$

For the direction $\mathbf{u}$, this equation relates the fields existing in the feed waveguide when the antenna is transmitting (term A) or receiving (term B). We deduce the power P_R effectively received by the antenna when it is placed in the plane-wave field

$$P_R = \frac{|B|^2}{2Z_g} = \frac{\lambda^4}{8\pi^2} \frac{Z_g |E_0|^2}{\eta^2 |A|^2} |\mathbf{p.e}(\mathbf{u})|^2 \qquad (4.13)$$

P_R is a maximum when the polarization of the incident wave, defined by the unit vector $\mathbf{p}$, is coincident with that of the wave transmitted by the antenna, defined by $\mathbf{e}(\mathbf{u})$. We have then $|\mathbf{p.e}(\mathbf{u})|^2 = |\mathbf{e}(\mathbf{u})|^2$ and

$$P_R = \frac{|B|^2}{2Z_g} = \frac{\lambda^4}{8\pi^2} \frac{Z_g |E_0|^2}{\eta^2 |A|^2} |\mathbf{e}(\mathbf{u})|^2 \qquad (4.14)$$

from which we deduce the relation between the power received by the antenna and the gain of the antenna (equation (4.7))

$$P_R = \frac{\lambda^2}{4\pi} \left[\frac{dP}{dS} \right]_{inc} G(\mathbf{u}) \qquad (4.15)$$

where $\left[\dfrac{dP}{dS} \right]_{inc} = \dfrac{|E_0|^2}{2\eta}$ is the power flux density of the incident plane wave.

This relationship shows that the gain, as defined on transmit, equally describes the properties of the antenna on receive.

4.2 ANTENNA EFFECTIVE RECEIVING AREA

4.2.1 Definition

An antenna on receive is located in the field of a plane wave incoming from the direction $\mathbf{u}$ (Fig. 4.3). This antenna is connected to the receiver by a line (or waveguide) which we suppose to be perfectly matched to the antenna. The power P_R effectively received by the antenna (which is delivered to the receiver) is proportional to the power density of the incident plane wave. This property allows us to define the effective receiving area $A_e(\mathbf{u})$ of the antenna by the relation

$$P_R = A_e(\mathbf{u})\left[\frac{dP}{dS}\right]_{inc} \tag{4.16}$$

The effective receiving area A_e not only depends on the direction of observation $\mathbf{u}$, but also on the geometry of the antenna, and the material from which it is made. It also depends, for that direction of observation, on the polarization characteristics of the incident wave. It is a maximum when these characteristics are coincident (matched polarization).

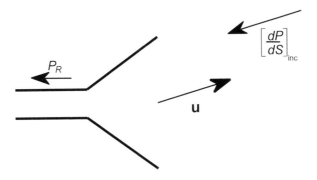

Fig. 4.3 Antenna effective receiving area.

4.2.2 Relationship between gain and effective receiving area

When the polarization of the antenna and the incident wave are coincident, comparison of equations (4.15) and (4.16) gives the relationship between the effective receiving area and the antenna gain in the direction $\mathbf{u}$

$$A_e(\mathbf{u}) = \frac{\lambda^2}{4\pi}G(\mathbf{u}) \tag{4.17}$$

This simple relationship allows us to use the transmission gain instead of the effective receiving area to describe the antenna properties on receive. Since the antenna maximum gain of the aperture cannot exceed $4\pi S/\lambda^2$, where S is the physical area of the aperture (§ 3.3.2), we have

$$A_e \leq S \tag{4.18}$$

The effective receiving area is at most equal to the physical area of the antenna.

4.3 ENERGY TRANSMISSION BETWEEN TWO ANTENNAS

4.3.1 The Friis transmission formula

Consider the transfer of power in a radio link between a transmitting antenna and a receiving antenna, separated by a distance r which is large with respect to the antenna dimensions and the wavelength (Fig. 4.4). These antennas are assumed to be correctly polarized and impedance matched to the transmission lines connecting both the receiver and transmitter. The link is defined by the following parameters

- P_T is the power accepted by the transmitting antenna.

- P_R is the power received at the terminal of the receiving antenna.

- G_T is the transmission antenna gain (in the direction of the receiving antenna).

- G_R is the receiving antenna gain (in the direction of the transmitting antenna).

- A_{eR} is the effective area of the receiving antenna (in the direction of the transmitting antenna).

The power received at the terminal of the receiving antenna is $P_R = A_{eR} \, dP/dS$ where $dP/dS = P_T G_T / (4\pi r^2)$ is the power flux density at the receiving antenna and $A_{eR} = G_R \lambda^2 / (4\pi)$ is the effective area of the receiving antenna.

From these we deduce the Friis transmission formula, which gives the ratio of the transmitted power to the received power

$$\frac{P_R}{P_T} = \frac{\lambda^2 G_T G_R}{(4\pi r)^2} \qquad (4.19)$$

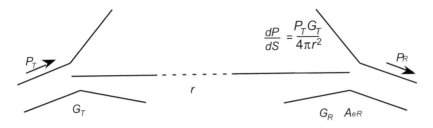

Fig. 4.4 Transmission between two antennas.

4.3.2 Radar equation

In a radar the transmitting antenna radiates a signal which reaches the receiving antenna after scattering from a target. We assume, at least initially, a bistatic configuration (Fig. 4.5). The distances are large compared with the wavelength and with the antenna and target dimensions.

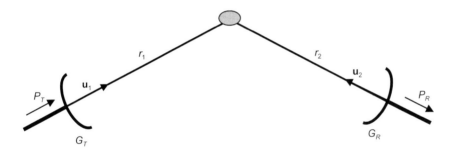

Fig. 4.5 Radar (bistatic configuration).

The power flux density at the target at a distance r_1 from the transmitting antenna is

$$\left[\frac{dP}{dS}\right]_{\text{target}} = \frac{P_T G_T}{4\pi r_1^2}$$

The power flux density at the receiving antenna, situated at a distance r_2 from the target, is proportional to the power density beyond the target. It decreases as $1/r^2$ and also depends on the geometry and orientation of the target with respect to the directions $\mathbf{u}_1$ and $\mathbf{u}_2$. The radar cross-section σ characterizes the behaviour of the target. It is defined by the relationship

$$\left[\frac{dP}{dS}\right]_{\text{receiving antenna}} = \frac{\sigma(\mathbf{u}_1, \mathbf{u}_2)}{4\pi r_2^{\,2}}\left[\frac{dP}{dS}\right]_{\text{target}} \tag{4.20}$$

The power received at the receiving antenna

$$P_R = A_{eR}\left[\frac{dP}{dS}\right]_{\text{receiving antenna}}$$

is then given by the radar equation

$$\begin{aligned}
P_R &= \frac{P_T G_T}{4\pi r_1^{\,2}}.\sigma.\frac{1}{4\pi r_2^{\,2}}.\frac{G_R \lambda^2}{4\pi} \\
&= \frac{P_T G_T G_R \lambda^2 \sigma}{(4\pi)^3 r_1^{\,2} r_2^{\,2}}
\end{aligned} \tag{4.21}$$

in which propagation losses have been ignored.

If the two antennas are co-located, which is the more usual monostatic configuration, their distance to the target is $r = r_1 = r_2$. The radar equation becomes

$$P_R = \frac{P_T G_T G_R \lambda^2 \sigma}{(4\pi)^3 r^4} \tag{4.22}$$

Moreover, if the same antenna is used for transmitting and receiving ($G_T = G_R = G$)

$$P_R = \frac{P_T G^2 \lambda^2 \sigma}{(4\pi)^3 r^4} \tag{4.23}$$

which is the standard form of the radar equation.

4.3.3 Antenna Radar Cross Section (RCS)[1,2,3,4]

In some cases we may be concerned not only with how an antenna radiates or receives signals, but also with its properties in scattering radiation. There is no point in designing a stealthy aircraft if its radar signature is dominated by scattering from its antennas !

Antenna RCS

Antenna RCS is made up of two components: the *structural mode RCS*, i.e. scattering from the antenna structure, and the *antenna mode RCS*.

The structural mode RCS can be defined as the RCS obtained when the antenna is terminated in a matched load.

The antenna mode RCS is defined in the following way. Suppose there is a power density p W/m^2 incident on the antenna in its boresight direction (Fig. 4.6). The power reaching the feedpoint of the antenna will be $pG\lambda^2/4\pi$, and the power reflected from the feedpoint will be

$$|\rho|^2 \, p \, \frac{G\lambda^2}{4\pi} \qquad (4.24)$$

where ρ is the voltage reflection coefficient at the antenna feedpoint.

This power is reradiated, so the antenna mode RCS in the direction of the main beam of the antenna is

$$\sigma_A = \frac{G^2\lambda^2 |\rho|^2}{4\pi} \qquad (4.25)$$

The result can be generalized for signals incident in other directions, giving an antenna mode RCS pattern proportional to the square of the antenna radiation pattern.

Whilst ρ, and hence σ_A, may be low in the operating band of the system connected to the antenna, outside of that band the antenna may present a very large RCS.

---

[1] Knott, E.F., Shaeffer, J.F. and Tuley, M.T., *Radar Cross Section* (second edition), Artech House, 1993.

[2] Bhattacharyya, A.K. and Sengupta, D.L., *Radar Cross Section Analysis and Control*, Artech House, 1991.

[3] Lynch, D. Jr, *Introduction to RF Stealth*, Scitech/Peter Peregrinus, 2003.

[4] Stimson, G.W., 'Antenna RCS reduction', chapter 39 in *Introduction to Airborne Radar* (second edition), SciTech/Peter Peregrinus, 1998.

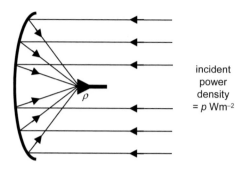

Fig. 4.6 Definition of antenna mode RCS.

There are several techniques that can be used to control the RCS of antennas. The structural RCS can be minimised by using absorbing material, and the antenna mode RCS by ensuring a good match at the feedpoint, though this may not be possible over a broad bandwidth. Frequency Selective Surfaces (FSSs)[5] can therefore be used in a radome to pass only signals within the desired operating band.

Consideration of the structural mode and antenna modes together shows that the two can be made to cancel, leading to the concept of the *minimum scattering antenna*[6,7].

4.4 ANTENNA BEHAVIOUR IN THE PRESENCE OF NOISE

Thermodynamics shows that all bodies radiate, and that the radiated power depends particularly on the frequency and on the absolute temperature of the body. This radiation is unpolarized (i.e. the two linear polarizations exist simultaneously). When surrounded by bodies at different temperatures, the antenna receives their thermal radiation, which is delivered in the form of noise to the receiver.

4.4.1 Power radiated by a body at absolute temperature *T*

We assume that the bodies located around the antenna radiate as black bodies. The power dP_B radiated (Fig. 4.7) in the solid angle $d\Omega'$ and in the frequency bandwidth

---

[5] Munk, B.A. *Frequency Selective Surfaces: Theory and Design*, Wiley, 2000.

[6] Kahn, W.K. and Kurss, H., 'Minimum scattering antennas', *IEEE Trans. Antennas and Propagation*, Vol.AP-13, September 1965, pp671-675.

[7] Pozar, D., 'Scattered and absorbed powers in receiving antennas', *IEEE Antennas and Propagation Magazine,* Vol.146, No.1, February 2004.

$(f, f + \Delta f)$, by a surface element dS' of a body at absolute temperature T, is given by the Rayleigh approximation for a black body

$$dP_B = \frac{2kT}{\lambda^2} \cos\theta' \, d\Omega' \, dS' \, \Delta f \qquad (4.26)$$

This approximation is valid when

$$\frac{hf}{kT} = 4.75 \times 10^{-11} \frac{f}{T} << 1 \qquad (4.27)$$

($h = 6.63 \times 10^{-34}$ Js is Planck's constant and $k = 1.38 \times 10^{-23}$ JK^{-1} is Boltzmann's constant).

The error involved in this approximation is negligible at microwave frequencies.

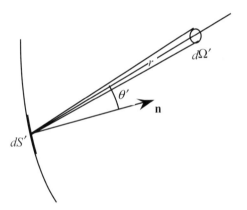

Fig. 4.7 Thermal radiation for a surface element.

The surface element dS' acts as an antenna that radiates equally in both polarizations. The power is radiated according to a $\cos\theta'$ function. In agreement with the results of Table 3.1, the total radiated power over the half-space is

$$dP_{rad} = \frac{2\pi kT}{\lambda^2} dS' \Delta f$$

and the directivity is $D(\theta') = 4\cos\theta'$. The receiving antenna, whose gain is $G(\mathbf{u})$, situated at a distance r, will only receive half the thermal power of that predicted by the Friis formula (4.19), since it is only sensitive to one of the polarizations.

Thus, the power dP_{noise} received in the form of thermal noise by the antenna is

$$dP_{\text{noise}} = \frac{\lambda^2 D(\theta') G(\mathbf{u})}{2(4\pi r)^2} dP_{\text{rad}} = \frac{kT}{4\pi} \frac{dS' \cos \theta'}{r^2} G(\mathbf{u}) \Delta f$$

$$dP_{\text{noise}} = \frac{k\Delta f}{4\pi} T(\mathbf{u}) G(\mathbf{u}) d\Omega \qquad (4.28)$$

- $T(\mathbf{u})$ is the absolute temperature in the direction $\mathbf{u}$,

- $G(\mathbf{u})$ is the gain in the direction $\mathbf{u}$.

- $d\Omega = \dfrac{dS' \cos \theta'}{r^2}$ is the solid angle element around the direction $\mathbf{u}$ (the solid angle subtended by the surface element dS' at the antenna) (Fig. 4.8).

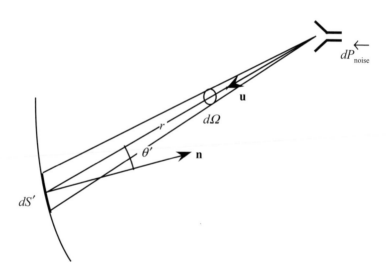

Fig. 4.8 Noise power received from a surface element.

The total noise power received by the antenna is

$$P_{\text{noise}} = \frac{k\,\Delta f}{4\pi} \int_{(4\pi)} T(\mathbf{u}) G(\mathbf{u}) d\Omega \qquad (4.29)$$

4.4.2 Noise temperature of the antenna

For a lossless antenna (gain = directivity) surrounded by a distribution of bodies or sources at the same temperature T, we have $\int_{(4\pi)} D(\mathbf{u})d\Omega = 4\pi$ and

$$P_{noise} = kT\Delta f$$

In the general case, by analogy, we define the noise temperature T_A of the antenna by

$$P_{noise} = kT_A \Delta f \qquad (4.30)$$

Eliminating the term P_{noise} between equations (4.29) and (4.30) yields the expression for the noise temperature of the antenna

$$T_A = \frac{1}{4\pi} \int_{(4\pi)} T(\mathbf{u})G(\mathbf{u})d\Omega \qquad (4.31)$$

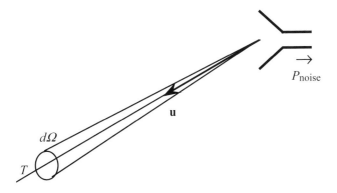

Fig. 4.9 Total noise power received by the antenna. Illustration of equation (4.31).

The noise temperature is the average, weighted by the gain, of the temperatures surrounding the antenna.

The thermal noise received by an antenna depends on the temperature in the directions from which the antenna is able to receive electromagnetic waves. This temperature is ~ 3 K for the sky at the zenith and in the absence of the sun or stars, ~ 290 K for the Earth and ~ 310 K for the human body. The thermal noise also depends on the directional properties of the antenna. Thus, the presence of a sidelobe will significantly increase the thermal noise if its level is not sufficiently small and if it is in a direction where the temperature is high.

4.4.3 Noise temperature of the receiving system

The receiving system is composed of the antenna and the receiver, connected by a transmission line (or waveguide). The noise temperature of the system is given by the formula

$$T_{\text{sys}} = T_A + T_{AP}\left(\frac{1}{\varepsilon_1} - 1\right) + T_L\left(\frac{1}{\varepsilon_2} - 1\right) + \frac{1}{\varepsilon_2}T_R \tag{4.32}$$

- T_A is the noise temperature (K) of the antenna,

- T_{AP} is the physical temperature (K) of the antenna,

- ε_1 is the thermal efficiency coefficient of the antenna ($0 \le \varepsilon_1 \le 1$), which is dimensionless and very close to 1 for a good antenna (low losses),

- T_L is the physical temperature (K) of the transmission line,

- ε_2 is the efficiency coefficient of the transmission line, related to its attenuation $\varepsilon_2 = \exp(-2\alpha\ell)$ (where α is the attenuation coefficient and ℓ is the line length),

- T_R is the noise temperature (K) of the receiver, given by

$$T_R = T_1 + \frac{T_2}{G_1} + \frac{T_3}{G_1 G_2} + \dots \tag{4.33}$$

- T_n and G_n are respectively the noise temperature and the gain of the nth stage of the receiver amplification chain.

In the case where ε_1 and ε_2 are close to 1, we have $T_{\text{syst}} = T_A + T_R$. If the losses in the line cannot be neglected then the noise temperature, evaluated at the antenna, is

$$T_{\text{sys}} = T_A + T_L\left(\frac{1}{\varepsilon_2} - 1\right) + \frac{1}{\varepsilon_2}T_R \tag{4.34}$$

The noise figure F is also used, defined from the noise temperature T by

$$F = 1 + T/T_0 \text{ with } T_0 = 290 \text{ K} \tag{4.35}$$

or in decibels

$$F_{(\text{dB})} = 10\log F = 10\log\left(1 + \frac{T}{T_0}\right) \tag{4.36}$$

FURTHER READING

1. Collin, R.E., *Antennas and Radiowave Propagation*, McGraw-Hill, New York, 1985.
2. Collin, R.E. and Zucker, F.J. (eds), *Antenna Theory*, Parts 1 and 2, McGraw-Hill, New York, 1969.
3. Munk, B.A., *Finite Antenna Arrays and FSS*, Wiley, 2003.
4. Schelkunoff, S.A. and Friis, H.T., *Antenna Theory and Practice*, Wiley, New York, 1952.
5. Vardaxoglou, J.C., *Frequency Selective Surfaces: Analysis and Design*, Research Studies Press, 1997.

EXERCISES

4.1 Earth-Moon radio link

A radio transmitter placed on the Moon, and operating at a frequency $f = 2.5$ GHz transmits a power $P_R = 2$ W, using an antenna whose gain is $G_T = 1000$. What should be the gain G_R of the receiving antenna on the Earth, such that the received power should be greater than the detection threshold $P_0 = 10^{-14}$ W? The propagation delay from the Moon to the Earth is 1.27 seconds.

4.2 Radiocommunication link

A radiocommunication link between two points 50 km apart operates at 12 GHz using antennas of gain 45 dB.

Q1 What is the order of magnitude of the area of the antennas?

Q2 What is the ratio between the transmitted power and the received power?

4.3 DBS TV

Consider a link at a frequency $f = 12$ GHz between a DBS TV satellite in geostationary orbit and a parabolic receiving antenna on the surface of the Earth, at a distance $d = 36,000$ km from the satellite. The transmit power is $P_T = 100$ W, and the transmit antenna gain is $G_T = 40$ dB.

Q1 Determine the power density dP/dS at the receive antenna.

Q2 The TV picture quality is acceptable if the receive antenna (assumed lossless) receives a power P_R which exceeds a threshold $P_S = 2 \times 10^{-11}$ W. What should the antenna area S_A and the antenna gain G_A be to achieve this ? Give an order of magnitude for the antenna dimensions.

4.4 Radio link

Consider a radio link at a frequency $f = 4$ GHz between two antennas each of gain $G = 25$ dB at a distance $r = 50$ km (Fig. 4.8).

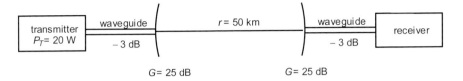

Fig. 4.10 Radio link.

Q1 Determine the total power P_T' radiated by the antenna. Hence deduce the power density at the receive antenna.

Q2 What is the power P_R at the input to the receiver ?

5

Antennas of simple geometry

The antenna used to broadcast the BBC's Radio 4 transmission on 198 kHz, from Droitwich, just to the south of Birmingham, England. Despite the enormous physical size of this antenna (some 200 m high), compared to the wavelength it is a 'small antenna' (photo: H. Griffiths).

5.1 APERTURE ANTENNAS

5.1.1 Parabolic antennas

Parabolic antennas are a particular case of the axially-symmetric systems described in Chapter 9. Here, we give a simplified theory of their operation.

Fig. 5.1 A telecommunications tower installation, with parabolic reflector antennas (protected from the weather by radomes), folded dipoles and slot antennas. Photo: Aerial Facilities Ltd.

Geometry of a parabola

A paraboloid of revolution having a focal length f between the focus F and the apex A can be described in polar coordinates (r', θ'), taking the origin at F, by the equation (Fig. 5.2)

$$r' = \frac{2f}{1 + \cos\theta'} \qquad (5.1)$$

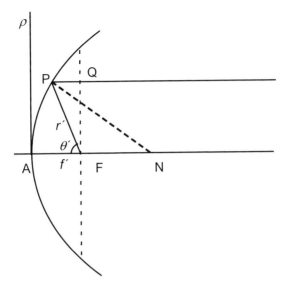

Fig. 5.2 Paraboloid of revolution.

Field in the aperture

A primary source, for example a horn, is placed at the focus F (Fig. 5.3). The electromagnetic field component tangential to the aperture plane is assumed to be linearly polarized and the x' axis is chosen parallel to the field polarization in a rectangular coordinate system (x', y', z') having its origin at the focus F[1]. We further assume that the reflector is placed in the far field of the primary source. The field $\mathbf{E}^i$ at a point P on the reflector is given by (equation (2.25))

$$\mathbf{E}^i = 2\pi j \, \frac{\exp(-jkr')}{kr'} \, F(\alpha', \beta') \, (\mathbf{u}_{x'}\gamma' - \mathbf{u}_{z'}\alpha') \qquad (5.2)$$

where $F(\alpha', \beta')$ is the angular spectrum of the primary source ($\alpha' = \sin \theta' \cos \phi'$, $\beta' = \sin \theta' \sin \phi'$, $\gamma' = \cos \theta'$). The plane wave spectrum method, associated with a model based on geometrical optics, will be used to calculate the radiation from the parabolic reflector.

---

[1] In this example, the primary feed is *not* a Huygens source (as defined in §7.4.5 and §8.12).

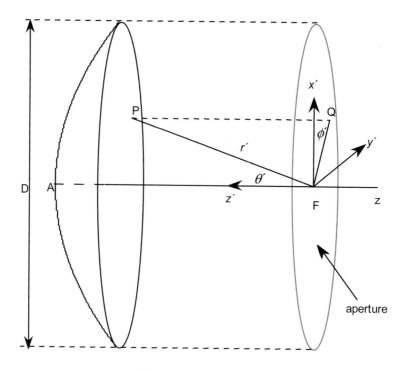

Fig. 5.3 Parabolic antenna.

- The radiating aperture is a circular aperture of diameter D in the focal plane. This aperture is illuminated by the rays from the primary feed, which is placed at the focal point.
- The field is assumed to be negligible outside the circle of diameter D in the focal plane. At all points Q within the circle, the field is expressed as $\mathbf{E}(Q)$.
- The phase of $\mathbf{E}(Q)$ is independent of Q, i.e. is the same for all points Q within the circle of diameter D, because the equation of the parabola tells us that the distance from the source to the point Q via the reflector is the same for all points Q (FP + PQ = $2f$). The phase difference between the source at the focal point F and the point Q is $2kf$, where $k = 2\pi/\lambda$.
- The amplitude of the field varies as $1/r'$, where r' is the distance between the focal point F and a point P on the reflector. The amplitude of the field is then modified by reflection. If we assume that the reflector is a perfect conductor, then the tangential component of the field at the point P may be written

$$\mathbf{E}_{\text{tangential}}^{\text{incident}} = -\mathbf{E}_{\text{tangential}}^{\text{reflected}} \, , \qquad \mathbf{E}_{\text{normal}}^{\text{incident}} = \mathbf{E}_{\text{normal}}^{\text{reflected}}$$

Thus

$$\mathbf{E}^{\text{reflected}} = -\mathbf{E}^{\text{incident}} + 2\left(\mathbf{n}.\mathbf{E}^{\text{incident}}\right)\mathbf{n} \tag{5.3}$$

where **n** is the normal vector at point P, pointing towards the focus F, and whose components are

$$\mathbf{n}_{x'} = -\sin\frac{\theta'}{2}\cos\phi' \quad \mathbf{n}_{y'} = -\sin\frac{\theta'}{2}\sin\phi' \quad \mathbf{n}_{z'} = -\cos\frac{\theta'}{2} \quad (5.4)$$

Considering the radiation from the primary source according to the model we have adopted, the field **E**(Q) in the interior of the aperture of diameter D is given by

$$\mathbf{E}(Q) = -2\pi j \frac{\exp(-2jkf)}{kr'} F(\alpha',\beta') \left[\mathbf{u}_{x'}(\cos\theta'+2\sin^2\frac{\theta'}{2}\cos^2\phi')+\mathbf{u}_{y'}\sin^2\frac{\theta'}{2}\sin 2\phi' \right]$$

$$(5.5)$$

In this equation, note the absence of any z component (because the field behaves as a plane wave), and the presence of a y component, in spite of the fact that we have assumed a pure polarization parallel to the x-axis at the feed[2]. Fig. 5.4 shows the electric field lines in the aperture plane (focal plane) for the case of a uniform angular spectrum over the entire reflector.

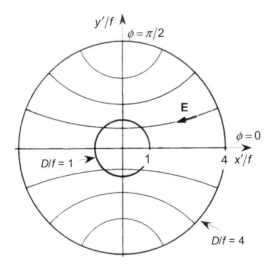

Fig. 5.4 Field lines in the focal plane for a uniform illumination of the reflector. For the case $D/f = 4$ (i.e. $f/D = 0.25$), the edge of the reflector is in the focal plane.

---

[2] It will be shown (§9.2) that there is zero cross-polarization in the aperture plane of an axially-symmetric antenna (e.g. paraboloid) illuminated by a Huygens source.

The field at a point Q in the aperture may be expressed in terms of the rectangular coordinate system (x', y') having its origin at the primary feed F by using the relations $x' = r' \sin \theta' \cos \phi'$, $y' = r' \sin \theta' \sin \phi'$ and equation (5.1).

We find

$$E(Q) = -2\pi j \frac{\exp(-2jkf)}{kr'} F\left(\frac{x'}{r'}, \frac{y'}{r'}\right)\left[\mathbf{u}_{x'}\left(\frac{2f-r'}{r'} + \frac{x'^2}{2fr'}\right) + \mathbf{u}_{y'}\frac{x'y'}{2fr'}\right] \qquad (5.6)$$

with

$$r' = f\left(1 + \frac{x'^2 + y'^2}{2f^2}\right)$$

Introducing a rotation of the Cartesian coordinate system in order to describe the radiation: $x = x'$, $y = -y'$, $z = -z'$, the electric field at a point Q of the aperture may be written

$$E(x,y) = -2\pi j \frac{\exp(-2jkf)}{kr'} F\left(\frac{x}{r'}, \frac{y}{r'}\right)\left[\mathbf{u}_x\left(\frac{2f-r'}{r'} + \frac{x^2}{2fr'}\right) + \mathbf{u}_y\frac{xy}{2fr'}\right] \qquad (5.7)$$

with

$$r' = f\left(1 + \frac{x^2 + y^2}{2f^2}\right)$$

Now introducing the polar coordinates (ρ, ϕ) in the aperture plane xFy, which are related to the Cartesian coordinates by the transformation $x = \rho \cos\phi$ and $y = \rho \sin\phi$, the electric field at a point Q of the aperture may be written

$$E(\rho,\phi) = -2\pi j \frac{\exp(-2jkf)}{kr'} F\left(\frac{\rho\cos\phi}{r'}, \frac{-\rho\sin\phi}{r'}\right)\left[\mathbf{u}_x\left(\frac{2f-r'}{r'} + \frac{\rho^2 \cos^2\phi}{2fr'}\right) + \mathbf{u}_y\frac{\rho^2 \sin 2\phi}{4fr'}\right]$$

$$(5.8)$$

with

$$r' = f\left(1 + \frac{\rho^2}{2f^2}\right)$$

Far field radiation in the case of a uniform illumination of the reflector by a primary feed

We assume $F(\alpha, \beta)$ is constant, i.e. independent of α and β, and we write the illuminating field amplitude as

$$E_0 = -2\pi j \frac{\exp(-2jkf)}{kf} F(0,0) \tag{5.9}$$

The aperture illumination function may then be written as

$$\mathbf{E}(\rho,\phi) = \mathbf{u}_x E_{0x}(\rho,\phi) + \mathbf{u}_y E_{0y}(\rho,\phi) \tag{5.10}$$

$$E_{0x}(\rho,\phi) = E_0 \frac{f}{r'} \left(\frac{2f - r'}{r'} + \frac{\rho^2 \cos^2 \phi}{2fr'} \right) \tag{5.11a}$$

$$E_{0y}(\rho,\phi) = E_0 \frac{\rho^2 \sin 2\phi}{4r'^2} \tag{5.11b}$$

5.1.2 Rectangular horns (see also §8.2)

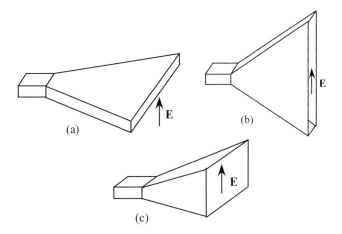

Fig. 5.5 Different types of rectangular horn: (a) H-plane sectoral horn; (b) E-plane sectoral horn; (c) pyramidal horn.

Rectangular horns consist of a flared section of rectangular waveguide, and generally operate in the TE_{10} mode of the waveguide. There are three basic types of horn, distinguishable by their respective geometries: the E-plane sectoral horn, the H-plane sectoral horn, and the pyramidal horn.

H- plane sectoral horn

The geometry of an H-plane sectoral horn, as shown in Fig. 5.6, leads to the relationships

$$\ell_H^2 = R_1^2 + \frac{A^2}{4}, \quad \tan\alpha_H = \frac{A}{2R_1}, \quad R_H = (A-a)\sqrt{\frac{\ell_H^2}{A^2} - \frac{1}{4}} \tag{5.12}$$

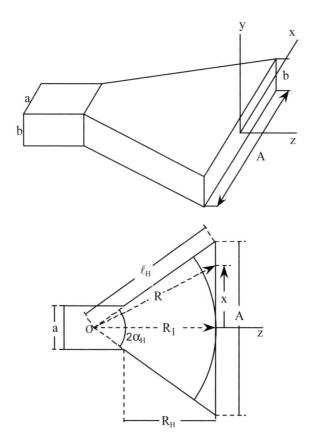

Fig. 5.6 Geometry of H-plane sectoral horn.

In a rectangular waveguide excited in a TE_{10} mode, the electric field has no x component, but only a y component E_y which varies as $\cos(\pi x/a)\exp(-j\beta z)$. The phase of the field is thus the same at all points of a plane perpendicular to the axis of the waveguide. The horn is obtained by increasing the dimension A in the plane of the aperture. We can then make the following reasonable approximations for the field in the horn aperture:

- The x component of the electric field is zero;
- The amplitude of the y component varies as $\cos(\pi x/A)$ in the same way as within the rectangular waveguide
- The phase at a point in the aperture depends on the position of the point, because the distance R is expressed as a function of its position. The phase of the component E_y at a point of the abscissa x is

$$\delta = \frac{2\pi}{\lambda}(R - R_1) \approx \frac{\pi}{\lambda R_1} x^2$$

The maximum $\delta_M = \dfrac{\pi A^2}{4R_1\lambda}$ occurs for $x = A/2$.

- E_y, the y component of the electric field in the rectangular aperture may be expressed as

$$E_y = E_0 \cos\frac{\pi x}{A}\exp\left(-j\frac{k}{2R_1}x^2\right) \tag{5.13}$$

The angular spectrum $F_y(\alpha, \beta)$ is obtained by integration over the rectangular aperture (A, b)

$$F_y(\alpha,\beta) = \frac{E_0}{\lambda^2}\iint\cos\frac{\pi x}{A}\exp\left(-j\frac{k}{2R_1}x^2\right)\exp\left[jk(\alpha x + \beta y)\right]\,dx\,dy \tag{5.14}$$

It can be seen that the angular spectrum may be expressed as a product of two terms $f_H(\alpha).f_E(\beta)$

- The term $f_E(\beta) = \dfrac{\sin(\pi\beta b/\lambda)}{(\pi\beta b/\lambda)}$ resulting from the integral with respect to y describes the antenna pattern observed in the E plane (the plane yOz with $\alpha = 0$, $\beta = \sin\theta$);

- The other term $f_H(\alpha)$ resulting from the integral with respect to x describes the antenna pattern observed in the H plane (the plane xOz with $\beta = 0$, $\alpha = \sin\theta$). This term may be calculated by introducing the Fresnel integral

$$F(v) = \int_0^v \exp\left(-j\pi u^2/2\right) du \tag{5.15}$$

The result is given by the universal curves shown in Fig. 5.7, shown for different values of the parameter

$$t = \frac{A^2}{8\lambda R_1} = \frac{1}{8}\frac{(A/\lambda)^2}{R_1/\lambda} \tag{5.16}$$

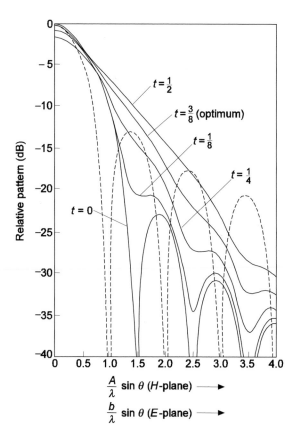

Fig. 5.7 Radiation pattern of an H-plane sectoral horn (———— H-plane – – – E-plane).

The directivity of the horn is shown in Fig. 5.8. For each value of R_1 there is an optimum dimension of the horn $A_{opt} = \sqrt{3\lambda R_1}$ corresponding to $t_{opt} = 3/8$. The directivity is maximum for this optimum value and the −3 dB half-beamwidth is 0.68 λ/A radians.

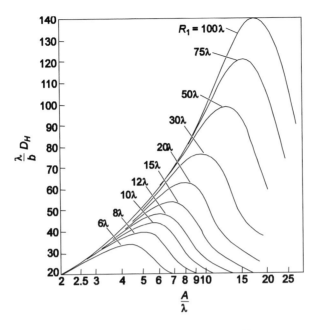

Fig. 5.8 Directivity of an H-plane sectoral horn.

E- plane sectoral horn

The geometry of an E-plane sectoral horn shown in Fig. 5.9 gives rise to the following relations

$$\ell_E^2 = R_2^2 + \frac{B^2}{4}, \quad \tan\alpha_E = \frac{B}{2R_2}, \quad R_E = (B-b)\sqrt{\frac{\ell_E^2}{B^2} - \frac{1}{4}} \qquad (5.17)$$

The method used for the calculations is analogous to that shown above for the H-plane sectoral horn. The tangential field in the aperture is oriented parallel to the y-axis. The component E_y may be written

$$E_y = E_0 \cos\frac{\pi x}{a}\exp\left(-j\frac{k}{2R_2}y^2\right) \qquad (5.18)$$

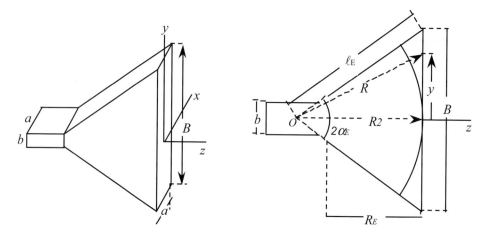

Fig. 5.9 Geometry of an E-plane sectoral horn.

The angular spectrum $F_y(\alpha, \beta)$ is obtained by integration over the rectangular aperture (a, B)

$$F_y(\alpha, \beta) = \frac{E_0}{\lambda^2} \iint \cos\frac{\pi x}{a} \exp\left(-j\frac{k}{2R_2} y^2\right) \exp[jk(\alpha x + \beta y)]\, dx dy \qquad (5.19)$$

As before, the angular spectrum may be expressed as the product of two terms $g_H(\alpha) \cdot g_E(\beta)$:

- The term $g_H(\alpha) = \dfrac{\cos\pi(\alpha a/\lambda)}{1 - [\pi\alpha a/\lambda]^2}$ resulting from the integral with respect to x

  describes the antenna pattern observed in the H plane (the plane xOz with $\beta = 0$, $\alpha = \sin\theta$).

- The other term $g_E(\beta)$ resulting from the integral with respect to y describes the antenna pattern observed in the E plane (the plane through yOz with $\alpha = 0$, $\beta = \sin\theta$). This term may be calculated by means of the Fresnel integral. The result is given by the universal curves shown in Fig. 5.10, shown for different values of the parameter s

$$s = \frac{B^2}{8\lambda R_2} = \frac{1}{8}\frac{(B/\lambda)^2}{R_2/\lambda} \tag{5.20}$$

The directivity of the horn is represented in Fig. 5.11. For each value of R_2 there is an optimum dimension of the horn $B_{opt} = \sqrt{2\lambda R_2}$ corresponding to $s_{opt} = 1/4$. The directivity is optimum for this optimum value and the -3 dB half-beamwidth is 0.47 λ/B radians.

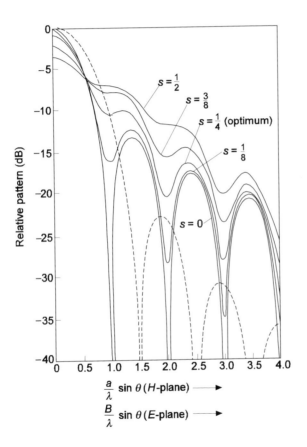

Fig. 5.10 Radiation pattern of an E-plane sectoral horn. (——— E-plane – – – H-plane).

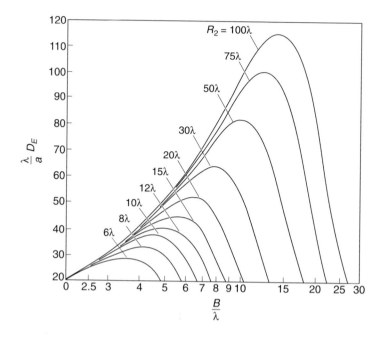

Fig. 5.11 Directivity of an E-plane sectoral horn.

Pyramidal horn

The geometry of the pyramidal horn is shown in Fig. 5.12. The aperture field is of the form

$$E_y = E_0 \cos\frac{\pi x}{A}\exp\left[-j\frac{k}{2}\left(\frac{x^2}{R_1^2}+\frac{y^2}{R_2^2}\right)\right] \qquad (5.21)$$

Comparing this expression to those obtained for the sectoral horns above, we observe that the angular spectrum of the pyramidal horn involves the same integrals. As a result, the universal curves of Figs 5.7, 5.8, 5.10 and 5.11 may be used to determine the radiation pattern and the directivity of the pyramidal horn (by replacing a with A and b with B).

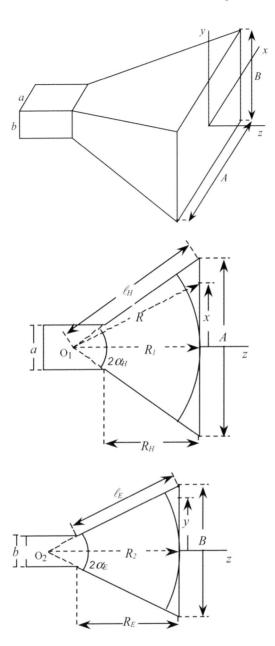

Fig. 5.12 Geometry of pyramidal horn.

- The E-plane radiation pattern is the same as that of the E-plane pattern of the E-plane sectoral horn.
- The H-plane radiation pattern is the same as that of the H-plane pattern of the H-plane sectoral horn.
- The directivity is

$$D = \frac{\pi}{32}\left[\frac{\lambda}{A}D_E\right]\left[\frac{\lambda}{B}D_H\right] \qquad (5.22)$$

- The optimum directivity is achieved with

$$A_{opt} = \sqrt{3\lambda R_1} \quad \text{and} \quad B_{opt} = \sqrt{2\lambda R_2} \qquad (5.23)$$

5.2 WIRE ANTENNAS

5.2.1 Electric dipoles

We define the term dipole as those antennas (Fig. 5.13a) formed by a portion of a rectilinear circuit with vanishing thickness and length ℓ in which a current $I(z')$ flows. The antenna may be excited by a transmission line such as that shown in Fig. 5.13b whose two conductors are sufficiently close to one another to avoid any perturbation of the dipole radiation.

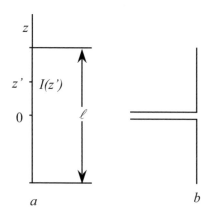

Fig. 5.13 Electric dipole.

Hertzian dipole (elementary electric dipole)

A Hertzian dipole is a theoretical object having a length ℓ which is assumed very small compared to the wavelength λ (for example $\ell < \lambda/100$), and in which the current I is assumed constant. Even though these assumptions are never fully obtained in a real physical antenna, such an antenna has a great theoretical importance for two reasons:

(1) The Hertzian dipole serves as a reference for comparison with the properties of more complex antennas, and
(2) Any arbitrary current distribution may be considered as the linear superposition of a set of Hertzian dipoles.
    A thorough study of an elementary electric dipole oriented along the unit vector $\mathbf{u}_D$ has been given in Chapter 3. We recall below the main results:

• The radiated electromagnetic field at large distance in the direction $\mathbf{u}$ is given by

$$\mathbf{E} = \frac{jk\eta}{4\pi} \frac{\exp(-jkr)}{r} \, I\ell \, \mathbf{u} \times (\mathbf{u} \times \mathbf{u}_D) \tag{5.24a}$$

$$\mathbf{H} = \frac{jk}{4\pi} \frac{\exp(-jkr)}{r} \, I\ell \, (\mathbf{u}_D \times \mathbf{u}) \tag{5.24b}$$

• The total radiated power is given by

$$P_{rad} = \frac{\pi}{3\lambda^2} \eta |I\ell|^2 \tag{5.25}$$

• The radiation resistance is given by

$$R_{rad} = \frac{2\pi}{3} \eta \left[\frac{\ell}{\lambda}\right]^2 = 800 \left[\frac{\ell}{\lambda}\right]^2 \quad (\Omega) \tag{5.26}$$

• The directivity is given by

$$D = \frac{3}{2} \sin^2 \theta \tag{5.27}$$

Electric dipoles of arbitrary length

In the case where the length ℓ cannot be considered as very small compared to the wavelength λ, the current distribution may be considered to depend on ℓ/λ as shown schematically in Fig. 5.14.

A simplified radiation study of dipoles may be based on an approximate expression for the current

$$I(z') = I_{\mathrm{M}} \sin\left[\frac{2\pi}{\lambda}\left(\frac{\ell}{2} - |z'|\right)\right] \tag{5.28}$$

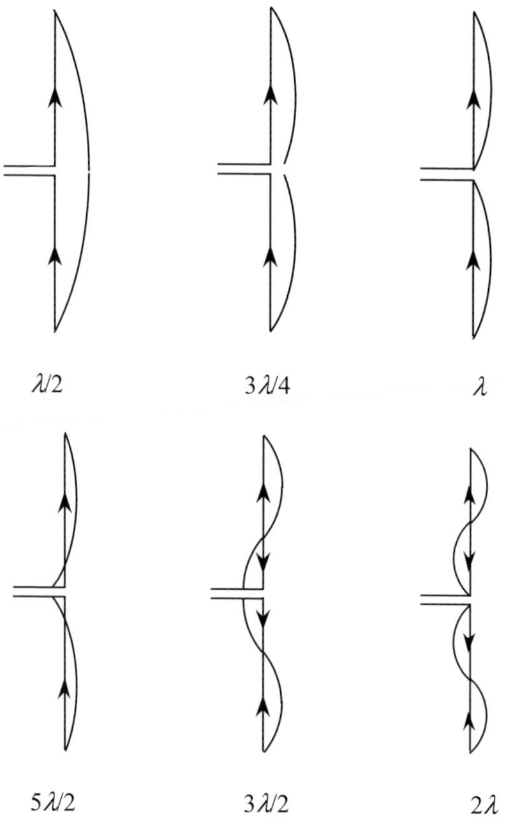

Fig. 5.14 Approximate current distribution along the length of dipoles of different lengths.

The current is zero at the extremities $z' = \pm\ell/2$ and takes a value $I(0) = I_M \sin(k\ell/2)$ in the centre. The radiated field results from the superposition of the contribution from all of the elementary dipoles characterized by $I(z')dz'$ and oriented along Oz. After integration of the contribution of the set of elementary dipoles which make up the antenna, the far field radiated electric field may be written

$$\mathbf{E} = \frac{j\eta I_M}{2\pi} \frac{\exp(-jkr)}{r} \frac{\left[\cos\left(\frac{1}{2}k\ell\cos\theta\right) - \cos\left(\frac{1}{2}k\ell\right)\right]}{\sin\theta} \qquad (5.29)$$

The factor

$$\frac{\cos\left(\frac{1}{2}k\ell\cos\theta\right) - \cos\left(\frac{1}{2}k\ell\right)}{\sin\theta}$$

characterises the far field radiation as a function of the observation direction θ. This term is simplified when the antenna has a length which is a multiple of $\lambda/2$, as shown in table 5.1.

ℓ	$\dfrac{\cos[(k\ell\cos\theta)/2] - \cos(k\ell/2)}{\sin\theta}$
$\ell = \lambda/2$	$\dfrac{\cos[(\pi/2)\cos\theta]}{\sin\theta}$
$\ell = \lambda$	$\dfrac{\cos(\pi\cos\theta) + 1}{\sin\theta}$
$\ell = 3\lambda/2$	$\dfrac{\cos[(3\pi/2)\cos\theta]}{\sin\theta}$

Table 5.1 Radiated field factor for $\ell = \lambda/2,\ \lambda,\ 3\lambda/2$.

The power density varies as the square of the radiation field

$$\left|\frac{\cos\left(\frac{1}{2}k\ell\cos\theta\right) - \cos\left(\frac{1}{2}k\ell\right)}{\sin\theta}\right|^2$$

Fig. 5.15 shows the form of the radiation pattern (field intensity) for $\ell = \lambda/2$, λ and $3\lambda/2$. There is only a single lobe when $\ell \leq \lambda$, but multiple lobes appear when $\ell > \lambda$ because the power density has zeros for angles of observation θ_0 defined by

$$\cos\theta_0 = \pm 1 + 2n\lambda/\ell \qquad \text{(where } n \text{ is a positive integer)} \qquad (5.30)$$

The calculation of the radiated power P_{rad}, the directivity D and the radiation resistance R_{rad} may be performed using the functions $\mathrm{Si}(x)$ and $\mathrm{Ci}(x)$. For the half-wave dipole ($\ell = \lambda/2$) we obtain

$$P_{rad} = \frac{\eta}{8\pi}\mathrm{Ci}(2\pi)|I_M|^2 = 15\,\mathrm{Ci}(2\pi)|I_M| \qquad (5.31)$$

and using the values $\eta = 120\pi$ and $\mathrm{Ci}\,(2\pi) = 2.44$

$$P_{rad} = 36.6\,|I_M|^2 \qquad (5.32)$$

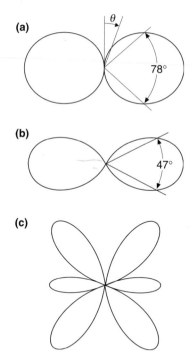

Fig. 5.15 Field radiation pattern for different dipole antenna lengths:
(a) $\ell = \lambda/2$ (b) $\ell = \lambda$ (c) $\ell = 3\lambda/2$.

Using this result we can thus express the directivity as

$$D(\theta) = 1.64 \left| \frac{\cos\left[(\pi/2)\cos\theta\right]}{\sin\theta} \right|^2 \tag{5.33}$$

for which the maximum value $D_M = 1.64$ is obtained for $\theta = \pi/2$, and the radiation resistance

$$R_{rad} = \frac{2P_r}{|I_M|^2} = 73.2\ \Omega \tag{5.34}$$

which represents the real part of the input impedance of the antenna. A thorough study of the half-wave antenna shows that the input impedance has an imaginary part equal to 42.5 Ω which may be cancelled by slightly shortening the length of the antenna.

The plane $z = 0$ is a plane of symmetry for the current distribution, and may thus be replaced by a ground plane without modifying the current distribution in the region $z > 0$ (as shown in Fig. 5.16). Thus a quarter-wave antenna ($\ell = \lambda/4$) placed above a ground plane will have the same radiation field in the region $z > 0$ as a half-wave dipole antenna in free space. The radiated power, the radiation resistance, and the input impedance are half that of the half-wave dipole, whereas the directivity is doubled.

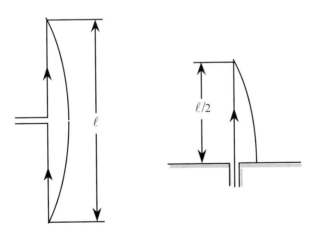

Fig. 5.16 Comparative geometry of a half-wave dipole antenna and a quarter-wavelength wire perpendicular to a ground plane.

Numerous antennas are made up of arrays of dipoles. The Yagi-Uda antenna is the best-known antenna of this type, widely used for TV reception[3]. The Yagi antenna is composed of several dipoles of which only one is actually fed by the transmission line, the other elements being director elements in the principal direction of radiation, and a single reflector element behind the driven element.

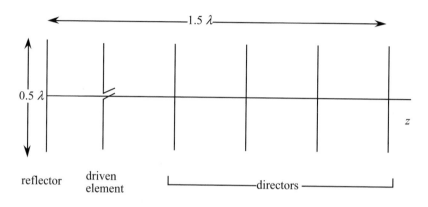

Fig. 5.17 Example of the geometry of a Yagi-Uda antenna.

5.2.2 Travelling wave rectilinear antennas

We consider a rectilinear antenna (as shown in Fig. 5.18) of length ℓ, upon which the current propagates with a phase velocity $v_\phi = \omega/\beta$

$$I(z') = I_0 \exp(-j\beta z') \tag{5.35}$$

[3] Yagi, H., 'Beam transmission of ultra short waves', *Proc. IRE*, Vol. 26, pp715-741, June 1928; also *Proc. IEEE*, Vol. 72, No. 5, pp634-645, May 1984.

Uda, S., 'Wireless beam of short electric waves', *J.IEE* (Japan), pp273-282, March 1926, and pp1209-1219, November 1927.

The original Yagi antenna is on display at Sendai University in Japan.

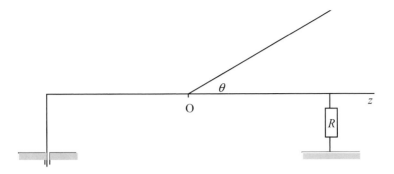

Fig. 5.18 Travelling wave rectilinear antenna.

The radiated field is obtained by integration of the contributions of the elementary dipoles which make up the antenna

$$\mathbf{E} = \frac{jk\eta}{4\pi}\frac{\exp(-jkr)}{r}I_0\ell\sin\theta\;\frac{\sin\left[\frac{1}{2}k\ell\left(\cos\theta - \beta/k\right)\right]}{\frac{1}{2}k\ell\left(\cos\theta - \beta/k\right)}\mathbf{u}_\theta \qquad (5.36)$$

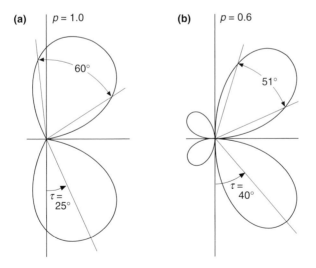

Fig. 5.19 Radiation patterns of travelling wave antennas of length $\ell = \lambda/2$:
(a) $v_\phi/c = 1$ (b) $v_\phi/c = 0.6$.

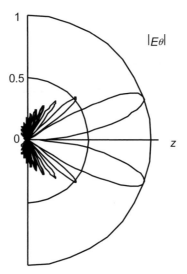

Fig. 5.20 Radiation pattern of a travelling wave antenna of length $\ell = 6\lambda$ and phase velocity
$$v_\phi = c.$$

The radiation pattern has symmetry of revolution about the z-axis Oz. It is always zero in the z direction ($\theta = 0$) and in other directions which depend on ℓ/λ and $v_\phi/c = k/\beta$. Fig. 5.21 shows the patterns which may be obtained from an array of travelling wave antennas of length 6λ.

Fig. 5.21 Radiation from an array of four antennas each of length 6λ, arranged in a rhombic configuration terminated with a resistance R: (a) arrangement of the array, showing the radiation pattern of each leg of the rhombic; (b) resulting radiation pattern of the array.

5.2.3 Loops and helical antennas

Many different geometries have been proposed for wire antennas in order to obtain particular electromagnetic properties, for example circular polarization, high gain, a specific form of the radiation diagram, etc. The study of such special antennas is not within the scope of this book, but has been explored in numerous books and other scientific publications. The reader is referred to Kraus[4] which includes further bibliographic references. In what follows we will present only the most widely used wire antennas.

Loops

When the dimensions of the loop are small compared to the wavelength, the current in the loop may be considered as constant, and the loop radiates as a magnetic dipole (cf. Chapter 3). When the dimensions are not negligible compared to the wavelength, the radiation field may only be calculated for very simple loop geometries.

An example is shown in Fig. 5.22: a simple circular loop of radius a and axis Oz, with a constant amplitude current I of constant phase.

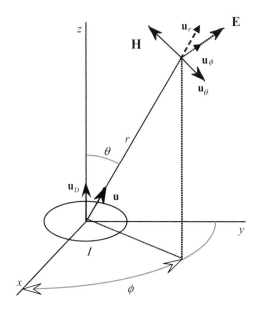

Fig. 5.22 Circular loop of radius a.

---

[4] Kraus, J.D., *Antennas*, McGraw-Hill, New York, 1988.

Recalling the result of §3.1.3

$$\mathbf{H} = \frac{-jk}{4\pi} \frac{\exp(-jkr)}{r} \mathbf{u} \times \iint_S \mathbf{J}(\mathbf{r}') \exp(jk\,\mathbf{u.r}')\,dV$$

The integral is performed over the circle $r' = a$ and $\mathbf{J(r')}dV$ becomes $Ia\,d\phi'\mathbf{u}_{\phi'}$. Making use of the relation $\mathbf{u.r}' = a\cos(\phi - \phi')\sin\theta$, we obtain

$$\mathbf{H} = \frac{-jk}{4\pi} \frac{\exp(-jkr)}{r} Ia\,\mathbf{u} \times \int_0^{2\pi} \mathbf{u}_{\phi'} \exp\left[jka\cos(\phi' - \phi)\sin\theta\right]d\phi' \qquad (5.37)$$

Using the formula

$$\int_0^{2\pi} \cos(\phi' - \phi)\{\exp[jz\,\cos(\phi' - \phi)]\}\mathbf{u}_{\phi'}\,d\phi' = 2\pi j\,J_1(z)$$

we finally obtain

$$\mathbf{H} = \frac{\pi a}{\lambda}\,I\,\frac{\exp(-jkr)}{r}\,J_1\left(2\pi\frac{a}{\lambda}\sin\theta\right)\mathbf{u}_\theta, \qquad \mathbf{E} = \frac{1}{\eta}\mathbf{u}\times\mathbf{H} \qquad (5.38)$$

where $J_1(x)$ is the Bessel function of first order.

For $a \gg \lambda$, the approximation $J_1(x) \approx x/2$ is valid for small x, which allows us to estimate the radiation of an elementary magnetic dipole. In the case where the radius a of the loop is of an arbitrary value, the function $J_1\left(2\pi\dfrac{a}{\lambda}\sin\theta\right)$ as represented in Fig. 5.23 allows us to find the radiation pattern, represented in Fig. 5.24 for values of radius $a = \lambda/5$, $\lambda/2$ and $3\lambda/4$.

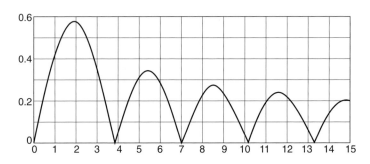

Fig. 5.23 The function $J_1\left(2\pi\dfrac{a}{\lambda}\sin\theta\right)$.

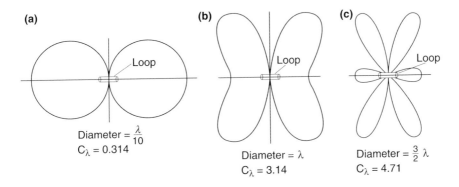

Fig. 5.24 Radiation from a constant current circular loop with (a) $a = \lambda/5$, (b) $a = \lambda/2$, and (c) $a = 3\lambda/4$.

The radiated power is calculated from the field by integration of the Poynting vector. From this result, we may also calculate the radiation resistance and the directivity. The maximum directivity is 3/2 for a small loop, and it can be shown that it is of the order of $4.25\ a/\lambda$ when $a/\lambda > 1$.

Helical antennas

A helix, as shown in Fig. 5.25, is characterized by

D = the diameter of the helix,
p = the distance between successive turns of the helix,
n = the number of turns of the helix

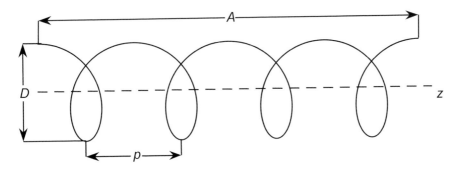

Fig. 5.25 Geometry of the helix.

The helix has several modes of radiation. It is most often used in its axial mode, which has its maximum radiation along the axis of the helix. When the helix dimensions are chosen appropriately, the helix radiates with circular polarization.

A detailed study of helical antennas and their radiation may be found in reference [9].

FURTHER READING

1. Balanis, C.A., *Antenna Theory and Design*, John Wiley & Sons, New York, 2nd edition, 1997.
2. Collin, R.E. and Zucker F.J., *Antenna Theory*, McGraw-Hill, New York, 1969.
3. Collin, R.E., *Antennas and Radiowave Propagation*, McGraw-Hill, New York, 1985.
4. Elliott, R.S., *Antenna Theory and Design*, Prentice Hall Inc., Englewood Cliffs, N.J., 1981
5. Fujimoto, K., Henderson, A., Hirasawa, K. and James, J.R., *Small Antennas*, Research Studies Press Ltd, Letchworth, England, 1987.
6. Jasik, H., *Antenna Engineering Handbook*, McGraw-Hill, New York, 1961.
7. Johnson, R.C. and Jasik, H., *Antenna Engineering Handbook*, McGraw-Hill, New York, 1984.
8. Johnson, R. C., *Antenna Engineering Handbook*, McGraw-Hill, New York, 1993.
9. Kraus, J.D., *Antennas*, McGraw-Hill, New York, 1988.
10. Kraus, J.D. and Marhefka, R.J., *Antennas for all Applications*, McGraw-Hill, New York, 2002.
11. King, R.W.P., *Table of Antenna Characteristics*, Plenum Press, New York, 1971.
12. Lo, Y.T. and Lee, S.W., *Antenna Handbook: Theory, Applications and Design*, Van Nostrand Reinhold, New York 1988.
13. Lowe, W., *Electromagnetic Horn Antennas*, IEEE Press, New York 1976.
14. Milligan, T. A., *Modern Antenna Design*, McGraw-Hill, New York, 1985.
15. Olver, A.D., Clarricoats, P.J.B., Kishk, A.A. and Shafai, L., *Microwave Horns and Feeds*, IEEE Press, New York 1994.
16. Rudge, A.W., Milne, K., Olver, A.D. and Knight, P. (eds), *The Handbook of Antenna Design*, Peter Peregrinus, Stevenage, 1982.
17. Schelkunoff, S.A., *Advanced Antenna Theory*, John Wiley & Sons Inc, New York, 1952.
18. Schelkunoff, S.A. and Friis, H.T., *Antennas: Theory and Practice*, John Wiley & Sons Inc, New York, 1952.
19. Silver, S. (ed), *Microwave Antenna Theory and Design*, MIT Radiation Lab. Series, Vol. 12, McGraw-Hill, New York, 1949, reprinted by Peter Peregrinus, Stevenage, 1984.
20. Stutzman, W.L. and Thiele, G.A., *Antenna Theory and Design*, John Wiley & Sons Inc, New York, 1981.
21. Wait, J.R., *Electromagnetic Radiation from Cylindrical Structures*, Pergamon Press, New York, 1959.
22. Walter, C.H., *Traveling Wave Antennas*, McGraw-Hill, New York, 1965.

EXERCISES

5.1 *Half-wave dipole*

A wire antenna of length $\lambda/2$ lies along the axis Oz, between $z' = -\lambda/4$ and $z' = +\lambda/4$. It carries a current

$$I(z') = I_0 \cos 2\pi \frac{|z'|}{\lambda}$$

Q1 Determine the electromagnetic field at a point whose spherical coordinates are (r, θ, ϕ).

Q2 Determine the power density dP/dS radiated in the direction (θ, ϕ). Sketch the shape of the radiation pattern of the antenna in the plane xOz.

Q3 The antenna is lossless. Its input impedance is $Z_i = 73 + j42.5\ \Omega$. Determine the total radiated power P_r.

Q4 Determine the directivity $D(\theta)$ of the antenna. Compare its maximum directivity D_M to that of a Hertzian dipole.

Q5 This dipole radiates a power $P_t = 300$ W at a frequency $f = 300$ MHz. A second dipole, parallel to the first is placed with its centre at a point whose coordinates are $r = 100$ m, $\theta = 90°$ and $\phi = 30°$. What is the power P_R that it receives ?

Q6 Suppose now that the dipole radiates a power $P_t = 1$ mW. How should the second antenna be oriented so as to receive maximum power, and what is the power P that it receives ?

5.2 *Parabolic reflector of diameter D small compared with the wavelength λ, uniformly illuminated by a primary feed*

Q1 Using equations (5.11), show that inside the circle of diameter D, the components of the electric field in the focal plane are

$$E_{0x} \approx E_0 \qquad E_{0y} \approx E_0 \frac{\rho^2}{4f^2}\sin 2\phi$$

Q2 Using the following Bessel function integral relationships

$$\int_0^{2\pi} \exp[ju\cos(\phi-\phi')]\exp(jn\phi')\,d\phi' = 2\pi\, j^n J_n(u)\exp(jn\phi)$$

$$\int u^{n+1} J_n(u)\,du = u^{n+1} J_{n+1}(u)$$

show that the Fourier transforms of the components can be written

$$E_{0x}(\alpha,\beta) = E_0 \frac{\pi D^2}{4\lambda^2}\, \frac{2J_1\left(\dfrac{\pi}{\lambda}D\sin\theta\right)}{\dfrac{\pi}{\lambda}D\sin\theta}$$

$$E_{0y}(\alpha,\beta) = -E_0 \frac{\pi D^4}{32\,\lambda^2 f^2}\, \frac{J_3\left(\dfrac{\pi}{\lambda}D\sin\theta\right)}{\dfrac{\pi}{\lambda}D\sin\theta}\sin 2\phi$$

where $\sin\theta = \sqrt{\alpha^2 + \beta^2}$

Q3 Deduce from this the radiated field in the far field.

5.3 *Pyramidal horn*

Determine the optimal geometry of a pyramidal horn fed by a rectangular waveguide of dimensions $a = 2.25$ cm, $b = 1$ cm, and whose directivity is 20 dB at 10 GHz.

5.4 *Rectilinear travelling wave antenna*

A travelling wave propagates with phase velocity v_ϕ along an antenna of length $\ell = \lambda/2$. Determine the far field radiation pattern in the following cases

$$\text{(i) } v_\phi = c \qquad \text{(ii) } v_\phi = c/2 \qquad \text{(iii) } v_\phi = c/10$$

5.5 *Power radiated by a loop carrying a constant current*

Q1 Show that the total power radiated by a loop of radius a carrying a constant current I_0 is

$$P=30\pi^2(2\pi a/\lambda)^2\left|I_0\right|^2\int_0^\pi J_1\left[\frac{2\pi a}{\lambda}\sin\theta\right]\sin\theta\,d\theta$$

Q2 Deduce from this expressions for the radiation resistance R_r and the directivity D of the loop.

Q3 What do these expressions become when the dimensions of the loop are very small compared with the wavelength ?

6

Printed antennas

An active printed patch array, designed and realized at the University of
Nice–Sophia Antipolis, France. Each patch is fed by its own
transistor amplifier, integrated as part of the array.

6.1 INTRODUCTION

Printed antennas originated from the use of planar microwave technologies such as
microstrip, slot lines, coplanar lines, etc. Although the possibility of using the
radiation from such lines (particularly important at discontinuities) was suggested as
early as the 1950s, the first practical printed antennas appeared in the mid-1970s[1]
and have been widely developed since then. They are realized with printed circuit
technology, and since their introduction they have accompanied the general trend of

---

[1] Munson, R.E., 'Conformal microstrip antennas and microstrip phased arrays',
IEEE Trans. Antennas & Propagation, Vol. AP-22, pp74-78, 1974.

electronic miniaturization. The different types of printed antennas are distinguished by the geometry of the radiating element and by the feeds or ports which allow its excitation.

Printed antennas have numerous advantages, including:

- light weight and small overall dimensions;
- easy manufacturing using printed circuit technology;
- flush planar technology, aerodynamic characteristics suitable for aeronautical applications;
- easily integrated with electronic components;
- possibility of printing on curved surfaces to make conformal antennas;
- easily integrated into arrays.

However they also have some drawbacks which limit their use:

- relatively narrow bandwidth in general, except for more complicated geometries;
- several effects may lead to low overall efficiency, including dielectric losses, generation of surface waves at the dielectric-air interface, and losses in the coplanar feed lines;
- difficulties to obtain high polarization purity.

Printed antennas have found use in most classical microwave applications, including for example radars, telecommunications, satellites for television broadcast, mobile communications, aeronautical applications, the space industry, GPS, systems for detection and identification, medical applications, etc. They are used at frequencies ranging from UHF to millimetre waves. They are increasingly associated with MMIC components, giving rise to integrated active antenna arrays.

Printed antennas generally behave as resonators. This naturally leads to a fairly narrow bandwidth, typically a few percent. However some printed antennas behave differently and may possess a wider passband.

6.2 DIFFERENT TYPES OF PRINTED RADIATING ELEMENTS

The simplest type of printed antenna is a metallic patch on one side of a dielectric substrate with the other side completely metallized (Fig. 6.1).

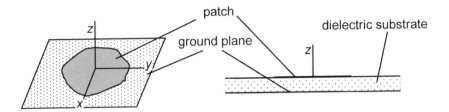

Fig. 6.1 Printed radiating element.

The patches may have various different forms, some of which are shown in Fig. 6.2: rectangular, circular, triangular, elliptical, annular, etc.

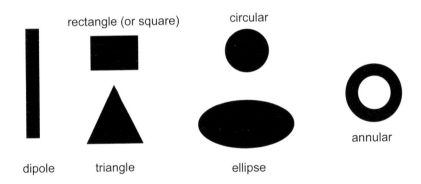

Fig. 6.2 Different forms of patches.

Printed antennas may be excited or 'fed' by different types of transmission lines, for example coaxial, microstrip, or coplanar. The radiating elements may be fed directly, with electrical continuity between the conductor of the transmission line and the conducting patch. Or, on the other hand, the radiating elements may be fed by electromagnetic coupling, in which case there is no direct contact between the transmission line and the patch, which is placed close enough to the transmission line to be excited by it. Several different types of feed, for the case of a rectangular patch, are shown in Figs 6.3 to 6.6.

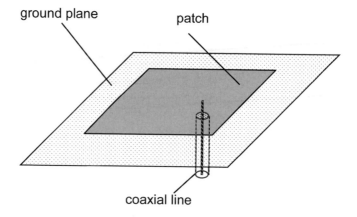

Fig. 6.3 Rectangular patch with coaxial feed. The central conductor is electrically connected to the patch, whereas the outer conductor is connected to the ground plane.

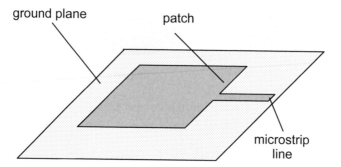

Fig. 6.4 Direct feed by microstrip line.

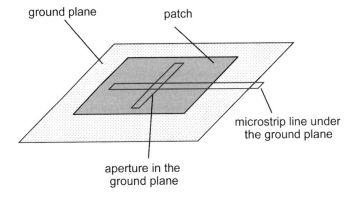

Fig. 6.5 Feed across an opening in the ground plane.

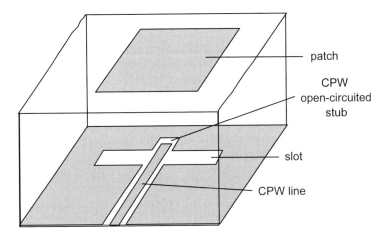

Fig. 6.6 Feeding of a rectangular patch by a coplanar line and slot.

6.3 FIELD ANALYSIS METHODS

There are numerous methods to analyse the fields radiated by printed antennas. Certain methods are based on numerical calculations generally used in electromagnetism problems, for example finite elements or finite differences, using harmonic (Fourier) or time-domain analysis, or the transmission line matrix (TLM) method. These methods yield precise results at the expense of heavy computing power or long calculation times.

6.3.1 Methods of analysis of printed antennas

Other methods are more specifically used for printed antennas:

- the transmission line method, in which a rectangular printed antenna is considered as a portion of microstrip line which radiates at its two extremities;
- the cavity method, in which the space between the patch and the ground plane is considered as a resonant electromagnetic cavity which is leaky (and hence radiates) via its lateral surface;
- the method of moments used to resolve a two-dimensional integral equation, which introduces the use of Green's functions in a planar geometry, and which may be extended to multilayer structures.

6.3.2 The cavity method

The cavity method yields good results for an approximation, but above all leads to a better understanding of the physical operation of the antenna. We shall use this method for the description of antennas having simple geometry.

The patch is assumed to be a perfectly conducting metallic surface S having a closed contour C. The patch is disposed on a dielectric substrate having the relative permittivity ε_r and thickness (or height of the cavity) h, assumed small compared to the wavelength λ_d in the substrate. The cavity method is based on the following physical model:

- the electric field is mainly localized in the cylinder of height h between the patch and the ground plane (Fig. 6.7);
- the radiation is the result of leakage from the cylindrical cavity via its lateral walls (as the cavity ends are perfectly conducting).

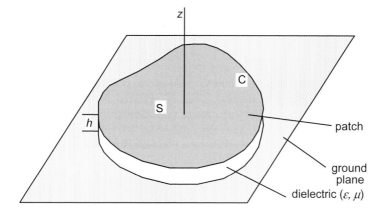

Fig. 6.7 The printed radiating element is considered as an electromagnetic cavity.

Simplifying hypotheses

The small thickness h of the dielectric substrate allows the following approximations to be made:

• The fields **E** and **H** are independent of z;
• The electric field **E** in the cavity is oriented parallel to $\mathbf{u}_z$.

Applying Maxwell's equations under these approximations yields a single component $E_z(x, y)$ of the electric field within the cavity

$$\mathbf{E} = E_z\,\mathbf{u}_z \tag{6.1}$$

Therefore

$$\mathbf{H} = -\frac{j}{\omega\mu}\,\mathbf{u}_z \times \operatorname{grad} E_z \tag{6.2}$$

The surface current density $\mathbf{J}_s$ on the lower (dielectric) side of the conducting patch is given by

$$\mathbf{J}_s = \mathbf{H} \times \mathbf{u}_z = -\frac{j}{\omega\mu}\operatorname{grad} E_z \tag{6.3}$$

Calculation of E_z

The component E_z satisfies the Helmholtz equation within the cavity

$$\nabla^2 E_z + \omega^2 \varepsilon \mu E_z = 0$$

However this relation is not sufficient to calculate E_z: we must also impose the boundary conditions at the lateral walls of the cylindrical cavity. We observe that the surface current density $\mathbf{J}_s$ on the internal side of the patch must be tangent to the contour C (or vanish). The magnetic field $\mathbf{H}$ obtained from the surface current density $\mathbf{J}_s$ via the relation $\mathbf{H} = \mathbf{u}_z \times \mathbf{J}_s$ is thus normal to the lateral surface (or vanishes). The component E_z is thus the solution to the following two-dimensional problem (ρ, ϕ in cylindrical coordinates, or x, y in Cartesian coordinates).

$$\nabla^2 E_z + \omega^2 \varepsilon \mu E_z = 0 \text{ on the surface S of the patch} \tag{6.4}$$

$$\mathbf{n}.\text{grad } E_z = 0 \text{ on the contour C which surrounds the patch} \tag{6.5}$$

where $\mathbf{n}$ represents a normal vector with respect to the lateral surface and to the contour C (Fig. 6.8).

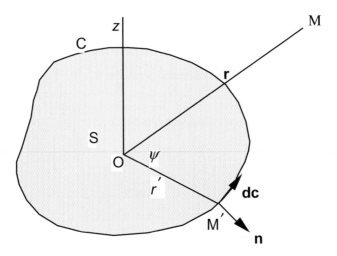

Fig. 6.8 Geometry for the calculation of the electromagnetic field.

Calculation of the radiated field

The cavity radiates via its lateral surface, and we may apply the equations governing radiation of apertures. The far field approximation of the radiated electric field $\mathbf{E}(\mathbf{r})$, at a point M situated in the direction $\mathbf{u}$ may be expressed as

$$\mathbf{E}(\mathbf{r}) = -\frac{jh}{\lambda}\frac{\exp(-jkr)}{r}\int_{C} E_z(\mathbf{r'})\exp(jkr')\cos\psi \ \mathbf{dc} \qquad (6.6)$$

where

$\mathbf{r}$ = OM defines the position of the point of observation,

$\mathbf{r'}$ = OM′ defines a point M′ on the contour C,

$E_z(\mathbf{r'})$ is the field on the lateral surface of the cavity at the point M′, and

$\mathbf{dc}$ is the element of contour C at point M′.

As an exercise, the reader is invited to demonstrate equation (6.6) using the results obtained in Chapter 2.

6.3.3 Application to a rectangular patch

The simplest printed antenna geometry is that of a rectangular patch having dimensions (a, b) disposed on a substrate characterized by its dielectric properties (ε, μ) and thickness h (Fig. 6.9)

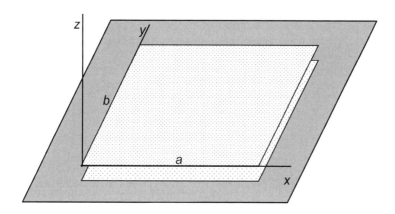

Fig. 6.9 Rectangular patch antenna.

Expression for the electromagnetic field

The component E_z satisfies the Helmholtz equation within the cavity

$$\nabla^2 E_z + \omega^2 \varepsilon\mu \, E_z = 0 \tag{6.7}$$

with boundary conditions

$$\frac{\partial E_z}{\partial y} = 0 \;\; \text{for } x = 0 \text{ and } x = a \quad \text{and} \quad \frac{\partial E_z}{\partial x} = 0 \;\; \text{for } y = 0 \text{ and } y = b \tag{6.8}$$

E_z may be easily calculated by the method of separation of variables

$$E_z = E_0 \cos\frac{m\pi x}{a} \cos\frac{n\pi y}{b} \tag{6.9}$$

where E_0 is a constant of integration which gives the amplitude and the phase of the field, and (m, n) are integers related to the angular frequency ω by the relation

$$\left[\frac{m\pi}{a}\right]^2 + \left[\frac{n\pi}{b}\right]^2 = \omega^2 \varepsilon\mu \tag{6.10}$$

Resonant modes

Each pair (m, n) denotes a resonant mode of the cavity. In view of the above relation, the resonant frequencies f_{mn} are given by the equation

$$f_{mn} = \frac{v}{2}\sqrt{\frac{m^2}{a^2} + \frac{n^2}{b^2}} \tag{6.11}$$

where $v = \dfrac{1}{\sqrt{\varepsilon\mu}}$ is the velocity of plane waves in the substrate.

The mode $(0, 0)$ may be eliminated because it corresponds to the electrostatic field inside the capacitor formed by the patch and the ground plane. The first modes to appear are thus:

- the mode $(1, 0)$ whose resonant frequency is $f_{10} = v/2a$. This is the fundamental mode when $a > b$;
- the mode $(0, 1)$ whose resonant frequency is $f_{01} = v/2b$. This is the fundamental mode when $a < b$.

Magnetic field and surface current on the patch

The magnetic field in the cavity and the surface current density on the patch are obtained from the relations

$$H_x = -J_{sy} = -\frac{j}{\omega\mu}\frac{n\pi}{b}E_0\cos\frac{m\pi x}{a}\sin\frac{n\pi y}{b} \tag{6.12}$$

$$H_y = -J_{sx} = -\frac{j}{\omega\mu}\frac{m\pi}{a}E_0\sin\frac{m\pi x}{a}\cos\frac{n\pi y}{b} \tag{6.13}$$

Properties of the fundamental mode

We choose $a > b$ and examine the fundamental mode (1, 0). The resonant frequency is $f_{10} = v/2a$, which implies that the dimension $a = \lambda_d/2$ where λ_d is the wavelength in the substrate. The distribution of fields and surface current is given (Fig. 6.10) by

$$E_z = E_0\cos\frac{\pi x}{a} \tag{6.14}$$

$$H_x = -J_{sy} = 0 \tag{6.15}$$

$$H_y = J_{sx} = \frac{j}{\omega\mu}\frac{\pi}{a}E_0\sin\frac{\pi x}{a} \tag{6.16}$$

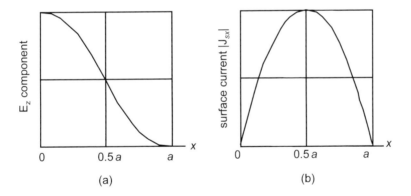

Fig. 6.10 (a) Electric field, and (b) surface current on a rectangular printed antenna.

The radiated field in the fundamental mode is obtained from equation (6.6) which allows the field components to be expressed in spherical coordinates. We show below the results in the planes $\phi = 0$ and $\phi = \pi/2$.

For the E-plane (which contains the electric field vector), $\phi = 0$ and

$$E_\theta = \eta H_\phi = \frac{2jbh}{\lambda} E_0 \frac{\exp(-jkr)}{r} \exp\left[j\frac{ka}{2}\sin\theta\right]\cos\left[\frac{ka}{2}\sin\theta\right] \quad (6.17)$$

$$E_\phi = -\eta H_\theta = 0 \quad (6.18)$$

For the H-plane (which contains the magnetic field vector), $\phi = \pi/2$ and

$$E_\theta = \eta H_\phi = 0 \quad (6.19)$$

$$E_\phi = -\eta H_\theta = -\frac{2jbh}{\lambda} E_0 \frac{\exp(-jkr)}{r} \exp\left[j\frac{kb}{2}\sin\theta\right]\frac{\sin\left[\frac{kb}{2}\sin\theta\right]}{\frac{kb}{2}\sin\theta} \quad (6.20)$$

6.3.4 Application to a circular patch

The circular printed antenna geometry is that of a circular patch of radius a, disposed on a substrate characterized by its dielectric properties (ε, μ) and thickness h (Fig. 6.11).

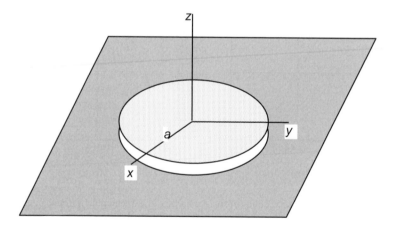

Fig. 6.11 Circular printed antenna.

The cavity is thus a right circular cylinder. The electric field E_z obeys the Helmholtz equation within the cavity, which may be written in polar coordinates (ρ, ϕ) as

$$\frac{\partial^2 E_z}{\partial \rho^2} + \frac{1}{\rho}\frac{\partial E_z}{\partial \rho} + \frac{1}{\rho^2}\frac{\partial^2 E_z}{\partial \phi^2} + \omega^2 \varepsilon \mu E_z = 0 \qquad (6.21)$$

The solutions which are thus obtained must also satisfy the boundary conditions $\frac{\partial E_z}{\partial \rho} = 0$ at $\rho = a$, and may be written:

$$E_z = E_0 \frac{J_m\left(\omega\sqrt{\varepsilon\mu}\,\rho\right)}{J_m\left(\omega\sqrt{\varepsilon\mu}\,a\right)}\cos m\phi \qquad (6.22)$$

where E_0 is a constant of integration which gives the amplitude and the phase of the field on the lateral surface of the cavity. $J_m(x)$ is the Bessel function of order m.

The boundary conditions $\frac{\partial E_z}{\partial \rho} = 0$ at $\rho = a$ are satisfied for $J'_m\left(\omega\sqrt{\varepsilon\mu}\,a\right) = 0$.

The resonant frequency f_{mn} of the mode TM_{mn} of the antenna thus contains a factor x'_{mn} of the nth root of the derivative $J'_m(x)$ of the Bessel function

$$f_{mn} = \frac{x'_{mn}}{2\pi a\sqrt{\varepsilon\mu}} \qquad (6.23)$$

The table below gives the first few roots in increasing order

m, n	x'_{mn}
1, 1	1.841
2, 1	3.054
0, 1	3.832

The frequency of the fundamental mode (1,1) is given by

$$f_{11} = \frac{0.293}{a\sqrt{\varepsilon\mu}} \qquad (6.24)$$

The radius a is thus related to the wavelength in the substrate λ_d by the relation $a = 0.293\,\lambda_d$. The electromagnetic field and the current density on the patch are given by

$$E_z = E_0 \frac{J_1\left(\omega\sqrt{\varepsilon\mu}\rho\right)}{J_1\left(\omega\sqrt{\varepsilon\mu}a\right)}\cos\phi \tag{6.25}$$

$$H_\rho = -J_\phi = \frac{j}{\omega\mu}\frac{1}{\rho}\frac{\partial E_z}{\partial\phi} = -\frac{j}{\omega\mu}E_0\frac{J_1\left(\omega\sqrt{\varepsilon\mu}\ \rho\right)}{J_1\left(\omega\sqrt{\varepsilon\mu}\ a\right)}\sin\phi \tag{6.26}$$

$$H_\phi = J_\rho = -\frac{j}{\omega\mu}\frac{\partial E_z}{\partial\rho} = -\frac{j}{\omega\mu}E_0\frac{J_1'\left(\omega\sqrt{\varepsilon\mu}\ \rho\right)}{J_1\left(\omega\sqrt{\varepsilon\mu}\ a\right)}\cos\phi \tag{6.27}$$

The radiated field may be obtained from the expression for E_z on the edge of the disk by using equation (6.6).

6.4 INPUT IMPEDANCE, BANDWIDTH AND RADIATION PATTERN

6.4.1 Input impedance

The antenna is fed by a transmission line having a characteristic impedance Z_c, and behaves as a complex impedance $Z_a = R_a + jX_a$, connected to the transmission line. The input impedance Z_a depends mainly on the geometry of the coupling between the transmission line and the antenna. For direct feeds, the input impedance depends strongly on the point of contact with the patch. The input admittance $Y_a = G_a + jB_a$ is the inverse of the input impedance.

The cavity method does not give exact results in calculations of input impedance. Other methods, such as the method of moments, give results of a better accuracy. We shall present some general properties of the input impedance of thin printed antennas of simple geometry (rectangular, circular, etc.). Such antennas behave as resonators and may be modelled as a parallel resonant circuit characterized in part by a quality factor Q. The input impedance of such a circuit near resonance is given by

$$Z_a = R_a + jX_a = \frac{R}{1 + 2jQ\left(\dfrac{f - f_r}{f_r}\right)} \tag{6.28}$$

where f_r is the resonant frequency and R is the resonance resistance. The variation of the input impedance as a function of frequency is shown in Fig. 6.12.

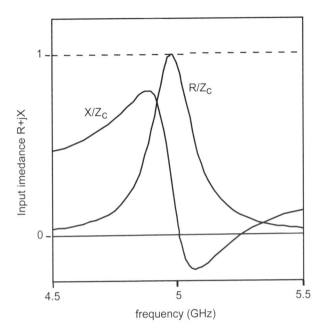

Fig. 6.12 Input impedance of a square patch of side 20 mm on a substrate of thickness 0.8 mm and of relative permittivity 2.25. Fed by a coaxial feed (Z_c =50 Ω) at 6.55 mm from one edge.

6.4.2 Bandwidth

The impedance matching between the feed line and the antenna may be accomplished by any of the standard matching techniques; however it is often advantageous to obtain impedance matching at the resonant frequency f_r such that Z_a = $R = Z_c$ by adjusting the geometry of the antenna and its coupling with the feed line. Under these conditions the bandwidth Δf for which the Voltage Standing Wave Ratio (VSWR) at the antenna input is less than a value S is given by

$$\frac{\Delta f}{f_r} = \frac{S-1}{Q\sqrt{S}}$$

(6.29)

This formula shows that in order to increase the bandwidth, one must lower the Q factor, thus increasing the losses. Increase of heat losses, either in the metal or in the dielectric, decreases the antenna efficiency. Therefore, in order to increase the bandwidth, it is preferable to increase the radiation losses. This may be achieved by increasing the substrate thickness, which results in an increased area of the lateral radiating surface of the cavity. In practical realizations, thin printed antennas having simple geometries have bandwidths of the order of 0.5% to 4%. In order to increase the bandwidth significantly, several resonators must be used together (patches, slot radiators, etc.) in each radiating element. By coupling several resonators in this manner, bandwidths of up to 50% may be achieved. Fig. 6.13 shows several structures which obtain wider bandwidths by coupling between patches or with slots. The method of moments is better suited than the cavity method for calculating the geometry of such radiating elements. The presence of several elements requires a larger overall size of the antenna, either in thickness or in lateral dimensions, and results in a more complicated assembly.

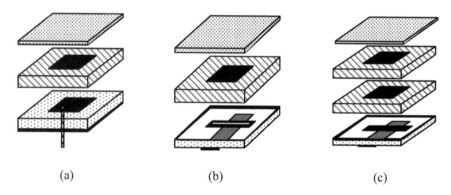

(a)   (b)   (c)

Fig. 6.13 Wideband antennas having coupled resonators: (a) two patches one above the other; (b) resonating slot and patch; (c) resonating slot and two patches.

With special design, some printed antennas may achieve bandwidths of several octaves. Fig. 6.14 shows two types of antenna with exponentially tapered slots: the Vivaldi antenna[2] and the Exponentially Tapered Slot Antenna (ETSA) in a triplate structure[3]. Fig. 6.15 shows the printed spiral antenna.

---

[2] Gibson, P.J. 'The Vivaldi aerial', *Proc. IX European Microwave Conference*, Brighton, Microwave Exhibitions and Publishers Ltd., pp101–105, September 1979.

[3] Guillanton, E. Dauvignac, J.Y., Pichot, Ch. and Cashman, J., 'A new design tapered s lot antenna for ultra-wideband applications', *Microwave Opt Technol. Letters*, Vol.19, pp286-289.

(a)

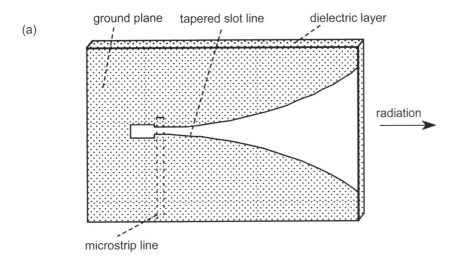

ground plane   tapered slot line   dielectric layer

radiation

microstrip line

(b)

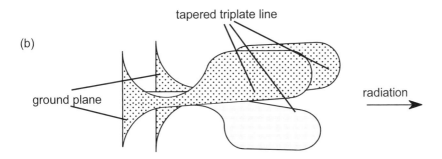

tapered triplate line

ground plane

radiation

Fig. 6.14 (a) Vivaldi antenna; (b) triplate Exponentially Tapered Slot Antenna.

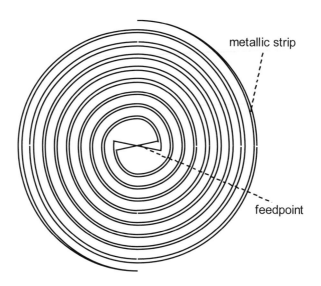

Fig. 6.15 Printed spiral antenna.

6.4.3 Radiation pattern

The radiated field has been calculated in the perpendicular planes xOz and yOz for the rectangular patch (equations (6.17) and (6.19)). The corresponding radiation patterns are shown in Fig. 6.16.

The gain of printed antennas depends strongly on the antenna geometry and the quality of the materials used. For simple geometries, antenna gain is of the order of 4 to 5 dB. The gain may be as high as 8 to 9 dB for multilayer antennas with stacked patches.

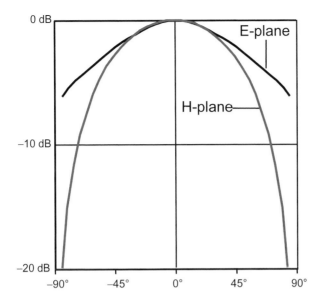

Fig. 6.16 Example of the theoretical radiation pattern of a square patch on an infinite ground plane.

6.4.4 Polarization

Linear polarization

Printed antennas having simple geometries (rectangular, circular, etc.) and used in their fundamental mode radiate with a linear polarization in the direction of the axis Oz perpendicular to the patch. The direction of polarization is determined by the surface currents on the patch. Consequently, the position of the antenna feed is important in determining the polarization direction, as the polarization will be parallel to the general direction of the surface current on the patch. Fig. 6.17 shows antennas radiating in the direction Oz with linear polarization parallel to Ox. Because of the asymmetry of the radiating element and/or the feed, the quality of the linear polarization is often inadequate for applications which require a high degree of polarization purity. This purity can be improved by making the arrangement of patches and feeds symmetrical, for example by feeding two points symmetrically disposed from the centre of the patch in phase opposition.

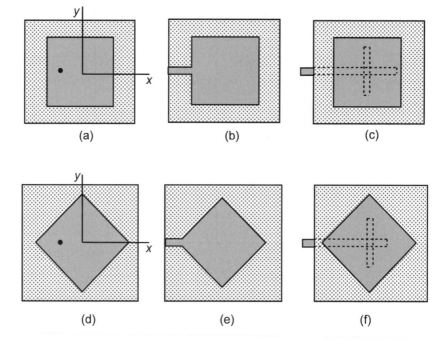

Fig. 6.17 Examples of antennas radiating with linear polarization, using a square patch
(radiation in the far field in the direction O*z* is polarized according to O*x*):
(a) coaxial feed in the middle of one side;
(b) direct feed by microstrip line at the middle of one side;
(c) feed by a microstrip line across a slot in the ground plane;
(d) coaxial feed on one diagonal of the square;
(e) direct feed by microstrip line at one corner of the square;
(f) feed by a microstrip line across a diagonal slot in the ground plane.

Circular polarization

In order to obtain circularly-polarized radiation, two equal-amplitude waves are
superposed with a phase difference of $\pi/2$, linearly polarized along the O*x* and O*y*
axes respectively. There are two ways to obtain these conditions:

- using radiating elements which operate in double linear polarization, fed at two
  points on the patch (Fig. 6.18). Two orthogonal modes (1, 0) and (0, 1) are thus
  excited. In this case, the two feeds are excited by equal-amplitude signals in
  quadrature, obtained for example from a 3dB coupler. The circularly-polarized

radiation will then have a bandwidth which is the same as that of the patch excited in linear polarization.

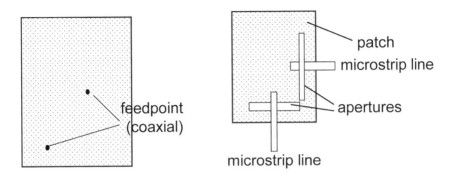

Fig. 6.18 Examples of radiating elements operating in double linear polarization: (a) rectangular patch fed by coaxial lines; (b) rectangular patch fed by slot apertures.

•using radiating elements having a single feed, but whose geometry has been calculated in such a manner that two orthogonal modes will be excited with equal amplitude and in phase quadrature, giving rise to radiation having two orthogonal linear polarizations. (Fig. 6.19). The cut corner of the square patch of Fig. 6.19a couples the two orthogonal modes (1, 0) and (0, 1) of the square patch. The geometry of the radiating element and the position of the feed point are calculated so that these modes are in quadrature and have equal amplitude at the operating frequency. The nearly square patch shown in Fig. 6.19b operates in a similar manner: the feed on the diagonal excites the two orthogonal modes (1, 0) and (0, 1) of the rectangle at two slightly different resonant frequencies. The operating frequency is chosen between these two frequencies such that the two modes are excited with equal amplitude and in phase quadrature. The notched disk shown in Fig. 6.19c operates like the square antenna with the cut corner. The notches couple two fundamental modes of the circular patch which are orthogonal polarizations. For these single-feed antennas, the geometry is critical and must be carefully calculated. Pure circular polarization is obtained only over a very narrow bandwidth because of the quadrature condition, much narrower than the bandwidth of the same patch excited in linear polarization.

The polarization purity from these radiating elements is generally poor, especially in off-boresight directions. The technique of sequential rotation (Fig. 19d) used in circularly-polarized arrays, gives considerably better polarization purity.

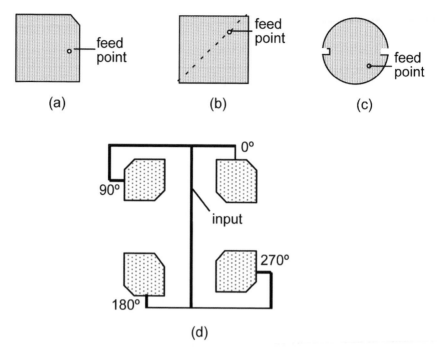

Fig. 6.19 Radiating elements operating with circular polarization and having a single feed point: (a) cut corner square patch; (b) nearly square patch; (c) notched circular patch; (d) 4-element array using sequential rotation.

6.5 LOW-PROFILE, WIDEBAND OR MULTIBAND ANTENNAS FOR MOBILE COMMUNICATIONS AND SHORT-RANGE APPLICATIONS

The relentless miniaturization of electronic equipment requires the design and realization of low-profile antennas. At the same time, antennas are required to operate simultaneously over several frequency bands, of bandwidth of the order of 10%. This is particularly the case for mobile telephony (GSM, DCS, PCS, UMTS) and for short-range wireless communications (Bluetooth, Wi-Fi, WLAN, ...). Antennas of very small dimensions generally possess narrow bandwidths. The techniques used to reduce the dimensions of antennas, as well as to increase the bandwidth and to realize multiband antennas, are presented below.

6.5.1 Miniaturization

Rectangular microstrip radiating elements can be considered as sections of microstrip line terminated in open-circuits at their two ends, thus forming a circuit which resonates when its electrical length is a multiple of $\lambda/2$. To reduce the dimensions of the antenna the following techniques can be used:

Increasing the permittivity. The electrical length $\lambda/2$ corresponds to a physical length which varies as $1/\sqrt{\varepsilon}$, where ε is the permittivity of the substrate that supports the radiating element. Using a substrate with a higher permittivity therefore allows a shorter antenna to be realized. However, this miniaturization generally results in a narrower bandwidth and a lower efficiency, so this method is only usable for antennas which operate only over a very narrow bandwidth.

Modification of the geometry of the radiating element. By altering the geometry (meander lines, folding, introduction of slots, …) it is possible to obtain an electrical length of $\lambda/2$ with physically smaller radiating elements.

Introduction of short-circuits. In a rectangular $\lambda/2$ resonant antenna a short-circuit plane can be introduced can be introduced in the median plane because the field there is zero. This results in a radiating element whose electrical length is $\lambda/4$, i.e. a factor of 2 smaller. The PIFA (Planar Inverted-F Antenna) is based on this principle (Fig. 6.20).

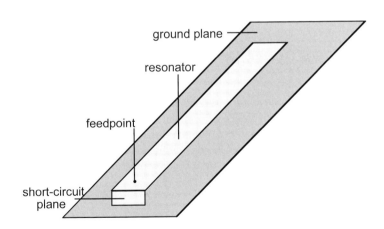

Fig. 6.20 Planar Inverted-F Antenna (PIFA).

More generally, short-circuits in the form of planes, strips, shorting pins or wires are used to reduce the dimensions of antennas, giving more or less complicated geometries. For example, the monopolar wire-patch antenna[4] (Fig. 6.21) has lateral dimensions of the order of $\lambda/7$.

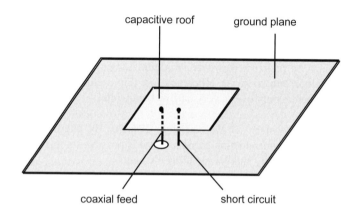

Fig. 6.21 The monopolar wire-patch antenna.

Capacitive loading. To artificially shorten the electrical length of the line it is possible also to load the non-short circuited end with a capacitance. Figure 6.22 shows several ways of realizing this capacitive loading.

6.5.2 Enlargement of the bandwidth and realization of multiband antennas

The method most often used is to introduce additional resonant structures into the radiating element. By suitable choice of resonant frequencies and coupling between different resonators, it is possible to obtain broadband antennas or multiband antennas. The principal ways of introducing additional resonators (parasitic radiating elements and slots), whilst at the same time keeping the transverse dimensions of the antenna small, are described below.

---

[4] Delaveaud, C., Leveque, P. and Jecko, B., 'The monopolar wire-patch antenna', *Electronics Letters*, Vol.30, No.1, pp1-2, 6 January 1994; see also comments in *Electronics Letters*, Vol.30, No.10, p745, 12 May 1994.

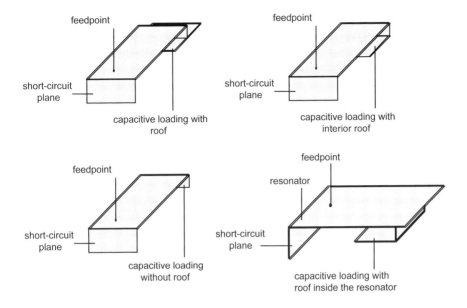

Fig. 6.22 Capacitive loading.

Parasitic resonators superposed or juxtaposed in proximity of the radiating element.
This simple method has already been mentioned (Fig. 6.12) in §6.4.2.

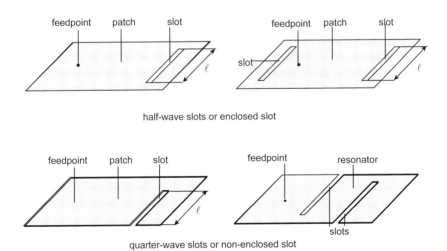

Fig. 6.23 Slots in radiating elements.

Slots in radiating elements. The insertion of slots in patches provides a simple technique for introducing additional resonances (Fig. 6.23). These slots also have the effect of increasing the pathlength of the electric currents, thereby contributing to the reduction in size of the antenna.

6.5.3 Example: multiband antenna for telecommunications[5]

We now describe a number of different antennas derived from the same basic geometry, obtained by adding successively slots, short-circuits and additional radiating elements.

Basic antenna. This is derived from the planar rectangular antenna to which has been added (Fig. 6.24) a short-circuit and an angled slot. This has the effect of increasing the pathlength of the surface electric currents, as well as a capacitive effect between its ends. A suitable choice of dimensions allows the realization of antennas to operate in the GSM band and in bands around 2 GHz.

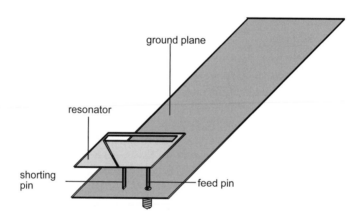

Fig. 6.24 Basic structure of multiband antennas.

Multiband antennas. The juxtaposition of a parasitic element allows an additional band to be covered. In the antenna shown in Figure 6.25 a short-circuited radiating element has been added, which allows operation in the GSM bands (880 MHz – 960 MHz) and UMTS (1920 MHz – 2170 MHz).

---

[5] The antennas presented in this section have been realized by P. Ciais, G. Kossavias, C. Luxey and R. Staraj at the Laboratoire d'Electronique, Antennes et Télécommunications de l'Université de Nice-Sophia Antipolis, France.

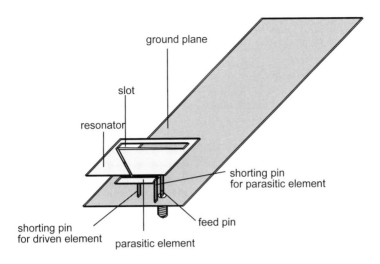

Fig. 6.25 GSM-UMTS antenna with a parasitic short-circuited element.

In the structure of Figure 6.26, a short-circuited parasitic element with capacitive loading has been added to the basic structure, in a different placement to that used in the previous antenna. The antenna realized in this way allows coverage of the GSM band (880 – 960 MHz), DCS band (1710 – 1880 MHz) and PCS band (1850 – 1990 MHz).

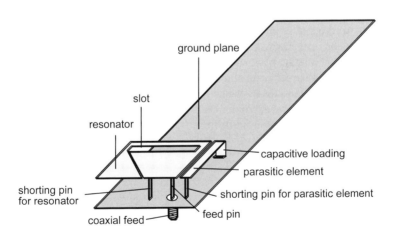

Fig. 6.26 GSM-DCS-PCS antenna with a parasitic short-circuited and capacitively-loaded element.

The simultaneous addition of these two types of parasitic element (Fig. 6.27) allows an antenna covering the DCS, PCS and UMTS bands, but only part of the GSM band.

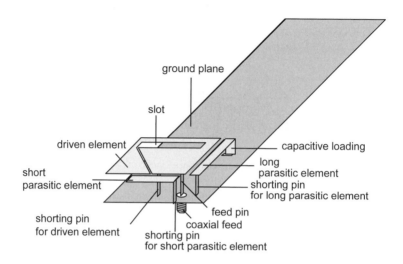

Fig. 6.27 GSM-PCS-UMTS antenna with two parasitic elements, with partial coverage of the GSM band.

An extra parasitic element (Fig. 6.28) of greater length, short-circuited and terminated in a capacitive load, allows coverage of the whole of the GSM band. In this way we can realize an antenna covering the GSM, DCS, PCS and UMTS bands.

Simultaneous coverage of the WLAN bands is possible, with the introduction of an additional slot in the basic radiating element. A bandwidth of 20% centred on 5.4 GHz has been obtained with the structure shown in Figure 6.29. As well as the GSM, DCS, PCS and UMTS bands this antenna covers the HYPERLAN/2 (indoor 5150 MHz – 5350 MHz, outdoor 5470 – 5725 MHz) and IEEE 80211a (indoor 5120 MHz – 5250 MHz, outdoor 5250 MHz – 5350 MHz and 5725 MHz – 5825 MHz) bands.

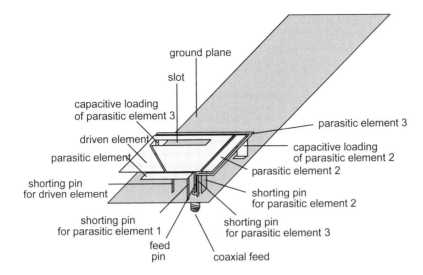

Fig. 6.28 GSM-DCS-PCS-UMTS antenna with three parasitic elements.

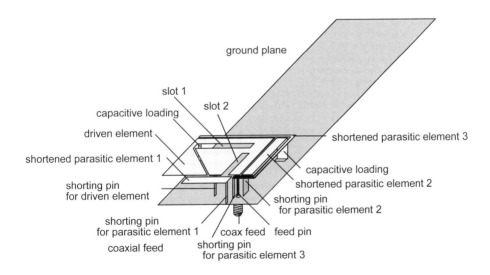

Fig. 6.29 GSM-DCS-PCS-UMTS-WLAN antenna.

FURTHER READING

1. Bahl, I.J. and Bhartia, P., *Microstrip Antennas*, Artech House, Dedham, MA., 1980.
2. Balanis, C.A., Antenna Theory and Design, John Wiley & Sons, New York, second edition, 1997.
3. Howe, H. Jr, *Stripline Circuit Design*, Artech House, Dedham, MA, 1974.
4. Fujimoto, K., Henderson, A., Hirasawa, K. and James, J.R., *Small Antennas*, Research Studies Press, 1987.
5. Fujimoto, K. and James, J.R., *Mobile Antenna Systems Handbook*, Artech House, 2nd edition, pp63-68, 2001.
6. James, J.R., Hall, P.S., Wood, C. and Henderson, A., 'Some recent developments in microstrip antenna design', *IEEE Trans. Antennas & Propagation*, Vol. AP-29, No. 1, pp124-128, January 1981.
7. James, J.R. and Hall, P.S., *Handbook of Microstrip Antennas*, Peter Peregrinus, Stevenage, 1989.
8. Lee, H.F. and Chen, W. (eds), *Advances in Microstrip Antennas*, Wiley, New York, 1997.
9. Mailloux, R.J., McIlvena, J.F. and Kernwels, N.P., 'Microstrip array technology', *IEEE Trans. Antennas & Propagation*, Vol. 29, pp25-37, 1981.
10. Mosig, J., 'Arbitrarily shaped microstrip structures and their analysis with a mixed potential integral equation', *IEEE Trans. Microwave Theory & Techniques*, Vol. 36, pp314-323, 1988.
11. Pozar, D.M., 'A reciprocity method of analysis for printed and slot coupled microstrip antennas', *IEEE Trans. Antennas & Propagation*, Vol. 34, No. 12, pp1439-1446, Dec 1986.
12. Pozar, D.M., 'Microstrip antennas', *Proc. IEEE*, Vol. 80, No. 1, pp79-91, January 1992.
13. Zurcher, J.F. and Gardiol, F., *Broadband Patch Antennas*, Artech House, Dedham, MA, 1995.

EXERCISES

6.1 Simplified model of a rectangular printed patch

An antenna consists of a rectangular patch (Fig. 6.20a) of sides $a = 2$ cm, $b = 4$ cm, printed on a substrate of relative permittivity $\varepsilon_r = 2.2$ and of thickness $h = 1.6$ mm. It can be modelled by the equivalent circuit of Fig. 6.20b, which consists of a section of line of characteristic impedance Z_A placed between two conductances G_1 and G_2, and fed by a line of characteristic impedance Z_C.

Q1 Give a simple physical justification for this model.

Q2 Determine the lowest resonant frequency of the antenna

Q3 Determine the feedpoint admittance Y_{in} of the antenna at the resonant frequency, and the width of the feedline in order to match the antenna, using the following approximate expressions:

Conductances representing the radiation from the antenna: $G_1 = G_2 = \dfrac{1}{90}\left(\dfrac{b}{\lambda}\right)^2$

Characteristic impedance of a microstrip line of thickness h and width w:

$$Z_C = \sqrt{\frac{\mu}{\varepsilon}}\,\frac{h}{w}$$

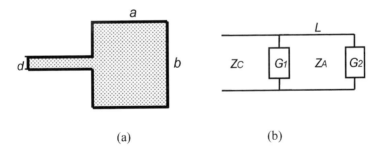

(a)   (b)

Fig. 6.30 Rectangular patch and simplified model.

6.2 *Semi-circular patch analyzed using the cavity method*

A semi-circular patch of radius $a = 4$ cm is printed on a substrate of relative permittivity $\varepsilon_r = 2.25$ and of height $h = 6$ mm (Fig. 6.31).

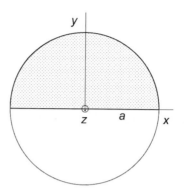

Fig. 6.31 Semi-circular patch.

Q1 Using the approximations of the cavity method, show that the electric field consists only of the component $E_z = E_0 \, J_m(kr)\cos m\phi$.

Q2 What are the boundary conditions imposed on the component E_z on the sides of the cavity ?

Q3 Deduce from this the resonant modes (m, n) of the antenna. What are the components of the electric and magnetic fields associated with each mode ?

Q4 What is the fundamental mode, and what is its resonant frequency ? What is the form of the surface currents on the patch, and their positions in the E- and H-planes. Where should a coaxial feed be placed in order to excite the fundamental mode ?

6.3 *Dual-frequency rectangular microstrip antenna*

A rectangular microstrip antenna of sides a and b ($a < b$) is printed on a substrate of relative permittivity $\varepsilon_r = 2.6$. This antenna is required to function with two modes of frequencies $f_1 = 1.5$ GHz and $f_2 = 2.2$ GHz at orthogonal polarizations, all in as compact a manner as possible.

Q1 How should a and b be chosen ?

Q2 Where should a coaxial feed be placed to excite (a) just the first mode; (b) just the second mode; (c) both modes simultaneously ?

7

Large antennas and microwave antennas

Dual-beam surveillance radar antenna. Photo: THALES (formerly Thomson-CSF).

7.1 INTRODUCTION

Large antennas and microwave antennas, which form the subject of the remaining chapters of this book, are characterized by the property that their useful dimensions are very large compared with the wavelength involved. The word 'large' refers to the physical dimensions of the antenna (radio telescopes, for example) compared with which the wavelengths involved are so small (centimetres or millimetres) that very specific technologies may be involved. In this way, a glass lens may be considered as a 'large antenna' whereas a radio transmitting tower may be considered 'small' in comparison with the wavelengths it transmits.

The main importance of large antennas, defined in this way, lies in the possibility to transmit or to receive radiation with high directivity and high gain, or with radiation patterns whose shape can be synthesized with great accuracy. In this way an antenna of diameter D can radiate or receive a signal of wavelength λ with a beamwidth of $\theta \approx \lambda/D$ radians (§7.4.4). Its maximum theoretical gain is $(\pi D/\lambda)^2$

(§7.4.8) and its radiation pattern can be synthesized with a sampling step of λ/D (§13.2.2) With array antennas, signal processing techniques may allow improvement of the angular resolution and spatial filtering of the incident signals (Chapter 14).

Nevertheless, in any antenna, *directivity* and *radiation pattern* cannot be optimized independently:

If the radiation, in a direction of unit vector **u** is characterized, both in amplitude, phase and polarization by any complex vectorial function $f(\mathbf{u})$, it results from eqn. 3.36b (§ 3.3.2) that the directivity (or gain in the absence of losses) is given by:

$$D(\mathbf{u}) = 4\pi \frac{\left|f(\mathbf{u})\right|^2}{\displaystyle\int_{4\pi}\left|f(\mathbf{u})\right|^2 d\Omega} \qquad (7.1)$$

where $d\Omega$ is the elementary solid angle within the direction **u** .

Applications of this important relationship are given further in §7.2.2 and in exercise 7.1.

The possibilities we have evoked so far allow us to consider a number of applications of large antennas: terrestrial or space telecommunications, radar and radioastronomy.

3-D electromagnetic problems related to such antennas can be approached using mathematical models whose complexity depends on the accuracy that is required: geometrical optics, physical optics either in scalar form, or in vector form if polarization is to be taken into account. Many problems are equivalent to the radiation from an aperture and may be solved using the approximations of Kirchhoff, Fresnel or Fraunhofer.

The choice of the mathematical model to be used depends on the requirements of the project. Generally, the engineering work consists, at the start, of the definition of the type of antenna which must fulfil not only the electrical specifications but also the mechanical constraints and the technological and economical requirements.

During this initial phase, only the simplest models - geometrical optics, scalar physical optics - allow the choice of the appropriate type of antenna for the preliminary design.

During a second phase a more detailed study is carried out. Making use of the powerful computational techniques now available, it is possible to obtain relatively rigorous results. But computation time can often be so long that it may limit the accuracy desired by the designer. Furthermore, a number of parasitic (poorly-identified) phenomena, difficult to calculate, may cause errors which are greater than those due to the approximations on which the calculations are based. Nevertheless, sometimes it may be possible to carry out an experimental evaluation - automatic measurement techniques allow digitized data to be recorded and processed by computer.

In summary, an important aspect of the art of the engineer lies in the choice of the appropriate models (in the phases of synthesis and analysis) and in the use of experimental measurements (Chapter 15), which all need to be taken into account in a successful design.

7.2 STRUCTURES AND APPLICATIONS

7.2.1 Structures

In considering the choice of a radiating structure for a specific application, many different types of antennas can be considered. Nevertheless, most large antennas may be derived from two main categories: *focused systems* and *arrays*.

A focused system requires a primary feed, generally associated with one or more reflectors, sometimes a dielectric lens, that transforms the incident wave phase - and in a less important manner the amplitude - so that the whole radiation should have the required characteristics. One important class of focal systems, rotationally symmetric, are the so called axially-symmetric systems (Chapter 9). The parabolic reflector and the Cassegrain antenna are two important examples (see Fig. 7.1). When the radiation characteristics are not axially-symmetric, the focal system itself is not centred. This is the case with offset reflectors or double curvature reflectors used in surveillance radars (§10.3).

Fig. 7.1 Cassegrain antenna for space communications, with equatorial mount.
Photo: THALES (formerly Thomson-CSF).

Seen from the point of view of reception, the focused system concentrates all the received energy from a plane wave into a 'focal' zone forming a diffraction pattern (§9.8). A knowledge of the structure of this 'diffraction spot' pattern is important since its transfer conditions into the primary feed determine the overall behaviour of the system in reception (§9.9): gain, noise temperature, and so on.

Fig. 7.2 Stacked-beam radar antenna. Photo: THALES (formerly Thomson-CSF).

The primary feed properties play, naturally enough, an important part in the whole system (Chapter 8). They can fulfil several functions: transmission, reception, angular tracking, polarization and so on. Their study is quite tricky since dimensions are not very large compared with the wavelength and the chosen approximation methods must be accurate enough and used with sufficient care to prevent significant errors. During the development phase, experimental measurements play an important role.

An array is composed of a group of identical radiating elements placed at the nodes of a regular grid whose lattices are square or triangular, and (usually) regularly spaced. The inter-element spacing (or 'step') is a basic parameter of the array. The radiating elements are fed by a feed network (or power divider) which defines an array illumination, generally fixed in amplitude. The radiation phases of the elements are controlled by a set of phase shifters, usually controlled by computer.

The interest in array antennas is that they provide a quasi-instantaneously steerable beam. This is obtained by imposing a constant phase gradient between the array elements, according to the desired pointing direction (Fig. 7.4).

Fig. 7.3 94 GHz horn antenna ($\lambda = 3$ mm). In contrast to the antenna shown at the beginning of Chapter 5, this is a 'large antenna'. (Photo: Dr Paul Brennan).

Fig. 7.4 Secondary radar array. Photo THALES (formerly Thomson-CSF).

Active arrays represent the most sophisticated type of antenna. Using a separate amplifier for each individual element of the array is essential since it improves the transmitted power efficiency and also allows control not only of the phase but also the *delay* and the *amplitude* of the array excitation. This permits great possibilities in using wide bandwidths and in respect of adaptive suppression of jamming signals (§11.8).

The combination of active antennas with digital processing in real time opens up enormous possibilities in imaging and high resolution (Chapter 14).

With the same basic structures, large antennas come in many shapes and sizes; a great diversity of antennas may be obtained by combining focal structures and arrays.

Fig. 7.5 Phased array antenna. Photo THALES (formerly Thomson-CSF).

7.2.2 External characteristics required in applications

By external characteristics we mean the antenna properties considered as a whole, possibly combined with signal processing systems. Internal characteristics are those that are chosen to fulfil the external ones. We shall consider, as examples, two important areas: radar and space telecommunications.

Generalities

A radar antenna is not only directed to illuminate useful targets and to receive echoes from them, but it is also designed to give accurate angular discrimination. In this sense it can be considered as an angular information channel, and it will be shown that in such a case it is equivalent to a filter (Chapter 13).

The main desired qualities in reception and transmission are the gain and the shape of the pattern. In the case of a surveillance antenna with a scanning beam, we desire a well-defined pattern in the horizontal plane (azimuth) giving a good resolution, and with a particular shape in the vertical plane to give a particular elevation coverage. We are, then, confronted with the important problem of pattern synthesis (Chapter 13).

Regarding tracking antennas, the beamwidth is not sufficient to determine the target direction with the required accuracy. For this reason interpolation processes and multiple angular measurements, either simultaneous (monopulse) or successive (conical scanning), have been developed (§10.2).

These methods interpret 'angular pointing errors' by means of tracking relationships whose slope and linearity are the key characteristics. The abovementioned methods are examples where the antenna directive properties are based on a signal processing system.

In any case, clutter is to be avoided as well as naturally-occurring noise and jamming sources. To do that sidelobes must be kept to a minimum. Adaptive antennas have adaptive feedback loops that give nulls in the jammer directions, improving the useful signal / jammer ratio (Chapter 14).

Echoes from clouds and rain can be largely suppressed using circular polarization. In effect, spherical obstacles such as rain droplets reverse the sense of the circular polarization on reflection. This is not the case with echoes from aircraft which, owing to their complexity, often depolarize the echo signal (§10.2, Chapter 12). Only the unpolarized component is detected by the circularly polarized radar - the antenna *ellipticity ratio* is an indication of the radar performance.

Let us now specify some points concerning various types of radar antennas.

Surveillance antenna pattern

What is the optimum radiation pattern in the vertical plane of a surveillance radar antenna, such as those that can be seen at many airports ? Consider an aircraft target of radar cross section σ approaching radially at a constant altitude h. The target range is r, and it makes an elevation angle β with the horizon.

It is not necessary that the echo signal should increase as the aircraft approaches: it is better to have maximum gain towards a long range echo from a target at r_0 when the aircraft is seen at a small angle β_0. This is possible by appropriate choice of the antenna vertical pattern $F(\beta)$. In effect, the received power P_r is given by (§4.3.2, eqn. 4.21b)

$$P_r = \frac{P_t \left[G(\beta) \right]^2 \lambda^2 \sigma}{(4\pi)^3 r^4}$$

where P_t is the transmit power, and $G(\beta)$ is the antenna gain, which is proportional to $|F(\beta)|^2$.

The power P_r will be constant if

$$\frac{G(\beta)}{r^2} = \frac{G(\beta_0)}{r_0^2} = \text{constant}$$

We see that $r = \dfrac{h}{\sin \beta} = h \, \text{cosec} \, \beta$.

Thus

$$G(\beta) = G(\beta_0) \sin^2 \beta_0 \, \text{cosec}^2 \beta \tag{7.2}$$

The power pattern varies as $\text{cosecant}^2 \beta$, and is called a 'cosec squared' pattern.

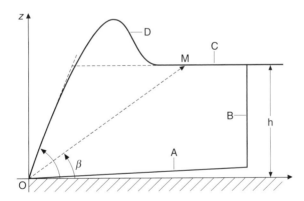

Fig. 7.6 Shaped beam for ground-based radar.

In polar coordinates the amplitude pattern corresponds to the part (C) of the pattern represented in Fig. 7.6. To avoid echoes from the ground, the pattern is shaped with an abrupt slope in the direction of the horizon. The dimensions of the antenna make up a limitation in that direction, part (A) of the preceding figure. At long range, ground clutter echoes are reduced and various filtering techniques are used to suppress those that remain.

At short range, they are strong and undesirable, so often the useful echo is strengthened by shaping the elevation pattern of the feed horn at high elevations, corresponding generally to short ranges (D).

In fact, optimization of the antenna dimensions alone suffices to obtain these conditions, at least approximately (Chapter 13). The pattern shape itself limits the maximum gain that may be obtained (B) and it must take into account the normalization condition that is inherent in the gain definition (eqn 7.1).

$$\frac{1}{4\pi} \iint_{(\Omega)} G(\mathbf{u})d\Omega = \rho \leq 1 \tag{7.3}$$

where Ω is the solid angle and ρ is the efficiency, due to losses in the various parts of the power and diplexing circuits (exercise 7.2).

Multibeam antenna for 3-D radar (§11.7).

This is an antenna having several superposed receiving radiation patterns in the vertical plane, the whole rotating around the vertical axis. This type of antenna possesses several advantages (Fig. 7.7):
- only the lowest beam is concerned with the ground clutter;
- the beam number corresponding to a detected target gives an indication of the target elevation. If it is detected on two adjacent beams, the target elevation can be determined by amplitude comparison (§10.2.4, §11.7);
- since each beam is relatively narrow, its gain is high and the range is increased, as well as the detection probability.

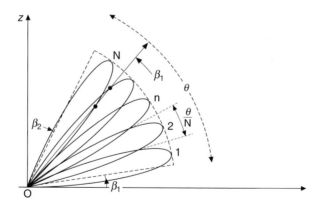

Fig. 7.7 Multibeam antenna for sectoral coverage with a 3-D radar.

The improvement in signal-to-noise ratio is, all other things being equal, given by the number of receive beams.

Multiple beams can be obtained by using a beamforming matrix (§11.7).

Electronically-scanned arrays (Chapter 11)

The main interest in electronic scanning is the ability to instantaneously point a beam in the wanted direction without any waste of time. In practice, this waste of time is quite important with mechanically scanned antennas. Only an electronically-scanned radar allows several targets to be tracked simultaneously and the information from each updated continuously.

Regarding angular sector surveillance, it is easy to show that an antenna scanning with a narrow beam presents, compared with a single large beam, the same advantages as those obtained with a multibeam antenna.

Low noise antennas for space communication links

Space communication links used in different applications (space telecommunications, space probes and so on) may involve very long ranges, much greater than terrestrial communication links (altitude of a geostationary satellite: 36,000 km, Earth–Moon distance: 360,000 km, Earth–Mars mean distance: 150 million km).

On the other hand, for economical reasons the mass of spacecraft is constrained. Consequently, the gain and the radiated power are also limited. The power is often split into a large range of frequencies in order to transmit a large number of channels simultaneously.

Therefore, the fields received on ground (power density in $W.m^{-2}\,Hz^{-1}$) are very small compared with noise and other signals. In order to obtain a sufficient signal-to-noise ratio, terrestrial stations are designed according to two criteria:
- use large area antennas, which means large gains, to maximize the incoming signal (diameters of several metres);
- use low noise temperature receivers and antennas which minimize external and internal noise ('cold antennas').

Gain problem: What are the factors at our disposal to obtain large antenna gains ?

The first is the antenna area. A circular antenna whose diameter and wavelength of operation are respectively D and λ has a maximum gain of

$$G = K\left(\pi\frac{D}{\lambda}\right)^2$$

where $K \leq 1$ is the antenna efficiency.

At the C-band frequencies normally used for space telecommunications (4 GHz satellite-to-Earth; 6 GHz Earth-to-satellite), with antenna diameters of the order of 12 metres it is possible to reach gains of the order of 50 to 60 dB. The cost of such antennas is, of course, very high but improvements, with great success, have been made on the *K* factor. At present, Cassegrain antennas can reach efficiencies from 70 to 80% and even more (Fig. 7.1).

The factor *K* depends on the following parameters.

- *illumination function of the antenna*: a quasi-uniform illumination may be obtained by means of a combination of appropriate reflectors with special shapes (conformal Cassegrain techniques) (§10.1.5).
- *primary feed*: this has several functions: reception, transmission, width detection. Its patterns are related to the antenna illumination function as well as the spillover losses (§9.5.3).
- *antenna fabrication accuracy*: this point is essential since it determines the cost of the antenna.

A thorough study of the effect of fabrication tolerances on antenna performance allows us to reach a compromise between cost and efficiency.

Noise temperature problem: Parasitic noise on reception has both internal and external origins.

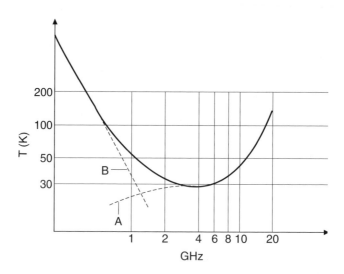

Fig. 7.8 Noise temperature of the sky versus frequency. A: cosmic noise. B: atmospheric noise.

Noise of external origin is mainly due to terrestrial thermal radiation: the Earth behaves as a black body at room temperature, approximately 300 K. Atmospheric thermal noise is due to wave attenuation in the atmosphere. Attenuation can be important when clouds or rain are present: note that 0.1 dB attenuation corresponds to an increase of approximately 7 K of noise temperature. Atmospheric noise increases with frequency (Fig. 7.8).

Cosmic noise is often low at centimetre wavelengths. It decreases with frequency (Fig. 7.8)[1].

The inverse variation of atmospheric and cosmic noise with frequency justifies the frequency choice between 4 and 6 GHz which has generally been taken in telecommunications. Nevertheless, congestion in this frequency range is such that communications links are now using higher Ku-band frequencies (11 to 15 GHz) in spite of the noise increase which results.

Finally, man-made noise sources have also to be taken into account.

All these noises may be intercepted by the antenna sidelobes giving rise to a global antenna temperature T_A. We have seen that this is the average angular distribution $T(\mathbf{u})$ of the external noises, weighted by the antenna gain function $G(\mathbf{u})$ (§4.3.2)

$$T_A = \frac{1}{4\pi} \iint_{(\Omega)} T(\mathbf{u}) G(\mathbf{u}) d\Omega \qquad (7.4)$$

where $d\Omega$ is the solid angle element around the vector direction $\mathbf{u}$.

The low noise antenna problem amounts to minimizing the $G(\mathbf{u})$ function (that is to say the diffuse radiation) particularly in the directions where $T(\mathbf{u})$ is important (the ground in particular). The problem is essentially solved by the choice of antenna and certain practical precautions. Regarding internal noise, the most important is due to the receiver temperature noise T_r. Masers are not commonly used today owing to their complexity. Generally, cooled parametric amplifiers are preferred (T_r between 30 K and 40 K) or even uncooled (T_r between 50 K and 120 K).

Another source of internal noise is ohmic losses that may intervene before amplification, in the radome, polarizer, primary feeds, frequency filters, mismatching, parasitic coupling and so on. All these factors, although almost negligible individually, may have an important overall effect not only on noise temperature (7 K per 0.1 dB of loss)[2] but also on the total gain of the antenna.

---

[1] Recall the discovery by Penzias and Wilson in 1967 of the cosmic microwave background at 2.7 K, which confirmed so emphatically the 'Big Bang' theory of the origin of the Universe.

[2] To minimize this effect, in some radiotelescopes such as the Nançay radiotelescope (France), the entire primary feed is cooled.

Receiving system factor of merit: The overall signal-to-noise ratio of the receiving system is proportional to the 'factor of merit', defined as the ratio of the maximum antenna gain (taking into account all the different losses) to the total noise temperature of the system (antenna temperature, receiver temperature, temperature due to losses)

$$M = \frac{G}{T}$$

This factor is expressed in decibels as follows.

Example: $G = 5 \times 10^5 = 57\text{dB}$
$T = 50$ K, that is to say +17 dB with respect to 1 K, then:

$$\left(\frac{G}{T}\right)_{\text{dB}} = 40$$

The choice of antenna parameters must be made so that the total cost should be minimum for a given factor of merit (see exercise 7.1).

7.3 FUNDAMENTAL PROPAGATION LAWS

The physical laws of radiation and propagation are treated in Chapter 2 and in numerous books on electromagnetism.

In considering microwave antennas whose dimensions are generally large compared with the wavelength, it is possible to present the propagation laws in an approximate and simplified form.

This form relies on concepts derived from physical and geometrical optics: wavefronts, rays and the Huygens and Fresnel principles.

7.3.1 Wavefronts

If at any point the electrical and magnetic fields oscillate with the same phase, it is then possible to define constant phase surfaces or *wavefronts*, some of whose properties related to spherical and plane waves can be applied to any surface. The case where the wavefront has a local character of plane wave will be often considered, then the Poynting vector will allow us to define a power flow, in W.m⁻², stored in the wavefront. This is the basis of the ray concept in terms of geometrical optics.

7.3.2 The Huygens–Fresnel principle of wave propagation

This principle is in fact a result of Maxwell's equations. Consider a homogeneous and isotropic medium, that is to say, a medium where the refractive index *n* may be considered as constant (at least locally). *It can be shown that each point M of a wavefront may be considered as a secondary source giving rise to an elementary spherical wave.* The field at a point P is obtained by superposing these elementary waves, taking into account their phase differences when they reach their destination point: this the result from the Kirchhoff integral (§2.2.2). Since the surface S is a wavefront it can be shown that this integral takes a simplified form (see Appendix 7A).

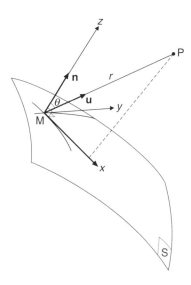

Fig. 7.9 Huygens–Fresnel principle.

Let $E(M)$ be any scalar component of the electric field at a point M of the wavefront (Fig. 7.9). The corresponding component of the electrical field $E'(P)$, at a point P whose distance $r = MP$, is given by the expression

$$E'(P) = \frac{j}{\lambda} \iint_{(S)} \gamma\, E(M) \frac{\exp\left(-j\dfrac{2\pi r}{\lambda}\right)}{r}\, dS \qquad (7.5)$$

The factor γ, called the 'obliquity factor', is quite close to unity. If the wavefront is locally plane, then we have

$$\gamma = \frac{1 + \cos\theta}{2}$$

This can be represented in polar coordinates by a cardioid. It represents the directivity of each element of the wavefront. It shows that the directivity decreases slowly as a function of θ, which is the angle of the vector **n** normal to the surface at the point M with the direction **u** defined by the direction MP. We may assume that $\gamma \approx 1$ since this factor is generally not too critical, as will be shown later. Let us observe that an identical relationship to equation (7.5) exists for the magnetic fields.

Remark: if S is not an equiphase surface, the same formulae may be applied provided that the phase variation is included in the $E(M)$ expression which becomes a complex scalar. The obliquity γ changes, but remains close to unity.

7.3.3 Stationary phase principle

This principle (§2.1.3) allows the transition between physical optics and geometrical optics[3] (Fig. 7.10).

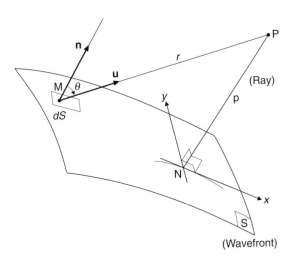

Fig. 7.10 Stationary phase principle.

---

[3] Silver, S. (ed), *Microwave Antenna Theory and Design*, p119, MIT Radiation Lab. Series Vol. 12, McGraw-Hill, New York, 1949; reprinted by Peter Peregrinus, Stevenage, 1984.

Let PN be an orthogonal line from the point P to a wave surface S. Assume that this normal is unique, in which case P is not a focal point. If NP = p is the vertical z axis and Nx and Ny are tangential to the principal cross sections of the surface, from equation (7.5) we obtain

$$E'(\mathrm{P}) = \frac{j}{\lambda}\exp\left(-j\frac{2\pi}{\lambda}p\right)\iint_{(S)}\gamma E(\mathrm{M})\frac{\exp\left[-j\dfrac{2\pi}{\lambda}(r-p)\right]}{r}dS \qquad (7.6)$$

This integral may be represented in the complex plane as a sum of elemental vectors δV. The amplitude of each vector is

$$\delta V = \gamma\frac{E(\mathrm{M})}{r}dS$$

and its phase difference compared with that of the corresponding vector at the point N is

$$\delta\alpha = \frac{2\pi}{\lambda}(r-p)$$

Let us consider the contribution of any surface element S that does not include the normal. If λ is large, the angles $\delta\alpha$ between the vectors are small. The vector sum diagram appears as a curve and the resultant is generally significantly different from zero (Fig. 7.11).

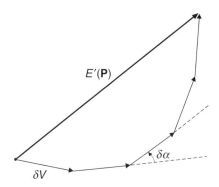

Fig. 7.11 Vector representation of the Huygens–Fresnel integral. Long wavelength case. Contribution of elements close to the normal.

If λ is small, the angles $\delta\alpha$ are large and the vector diagram appears as a tightly-coiled spiral. In such a case $E'(\mathrm{P})$ is almost zero (Fig. 7.12). Thus, in regions where the component wave phase varies rapidly, there is a kind of cancellation of the resulting interference field.

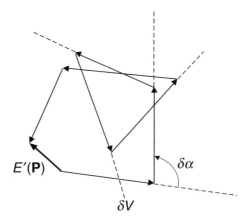

Fig. 7.12 Vector representation of the Huygens–Fresnel integral. Small wavelength case.

The situation is quite different in the vicinity of the origin of the normal at the point N where the phase function is stationary

$$\phi(x,y) = \frac{2\pi}{\lambda}(r-p) = \left[\sqrt{x^2 + y^2 + (p-Z)^2} - p \right] \frac{2\pi}{\lambda}$$

Therefore, close to this point, the phase varies slowly even if λ is small. *Then, only the contribution of the points close to the normal is important* in the expression of the field at P. Note that for these points, θ is small, thus $\gamma \approx 1$.

We can now formulate the stationary phase principle: *for short wavelengths, at a point P far away, the electrical quantities are, approximately, only dependent on the electrical state of the points on the wave surface near the foot of the normal passing through P.*

A simple mathematical relationship linking the electrical quantities at N to those at P, involving the wave surface radii of curvature, can be given.

The integral (7.6) reduces to

$$E'(P) = \frac{j}{\lambda} E_N \frac{\exp\left(-j\frac{2\pi}{\lambda}p\right)}{p} \iint_{(S)} \exp\left[-j\phi(x,y)\right] dx dy$$

Taking into account the surface equation in the vicinity of N and extending the integral to infinity, which the stationary phase principle permits us to do, we find

$$E'(\mathrm{P}) = E_N \sqrt{\frac{R_1 R_2}{(R_1 + p)(R_2 + p)}} \exp\left(-j\frac{2\pi}{\lambda} p\right) \qquad (7.7)$$

where R_1, R_2 are the principal radii of curvature.

7.3.4 Geometrical optics ray theory

Rays

The stationary phase principle establishes a correspondence from one point to another point of two successive wavefronts (Fig. 7.13).

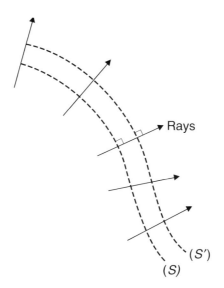

Fig. 7.13 Wavefront and rays.

The points involved are located on the common normal NN′ of two wave surfaces. This family of normals constitutes *optical rays*. This notion has a purely mathematical meaning: it does not suppose the possibility of physically isolating such rays, but it allows us to extend within the limits of validity of the stationary phase principle to the principles and methods of geometrical optics.

When ultra-short waves are involved we shall use two main methods based either on *geometrical optics* or on *physical optics*. The first method will be used when the wavelength is short enough and when we study the relationships between the electrical quantities at *non focal points*, since in this case we are allowed to apply the stationary phase principle. When *focal points* are involved *Huygens diffraction formulae* will be used based on physical optics (see equation (7.5)).

Flow of electromagnetic energy and geometrical optics

It can be shown that the rays defined above are the envelopes of the electromagnetic Poynting vectors.

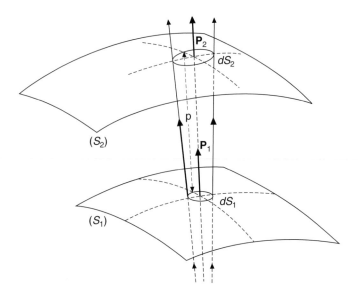

Fig. 7.14 Energy flow in ray bundles.

Rays, then, are lines of energy propagation. Let us consider two successive wave surfaces S_1, S_2 and a conical ray bundle intercepting them (Fig. 7.14). The energy conservation principle applied between the Poynting vectors (§1.2.3) P_1 and P_2 gives the following relation

$$|\mathbf{P}_1| \, dS_1 = |\mathbf{P}_2| \, dS_2$$

$$\frac{1}{2\eta} \mathbf{E}_1^2 dS_1 = \frac{1}{2\eta} \mathbf{E}_2^2 \, dS_2 \qquad (7.8)$$

This equation allows us to establish the relationship (7.7) that we obtained using the Huygens–Fresnel principle.

Optical path

A fundamental principle of geometrical optics is Fermat's principle, often taken as a postulate. Before formulating it is necessary to define the *'optical path'*.

The *optical path* L along a curve (Γ) between two points P_1P_2 is defined according to the line integral

$$L = \int_{(\Gamma)} nds \qquad (7.9)$$

where n is the refractive index of the medium involved and ds an element of length.

Since we have

$$\phi = \frac{2\pi L}{\lambda} \qquad (7.10)$$

the optical path corresponds to a phase difference between the two points P_1P_2.

Example: If we consider two closely-spaced wave surfaces, the optical paths followed by the rays are equal since the rays are orthogonal to the two wave surfaces.

Fermat's principle: If P_1 is an illuminating source and P_2 an observation point, the path followed by the light between P_1 and P_2 is such that the path $P_1 P_2$ should be stationary (generally a minimum).

Consequences are:

- in a homogeneous medium rays are straight lines,
- Consider a ray incident on a surface separating two different media. Snell's law[4] is

$$n_1 \sin i_1 = n_2 \sin i_2$$

where i_1, i_2 are, respectively, the incident and emerging angles with the normal to the surface and n_1, n_2 are the refractive indices of the two media considered.

---

[4] Snell Van Royen (1550–1626) (Netherlands). This law was formulated by the French mathematician and philosopher René Descartes (1596–1650) at approximately the same time.

Optical path law - applications

Consider now any reflective or refractive medium, for example a lens. Consider the optical paths separating a wave surface S_1, related to the incident rays and a wave surface S_2 related to the emerging rays. Consider two rays ABC, A′B′C′ going from one surface to the other (Fig. 7.15). The 'optical path law' may be demonstrated using the Fermat principle

$$\int_{ABC} nds = \int_{A'B'C'} nds \qquad (7.11)$$

Moving along any light path from any incident wave surface to an emerging wave surface, the optical path, and thus the phase variation, are independent of the path chosen.

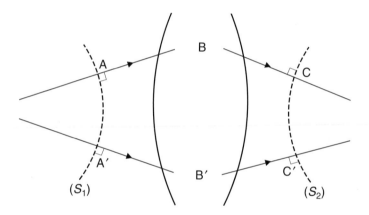

Fig. 7.15 Optical path law.

Application of the law - parabolic reflector: The optical path law that we have just described often gives a simpler and more direct means to determine the reflection and refraction surfaces than Snell's law.

Example: Let us derive the shape required for a reflector to transform a spherical wave into a plane wave. Let F be the centre of the spherical wave, i.e. 'focus' of the radiation. Let FX be the direction of the emerging plane wave radiation. The reflector we are looking for should be rotationally symmetric. Let us take the plane of the figure as a meridian plane (Fig. 7.16). Let S be the apex of the reflector, then, the focal distance is $f = $ SF.

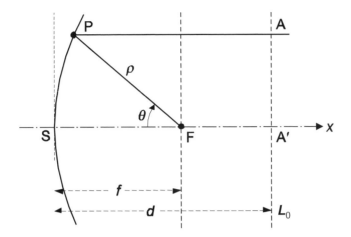

Fig. 7.16 Parabolic reflector synthesis.

Let L_0 be the section, in the meridian plane, of an emerging plane wave, at a distance d from the apex S (SA = d). Consider the diverging rays FP and FS (axial radius) associated with the spherical waves and the parallel rays PA and SA′ associated with the plane waves. The optical path equality can be written as follows:

$$FP + PA = \text{constant} = FS + SA'$$

or

$$\rho + \rho \cos \theta + d - f = f + d$$

$$\rho = \frac{2f}{1 + \cos \theta} = \frac{f}{\cos^2 \dfrac{\theta}{2}} \tag{7.12}$$

which is the equation of a parabola.

In contrast to the preceding calculation, if Snell's law is applied a differential equation for the surface is obtained.

7.3.5 Ray theory in quasi-homogeneous media

Such media can be found in nature: for example, the Earth's atmosphere, where the refractive index decreases with altitude. It is also the case with dielectric lenses with varying refractive index.

They are defined as media where the refractive index $n(M)$ varies 'slowly' as a function of wavelength, so they can therefore be considered as locally 'homogeneous' and the light rays may be defined as the orthogonal trajectories to the wave surfaces.

(a) Wave surface and ray differential equations

The wave surfaces are equiphase surfaces characterized by relationships of the form

$$\Phi(M) = \text{constant}$$

or, according to (7.10)

$$L(M) = \text{constant}$$

Let us consider two closely-spaced wave surfaces (Fig. 7.17)

$$L_1(M) = L_0 \qquad L_2(M) = L_0 + dL$$

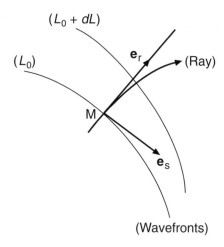

Fig. 7.17 Curvature of rays and index gradient.

In the vicinity of point M where the refractive index is $n(M)$, these two surfaces are separated by distance $ds = dL/n(M)$. The normal derivative of the optical path is the refractive index

$$\frac{dL}{ds} = n(M) \tag{7.13}$$

The unit vector normal to the wave surface (L) at M is equally the vector $\mathbf{e}_r$ tangential to the ray passing through M, thus

$$\operatorname{grad} L = n(\mathrm{M})\mathbf{e}_r \qquad (7.14)$$

Differential equation of rays: Differentiating equation (7.13), we obtain

$$\operatorname{grad} n = \operatorname{grad}\left(\frac{\partial L}{\partial s}\right) = \frac{\partial}{\partial s}\operatorname{grad} L$$

and taking into account equation (7.14), we have

$$\operatorname{grad} n = \frac{\partial\left[n(\mathrm{M})\mathbf{e}_r\right]}{\partial s} \qquad (7.15)$$

(b) Curvature of rays

Let $\mathbf{e}_s$ be the unit vector at M of the principal normal of the ray. Consider a local basis $(\mathbf{e}_s, \mathbf{e}_r)$ in a plane defined by an orthogonal system $\mathbf{e}_x$, $\mathbf{e}_y$. Let φ be the angle between $\mathbf{e}_r$ and $\mathbf{e}_x$ (Fig. 7.18).

The radius of curvature of the ray at M is the derivative of the curvilinear abscissa with respect to

$$R = \frac{ds}{d\varphi}$$

Since

$$\mathbf{e}_r = \mathbf{e}_x \cos\varphi + \mathbf{e}_y \sin\varphi$$

then

$$\frac{d\mathbf{e}_r}{ds} = \frac{d\varphi}{ds}\left(-\mathbf{e}_x \sin\varphi + \mathbf{e}_y \cos\varphi\right) = \frac{1}{R}\mathbf{e}_s$$

and using equation (7.15), we obtain

$$\operatorname{grad} n = \frac{\partial\left(n\mathbf{e}_r\right)}{\partial s} = \frac{n}{R}\mathbf{e}_s + \mathbf{e}_r\frac{dn}{ds}$$

$$\operatorname{grad} n.\mathbf{e}_s = n\frac{1}{R}$$

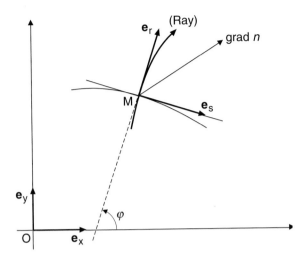

Fig. 7.18 Transversal gradient component effect on ray curvature.

Therefore, the curvature of rays is given by the relationship

$$\frac{1}{R} = \frac{1}{n} \mathbf{e}_s \cdot \operatorname{grad} n \qquad (7.16)$$

$\mathbf{e}_s$ being the tangential vector already defined. We see that the more the transversal gradient component with the respect to the ray is important the more the curvature is important. This results in a bending of the rays toward the increasing indices in the same way as with a prism. On the other hand, a longitudinal variation of the dielectric constant does not cause any deviation of the rays.

(c) Important particular case: radial variation of the index - Bouguer's law (Fig. 7.19)

If the refractive index $n(M)$ is only dependent on the distance $\rho = OM$, we have

$$\operatorname{grad} n = \frac{dn}{d\rho} \mathbf{e}_\rho$$

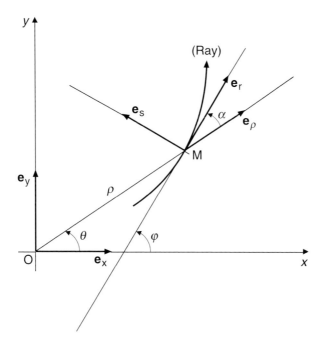

Fig. 7.19 Bouguer's law.

Taking into account equation (7.16), we obtain

$$\frac{n}{R} = \frac{dn}{d\rho}\mathbf{e}_\rho.\mathbf{e}_s = -\frac{dn}{d\rho}\sin\alpha$$

Since

$$\frac{1}{R} = \frac{d\varphi}{ds} = \frac{d(\theta+\alpha)}{ds} = \frac{d\theta}{ds} + \frac{d\alpha}{ds}$$

and together with the classic geometrical relationships

$$\rho\frac{d\theta}{ds} = \sin\alpha \qquad \frac{d\rho}{ds} = \cos\alpha$$

we deduce

$$n\sin\alpha\, d\rho + n\rho\cos\alpha\, d\alpha = -\rho\, dn\sin\alpha$$

Finally, after integration we obtain Bouguer's law

$$\rho n \sin\alpha = K \qquad\qquad (7.17)$$

where K is a constant.

(d) Case of a laminar medium: see exercise 7.3

Applications of Bouguer's law - media where the index is constant (Fig. 7.20):

In this case $\rho\sin\alpha = $ const. $= $ OH. Thus, the rays are straight lines.

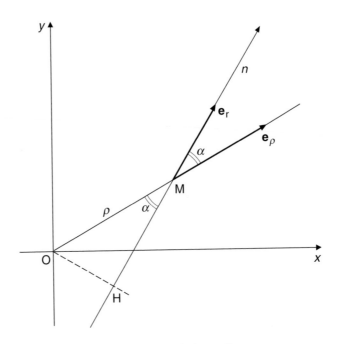

Fig. 7.20 Constant index medium.

Media where the index is_inversely proportional to the distance to the centre: Thus α is a constant. The rays are logarithmic spirals.

General case – ray differential equation: We start from the classical geometric relationship in polar coordinates

$$\tan \alpha = \frac{\rho d\theta}{d\rho}$$

Then, using Bouguer's law (7.17), we find

$$\frac{d\theta}{d\dfrac{K}{\rho}} = -\frac{1}{\sqrt{n^2 - \left(\dfrac{K}{\rho}\right)^2}} \tag{7.18}$$

Spherical lenses: The first example ('fish eye') was given by Maxwell using the following refractive index:

$$n(\rho) = \frac{2}{1 + \dfrac{\rho^2}{R^2}}$$

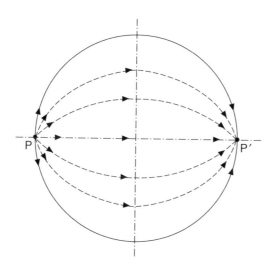

Fig. 7.21 'Fish eye' lens.

Each point of the sphere surface $r = R$, is the stigmatic image of the diametrically opposite point (Fig. 7.21).

Luneburg lens: The variation of refractive index is as follows

$$n(\rho) = \sqrt{2 - \frac{\rho^2}{R^2}}$$

The rays from any point F of the surface emerge parallel to the axis FO, therefore the emerging wave is plane (Fig. 7.22). This property is used to realize multiple-beam or array antennas by switching the primary feeds located at the sphere surface. A retro-reflector can be realized by coating a portion of the sphere surface with reflective material.

All these lenses can be made by means of concentric layers of dielectric whose refractive indices follow the appropriate law according to successive levels.

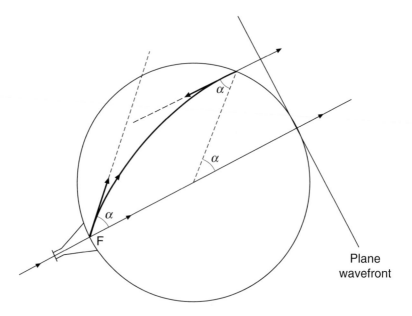

Fig. 7.22 Luneburg lens.

Geodesic lenses: These make use of guided propagation between two conducting parallel spherical shells with appropriate curvature (Fig. 7.23). Rays are the geodesic lines of the median surface. They are obtained from classical differential geometry. A Luneburg lens allowing large angle scanning can be realized by means of a movable primary feed or switchable sources.

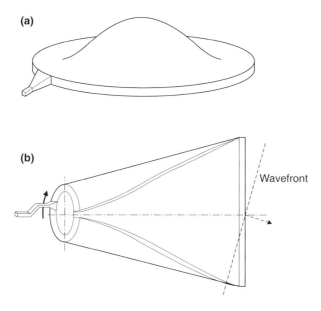

Fig. 7.23 (a) Geodesic lens; (b) Robinson antenna.

A surface with a circular side and an opposed straight one allows us to convert the excitation obtained from a primary feed rotating around the circular side into an alternative scanning in the straight side.

Exercise: Using the analogy of media of refractive index central variation discussed in the preceding section analyse laminar media such that $n = n(z)$. Discuss the different types of rays.

Fresnel zone plate lenses: One of the simplest types of microwave lens antenna is the Fresnel zone plate lens[5,6]. In its most basic form this consists of alternate transparent and opaque annuli (Figure 7.24). The radii of the boundaries of the annuli are given by

5 Van Buskirk, L.F. and Hendrix, C.E., 'The zone plate as a radio frequency focusing element', *IRE Trans. Antennas and Propagation*, Vol.AP-9, pp319–320, May 1961.

6 Wiltse, J.C. and Garrett, J.E., 'The Fresnel zone plate antenna', *Microwave Journal*, Vol.34, pp101–114, Jan. 1991.

$$r_n = \sqrt{\left(f + n\frac{\lambda}{2}\right)^2 - f^2}$$ (7.19)

where *n* is an integer. The annuli may be conveniently realized by printed metallization or by strips of aluminium foil, on a transparent background. Thus the lens might be formed on a window, with the feedhorn indoors, giving a notably simple, cheap and unobtrusive antenna, for applications such as DBS TV reception.

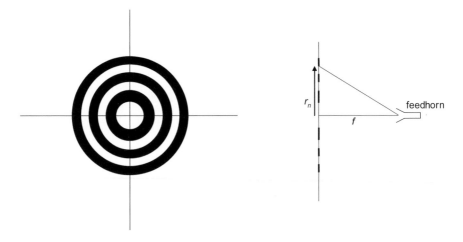

Fig. 7.24 Basic form of Fresnel zone plate lens, in transmission.

The efficiency of such a system is only about 20%, since approximately half the incoming energy is lost and the rest is only focused with coarse phase increments (180°). The bandwidth is approximately 21% of the centre frequency, and the radiation pattern is found to be reasonably well-behaved, with a mean sidelobe level of about –20 dB[7].

Refinements to the basic scheme include:

- Squinting the beam off boresight, in azimuth and/or elevation, and locating the feedhorn off-axis. In this case (Fig. 7.25) the boundaries of the zones are no longer circular, but are given by the locus of the points (*x*, *y*) such that:

[7] Guo, Y.C., Barton, S.K. and Wright, T.M.B., 'Focal field distribution of Fresnel zone plate antennas', *Proc. 7th IEE Intl. Conference on Antennas and Propagation*, York, IEE Conf. Publ. No. 333 Part I, pp6–8, 15–18 April 1991.

$$\left(\sqrt{\left(f\cos\phi_f \sin\theta_f + x \right)^2 + \left(f\sin\phi_f + y \right)^2} - f \right) - x\sin\theta_i - y\sin\phi_i = n\frac{\lambda}{2} \quad (7.20)$$

- Replacing the opaque zones by additional $\lambda/2$ pathlengths through a suitable dielectric (which will double the efficiency)[8];
- Operation in reflection mode rather than transmission, perhaps backed with a quarter-wavelength substrate and a reflecting layer to increase the efficiency.

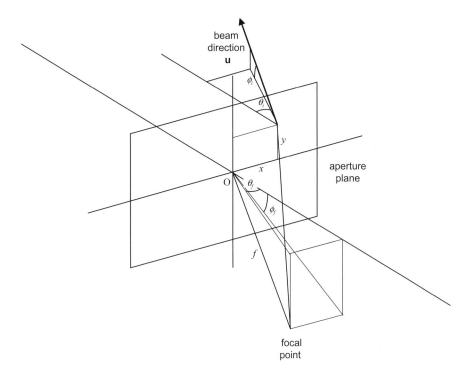

Fig. 7.25 General form of Fresnel zone plate lens with off-boresight beam and offset feed. The signal is incident from angle (θ_i, ϕ_i) and the focal length is f, offset from the axis in direction (θ_f, ϕ_f). The boundaries of the zones are no longer circular, but are defined by the locus of points (x, y) given by equation. (7.20).

---

[8] Black, D.N. and Wiltse, J.C., 'Millimeter-wave characteristics of phase-correcting Fresnel zone plates, *IEEE Trans. Microwave Theory and Techniques*, Vol.35, No.12, pp1122–1129, December 1987.

7.4 ANTENNAS AS RADIATING APERTURES

7.4.1 Antenna radiation and equivalent aperture method

Consider any antenna A and its feed, and a surface S enclosing the whole (Fig. 7.26). The following theorem can be demonstrated (see Chapter 2): 'If a component (electric or magnetic) of the electromagnetic field is known at any point M of S, then the radiated field at any point P outside the surface can be calculated'.

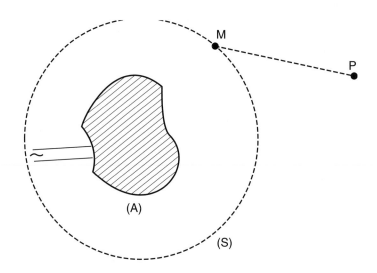

Fig. 7.26 Surface (S) enclosing an antenna and its feed.

If the surface S is a wavefront (or approximately a wavefront) the field at P can be deduced by applying Huygens's principle given in the form of equation (7.5) in section 7.3.

The above result is quite interesting for the following reasons:
• In most microwave antennas, the surface S is reduced to a plane where the electromagnetic field is negligible outside a certain domain called the 'equivalent aperture' of the antenna. A few examples will follow.

- The electromagnetic field on this surface can often be determined, to a first approximation, from methods based on geometrical optics.
- The radiated field can be calculated from methods derived from Huygens's principle.

7.4.2 Examples of microwave antennas and equivalent apertures

Electromagnetic horn (§5.1.2 and §8.2)

The horn is perhaps one of the simplest types of microwave antenna, but it can take many different forms.

It is a structure derived from a waveguide whose cross-sectional area increases progressively along the guide until a plane (S) where the section becomes an equivalent radiating aperture (Fig. 7.27).

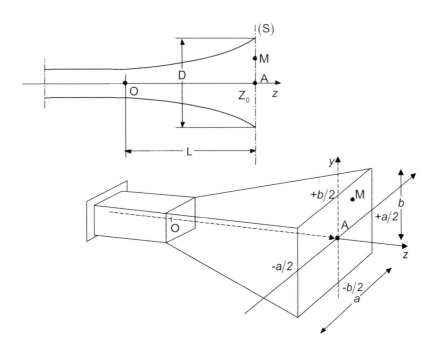

Fig. 7.27 Geometry of a horn.

The cross sectional area of the horn may be of arbitrary form, but very often it is rectangular or circular.

The absence of discontinuities in the horn avoids parasitic higher-order propagation modes, therefore the field at any point M of the aperture is only dependent on the propagation modes created by the feed power excitation. Nevertheless, it is possible to intentionally create discontinuities (multimode horns) or to design the horn longitudinal profile so that the field distribution in the aperture is modified and, hence the radiation pattern.

Parabolic reflector

The parabolic reflector will be analysed in greater detail later on in §9.3. It has been shown (§7.2.4) that a parabolic reflector allows us to transform a spherical wave emitted by a radiating source (a horn for instance) into a plane wave. If a plane (P) passing through the focus F is considered, then at any point M of (P) the electromagnetic field may be evaluated, approximately, applying the principle of conservation of energy to the energy transmitted in the cones and ray bundles (Fig. 7.28).

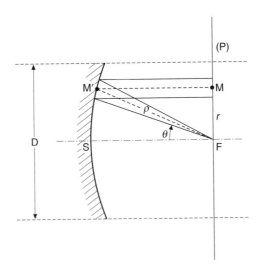

Fig. 7.28 Parabolic reflector. Cone and ray bundles.

In the plane (P) let us define a radiating aperture whose diameter D is approximately the same as that of the reflector. Moving along a ray FM′ M, we can see that the field is only attenuated in the conical part FM′ $= \rho$ and it remains constant along MM'. If the feed at F is a point source, then the field at M is of the form

$$E_M = E_0 \frac{\text{SF}}{\text{M'F}} = E_0 \frac{f}{\rho} \qquad (7.21)$$

where E_0 is the field at the apex S of the reflector and f the focal length. Since the radius vector ρ is given by equation (7.12), the position of M is given by

$$\text{FM} = r = \rho \sin \theta \qquad (7.22)$$

where $r \leq D/2$, then we can write

$$E_M(r) = \frac{E_0}{1 + \left(\dfrac{r}{2f}\right)^2} \qquad (7.23)$$

The equivalent aperture possesses an axially symmetric illumination which tapers towards the edges (Fig. 7.29). This is an example of 'transfer functions', whose general properties are analyzed in §9.4.

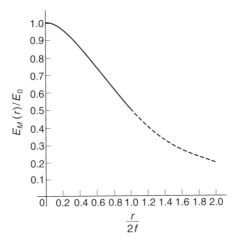

Fig. 7.29 Aperture illumination of a parabolic reflector with a point source primary feed.

Array antenna

Array antennas are treated in Chapter 11. In its most classical form, an antenna array is composed of N regularly spaced antenna elements arranged in a straight line or on a plane (P) (Fig. 7.30).

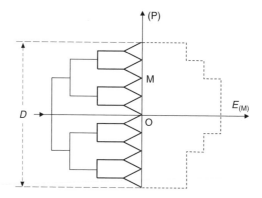

Fig. 7.30 Example of an array feeding system and of the resulting E(M) of the equivalent aperture.

This plane is that of the equivalent aperture and it is limited to the length D of the array. The antenna elements may be fed in various ways, for instance by means of line sources and couplers, possibly having phase shifters, amplifiers and attenuators. The illumination function $E_{(M)}$ - and hence its directive properties - may be controlled according to the excitation function of its elements. This is the main attraction of array antennas with respect to those previously considered.

7.4.3 Far-field radiation from an aperture

Introduction

The field existing in the aperture plane is not always cophasal: phased arrays use such phase variations in order to direct the maximum radiation in a desired direction.

Nevertheless, the radiation from an aperture may be obtained, with good accuracy, by applying the Huygens–Fresnel principle (§7.2) at all the points M of the aperture, provided that the field phase at M is taken into account. The field at an external point P may be considered as the sum of elementary spherical waves from all points M of the aperture, that is to say weighted in amplitude, polarization and phase existing at M.

In order to calculate the radiation from an aperture it is assumed that the field at each point of the aperture is known. In practice, only the 'applied' field at the aperture is known. The calculation of the resultant field is related to a diffraction problem, so rigorously it should be carried out by means of an integral equation. It can be shown that the far field wave is spherical and that the field has the local character of a plane wave, in particular, the polarization vector is orthogonal to the propagation direction **u**.

Thus, it is not possible to arbitrarily fix the field in the aperture, but only a transverse component (contained in the aperture plane) of either the electric field or the magnetic field. We shall proceed as follows.

(1) We assume that the transverse component $\mathbf{E}_T(M)$ in the plane of the aperture is known.
(2) From the *Kirchhoff* integral (§2.2.2) we deduce the corresponding component $\mathbf{E}'_T(P)$ (parallel to the aperture) of the radiated far field at P in the direction of the unit vector **u.**
(3) Since we know that the total field $\mathbf{E}'_T(P)$ is orthogonal to the direction **u** we deduce it from the component $\mathbf{E}'_T(P)$.

Remark: we may start from the magnetic field in the aperture instead and deduce the radiated magnetic field using the same formulae, then determine the electric field since the radiated field has the local character of a plane wave.

Calculation of the transverse component $\mathbf{E}_T(P)$ of the field radiated by an aperture from the transverse component $\mathbf{E}_T(M)$ of the electric field in the aperture.

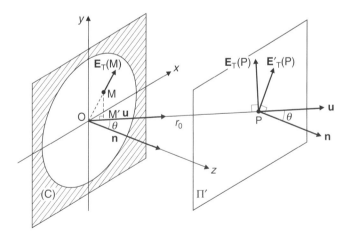

Fig. 7.31 Calculation of the transverse component of the field radiated by an aperture.

Consider an aperture (S) (Fig. 7.31), in an *xy* plane, with the unit vector **n** of the axis Oz orthogonal to the aperture. Consider a point P located at a large distance OP $= r_0$ in the direction of the unit vector **u** whose coordinates (or direction cosines) are α, β, γ

$$|\mathbf{u}|^2 = \alpha^2 + \beta^2 + \gamma^2 = 1$$

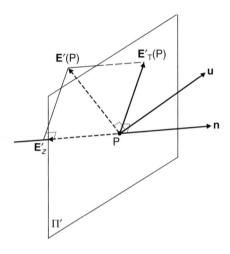

Fig. 7.32 Longitudinal (E'_z) and transverse (E'_T) components of the radiated field E(P).

The transverse field at P is obtained by application of the Kirchhoff integral at each component of $\mathbf{E}_T(M)$ in the *xy* plane (equation (2.41)). Thus, we have

$$\mathbf{E}'_T(P) = -\frac{1}{2\pi} \iint_S \mathbf{E}_T(M) \frac{d}{dz} \frac{\exp(-jkr)}{r} dS$$

At large distances, the derivative of the Green's function can be approximated as (see appendix 7A)

$$\frac{d}{dz} \frac{\exp(-jkr)}{r} \cong -\frac{\exp(-jkr)}{r} jk \frac{z}{r}$$

with

$$\frac{z}{r} = \cos\theta = \gamma$$

Thus

$$\mathbf{E}'_T(P) = \frac{j}{\lambda} \iint_S \gamma\, \mathbf{E}_T(M) \frac{\exp\left(-j\dfrac{2\pi}{\lambda}\right)}{r}\, dS \tag{7.24}$$

Let M′ be the orthogonal projection of M on OP. Since P is far away, we have

$$r \cong \mathrm{MP} = \mathrm{OP} - \mathrm{OM}' = \mathrm{OP} - \mathbf{OM.\,u} = r_0 - \mathbf{OM.\,u}$$

The attenuation factor $1/r$ is equivalent to $1/r_0$, so it is constant and can be removed from the integrand. Likewise the factor γ is constant since the angle $(\mathbf{n}, \mathbf{u})$ is the same for any point of the aperture. Therefore we have

$$\mathbf{E}'_T(P) = \frac{j\exp\left(-j2\pi\dfrac{r_0}{\lambda}\right)}{\lambda r_0}\, \gamma \iint_{(S)} \mathbf{E}_T(M)\exp\left(j2\pi\frac{\mathbf{OM.u}}{\lambda}\right) dS \tag{7.25}$$

Radiation characteristic function: It is useful to express the directive properties independently of feed power, by using a reference field, for instance the maximum field E_0 in the aperture. Then, we define the 'aperture illumination law' by a dimensionless vector function $\mathbf{f}_T(M) = \mathbf{E}_T(M)/E_0$.

The transverse radiated field at P may be written as

$$\mathbf{E}'_T(P) = j\frac{\exp\left(-2j\tau\dfrac{r_0}{\lambda}\right)}{\dfrac{r_0}{\lambda}}\, E_0 \gamma \mathbf{F}_T(\mathbf{u}) \tag{7.26}$$

$\mathbf{F}_T(\mathbf{u})$ is a dimensionless vector function which is characteristic of the antenna radiation

$$\mathbf{F}_T(\mathbf{u}) = \frac{1}{\lambda^2} \iint_{(S)} \mathbf{f}_T(M)\exp\left(2j\pi\frac{\mathbf{OM.U}}{\lambda}\right) dS \tag{7.27}$$

Using the direction cosines of $\mathbf{u}$ and the Cartesian coordinates (x, y) of M, we can write

$$\mathbf{F}_T(\alpha, \beta) = \frac{1}{\lambda^2} \iint_{(S)} \mathbf{f}_T(x, y)\exp\left(2j\pi\frac{\alpha x + \beta y}{\lambda}\right) dxdy \tag{7.28}$$

This expression may be simplified by taking λ as unity and setting

$$\frac{x}{\lambda} = v, \frac{y}{\lambda} = \mu \qquad (7.29)$$

Then

$$\mathbf{F}_{\mathrm{T}}(\alpha, \beta) = \iint_{(S)} \mathbf{f}_{\mathrm{T}}(v, \mu) \exp\left[2j\pi(\alpha v + \beta \mu)\right] dv d\mu \qquad (7.30)$$

from which we obtain the fundamental result:

'*The characteristic function of the aperture radiation is the two-dimensional Fourier transform of the illumination function*'.

In a similar way, from equation (7.26) and setting

$$\frac{r_0}{\lambda} = \rho$$

the transverse field becomes

$$\mathbf{E}'_{\mathrm{T}}(\mathrm{P}) = j\frac{\exp(-2j\pi\rho)}{\rho} E_0 \gamma \, \mathbf{F}_{\mathrm{T}}(\alpha, \beta) \qquad (7.31)$$

Calculation of the total field at P from its transverse component

The total field has locally the character of a plane wave, therefore it is orthogonal to the propagation direction $\mathbf{u}$, and

$$\mathbf{E}'(\mathrm{P}).\mathbf{u} = 0$$

The field is the sum of the transverse and longitudinal components

$$\mathbf{E}'(\mathrm{P}) = \mathbf{E}'_{\mathrm{T}}(\mathrm{P}) + E'_z \mathbf{n}$$

and we can deduce

$$\mathbf{E}'_{\mathrm{T}}.\mathbf{u} + \mathbf{n}.\mathbf{u} E'_z = 0$$

Finally, we obtain

$$\mathbf{E}'(\mathrm{P}) = \mathbf{E}'_{\mathrm{T}}(\mathrm{P}) - \frac{\mathbf{E}'_{\mathrm{T}}.\mathbf{u}}{\mathbf{n}.\mathbf{u}}\mathbf{n} = \frac{\mathbf{E}_{\mathrm{T}}(\mathbf{n}.\mathbf{u}) - \mathbf{n}(\mathbf{E}'_{\mathrm{T}}.\mathbf{u})}{\mathbf{n}.\mathbf{u}}$$

The double vector product identity can be seen in the numerator and the direction cosine in the denominator, so that

$$\gamma = \mathbf{n} \cdot \mathbf{u}$$

and

$$\mathbf{E}'(\mathrm{P}) = \frac{1}{\gamma}(\mathbf{n} \times \mathbf{E}'_\mathbf{T}) \times \mathbf{u} \tag{7.32}$$

The total field is obtained from the radiation characteristic function by means of the expression (7.31) for the transverse field

$$\mathbf{E}'(\mathrm{P}) = j\frac{\exp(-2j\pi\rho)}{\rho} E_0 \left[\mathbf{n} \times \mathbf{F}_\mathbf{T}(\alpha,\beta) \times \mathbf{u} \right] \tag{7.33}$$

where the characteristic function $\mathbf{F}_\mathbf{T}(\alpha,\beta)$ is given by the Fourier transform (7.30) of $\mathbf{f}_\mathbf{T}(\gamma,\mu)$.

Electric field expression obtained from the magnetic field

Similarly, we obtain

$$\mathbf{H}'(\mathrm{P}) = j\frac{\exp(-2j\pi\rho)}{\rho} H_0 \left[\mathbf{n} \times \mathbf{H}_\mathbf{T}(\alpha,\beta) \right] \times \mathbf{u} \tag{7.34}$$

where $\mathbf{H}_\mathbf{T}(\alpha,\beta)$ is the magnetic field illumination function

$$\mathbf{h}_\mathbf{T}(\mathrm{M}) = \frac{\mathbf{H}_\mathbf{T}(\mathrm{M})}{H_0}$$

and H_0 is a reference magnetic field.

Since the far field is always locally plane, we can deduce $\mathbf{E}'(\mathrm{P})$ directly from equation (7.34), that is to say from the transverse magnetic field in the aperture

$$\mathbf{E}'(\mathrm{P}) = \eta\,\mathbf{u} \times \mathbf{H}'(\mathrm{P}) \tag{7.35}$$

where in free space $\eta = 120\pi\ \Omega$.

7.4.4 Examples of radiating apertures

In many applications it is useful to consider the radiation characteristic function of an aperture in its scalar form (as a first approximation) without taking the polarization into account. That is what will be done in the two important examples that follow.

Rectangular apertures

Consider the radiation from a rectangular aperture whose illumination is separable (Fig. 7.33).

Let D and D′ be the dimensions of the aperture. In this case, the transverse field at a point M in the aperture is equal to the product of two functions whose normalized coordinates are

$$v = \frac{x}{\lambda}, \quad \mu = \frac{y}{\lambda}$$

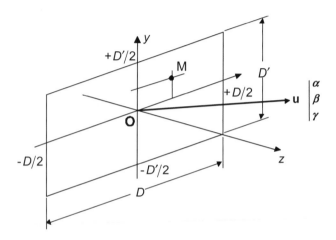

Fig. 7.33 Rectangular aperture.

If the polarization is fixed (vertical for example) the field is expressed in terms of a simple scalar function

$$E_T(v, \mu) = E_1(v) E_2(\mu) \tag{7.36}$$

Let us set

$$v_0 = \frac{D}{2\lambda}, \quad \mu_0 = \frac{D'}{2\lambda} \tag{7.37}$$

The integral (7.30) can be written in terms of the product of two integrals

$$F_1(\alpha) = \int_{-v_0}^{+v_0} E_1(v) \exp(2j\pi\alpha v) dv \tag{7.38}$$

$$F_2(\beta) = \int_{-\mu_0}^{+\mu_0} E_2(\mu)\exp(2j\pi\beta\mu)\,d\mu \tag{7.39}$$

If we assume that ν is a frequency, then the function $F_1(\alpha)$ is a signal in the time domain and its spectrum is $E_1(\nu)$. The spectrum is *band-limited* by the dimension D of the aperture and ν_0 may be considered as a *cut-off frequency*. That is why the variables ν and μ are often known as 'spatial frequencies'. All these analogies with the theory of band-limited signals will be developed later on (Chapter 13).

Exercise: consider a narrow slit for which $\mu_0 = D'/2\lambda$ is very small. If we set $E_2(\mu) = E_0$, we have

$$F_2(\beta) = 2\mu_0 E_0 = \text{constant}$$

Around the slit the radiation is omnidirectional

$$F(\alpha, \beta) = 2\mu_0 E_0 F_1(\alpha) \tag{7.40}$$

Let us calculate the characteristic function $F_1(\alpha)$ associated with a uniform illumination, $E_1(\nu) = E_0$. We find the classical result

$$F_1(\alpha) = 2\nu_0 E_0 \frac{\sin(2\pi\nu_0\alpha)}{2\pi\nu_0\alpha} = \frac{D}{\lambda} E_0 \frac{\sin\left(\pi\dfrac{D}{\lambda}\alpha\right)}{\pi\dfrac{D}{\lambda}\alpha} \tag{7.41}$$

This function is shown in Fig. 7.34. The radiation is maximum in the normal direction $(\alpha = 0)$. The relative level of the first sidelobe is 0.22, i.e. approximately -13 dB.

The angular half-width between adjacent zeroes is

$$\Delta\alpha = \frac{1}{2}\nu_0 = \frac{\lambda}{D} \tag{7.42}$$

This is approximately the 3 dB beamwidth, that is to say the angular width where the amplitude is reduced by 3 dB. This characteristic is general for radiating apertures (§7.1, Introduction).

This result is general: the angular beamwidth of the radiated beam is inversely proportional to the aperture D normalized to the wavelength λ.

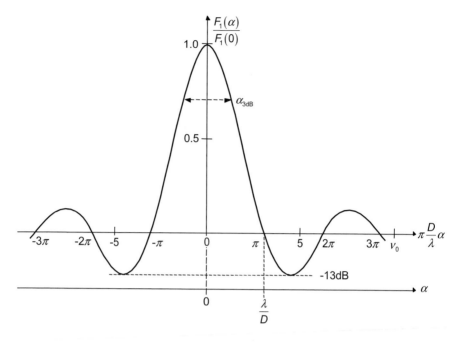

Fig. 7.34 Characteristic function of a linear aperture with uniform illumination.

Circular apertures

This problem is frequently found in many practical cases: circular horns, parabolic reflectors, Cassegrain antennas, etc.

Consider a circular aperture of diameter D (Fig. 7.35). Consider a point M of the aperture whose polar coordinates are r, ϕ. Let us assume that the polarization in the aperture is uniform, then a scalar function $f(r, \phi)$ is sufficient to describe the illumination. Let **u** be a direction whose spherical coordinates are (θ, ψ). The characteristic function of the radiation is the Fourier transform of the illumination $f(r, \phi)$. From equation (7.27), this can be written in the form

$$F(\mathbf{u}) = \frac{1}{\lambda^2} \int_0^{D/2} \int_0^{2\pi} f(r, \phi) \exp\left(2j\pi \frac{\mathbf{OM.u}}{\lambda}\right) dS \qquad (7.43)$$

or

$$F(\theta, \phi) = \frac{1}{\lambda^2} \int_0^{D/2} \int_0^{2\pi} f(r, \phi) \exp\left[2j\pi \frac{r}{\lambda} \sin\theta \cos(\phi - \psi)\right] r\, dr\, d\phi \qquad (7.44)$$

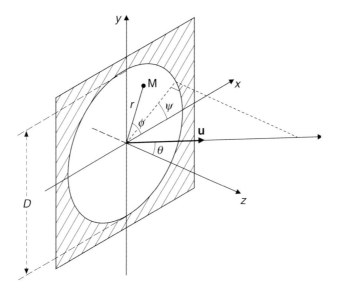

Fig. 7.35 Circular aperture.

Circularly symmetric illumination: In the case of a parabolic reflector, for instance, such an illumination can be obtained from a circularly symmetric primary Huygens source, for example a corrugated horn (§8.3). In this case it takes the form $f(r)$.
   The characteristic function depends only on θ

$$F(\theta) = \frac{1}{\lambda^2} \int\limits_0^{D/2} f(r) \left\{ \int\limits_0^{2\pi} \exp\left[2j\pi \frac{r}{\lambda} \sin\theta \cos(\varphi - \psi) \right] d\varphi \right\} r dr$$

The integral between brackets is independent of ψ. It represents the radiation contribution of an elementary annular aperture. It can be calculated by means of the Bessel function of order zero

$$\int_0^{2\pi} \exp\left[2j\pi \frac{r}{\lambda} \sin\theta \cos(\phi - \psi) \right] d\phi = 2\pi j \mathrm{J}_0\left(\frac{2\pi r}{\lambda} \sin\theta \right)$$

The characteristic function reduces, then, to a simple integral

$$F(\theta) = j\frac{2\pi}{\lambda^2} \int_0^{D/2} f(r) \mathrm{J}_0\left(\frac{2\pi r}{\lambda} \sin\theta \right) r dr \qquad (7.45)$$

Uniform illumination: $f(r) = f_0 = 1.$

We shall see (§7.4.9) that this illumination gives the aperture maximum gain in its axial direction. The characteristic function can be written

$$F_0(\theta) = j\frac{2\pi}{\lambda^2} \int_0^{D/2} J_0\left(\frac{2\pi r}{\lambda}\sin\theta\right) r\,dr$$

which can be expressed by means of a first order Bessel function

$$F_0(\theta) = j\frac{\pi D^2}{4\lambda^2} 2\frac{J_1\left(\frac{\pi D}{\lambda}\sin\theta\right)}{\frac{\pi D}{\lambda}\sin\theta} \qquad (7.46)$$

This expression can be normalized to the value in the axial direction. This yields the classical result given below

$$\boxed{H_0(\theta) = \frac{F_0(\theta)}{F_0(0)} = 2\frac{J_1\left(\frac{\pi D}{\lambda}\sin\theta\right)}{\frac{\pi D}{\lambda}\sin\theta}} \qquad (7.47)$$

which is very often written

$$H_0(\theta) = \Lambda_1(X), \quad \text{with } X = \frac{\pi D}{\lambda}\sin\theta \qquad (7.48)$$

The characteristic surface (angular spectrum - radiation pattern) of radiation is circularly symmetric. It can be represented by a meridian cross section, the aperture radiation pattern (Fig. 7.36 curve Λ_1).

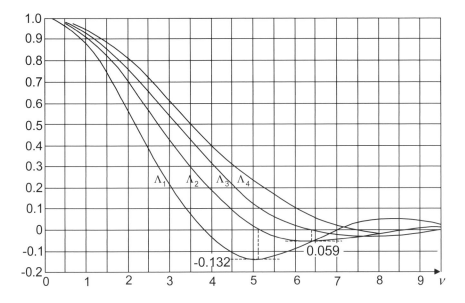

Fig. 7.36 Λ_1 function. Characteristic of a circular aperture.

We can note the following characteristics:

- 3 dB - beamwidth: $1.03 \dfrac{\lambda}{D}$ (radian);

- first sidelobe level: this corresponds to the first lobe either side of the main lobe: 0.132 or $N_{dB} = -17.6\,dB$.

Tapered illumination: Uniform illumination has the highest axial gain but it also has a rather high sidelobe level. In order to reduce sidelobes, a tapered illumination is often used (see Fig. 7.37). The more the illumination decreases toward the aperture edges, the broader is the main beam and the lower the sidelobe levels. Such an illumination may be obtained, for instance, from a Bessel, a Gaussian (§8.34) or a parabolic distribution.

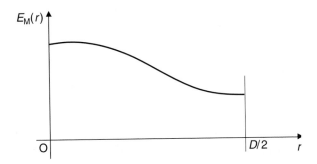

Fig. 7.37 Tapered illumination.

Let us consider an illumination having the form

$$f_1(r) = 1 - \left(\frac{2r}{D}\right)^2 \tag{7.49}$$

In this case the characteristic function is of the form

$$H_1(\theta) = 8 \frac{J_2\left(\dfrac{\pi D}{\alpha}\sin\theta\right)}{\left(\dfrac{\pi D}{\lambda}\sin\theta\right)^2} \tag{7.50}$$

denoted

$$H_1(\theta) = \Lambda_2(v)$$

where J_2 is the Bessel function of the second order. We obtain a reduction of the first sidelobe to $(-24.6$ dB$)$ at the cost of a 3 dB beamwidth of $1.27\,\lambda/D$ (radian) and a loss of gain (curve Λ_2 of Fig. 7.36).

Exercise: Calculate the loss of gain K from the expression of aperture gain (§7.4.8, equation 7.76).
  Answer: $K = 0.75$.

Remark: It is possible to combine a uniform illumination and a parabolic one, leading to a certain sidelobe level at the edges. This is often the case of parabolic reflector illuminated by a primary feed. In this case the pattern is a result of the combination of functions H_0 and H_1 with the appropriate distributions.

Generally, distributions of the form

$$f_n(r) = \left[1 - \left(\frac{2r}{D} \right)^2 \right]^n$$

lead to the curves Λ_{n+1} of Fig. 7.36 whose mathematical definition is given in §9.8.6 (equation (9.97)).

7.4.5 Polarization of the radiated field: case where the field in the aperture has the characteristic of a plane wave

To study the polarization of the radiated field it is necessary to consider the vector character of the electromagnetic field in the aperture. In cases where it has a local character of a plane wave we can define a 'reference polarization' that will allow us to characterize the polarization of any antenna.

Polarization of reference

In any case, the radiated field can be expressed from the electric and magnetic transverse fields, taking, for instance, half the sum of equations (7.32) and (7.35) and taking into account equation (7.34). Evaluating the Fourier transformations explicitly, we obtain

$$\mathbf{E}(\mathbf{u}) = -j \frac{\exp\left(-2j\pi \dfrac{R}{\lambda} \right)}{2\lambda R} \mathbf{u} \times \iint_{(S)} \left[\mathbf{n} \times \mathbf{E}_T + \eta (\mathbf{n} \times \mathbf{H}_T) \times \mathbf{u} \right] \exp\left(2j\pi \frac{\mathbf{OM.u}}{\lambda} \right) dS$$

$$(7.51)$$

This expression is known as the 'Kottler formula'. It should not be forgotten that it is not possible to specify both, independently and *a priori*, the transverse fields $\mathbf{E}_T$ and $\mathbf{H}_T$; one determines the other. Nevertheless, it is often useful to consider the case when these components have, at least locally, the characteristics of a plane wave. In this case they are related by the following equation.

$$\eta \mathbf{n} \times \mathbf{H}_T = -\mathbf{E}_T \qquad (7.52)$$

Then

$$\mathbf{E}(\mathbf{u}) = j \frac{\exp\left(-2j\pi \dfrac{R}{\lambda} \right)}{2\lambda R} \mathbf{u} \times \left[(\mathbf{n} + \mathbf{u}) \times \iint_{(S)} \mathbf{E}_T(M) \exp\left(2j\pi \frac{\mathbf{OM.u}}{\lambda} \right) dS \right] \quad (7.53)$$

Let us assume that the aperture is excited by a horizontal polarization

$$E_T(M) = E_T(M)e_x \tag{7.54}$$

e_x, e_y and e_z are the unit vectors of a Cartesian coordinate reference system.

The field then possesses a characteristic that we shall use as a reference for all the other cases (Fig. 7.38). We then have

$$E(u) = jp \frac{\exp\left(-2j\pi \dfrac{R}{\lambda}\right)}{2\lambda R} \iint_{(S)} E_T(M) \exp\left(2j\pi \frac{\mathbf{OM.u}}{\lambda}\right) dS \tag{7.55}$$

where **p** defines a polarization

$$p = u \times \left[(n+u) \times e_x \right] \tag{7.56}$$

or, according to the formula for the double vector product

$$p = (n+u)u.e_x - e_x.u(n+u)$$

or, using the direction cosines of **u**

$$p = (n+u)\alpha - e_x(1+\gamma) \tag{7.57}$$

Geometrical interpretation: Consider a sphere of centre O and of unit radius (Fig. 7.38). Let O′ be the intersection of this sphere with the axis Oz.

Thus, we have: **O′O = n**.

Let O′x′ be an axis parallel to Ox, thus the unit vector e_x is also associated with it. Let P be a point on the sphere such that

$$\mathbf{OP = u}$$

We can see that

$$\mathbf{O'P = n+u}$$

Then equation (7.57) can be written as

$$p = O'P\alpha - e_x(1+\gamma) \tag{7.58}$$

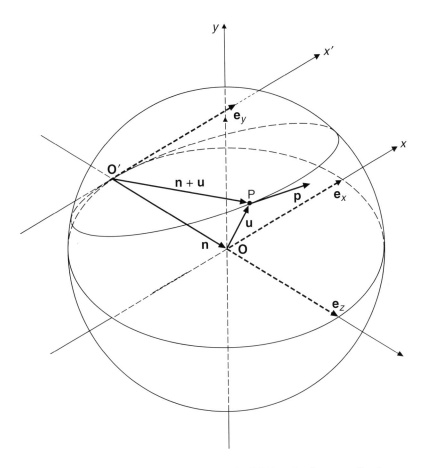

Fig. 7.38 Polarization of the radiated field when the field has the character of a plane wave.

This shows that the polarization **p** is in the plane $(O'P, e_x)$, that is to say, in the plane passing through the axis $O'x'$ and the point P.

Conclusion: The polarization lines of the electrical field on the sphere (S) are the *intersections of this sphere with the set of planes passing through the axis* $O'x'$ tangential to the sphere (Fig. 7.39). Evidently, the field magnetic lines are the intersection circles of (S) with the set of planes passing through the axis $O'x'$, tangential to the sphere and parallel to Oy.

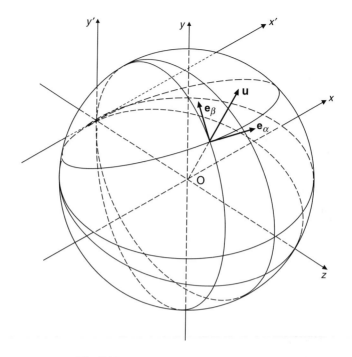

Fig. 7.39 Field lines of a Huygens source.

Huygens sources

Any antenna whose polarization possesses the above properties will be called a 'Huygens source'.

Most antennas are not Huygens sources, but certain types are good approximations to them, such as dual-mode horns (§8.2.4) and corrugated horns (§8.3).

The field lines defined in such a manner can be used as an orthogonal coordinate system allowing us to define on each point of the sphere (S) the polarization of any antenna by its components. Let us define the unit vectors of these coordinates, which for simplicity, we shall call 'Huygens coordinates'[9].

Unit vectors of Huygens coordinates

It is easy to calculate the modulus of the vector **p** and to deduce the unit vector of the polarization $\mathbf{e}_\alpha$

---

[9]Ludwig, A.C., 'The definition of cross polarization', *IEEE Trans. Antennas and Propagation*, Vol. AP-21, pp.116-119, January 1973.

$$\mathbf{e}_\alpha = \frac{1}{1+\gamma}\mathbf{p} = \mathbf{e}_x - \frac{\mathbf{n}+\mathbf{u}}{1+\gamma}\alpha \qquad (7.59)$$

$\mathbf{p}$ is given by equation (7.57).

Eventually, the field $\mathbf{E}(\mathbf{u})$ can be written from equation (7.55) in the following form:

$$\mathbf{E}(\mathbf{u}) = \mathbf{e}_\alpha j \frac{\exp\left(-2j\pi\dfrac{R}{\lambda}\right)}{\lambda R}\frac{1+\gamma}{2}\iint_{(S)} E_\mathrm{T}(\mathrm{M})\exp\left(2j\pi\frac{\mathbf{OM}.\mathbf{u}}{\lambda}\right)dS \quad (7.60)$$

In the same way we define the unit vector $\mathbf{e}_\beta$ of the *magnetic* field lines from equations similar to (7.58) and (7.59)

$$\mathbf{q} = (\mathbf{n}+\mathbf{u})\beta - \mathbf{e}_y(1+\gamma) \qquad (7.61)$$

$$\mathbf{e}_\beta = \frac{-\mathbf{q}}{1+\gamma} = \mathbf{e}_y - \frac{\mathbf{n}+\mathbf{u}}{1+\gamma}\beta \qquad (7.62)$$

Arbitrary aperture field

This case will be treated in §8.41 in the context of primary feeds, where this situation often occurs. We shall see that the field can be decomposed into several distributions where each is locally plane, allowing us to use the results of the preceding case.

7.4.6 Geometrical properties of the Huygens coordinates

Rotational symmetric properties

Consider a Huygens source. The following property plays an important part when the antennas possess rotational symmetry: 'on the sphere (S), the magnetic field lines are deduced, globally, from the field electric lines by a rotation through $\pi/2$ radians'.

The electric and magnetic field lines are both locally and globally orthogonal.

Relations with spherical coordinates

Consider the set of planes intercepting the sphere (S) according to the Huygens lines (Fig. 7.40).

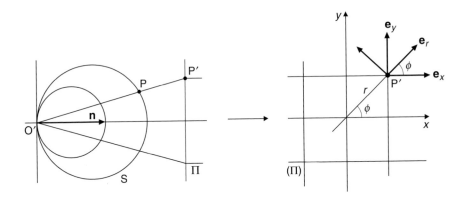

Fig. 7.40 Huygens field lines on a sphere (left) and the equivalent of Cartesian coordinates on the corresponding plane.

Let us perform the following geometrical transformation of any point P of the sphere (S) into a new point P′ such that

$$\mathbf{O'P.O'P'} = k \tag{7.63}$$

where k is any constant parameter.

The sphere (S) is transformed into a plane π normal to **n**. The set of planes intercept on π making a Cartesian coordinate system. Any point P′ of this plane may be referenced, either to a Cartesian coordinate system whose unit vectors are $\mathbf{e}_x, \mathbf{e}_y$, or to a polar coordinate system whose unit vectors are $\mathbf{e}_\Phi, \mathbf{e}_r$. We know that these two systems of unit vectors can be deduced from each other by a simple rotation through an angle Φ.

$$\mathbf{e}_x = \mathbf{e}_r \cos \Phi - \mathbf{e}_\Phi \sin \Phi$$
$$\tag{7.64}$$
$$\mathbf{e}_y = \mathbf{e}_r \sin \Phi + \mathbf{e}_\Phi \cos \Phi$$

If we come back to the sphere (S), we see that, apparently a point P can be referenced, either to Huygens coordinates, already defined ($\mathbf{e}_\alpha, \mathbf{e}_\beta$), or to spherical coordinates whose unit vectors are $\mathbf{e}_\Theta, \mathbf{e}_\Phi$. These spherical coordinates can be obtained directly from the polar coordinates of P′ by the same geometrical transformation (7.63). Therefore, between spherical and Huygens coordinates we find the *same* transformation as that between polar and Cartesian coordinates. The Huygens unit vectors can be deduced from the unit vectors of the spherical coordinates $(\mathbf{e}_\Theta \, \mathbf{e}_\Phi)$ from a simple rotation of angle Φ (Fig. 7.41).

$$\mathbf{e}_\alpha = \mathbf{e}_\Theta \cos \Phi - \mathbf{e}_\Phi \sin \Phi$$

$$(7.65)$$

$$\mathbf{e}_\beta = \mathbf{e}_\Theta \sin \Phi + \mathbf{e}_\Phi \cos \Phi$$

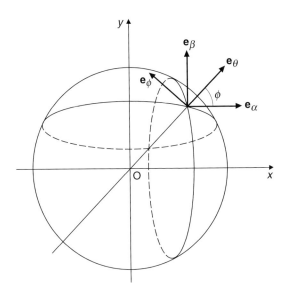

Fig. 7.41 Huygens field line unit vectors ($\mathbf{e}_\alpha$, $\mathbf{e}_\beta$) and spherical coordinates ($\mathbf{e}_\theta$, $\mathbf{e}_\phi$).

Polarization components of a field

It can be seen that a polarization field **p** may, at will, be expressed in terms of its spherical or Huygens coordinates as follows

$$\mathbf{p} = p_\Theta \mathbf{e}_\Theta + p_\Phi \mathbf{e}_\Phi$$

$$(7.66)$$

$$\mathbf{p} = p_\alpha \mathbf{e}_\alpha + p_\beta \mathbf{e}_\beta$$

They can be deduced from each other by means of a rotation matrix

$$\begin{bmatrix} p_\Theta \\ p_\Phi \end{bmatrix} = \begin{bmatrix} \cos \Phi & \sin \Phi \\ -\sin \Phi & \cos \Phi \end{bmatrix} \begin{bmatrix} p_\alpha \\ p_\beta \end{bmatrix}$$

$$(7.67)$$

7.4.7 Aperture radiation in the near field

A rigorous study of the near-field can easily be carried out using either the method of plane wave spectra (§2.2) or the Kirchhoff integral (§2.2.2).

In this paragraph, for simplicity, only the scalar aspect of the radiation will be considered, and the polarization is assumed to be known.

Consider a circular aperture of diameter D, whose illumination phase is almost constant (Fig. 7.42).

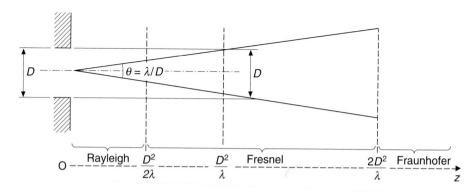

Fig. 7.42 Three radiation regions for the field diffracted from a radiating aperture.

As a first approximation, three radiation regions may be distinguished over an in-phase aperture.

Far-field or Fraunhofer region

This has just been studied. It extends out to many wavelengths beyond the aperture (see §15.2)

$$Z_3 \geq 2\frac{D^2}{\lambda}$$
(7.68)

Example: $\lambda = 10$ cm, $D = 10$ m, $Z_3 \geq 2$ km. This condition requires that the test towers should be located at a long distance from the place where the antenna measurements will be carried out (see Chapter 15).

Near-field or Rayleigh region

In this region, the stationary phase principle (equation (7.35)) allows us - at least as a first approximation - to use the geometric optics approach. It can be shown that this region is limited to distances Z_1 less than

$$Z_1 \le \frac{D^2}{2\lambda} \tag{7.69}$$

In this region, we can consider that the energy flow is almost dispersionless. The field remains approximately concentrated in a radiation beam whose diameter is the same as that of the aperture. In the preceding example, this distance reaches 500 m and at that distance the power density is comparable to that existing on the reflector.

Intermediate or Fresnel region

Consequently, this lies between the preceding regions. It is a transition region where the field gradually varies from the illumination law at the aperture to the resulting far-field pattern.

   To calculate the field at a point P in this region, one may use the Fresnel–Huygens principle (§7.3.2) but without, of course, applying the approximation valid only in the far field (equation 7.5). Nevertheless, a quadratic approximation can be made for the phase factor (Fresnel approximation)[10].

   If we set α, β, γ as the direction cosines in the direction **OP** = r**u**, where OP = r, then we have (Fig. 7.31)

$$\mathbf{MP} = \mathbf{OP} - \mathbf{OM} \; ; \quad \mathbf{MP}^2 = r^2 + \left(x^2 + y^2 \right) - 2\mathbf{OM.OP}$$

$$\mathbf{MP}^2 = r^2 \left(1 + \frac{x^2 + y^2 - 2\mathbf{OM.OP}}{r^2} \right); \quad \mathbf{MP} \approx r + \frac{x^2 + y^2}{2r} - \mathbf{OM.u}$$

If $f(x, y)$ is the illumination at a point M of the aperture, a scalar component of the field at P is then

---

[10] The plane wave spectrum method, which is equivalent to the Kirchhoff formula, gives the best results.

$$E_p\left(\alpha,\beta,r\right)=\frac{j}{\lambda}\frac{\exp\left(j\dfrac{2\pi r}{\lambda}\right)}{r} \times$$

$$\iint_{(S)} f\left(x,y\right)\exp\left[j\frac{2\pi}{\lambda}\left(\frac{x^2+y^2}{2r}\right)\right]\exp\left[-j2\pi\left(\alpha\frac{x}{\lambda}+\beta\frac{y}{\lambda}\right)\right]dxdy \tag{7.70}$$

The field is again the Fourier transform of the illumination, since the latter is modified by a quadratic phase factor. This phase factor vanishes at large distances, yielding the same far-field result as that found using the conventional method

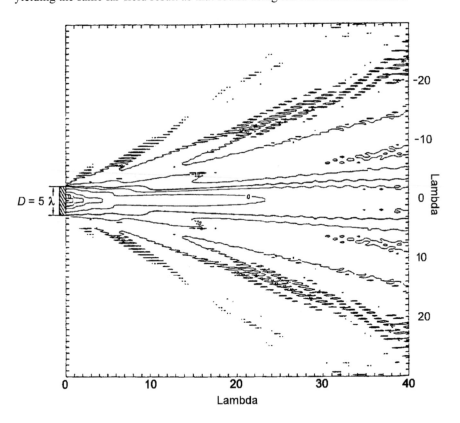

Fig. 7.43 Near-field distribution for an aperture with $D = 5\lambda$ (lines of equal amplitude).

Fig. 7.43 illustrates the field distribution at short distances of an aperture whose diameter is $D = 5\lambda$. Field contours at 0 dB, −3 dB, −5 dB, −10 dB are shown.

7.4.8 Gain factor of a radiating aperture

General expression

Let us consider an aperture (S) in the *xy* plane (Fig. 7.44). Let **n** be the unit vector of the normal O*z* to this aperture. In any direction **u,** the directivity (or the gain *G* if there is no loss) is given by (§3.3.2)

$$G(\mathbf{u}) = \frac{4\pi}{Wr} \frac{dW}{d\Omega} \tag{7.71}$$

where *dW* is the power radiated within the elementary solid angle *dΩ.*

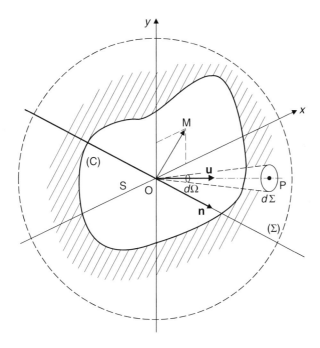

Fig. 7.44 Gain of a radiating aperture.

The radiated power W_r is the power crossing the aperture. If the field over it has, at least locally, a *plane wave character*, the power is evaluated from the integral of the square of the modulus of the electrical field over the aperture:

$$W_r = \frac{1}{2\eta} \iint_{(S)} |\mathbf{E}(M)|^2 \, dS \qquad (7.72)$$

with, in free space $\eta = 120\pi$ ohms.

The power density $dW/d\Omega$ in the direction $\mathbf{u}$ is obtained from the radiated field on a sphere (Σ) of large radius r_0

$$\frac{dW}{d\Omega} = \frac{r_0^2 \, dW}{d\Sigma} \qquad (7.73)$$

Since in this sphere the field is locally a plane wave, the power density is given by

$$\frac{dW}{d\Sigma} = \frac{|\mathbf{E}(\mathbf{u})|^2}{2\eta}$$

The field $\mathbf{E}(\mathbf{u})$ is that of a radiating aperture (see equations (7.25) and (7.32)). If $\mathbf{u}$ is close to $\mathbf{n}$, we may admit that the total field is given by the following relationship.

$$|\mathbf{E}(\mathbf{u})| = \frac{1}{r_0 \lambda} \left| \iint_{(S)} \mathbf{E}(M) \exp\left(2j\pi \frac{\mathbf{OM.u}}{\lambda}\right) dS \right| \qquad (7.74)$$

In conclusion, the aperture gain in the direction $\mathbf{u}$ is given by the expression

$$G(\mathbf{u}) = \frac{4\pi}{\lambda^2} \frac{\left| \iint_{(S)} \mathbf{E}(M) \exp\left(2j\pi \frac{\mathbf{OM.u}}{\lambda}\right) dS \right|^2}{\iint_{(S)} |\mathbf{E}(M)|^2 \, dS} \qquad (7.75)$$

Gain in the normal direction

Here the transverse field and the total field are identical, thus we have

$$G(\mathbf{u}) = \frac{4\pi}{\lambda^2} \frac{\left| \iint_{(S)} \mathbf{E} \, dS \right|^2}{\iint_{(S)} |\mathbf{E}|^2 \, dS} \qquad (7.76)$$

Remark: It may happen that only a fraction K_S of the radiated power crosses the aperture: this is the case with a parabolic reflector fed by a primary feed. A part of its pattern spills over around the reflector. In such a case the preceding expression should be multiplied by K_S.

Maximal theoretical gain

The question is 'What is the illumination $E(P)$ leading to the maximum normal gain?' In this Chapter the problem is limited to the general case where this illumination has, at least locally, the characteristics of a plane wave. More general conditions are considered in §13.4 (superdirective antennas). The answer can be obtained from the properties of Schwartz's inequality. Consider two complex vector functions **u** and **v**, integrable over a domain (S). It can be demonstrated that

$$\frac{\left|\iint_{(S)} \mathbf{u}.\mathbf{v}\, dS\right|^2}{\iint_{(S)} |\mathbf{u}|^2\, dS \iint_{(S)} |\mathbf{v}|^2\, dS} \leq 1 \tag{7.77}$$

Evidently, equality is obtained when $\mathbf{u} = \mathbf{v}^*$ (the complex conjugate of **v**).

In the present case, let us set $\mathbf{u} = \mathbf{E}_0$ (real and constant), $\mathbf{v} = \mathbf{E}(M)$. Then, the Schwartz inequality can be written as

$$\frac{\left|\iint_{(S)} \mathbf{E}_0.\mathbf{E}(M)\, dS\right|^2}{\iint_{(S)} |\mathbf{E}(M)|^2\, dS \iint_{(S)} |\mathbf{E}_0|^2\, dS} \leq 1 \tag{7.78}$$

$$\frac{1}{S} \frac{\left|\iint_{(S)} \mathbf{E}(M)\, dS\right|^2}{\iint_{(S)} |\mathbf{E}(M)|^2\, dS} \leq 1$$

Then the following fundamental result is obtained

$$\boxed{G(n) \leq \frac{4\pi S}{\lambda^2}} \tag{7.79}$$

Maximum gain is obtained when

$$\mathbf{E}(M) = \mathbf{E}_0$$

That is to say, *for a constant amplitude, phase and polarization illumination. Furthermore, the illumination should be locally plane.*

Illumination factor of an aperture

This is defined as the ratio of the actual gain to the maximum theoretical gain, therefore

$$K = \frac{G(\mathbf{n})}{\dfrac{4\pi S}{\lambda^2}} = \frac{1}{S} \frac{\left| \iint_{(S)} \mathbf{E}(M) \, dS \right|^2}{\iint_{(S)} |\mathbf{E}(M)|^2 \, dS} \tag{7.80}$$

Application

What is the minimum diameter of a circular radiating aperture for a gain of 50 dB ? The area of a circular aperture of diameter D is

$$S = \pi \frac{D^2}{4}$$

Its maximum gain is then

$$G_{\max} = \left(\frac{\pi D}{\lambda} \right)^2 \tag{7.81}$$

and we desire

$$\left(\frac{\pi D}{\lambda} \right)^2 \geq 10^5$$

Thus we must have, approximately: $D/\lambda \approx 100$.

For instance, for a frequency of 3 GHz, (wavelength = 10 cm), the diameter of the aperture should be of at least 10 metres

Exercise: The above aperture has parabolic radial illumination of the form

$$\mathbf{E}(r) = \mathbf{E}_0 \left[1 - \left(\frac{2r}{D} \right)^2 \right]$$

What is the illumination factor K of this aperture ?

Answer: $K = 0.75$

Appendix 7A: Deduction of the Huygens-Fresnel principle (§7.3.2) from the Kirchhoff integral

Let us start from the Kirchhoff integral (§2.2.2, equation (2.40)). Using the same notation as Fig. 7.9, we have

$$E(P) = \iint_S \left(G \frac{\partial E_M}{\partial z} - E_M \frac{\partial G}{\partial z} \right) dS$$

where G is the Green's function in free space

$$G = \frac{1}{4\pi} \frac{\exp(-jkr)}{r}, \qquad r^2 = (x - x_0)^2 + (y - y_0)^2 + (z - z_0)^2$$

Since the surface S is a wavefront, it is locally plane: the behaviour of the field, close to M, is then the same as that of a plane wave propagating in the direction of **n**. Thus

$$\frac{\partial E_M}{\partial z} = -jk E_M$$

The evaluation of $\partial G/\partial z$ is straightforward

$$\frac{\partial G}{\partial z} = \frac{\partial G}{\partial r} \frac{\partial r}{\partial z} = -\frac{1}{4\pi} \frac{z}{r} \frac{\exp(-jkr)}{r} \left(jk + \frac{1}{r} \right)$$

At large distances

$$\frac{\partial G}{\partial z} \approx -jk \frac{z}{r} G$$

and by inserting $z = r \cos\theta$ in the preceding equation, it follows that

$$\boxed{E(P) = \frac{-j}{\lambda} \iint_S \frac{1 + \cos\theta}{2} E_M \frac{\exp(-jkr)}{r} dS}$$

We find again equation (7.5) in a slightly different form.

FURTHER READING

1. Borgiotti, G., 'The Fourier transform method in aperture antenna problems', *Alta Frequenza*, Vol. 32, No. 11, November 1963.
2. Roubine, E., *Antennas - Vol.1: General Principles*, North Oxford Academic Press, 1987.
3. Schwering, F., 'On the range of validity of Fresnel-Kirchhoff's approximation formula', *IRE Trans. Antennas and Propagation*, Vol. AP-10, pp.99-100, 1962.

EXERCISES

7.1 *Factor of merit of a ground station antenna*

The factor of merit is defined as the ratio of the gain to the total noise temperature of the receiving system

$$M = \frac{G}{T}$$

where T is in Kelvins (K).

It is usually expressed in dB, in the form $M = 10 \log_{10} G/T$. In the following exercise we consider a very simple model for the antenna. The goal is to evaluate its performance.

Q1 Consider a directional antenna whose 'characteristic radiation surface' is defined by a cone of revolution of angle θ_0, limited by a spherical 'cap'. Evaluate the maximum gain G_0 of this antenna.

Note that the gain is zero if the direction of the unit vector **u** lies outside the cone, and constant and equal to G_0 if **u** lies inside the cone. Recall that the solid angle Ω_0 of such a cone is given by

$$\Omega_0 = 2\pi \left(1 - \cos \frac{\theta_0}{2} \right)$$

[Typical value: $\theta_0 = 10$ mrad - note that if θ_0 is small, Ω_0 can be approximated by $\Omega_0 \approx \pi \theta_0^2 / 4$]

Q2 Suppose now that this antenna has a mean absolute sidelobe level g. What is the new value G of its gain as a function of G_0 and g ? If the relative level L is denoted by the ratio G/g, what is G as a function of G_0 and L ?

[Typical values: $g_1 = 0.1$, $g_2 = 0.5$. Calculate the relative levels L_1 and L_2 and the corresponding gains G_1 and G_2. What do you conclude ?]

Q3 This antenna is now placed above a flat surface (ground) and pointed towards the sky. The sky is treated as a black body at a temperature $T_1 = 20$ K. The ground is equivalent to a black body at a temperature $T_0 = 300$ K. What is the noise temperature T_A of this antenna ?

Q4 The antenna is connected to a parametric amplifier whose noise temperature is $T_R = 50$ K. The loss between the antenna and the amplifier is defined by a coefficient α.

[Typical value: $\alpha = 10\%$ – i.e. $10 \log(1-\alpha) \approx 0.5$ dB]

What is the factor of merit M of this system ?

Note: In this exercise you can use the expressions (7.1), (7.3) and (7.4) for the gain and noise temperature.

ANSWERS

Q1 $\quad G_0 = \dfrac{4\pi}{\Omega_0}$

Q2 $\quad G = G_0(1-g) + g \; ; \; G = \dfrac{G_0}{1 + \dfrac{G_0 - 1}{L}}$

Q3 $\quad T_A = \dfrac{G}{G_0}T_1 + g\dfrac{T_0 + T_1}{2} - T_1\dfrac{g}{G_0} \approx (1-g)T_1 + \dfrac{g}{2}(T_0 + T_1)$

Q4 $\quad M = \dfrac{(1-\alpha)G}{(1-\alpha)T_A + \alpha T_0 + T_R}$

7.2 *Effect of cosecant-squared pattern on maximum gain*

Consider an antenna from which the radiation is specified with respect to a set of orthogonal axes Ox, Oy, Oz (with Oz vertical), and spherical coordinates (α, β), where α is the angle between the unit vector $\mathbf{u}$ and the xOz plane, and β is the angle of its projection onto this plane with the horizontal axis Ox (i.e. elevation angle). The radiation is confined to the neighbourhood of the vertical plane xOz in a cone of angle α_0. Suppose that the gain is constant in this plane, with unknown value G for an elevation angle β such that $0 \le \beta \le \beta_0$. For larger values of β we allow, for the sake of simplicity, the gain to vary not exactly as a cosecant-squared function, but rather as $1/\beta^2$, up to a maximum angle β_1.

We have therefore, for $|\alpha| \le \alpha_0/2$

$$\begin{cases} 0 \le \beta \le \beta_0 : & G(\alpha,\beta) = G \\ \beta_0 \le \beta \le \beta_1 : & G(\alpha,\beta) = G\beta_0^2/\beta^2 \end{cases}$$

Evaluate the gain for the following three cases:

Q1 With no cosecant-squared taper ($\beta_1 = \beta_0$) and a symmetrical pattern ($\beta_0 = \alpha_0$) [Typical value: $\alpha_0 = 10^{-2}$ radians]

Q2 With no cosecant-squared taper, but with a sectoral pattern up to an elevation β_0 = 1 (i.e. approximately 60°)

Q3 With a cosecant-squared taper up to an elevation angle $\beta_1 = 1$ (with $\beta_0 = \alpha$)

What conclusions do you draw ?

Note: an element of solid angle is written as: $d\Omega = \cos\alpha \, d\alpha \, d\beta$. If α is small we can make the approximation $d\Omega = d\alpha \, d\beta$. Also, you can use the expression (7.3) for the gain.

ANSWERS

Q1 $\qquad G_1 = \dfrac{4\pi}{\alpha_0^2}$

Q2 $\qquad G_2 = \dfrac{4\pi}{\alpha_0}$

Q3 $\qquad G_3 = \dfrac{4\pi}{\alpha_0^2 \left(2 - \dfrac{1}{\alpha_0} \right)}$

Conclusion: solution 3 (cosecant-squared) involves a loss of only 3 dB with respect to the maximum gain G_1, while solution 2 (sectoral pattern) involves a loss of 20 dB (with $\alpha_0 = 10^{-2}$ radians).

7.3 *Wave propagation in a laminar medium*

A rectangular coordinate system $Oxyz$ is used. The refractive index n is a function only of the z coordinate. Thus at any point M

$$n_M = n(z)$$

Consider rays propagating in the plane xOz. How do their trajectories behave ?

We make use of the results of §7.4.4 (Bouguer's law)

Q1 What is the expression for the gradient of the refractive index ?

Q2 Expression for the radius of curvature $1/R$ of a ray. If α denotes the angle between the tangent τ to a ray and a point M, derive the simple relationship between this angle and the refractive index.

Q3 Consider a ray passing through the origin O, and making an angle α_0 with the z-axis Oz. What can be said about the nature of such rays for $\alpha_0 = 0$? For what value of α_0 is the curvature maximum at this point ?

Q4 What is the differential equation describing the rays in Cartesian coordinates ?

Q5 Consider a refractive index which varies linearly with z

$$n(z) = n_0 (1 + kz)$$

Derive the explicit equation for the rays passing through O and y, making an angle $\alpha_0 = \pi/2$ with Oz.

ANSWERS

Q1 $\mathrm{grad}\, n = \dfrac{dn}{dz} \mathbf{e}_z$; $\dfrac{1}{R} = -\dfrac{1}{n}\dfrac{dn}{dz}\sin \alpha$

Q2 $n \sin \alpha = \text{const.} = A$

Q3   For $\alpha_0 = 0$, $\dfrac{1}{R} = 0$. The rays are parallel to Oz. The curvature is greatest

when the tangent at O is normal to Oz.

Q4   Using $\tan \alpha = \dfrac{dx}{dz}$ leads to

$\dfrac{dx}{dz} = \dfrac{1}{\sqrt{\left(\dfrac{n(z)}{A}\right)^2 - 1}}$

Q5   Setting $\dfrac{n(z)}{n_0} = 1 + kz = \cosh \varphi$, the integration leads, for rays passing

through O, to $z = \dfrac{\cosh kx - 1}{k}$. These are 'catenaries'.

Rays of arbitrary slope passing through O are given by the equation

$z = \dfrac{\cosh k (x - x_0) - \cosh kx_0}{k}$

with $\cot \alpha_0 = -\sinh x_0$

7.4 *Gain of an aperture with tapered illumination*

Consider a square aperture of side D, with a tapered illumination of the 'cosine on a pedestal' type. At each point M of the aperture, of coordinates (x, y), the aperture illumination is separable, in the form

$$F(x, y) = f(x).f(y)$$

with

$$f(x) = a + (1-a)\cos\frac{2x}{D}$$

We denote the *weighting factor* by the coefficient

$$\beta = \frac{\pi}{2}\frac{a}{1-a}$$

What is the gain factor K of the aperture as a function of β?

ANSWER

$$K = \left[\frac{1}{1+\dfrac{\dfrac{\pi^2}{8}-1}{(1+\beta)^2}}\right]^2$$

In the absence of weighting: $a = 1$, $\beta = \infty$, $K = 1$

In the absence of the 'pedestal': $a = 0$, $\beta = 0$, $K = \left(\dfrac{8}{\pi^2}\right)^2$
≈ 0.657, or -1.82 dB

8

Primary feeds

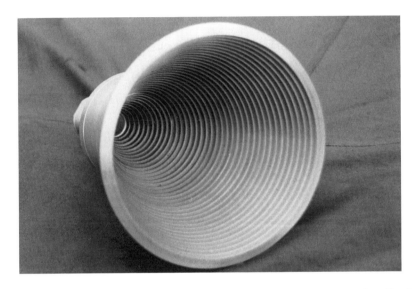

An X-band corrugated horn with a semi flare angle of 20°, designed and realized at
Queen Mary, University of London (photo: A.D. Olver).

8.1 GENERAL PROPERTIES

8.1.1 Introduction

A primary feed associated with a focusing system has as its purpose:

- In transmission, to illuminate the elements of the system (reflector, lens, and so on) with a characteristic function of well-defined amplitude, phase and polarization, in general stable over the required frequency band, and minimizing spillover losses (§9.5.3).
- In reception, to collect in an optimal manner the energy concentrated by the focusing system at its focus, in the form of a diffraction pattern (§9.8).

A primary feed may have several ports corresponding to different primary patterns and thus to several secondary patterns (§10.2.4). On the other hand, a primary feed is often associated with other elements: polarizers (§12.6), channel-splitters (polarization-splitters, frequency-duplexers) which are analysed using waveguide techniques. The performance of these elements directly influences that of the antenna. After examination of the general characteristics of primary feeds, we shall point out the main types that can be related to the fundamental structure of the horn, with particular emphasis on corrugated horns. In effect, apart from their own inherent interest, the *hybrid modes* that they propagate permit a practical description of the fields diffracted (in receiving) around the focus of axially-symmetric focusing systems (§9.8).

8.1.2 General characteristics of primary feeds

The principal characteristics to consider, in the frequency band of operation, are the following.

Matching of the feed to the feed guide

A primary feed can be any type of elementary antenna (see Chapter 5): dipole, slot, horn and so on.

The required matching conditions are sometimes difficult to realize [Voltage Standing Wave Ratio (VSWR) better than 1.1 for instance]. This is often the case with telecommunication antennas to avoid spurious reflections. This may also be the case with ultra-wideband radars for the same reason. If only the power transfer coefficient (return loss) is important, less stringent conditions may be acceptable.

If circular polarization is used, the ellipticity ratio of the radiated wave can be limited by the VSWR of the feed (if it is not possible to use a cross-polarization absorber).

Note that the focusing system itself interacts with the primary feed and may modify its matching and the ellipticity ratio of the wave radiated by the whole antenna.

Radiation pattern

The primary feed reference system is Oxyz, of unit vectors (e_x, e_y, n) (Fig. 8.1). Any point M for which the field is observed is defined by the unit vector $u = OM/OM$ and the distance $R = OM$. If R is sufficiently large, the field at M behaves locally as a plane wave: transverse field, the vectors (E, H, u) forming a direct orthogonal trihedral, the ratio of the magnitudes being $|E|/|H| = 120\pi \ \Omega$. The radiation is then defined by a knowledge of the electric field alone. It can be expressed by separating the information on polarization, amplitude and phase as follows:

$$\mathbf{E}_{O,R}(\mathbf{u}) = \mathbf{p}_{O,R}(\mathbf{u}) A_{O,R}(\mathbf{u}) \exp\left[j\phi_{O,R}(\mathbf{u})\right] \qquad (8.1)$$

where $\mathbf{p}$ is a unit vector normal to $\mathbf{u}$. The double subscript O, R expresses the fact that the field depends not only on the direction $\mathbf{u}$ but also, to a lesser extent, on the chosen point O and the distance R.

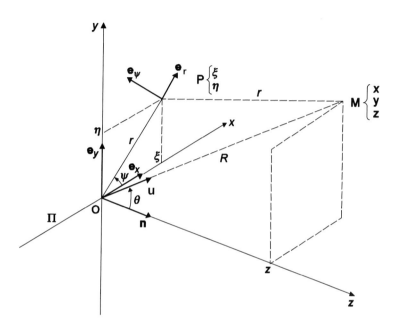

Fig. 8.1 Primary feed reference system.

Phase centre, astigmatism, caustic

The primary wave surfaces are defined by the condition

$$\phi_{O,R}(\mathbf{u}) = C \text{ (constant)} \qquad (8.2)$$

In the vicinity of the reference direction $\mathbf{n}$, they may be approximated by quadric surfaces (ellipsoids, in general). These quadrics possess two orthogonal planes of principal cross sections (P') and (P'') defining two centres of curvature O'_R and O''_R known as phase centres of the patterns located in these planes. These points often depend on the distance R, but if R is large enough, O'_R and O''_R become stable. If we express the variations of phase in the principal planes (P') and (P'') choosing these points as references, the phase is in principle constant. We know that for $R = \infty$

$$\phi_{O'}(\mathbf{u}) = \phi_O(\mathbf{u}) + 2\pi \frac{\mathbf{OO'.u}}{\lambda} \tag{8.3}$$

In fact, in general the phase is of the fourth order in terms of the angle $\theta = (\mathbf{n}, \mathbf{u})$. If ϕ_0 is constant, there is perfect stigmatism in the corresponding plane (Fig. 8.2).

The distance $O'O''$ measures the *astigmatism* of the antenna (the term is chosen by analogy with optics). Sectoral horns (§5.1.2) are examples of highly astigmatic antennas (Fig. 8.4).

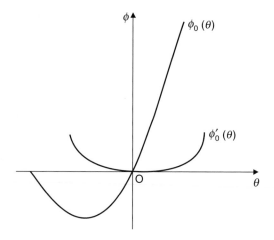

Fig. 8.2 Phase variation of the characteristic function of the primary feed with respect to the reference point O.

If O' and O'' are coincident, the antenna is stigmatic, and possesses a phase centre. This property is desired for primary feeds associated with stigmatic focal systems, that is to say with a well-defined focus. Certain types of feeds are almost perfectly stigmatic but we are often obliged to tolerate a certain amount of astigmatism at least within a part of the frequency band to be covered.

Even if the feed is perfectly stigmatic, the phase centre O may move as a function of frequency; the stability of the centre of the primary phase centre over the band of operation is another essential quality of focusing systems.

Caustic surface

Since a radiated primary wave surface is the locus of points of constant phase, the caustic surface is the envelope of the normal lines to the wave surface, or rays. If the feed is rigorously stigmatic, this envelope is the phase centre O′ itself. In case of astigmatism, the rays take support on two orthogonal straight lines passing through the centres O′ and O″ termed *focal straight lines* (*Sturm focal lines* in optics) (Figs. 8.3 and 8.4).

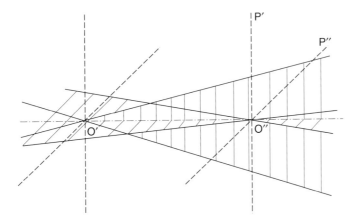

Fig. 8.3 Astigmatism.

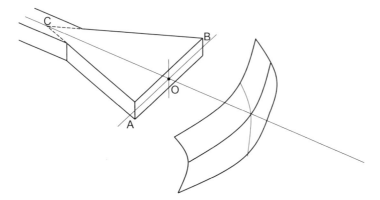

Fig. 8.4 Sectoral horn with apex C and quasi-linear aperture AB. The wavefront is toric in form: a cut with plane of symmetry perpendicular to AB has its phase centre at O, while the orthogonal cut in the plane CAB has its phase centre at C. The distance CO is the astigmatism of the antenna.

In general, the caustic surface can take a number of different forms. Fig. 8.5 shows the caustic meridian of an imperfectly stigmatic axially symmetric system.

Certain focusing systems are astigmatic; this is the case with spherical or toric reflectors. In these cases the form of the primary feed caustic should be matched to that of the focusing system to correct the astigmatic aberration.

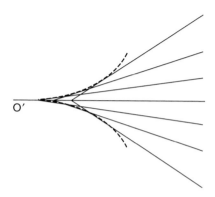

Fig. 8.5 Meridian of an imperfectly stigmatic axially symmetric radiation.

Polarization: vectorial expression of the radiated field

Huygens reference source: We have defined a reference aperture (§7.4.5) radiating a 'pure' polarization forming a 'Huygens source'. The field lines observed on a sphere of large radius (Fig. 8.6) are the intersections of this sphere with a set of planes. They define an orthogonal coordinate system allowing us to characterize any polarization by its components in this system.

To obtain a Huygens source it is sufficient[1] that the electromagnetic field over its aperture should be constant in polarization and phase, and that the field should be locally plane. Under these conditions the radiated field is given by the expressions (equations 7.53, 7.57) that we restate here

$$\mathbf{E}(\mathbf{u}) = \mathbf{e}_\alpha j \frac{\exp\left(-j2\pi\dfrac{R}{\lambda}\right)}{\lambda R} \frac{1+\cos\theta}{2} \iint_S E_r(M)\exp\left(j2\pi\frac{\mathbf{OM}}{\lambda}.\mathbf{u}\right)dS \quad (8.4)$$

To simplify let us set

---

[1] But not necessary: more general conditions will be given about radiation from circular apertures and hybrid waves (§8.1.3 and §8.3).

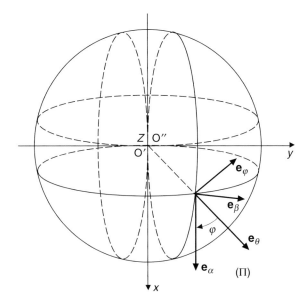

Fig. 8.6 Huygens polarization references.

$$\frac{R}{\lambda} = R', \qquad \mu = \frac{\mathbf{OM}}{\lambda}, \qquad d\Sigma = \frac{dS}{\lambda^2} \qquad (8.5)$$

Then, with an electric field of constant illumination and horizontal polarization $E_x(\mathbf{m})$, we have

$$\mathbf{E}(\mathbf{u}) = \mathbf{e}_\alpha j \frac{\exp(-j2\pi R')}{R'} \frac{1+\cos\theta}{2} F_\alpha(\mathbf{u}) \qquad (8.6)$$

where

$$F_\alpha(\mathbf{u}) = \iint_\Sigma E_x(\mu) \exp(j2\pi\mu.\mathbf{u}) d\Sigma$$

In practice, the electromagnetic field over an aperture is far from being locally plane because quite often it results from the mode (or modes) of propagation of the source. Nevertheless, we can always consider it as the result of superposition of plane waves, and in this case we still obtain the same results as before.

Illumination whose polarization is constant but whose transverse components are not locally plane: The fields $\mathbf{E}_T$ and $\mathbf{H}_T$ are parallel to the Ox and Oy axes respectively, but their ratio need not be constant (wave impedance), $\eta \neq \eta_0$ but rather may vary as a function of the point P considered over the aperture. The unit vectors of the axes are denoted by $\mathbf{e}_x$, $\mathbf{e}_y$:

$$\mathbf{E}_T = E_x \mathbf{e}_x, \quad \mathbf{H}_T = H_y \mathbf{e}_y \tag{8.7}$$

$$\frac{|\mathbf{E}_T|}{|\mathbf{H}_T|} = \frac{|E_x|}{|H_y|} = \eta(\mu) \tag{8.8}$$

Under these conditions, the calculation of the radiation by the Kottler formula (equation (7.49)) allows us to express the electric field in terms of its two polarization components in the form

$$E_{\alpha x} = j\frac{\exp(-j2\pi R')}{R'}\frac{1+\cos\theta}{2}F'_\alpha + j\frac{\exp(-j2\pi R')}{R'}\frac{1-\cos\theta}{2}\cos(2\varphi)G_x \tag{8.9a}$$

$$E_{\beta x} = j\frac{\exp(-j2\pi R')}{R'}\frac{1-\cos\theta}{2}\sin(2\varphi)G_x \tag{8.9b}$$

where

$$F'_\alpha = \iint_\Sigma \frac{E_x + \eta_0 H_y}{2}\exp(j2\pi\mu.\mathbf{u})d\Sigma \tag{8.10a}$$

$$G_x = \iint_\Sigma \frac{E_x - \eta_0 H_y}{2}\exp(j2\pi\mu.\mathbf{u})d\Sigma \tag{8.10b}$$

and $\eta_0 = 120\pi \ \Omega$.

We can verify that in the case where the wave is locally plane we have $G_x = 0$. Therefore, we are in the same situation as before. It is worth noting that the cross-polar field is small in the vicinity of the axis ($\theta = 0$) and over the principal planes ($\varphi = 0$ and $\varphi = \pi/2$) and even for $\eta \neq \eta_0$. This justifies the approximation often made, although not always valid, of an illumination which is locally plane.

General case: In the general case for which the polarization is not constant, we are led to decompose the field into two distributions of the preceding type associating the components (E_x, H_y) on one hand, and the components (E_x, H_x) on the other hand, and adding the corresponding radiated fields

$$E_\alpha = E_{\alpha x} + E_{\alpha y}; \quad E_\beta = E_{\beta x} + E_{\beta y} \tag{8.11}$$

We obtain the general expressions (8.12) for the radiating aperture. In these expressions the term associated with $(1+\cos\theta)/2$ is emphasized since it is most important in the vicinity of the normal to the aperture, and the term $(1-\cos\theta)/2$

which is zero in the normal direction can be neglected. Nevertheless, in certain cases this term is dominant (hybrid modes of the type EH, for example corrugated horns, §8.3.2).

$$\frac{E_\alpha}{j\psi_0} = \frac{1+\cos\theta}{2}F'_\alpha + \frac{1-\cos\theta}{2}\left[\cos(2\varphi)G_x + \sin(2\varphi)G_y\right] \qquad (8.12a)$$

$$\frac{E_\beta}{j\psi_0} = \frac{1+\cos\theta}{2}F'_\beta - \frac{1-\cos\theta}{2}\left[\cos(2\varphi)G_y - \sin(2\varphi)G_x\right] \qquad (8.12b)$$

where

$$\psi_0 = \frac{\exp(-j2\pi R')}{R'} \qquad (8.13)$$

$$F'_\beta = \iint_\Sigma \frac{E_v - \eta_0 H_x}{2}\exp(j2\pi\mu.\mathbf{u})d\Sigma \qquad (8.14a)$$

$$G_y = \iint_\Sigma \frac{E_v + \eta_0 H_x}{2}\exp(j2\pi\mu.\mathbf{u})d\Sigma \qquad (8.14b)$$

These expressions are of particular interest in the case of circularly symmetrical structures.

8.1.3 Radiation from radially-symmetric structures

General expressions for fields

Consider any structure (metallic, dielectric etc.) formed from surfaces of revolution about the same axis Oz (Fig. 8.7). This structure may be, for example, a monomode or multimode corrugated horn (§8.3), an axially symmetric system with reflectors and lenses and so on (Chapter 9).

Suppose that this structure is excited by means of a guide having the same axis Oz, transmitting only its dominant-propagation mode. Its polarization is represented by a unit vector **p**. Note that high order modes may exist in other parts of the structure.

Consider an arbitrary axisymmetrical surface (S), with the same axis Oz. This surface may be for instance the plane of the radiation aperture, a surface resting on the sides of the aperture, or a sphere (Γ) of large radius where the radiation patterns are observed.

We shall show that over such a surface the electromagnetic field exhibits a characteristic mathematical form. From this will be deduced a simple criterion for a zero cross-polarization radiated field. Some other conclusions will be drawn concerning 'hybrid' modes (§8.3) [5].

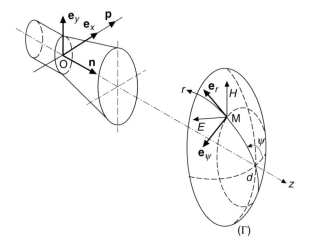

Fig. 8.7 Radiation from an axially symmetric primary feed.

Any point M of (S) is characterized by its polar coordinates (r, ψ) where r denotes the curvilinear abscissa along a meridian.

Consider for instance the case of an excitation of the structure in horizontal polarization

$$\mathbf{p} = \mathbf{e}_x$$

On any surface (S) an electromagnetic field distribution characterized by its radial and tangential components results.

$$\mathbf{E}_{(M)} \begin{cases} E_r(r,\psi) \\ E_\psi(r,\psi) \end{cases} \qquad \mathbf{H}_{(M)} \begin{cases} H_r(r,\psi) \\ H_\psi(r,\psi) \end{cases} \tag{8.15}$$

Note that the distribution has the same symmetries as the excitation **p**. Therefore,

$$E_r\left(r,\frac{\pi}{2}\right) = 0 \qquad H_r(r,0) = 0$$

$$\tag{8.16}$$

$$E_\psi(r,0) = 0 \qquad H_\psi\left(r,\frac{\pi}{2}\right) = 0$$

To express the field at a point $M(r,\psi)$ the excitation $\mathbf{p}=\mathbf{e}_x$ can be split into two orthogonal excitations $\mathbf{p}'$ and $\mathbf{p}''$ whose orientations are ψ and $\psi-\pi/2$ (Fig. 8.8).

$$\mathbf{p}=\mathbf{e}_x=\mathbf{p}'\cos\psi+\mathbf{p}''\sin\psi \tag{8.17}$$

The excitations $\mathbf{p}'$ and $\mathbf{p}''$ give rise, by symmetry, to the field distributions $(\mathbf{E}',\mathbf{H}')$ and $(\mathbf{E}'',\mathbf{H}'')$ deduced respectively from (8.15) by the rotations ψ and $\psi-\pi/2$. Their respective expressions are obtained by stating the conservation of the radial and tangential components in these rotations. The point $M(r,\psi)$ is therefore in a plane of symmetry of the two partial distributions that then satisfy the symmetric properties of equations (8.16); thus we have

$$E'_\psi(r,\psi)=E_\psi(r,0)=0 \qquad\qquad H'_r(r,\psi)=H_r(r,0)=0$$

$$E''_r(r,\psi)=E_r\left(r,\frac{\pi}{2}\right)=0 \qquad\qquad H''_\psi(r,\psi)=H_\psi\left(r,\frac{\pi}{2}\right)=0$$

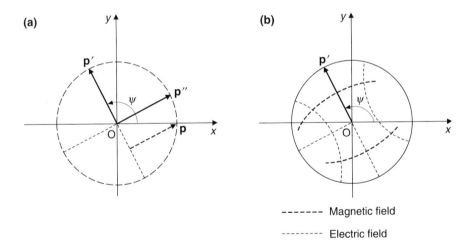

------- Magnetic field

--------- Electric field

Fig. 8.8 (a) Component of the incident field polarization; (b) Field distribution $(\mathbf{E}',\mathbf{H}')$ resulting from the $\mathbf{p}'$ component of the incident field.

Since the excitation is monomodal, their superposition gives the original distribution. We then obtain expressions of the form

$$E_r = a(r)\cos\psi \qquad\qquad H_r^+ = c(r)\sin\psi$$

$$E_\psi = -b(r)\sin\psi \qquad\qquad H_\psi^+ = d(r)\cos\psi$$

(8.18)

In Cartesian coordinates, by setting

$$a(r) = A(r) + B(r) \qquad\qquad b(r) = A(r) - B(r)$$

and using the rotation matrix (eqn. 7.65) we obtain

$$E_x = A(r) + B(r)\cos(2\psi) \qquad H_x^+ = C(r)\sin(2\psi)$$

$$E_y = B(r)\sin(2\psi) \qquad\qquad H_y^+ = D(r) - C(r)\cos(2\psi)$$

(8.19)

Note: For simplicity we set here and subsequently

$$H^+ = \eta_0 H$$

(8.20)

with $\eta_0 = 120\pi\ \Omega$

Properties of the radiation patterns

General form: If the surface (S) is a sphere Γ of large radius R (polar coordinates $\theta = r/R, \varphi$), the fields are the radiation patterns in the classical sense. They are of the form

$$E_\theta(\theta,\varphi) = E_\theta(\theta,0)\cos\varphi = E_\alpha(\theta,0)\cos\varphi$$

(8.21)

$$E_\varphi(\theta,\varphi) = E_\varphi\left(\theta,\frac{\pi}{2}\right)\sin\varphi = -E_\alpha\left(\theta,\frac{\pi}{2}\right)\sin\varphi$$

where $E_\alpha(\theta,0)$ and $E_\alpha(\theta,\pi/2)$ are the electric field patterns located in the principal planes E (electric) and H (magnetic). Since the wave is locally plane the magnetic fields can be deduced in a straightforward manner.

The radiation pattern expressions are deduced in both principal polarization and cross polarization $\mathbf{e}_\alpha$, $\mathbf{e}_\beta$ (§7.4.6):

$$E_\alpha(\theta,\varphi) = \frac{E_\alpha(\theta,0)+E_\alpha\left(\theta,\frac{\pi}{2}\right)}{2} + \frac{E_\alpha(\theta,0)-E_\alpha\left(\theta,\frac{\pi}{2}\right)}{2}\cos(2\varphi)$$

(8.22)

$$E_\beta(\theta,\varphi) = \frac{E_\alpha(\theta,0)-E_\alpha\left(\theta,\frac{\pi}{2}\right)}{2}\sin(2\varphi)$$

Conclusions (Fig. 8.9):

(1) The cross polarization is maximum in the diagonal planes ($\varphi = \pm \pi/4$). In these planes, the pattern obtained is the half difference between the E-plane and H-plane patterns.

(2) A criterion for the cross polarization to be zero is that the E-plane and H-plane radiation patterns should be identical in both amplitude and phase.

(3) In co-polarization, the pattern obtained in the diagonal plane is the average of E-plane and H-plane patterns.

(4) If the condition 2 is fulfilled, the pattern is cross-polarization free and axially symmetric (characteristic surface of revolution).

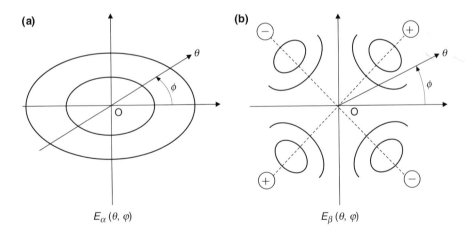

Fig. 8.9 (a) Copolar iso-level lines of the radiation pattern; (b) Cross polar iso-level lines of the radiation pattern.

Expressions from the fields over the aperture

The form of fields (equation (8.19)) over the aperture allows us, by means of the general equations (equation (8.12)), to express the radiation function of any circular aperture excited by a circular guide in its dominant mode

$$\frac{E_\alpha(\theta,\varphi)}{2\pi j\psi_0} = \frac{1+\cos\theta}{2\lambda^2}\int_0^a \frac{A+D}{2}J_0(ur)rdr \;-\; \frac{1-\cos\theta}{2\lambda^2}\int_0^a \frac{B+C}{2}J_2(ur)rdr$$

$$-\cos(2\varphi)\left[\frac{1+\cos\theta}{2\lambda^2}\int_0^a \frac{B-C}{2}J_2(ur)rdr \;-\; \frac{1-\cos\theta}{2\lambda^2}\int_0^a \frac{A-D}{2}J_0(ur)rdr\right]$$

$$(8.23)$$

$$\frac{E_\beta(\theta,\varphi)}{2\pi j\psi_0} = \left[\frac{1+\cos\theta}{2\lambda^2}\int_0^a \frac{B-C}{2}J_2(ur)rdr \;-\; \frac{1-\cos\theta}{2\lambda^2}\int_0^a \frac{A-D}{2}J_0(ur)rdr\right]\sin(2\varphi)$$

$$(8.24)$$

where $J_0(ur)$ and $J_2(ur)$ are Bessel functions of zero and second order respectively and of argument

$$ur = \frac{2\pi r}{\lambda}\sin\theta$$

with ψ_0 given by equation (8.13).

Polarization-purity condition: hybrid waves

The condition

$$E_\beta = 0$$

leads to

$$A = D \qquad B = C \qquad\qquad (8.25)$$

Thus we have

$$E_x = A + B\cos(2\psi) \qquad\qquad H_x^+ = B\sin(2\psi)$$

$$(8.26)$$

$$E_y = B\sin(2\psi) \qquad\qquad H_y^+ = A - B\cos(2\psi)$$

The set of illuminations fulfilling this condition (with the assumptions made) thus only depend on the parameters A and B. The plane wave character corresponds to the particular case $B = 0$.

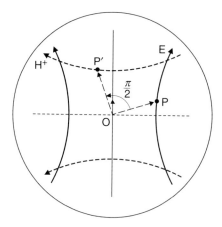

Fig. 8.10 Hybrid mode: magnetic field lines H are derived from electric ones (E) by a rotation of 90° about the z axis.

In polar coordinates (8.18) the condition found (8.25) leads to the following form:

$$E_r = a(r)\cos\psi \qquad\qquad H_r^+ = -a(r)\sin\psi$$

$$\tag{8.27}$$

$$E_\psi = b(r)\sin\psi \qquad\qquad H_\psi^+ = b(r)\cos\psi$$

This expresses the following fundamental property: *the magnetic field lines can be derived from the electric field lines by a rotation of $\pi/2$ about the Oz axis* (Fig. 8.10). It can be shown that this is also true for the longitudinal components. That is why these waves, which do not belong to the classical category of TE or TM waves, are termed '*hybrid waves*'.

The radiation pattern under these conditions then takes the following form:

$$\frac{E_\alpha}{2\pi j\psi_0} = \frac{1+\cos\theta}{2\lambda^2}\int_0^a AJ_0(ur)r\,dr - \frac{1-\cos\theta}{2\lambda^2}\int_0^a BJ_2(ur)r\,dr \qquad (8.28)$$

8.1.4 Primary aperture in an incident field

Introduction

The operation of a primary feed on reception can be deduced in principle from its transmission behaviour by means of the reciprocity theorem. Nevertheless, direct analysis of the operation on reception is of interest to avoid the calculation of the primary pattern and to separate the factors inherent to the primary feed from those of the focusing system. It is then relatively easy to define a criterion allowing the optimization of the complete system.

Consider a primary aperture S of contour (C) in the *xy* plane, where **n** is the unit vector normal to this plane (Fig. 7.42). Its illumination function on transmission is defined by the electromagnetic distribution **E**(M) and **H**(M). These components are supposed to be in-phase (only the 'active' powers of the radiation are considered). If the feed is illuminated by an incident field $E'(M)$ and $H'(M)$, it can be shown that the power-transfer coefficient of the incident wave toward the feed is given by the following expression[2] (see also Chapter 4):

$$
T = \frac{1}{4} \frac{\left| \iint_S (\mathbf{E} \times \mathbf{H}' - \mathbf{E}' \times \mathbf{H}).\mathbf{n}dS \right|^2}{\iint_S (\mathbf{E} \times \mathbf{H}^*).\mathbf{n}dS \iint_S (\mathbf{E}' \times \mathbf{H}'^*).\mathbf{n}dS}
\tag{8.29}
$$

where the integration is performed over the width of the aperture. Perfect matching with the incident field corresponds to the condition

$$
\mathbf{E}' = \mathbf{E}^*, \mathbf{H}' = -\mathbf{H}^*
\tag{8.30}
$$

in which case $T = 1$.

Gain factor of the focusing system

In the case for which the feed is associated with a focusing system, the incident field $(\mathbf{E}', \mathbf{H}')$ is that of the *diffraction pattern* of the system, as defined and calculated in §9.8.

Under these conditions, the above normalized transfer coefficient is identical to the directivity factor (or the gain factor in the absence of losses) of the complete focusing system. In particular, the matching condition $\mathbf{E}' = \mathbf{E}^*$, $\mathbf{H}' = -\mathbf{H}^*$ allows us to define a simple criterion for optimization of the primary feed, leading to

---

[2] Robieux, J., 'Lois générales de la liaison entre radiateurs d'ondes', *Annales de Radioélectricité*, Vol. XIV, No. 57, §1.1, p190, July 1959.

maximum gain. The illumination function *must reproduce, as well as possible, the form of the diffraction pattern.* (The change of sign of **H** with respect to that of **E** indicates a change of the direction of propagation). Other optimization criteria may be used: for instance, the tracking slope of a monopulse system (see §10.2). In the case of a focusing system with long focal length (Cassegrain), the character of the field in the diffraction pattern is that of a plane wave parallel to the focal plane. In this case, the central lobe of the diffraction pattern is, in general, several wavelengths in diameter and the radiating aperture is itself quite large. As a result, the illumination field $(\mathbf{E}', \mathbf{H}')$ is also locally plane. The above expression of the gain factor can then be simplified. Only the scalar products of the electric fields are involved

$$T = \frac{\left| \iint_S \mathbf{E}.\mathbf{E}' dS \right|^2}{\iint_S |\mathbf{E}|^2 dS \iint_S |\mathbf{E}'|^2 dS} \tag{8.31}$$

Primary feed gain

If the incident wave $(\mathbf{E}', \mathbf{H}')$ is a plane wave whose direction is **u**, the calculation of the transfer coefficient T gives the directivity factor of the primary feed in that direction. In particular, if we consider a normal incident wave whose polarization of unit vector **p** is that of the feed in that direction, we obtain the axial gain factor for a lossless primary feed (eqn. 7.78):

$$T = \frac{\left| \mathbf{p}.\iint_S \mathbf{E} dS \right|^2}{S \iint_S |\mathbf{E}|^2 dS} = \frac{\left| \iint_S \mathbf{E} dS \right|^2}{S \iint_S |\mathbf{E}|^2 dS} \tag{8.32}$$

8.2 HORNS[3]

8.2.1 General properties

A horn consists of a flared length of waveguide ending in a radiating aperture. The other end (connected to the feed guide) is the throat (Fig. 7.25). This kind of antenna is used extensively as a primary feed since it has many electrical and mechanical advantages. It is easy to protect from the weather by means of an electrically-transparent radome. It can also be pressurized to avoid condensation, which improves the power handling capability. In addition, a large range of radiation

---

[3] Rectangular horns are analysed in §5.1.2.

characteristics can be obtained according to the type of horn and the excitation conditions, such as polarization and propagation mode. In fact, there are several types of horns: monomode, multimode and corrugated. Moreover, in each category, horns have different forms: aperture angle, form of a transversal or longitudinal section. All these factors allow us to find the appropriate horn for the required application. To evaluate the radiation characteristics, we can often treat the horn as a radiating aperture whose illumination is expressed in terms of the excited propagation mode (or modes). It is sometimes necessary to take into account other higher order modes of diverse origin, such as discontinuities due to the aperture, parasitic reflections and so on. Nevertheless, if the horn is fed by a waveguide carrying its dominant mode, the longitudinal profile of the horn ensures a progressive transition to the aperture, avoiding the appearance of higher order modes. The illumination obtained is then a proper representation of the field distribution of the mode considered (monomode horn).

On the other hand, if the horn is discontinuous, high order modes are created that modify the illumination function and therefore the pattern (multimode horn). A horn may be fed with several feed ports, giving several patterns. If the arrangement of the horn is such that the feed-ports are isolated, the radiated patterns are orthogonal (§11.2.2). It is often useful to consider the operation of the horn on reception. If a primary feed is concerned, for instance, the horn is illuminated by in an incident field ($\mathbf{E'}, \mathbf{H'}$) whose form is the diffraction pattern of the focusing system (§8.1.4, §9.9.2, §10.2.4).

In the case of a monomode horn, for instance, the diffraction pattern generates in the vicinity of the aperture a set of diverse order modes that can propagate towards the throat of the waveguide until the cross-sections correspond to the cut-offs of the modes concerned; only the useful modes remain. The horn, therefore plays the role of a mode filter.

8.2.2 Small flare angle horns and open-ended guides

An open-ended guide is the limiting case of a horn whose angular aperture is zero. The study of its radiation can give a good approximation to that of a horn of finite angular aperture but sufficiently small so that the emerging wave is very close to a plane wave. In the case of a conical horn for instance, of apex angle 2α, length L and whose radiation aperture is a circle of diameter $\mathrm{AA'} = 2a$, the maximum longitudinal distance between the spherical wave of centre O′ and radius L = OO′ and the aperture plane is given by the relationship (Fig. 8.11)

$$\delta z = L(1 - \cos\alpha) \approx L\frac{\alpha^2}{2} = \frac{a^2}{2L} \qquad (8.33)$$

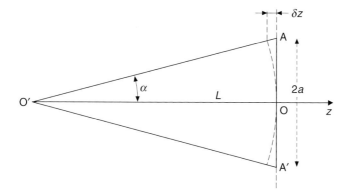

Fig. 8.11 Conical horn.

If we set a criterion such that this pathlength is less than or equal to $\lambda/4$, then

$$L \geq \frac{2a^2}{\lambda} \tag{8.34}$$

Under these conditions, we see that a horn whose aperture $2a$ exceeds a few wavelengths must necessarily be very long. Nevertheless, such horns are used as primary feeds to illuminate long focal length systems (Cassegrain). Besides, it is possible to completely cancel this phase error by adding a reflector or an auxiliary lens to the horn (§10.31).

8.2.3 Flared horns

When the angular aperture of a horn is not negligible, one must take into account the phase variations of the illumination function[4] (see also §5.1.2). A quadratic approximation to these variations gives the classical results summarized below. We shall encounter them again when we consider corrugated horns (§8.3.4):

- For a given aperture of a/λ, the principal lobe of the pattern remains close to that corresponding to uniform illumination as long as the phase error is less than $\pi/2$. The phase centre is close to the aperture centre. The beamwidth varies as $\lambda/(2a)$. It is therefore a function of frequency.

---

[4] Silver, S. (ed), *Microwave Antenna Theory and Design*, p186 and p334, MIT Radiation Lab. Series Vol. 12, McGraw-Hill, New York, 1949; reprinted by Peter Peregrinus, Stevenage, 1984.

- If the length of the horn decreases, this in turn fills in the nulls between the sidelobes and the main lobe is broadened and distorted. The phase function becomes complex. The phase centre moves away from the aperture towards the apex of the horn.

- For a given length L/λ of the horn, if the aperture a/λ increases, the beamwidth decreases, attains a minimum, then increases again and approaches the angle of the geometric aperture of the horn.

8.2.4 Multimode horns[5]

General characteristics

A multimode horn allows us to synthesize the illumination function of its aperture (and hence its pattern) by superposition of several modes of propagation, appropriately combined. A multimode horn can have several independent feed ports associated with orthogonal patterns (§11.7.3). This is the case with monopulse tracking antennas from which this technique originated (§11.3.4). Multimode horns can provide favourable primary feeds in respect of gain and noise temperature when the required bandwidth is not too high. The possibility to synthesize the illumination function of an open guide by the superposition of modes lies in the general property of orthogonality of the transverse field components associated with these modes. This provides an 'orthogonal base' permitting us to express linearly any illumination function with the coefficients obtained by 'projecting' the illumination to be synthesized onto this base. The required modes are generated by means of a mode generator.

Generation of modes

We shall use the term 'mode generator' to describe any guiding structure involving discontinuities intended to generate high order modes. The structure of a mode generator is in general as follows (Fig. 8.12). One or several 'excitation' guides, each one filtering their dominant propagation mode, lead to an input of a 'main' guide of length L, whose open end forms the radiation aperture. This main guide is capable of propagating various high order harmonic modes. The high order modes are generated in principle at the levels of the discontinuity planes where the

---

[5]Drabowitch, S., 'Multimode antennas', *Microwave Journal*, January 1966. Also in 'Théorie et applications des antennas multimodes', *Revue Technique C.F.T.H.*, No. 37, Nov. 1962.

excitation guides end. In effect, the boundary conditions imply the existence, at the front and at the rear of the main guide, of an infinite set of propagating and evanescent modes. The propagating dominant and harmonic modes that are generated propagate through the main guide at their characteristic phase velocity up to the end of the guide. Their relative phases may be controlled by means of this guide and in particular by its length *L*.

This result is only obtained in principle at a single frequency, but we can tolerate a certain phase error (a few percent) between the modes; this is generally acceptable in the radar domain but not always in that of telecommunications. This is one of the reasons for the development of corrugated horns, which do not have these limitations.

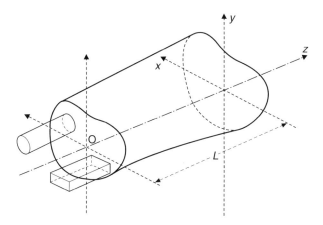

Fig. 8.12 Multimode horn. Mode generator.

Examples

Multimode horn, H-plane: Fig. 8.13 shows the schematic of a H-plane mode generator. If by means of the magic T, the inputs A and B are excited with a phase difference of π, then only the TE_{20} mode propagates up to the throat and produces a 'difference' (Δ) pattern utilized in monopulse-tracking antennas (10.2.4). If the inputs are exited in phase, the 'even' modes TE_{10} and TE_{30} are excited. Their superposition gives, as shown in Fig. 8.14, a 'sum' (Σ) illumination resembling closely the diffraction pattern of a focusing system (§9.8.7) and able to give an excellent on-axis system gain (§10.3.4).

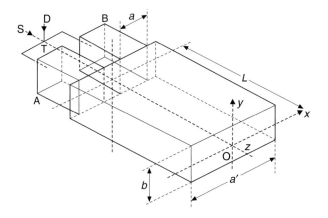

Fig. 8.13 H-plane multimode horn.

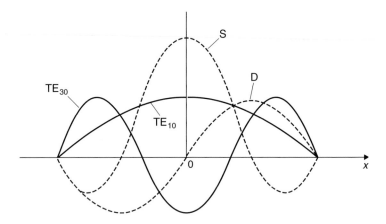

Fig. 8.14 Superposition of modes in the horn radiating aperture.

Dual-mode circular horn[6]: Fig. 8.15 shows how the superposition over the circular guide aperture of modes TE_{11} and TM_{11} allows us to obtain an approximately circularly-symmetric illumination with almost zero cross polarization. The generation of mode TM_{11} may be obtained by a simple discontinuity of the diameter of the guide, as shown in Fig. 8.16.

---

[6]Potter, P.D., 'A new horn antenna with suppressed side lobes and equal beamwidths', *Microwave Journal*, June 1963, pp71-78.

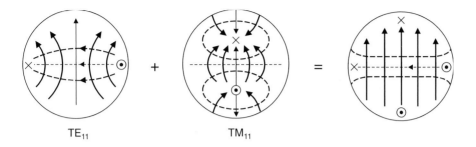

Fig. 8.15 Dual-mode circular horn.

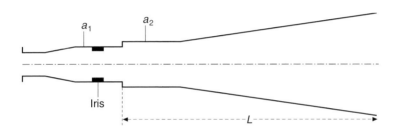

Fig. 8.16 Longitudinal section of a dual-mode circular horn.

8.3 HYBRID MODES AND CORRUGATED HORNS

8.3.1 Circular aperture radiating a pure polarization

We have established (§8.1.3) general expressions (equation (8.19)) for the field over the aperture of any circularly-symmetric structure with the only condition that the excitation guide should only transmit the dominant mode. Equation (8.23) for the radiated field has allowed us to formulate the condition for pure polarization (equation (8.25)), which leads to the more restrictive forms (equations (8.26) and (8.27)). These express the fact that *the field magnetic lines are deduced from the field electric lines by a $\pi/2$ rotation about the Oz axis*. Is it possible to imagine a waveguide capable of transmitting a mode having all these properties ?

8.3.2 Search for hybrid mode waves

Method used

On the sphere $\Gamma(O, R)$ the lines of polarization $(\mathbf{e}_\alpha, \mathbf{e}_\beta)$ are known (Fig. 7.37). We note that these plane sections of the sphere Γ are deduced ones from the others by a $\pi/2$ rotation about Oz. If we consider a conical spectrum of plane waves whose incidence directions all make *the same angle* θ_0 *with the* Oz *axis* and of electric polarization $\mathbf{e}_\alpha$ and magnetic polarization $\mathbf{e}_\beta$, we can expect the following results:

- Their interference will give a plane wave propagating along Oz with a propagation constant $\gamma = K \cos \theta_0$.

- The polarization lines of the interference field obtained in a plane of any cross section ($z = 0$ for example), will have the same symmetrical properties.

Once this is done it only remains to look for cylinders of axis Oz whose boundary conditions are compatible with such waves.

Hybrid mode generated by a conical spectrum of plane waves (Fig. 8.17)

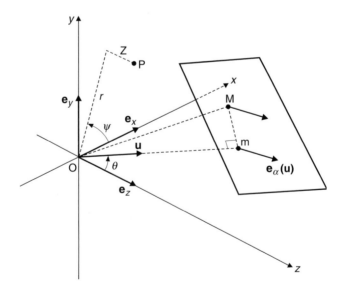

Fig. 8.17 Hybrid mode generated by a conical spectrum of plane waves.

A plane wave of direction **u** polarized according to the vector $\mathbf{e}_\alpha(\mathbf{u})$ can be written in the form

$$\mathbf{E}'_\mathbf{u}(M) = E_0 \mathbf{e}_\alpha(\mathbf{u}) \exp(jK\mathbf{OM.u}) \tag{8.35a}$$

$$\mathbf{H}'^+_\mathbf{u}(M) = H_0^+ \mathbf{e}_\beta(\mathbf{u}) \exp(jK\mathbf{OM.u}) \tag{8.35b}$$

where $\mathbf{H}_0^+ = \mathbf{E}_0$ and $K = \dfrac{2\pi}{\lambda}$.

Let (r,ψ,z) be the cylindrical coordinates at a point P for which we seek the resultant field; (θ_0, φ) the spherical coordinates of the direction **u**; and $(\mathbf{e}_x, \mathbf{e}_y, \mathbf{e}_z)$ the base vectors of the reference trihedral Oxyz.

We have

$$K\mathbf{OM.u} = Kr\sin\theta_0 \cos(\varphi-\psi) + Kz\cos\theta_0 \tag{8.36}$$

Then, the field sought is of the form

$$\mathbf{E}(r,\psi,z) = \mathbf{E}(r,\psi)\exp(-j\gamma z)$$
$$\tag{8.37}$$
$$\mathbf{H}^+(r,\psi,z) = \mathbf{H}^+(r,\psi)\exp(-j\gamma z)$$

with

$$\gamma = \frac{2\pi}{\lambda_g} = K\cos\theta_0 \tag{8.38}$$

and

$$\mathbf{E}(r,\psi) = E_0 \int_0^{2\pi} \mathbf{e}_\alpha(\theta_0,\varphi)\exp\left[jKr\sin\theta_0 \cos(\varphi-\psi)\right]d\varphi$$
$$\tag{8.39}$$
$$\mathbf{H}^+(r,\psi) = E_0 \int_0^{2\pi} \mathbf{e}_\beta(\theta_0,\varphi)\exp\left[jKr\sin\theta_0 \cos(\varphi-\psi)\right]d\varphi$$

Using expressions for $\mathbf{e}_\alpha$ and $\mathbf{e}_\beta$ (equation (7.62)), calculations show that the electric and magnetic fields are of the form (equation (8.26))

$$A = \frac{1+\cos\theta_0}{2}J_0(Kr\sin\theta_0); \qquad B = \frac{1-\cos\theta_0}{2}J_2(Kr\sin\theta_0) \tag{8.40}$$

Search for a cylindrical guide possessing the required boundary conditions - propagation conditions

A perfectly-conducting circular guide requires that the tangential electric field and the normal magnetic field should be zero. It can be verified in equations (8.26) and (8.40) that this condition is never satisfied for any value of r. Let us suppose that we can realize (we shall see how) a surface forcing the same boundary conditions of the tangential components of the electric and magnetic fields

$$\mathbf{E}_\psi = \mathbf{H}_\psi = 0$$

with

$$\mathbf{E}_z \neq 0, \mathbf{H}_z \neq 0$$

(8.41)

These express characteristics inherent to the nature of the surface considered introducing associated surface admittances: longitudinal Y_z and transverse Y_ψ. In effect, we have

$$Y_z = \frac{\mathbf{H}_\psi}{\mathbf{E}_z} = 0 ; \qquad Y_\psi \frac{\mathbf{H}_z}{\mathbf{E}_\psi} = \infty$$

(8.42)

Such a surface is then characterized by an anisotropic *surface admittance*: *infinite* in the transverse direction (as with a perfect conductor) and *zero* in the longitudinal direction. There are no longitudinal currents. The equations obtained show that such a cylindrical surface, of radius a, can propagate a hybrid mode with the condition

$$J_0 \left(Ka \sin \theta_0 \right) - \tan^2 \frac{\theta_0}{2} J_2 \left(Ka \sin \theta_0 \right) = 0$$

(8.43)

This propagation condition can be written, taking into account equation (8.38)

$$J_0 \left(\sqrt{(Ka)^2 - (\gamma a)^2} \right) - \frac{1 - \dfrac{\gamma a}{Ka}}{1 + \dfrac{\gamma a}{Ka}} J_2 \left(\sqrt{(Ka)^2 - (\gamma a)^2} \right) = 0$$

(8.44)

To each value of Ka corresponds a set of values $\gamma_{1m} a$ of γa defining guide wavelengths of a set of 'hybrid' modes denoted HE_{1m}. The cut-off conditions correspond to $\gamma_{1m} = 0$, that is to say $\theta_0 = \pi/2$ and $Ka = u_{1m}$, such as

$$J_0 \left(u_{1m} \right) - J_2 \left(u_{1m} \right) \equiv 2J_1' \left(u_{1m} \right) = 0$$

(8.45)

$$\frac{2\pi a}{\lambda_{c1m}} = u_{1m} \tag{8.46}$$

For the mode HE_{1m}, $u_{11} = 1.841$, $\lambda_{c_{11}} = 3.41a$.

Note that the cut-off condition of the mode HE_{1m} is the same as that of the dominant mode TE_{110} of the circular guide with smooth walls. The diagram expressing this relation (equation (8.44)) constitutes the 'Brillouin diagram' (K, γ) (Fig. 8.18).

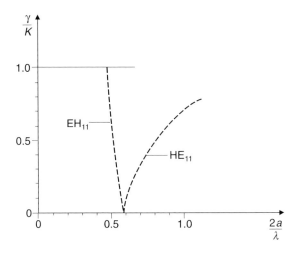

Fig. 8.18 Brillouin diagram for the EH_{11} and HE_{11} hybrid circular modes.

Field line structure of modes HE_{1m}

The fields may be written in the form (8.26) taking into account equation (8.38) as follows:

$$A = \frac{1+\dfrac{\gamma_{1m}}{K}}{2} J_0\left(v_{1m}\frac{r}{a}\right); \qquad B = \frac{1-\dfrac{\gamma_{1m}}{K}}{2} J_2\left(v_{1m}\frac{r}{a}\right) \tag{8.47}$$

with

$$v_{1m} = \sqrt{(Ka)^2 - (\gamma_{1m}a)^2} \tag{8.48}$$

These equations allow us to define the transverse electric and magnetic lines. It can be seen, as expected, that they can be deduced from each other by a rotation through

$\pi/2$ (Fig. 8.15). In contrast to the case of classical guides, their configuration changes in terms of the ratio $K/K_c = \lambda_c/\lambda$. Far away from cut-off, we can neglect terms in J_2 and we have, approximately

$$\mathbf{E}_x = E_0 J_0\left(u_{1m}\frac{r}{a}\right); \qquad \mathbf{H}_x = 0 \tag{8.49}$$

$$\mathbf{E}_y = 0; \qquad \mathbf{H}_y = \frac{E_0}{\eta}J_0\left(u_{1m}\frac{r}{a}\right) \tag{8.50}$$

with

$$J_0\left(u_{1m}\right) = 0$$

We observe an amplitude illumination that decreases towards the edge.

The field is locally plane, and the polarization is practically constant. We shall see that this field structure is of the same kind as those of the diffraction patterns of the axially symmetric focusing systems (§9.8.3). From this, we can see the advantage of using primary feeds radiating a hybrid mode to illuminate such systems.

Realization of the anisotropic surface admittance

Such an admittance is obtained by means of conductive surfaces with grooves transverse to the propagation direction (Fig. 8.19). The depth d of the grooves is in principle a quarter wavelength so that the surface admittance is close to zero. Since a corrugation is effectively a 'radial' guide, the depth condition is slightly greater than a quarter wavelength. The corrugations should be in practice narrow and close-spaced with respect to the wavelength to simulate a continuous superficial admittance, but experience shows that in practice only a small number of corrugations per wavelength are sufficient (three as a minimum).

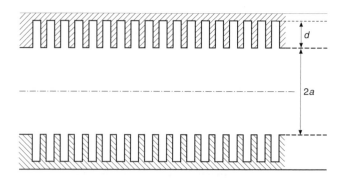

Fig. 8.19 Corrugated waveguide. Longitudinal section.

The conditions for a longitudinal admittance to be zero are the 'balanced hybrid conditions'. An exhaustive study of modes conveyed by such structures shows that to avoid the appearance of parasitic surface waves (EH_{11} for instance), the admittance due to the grooves should not be inductive in the frequency band to be covered. In other respects, grooves of depth equal to half a wavelength are equivalent to a smooth surface. As a general rule, we must have

$$\frac{\lambda}{4} < d < \frac{\lambda}{2} \tag{8.51}$$

8.3.3 Radiation pattern

Radiation from an open-ended corrugated waveguide

When the balanced hybrid conditions are satisfied, the patterns are given by equation (8.28). The integrals are explicitly calculated by means of the *Lommel* formulas. The pattern obtained is independent of φ, as expected, and with zero cross polarization. If the aperture is large enough, the illumination follows, as we have seen, an amplitude taper function that decreases towards the aperture periphery. This results in a very low sidelobe level as calculation and experiment shows.

Radiation from a corrugated horn

Introduction: The radiation of corrugated horns excited with mode HE_{11} have the same special symmetrical properties as open corrugated guides: in the vicinity of the hybrid matching conditions, the aperture illumination satisfies the condition for zero cross-polarization. With the classical approximations of the horns, its expression can be given, the same form as in a guide of the same diameter transmitting the same mode, with a quadratic-phase function corresponding to a pseudo-spherical wave of radius L (generatrix line length of the cone defining the horn). Moreover, as the horn aperture is assumed large enough, we shall admit that the field is locally plane. Thus, this is determined by the electric field alone. At any point M of the aperture of radius a (polar coordinates ρ, ψ, Cartesian coordinates ξ, η) is placed a source of illumination defined by the scalar

$$E_M = J_0\left(u_{11}\frac{\rho}{a}\right)\exp\left(-j\frac{2\pi}{\lambda}\frac{\rho^2}{2L}\right) \tag{8.52}$$

The radiation at any distance may be obtained from the Kirchhoff formulas with the Fresnel–Huygens approximation (§7.4.7, equation (7.68)). A simple 'radial' integral is thus found. The field polarization is that of a 'Huygens source'. A particularly simple and accurate method of calculation consists of decomposing the illumination into a series of Laguerre–Gauss functions[7] (Fig. 8.20). The particular form of the function J_0 allows us to limit the expansion, to a first approximation, to the first term

$$J_0\left(u_{11}\frac{\rho}{a}\right) \approx \exp\left(-\frac{\rho^2}{w^2}\right) \tag{8.53}$$

The dispersion w can be chosen by expressing the conservation of the energy enclosed in the aperture in these two forms. Then, we find $w \approx a/\sqrt{2}$. This approximation gives simple and complete results, adequately rigorous, and that have at least a qualitative value for other types of horns [circular dual mode horn, for instance (§8.2.4)].

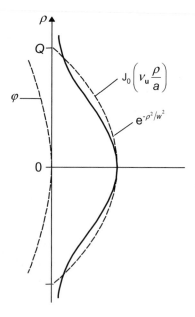

Fig. 8.20 Bessel function approximated by the Laguerre–Gauss function.

[7]Kildal, P.S., 'Gaussian beam model for aperture-controlled and flare angle-controlled corrugated horn antennas', *IEE Proc*, Vol. 135, Pt. H, No. 4, August 1988.

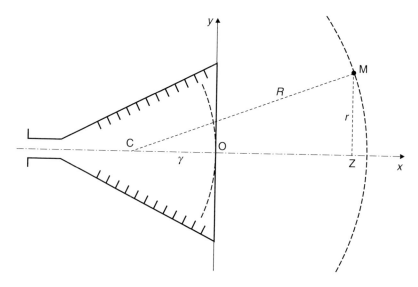

Fig. 8.21 Longitudinal section of a corrugated horn.

Radiation in the Fresnel region: A knowledge of the radiation in the near field is necessary when a horn is to illuminate an auxiliary reflector (Cassegrain antenna, for example). At a point M (coordinates x, y, z) (Fig. 8.21), the field is given by the Fresnel formula which is in fact a form of the Fourier transform. In this case integration can be carried out completely. Since the Fourier transform of a Gaussian function is itself a Gaussian, the field is also Gaussian in form

$$E_p(r, z) = A\exp\left(-j2\pi\frac{r^2}{2\lambda R}\right)\exp\left(-\frac{r^2}{w'^2}\right) \tag{8.54}$$

This result is confirmed by experimental results which give very low sidelobe levels (Fig. 8.22). The parameters w' and R characterize respectively the dispersion and the position of the phase centre.

Dispersion:

$$w'^2 = \frac{2}{\pi^2}\frac{z^2\lambda^2}{a^2} + \left(1 + \frac{z}{L}\right)^2\frac{a^2}{2} \tag{8.55}$$

- In the vicinity of the aperture $z \approx 0$, $w'^2 = \frac{a^2}{2} = w^2$

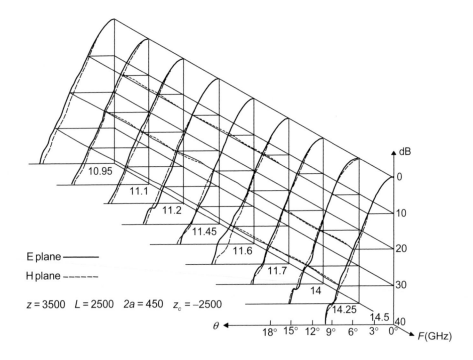

Fig. 8.22 Radiation patterns of a corrugated horn.

The dispersion is the same as the illumination.

• At long distances

$$\frac{w'^2}{z^2} \approx \frac{2}{\pi^2} \frac{\lambda^2}{a^2} + \frac{a^2}{2L^2} \tag{8.56}$$

For $L = \infty$ (equiphase aperture), the angular aperture of the beam is defined in radians as

$$\theta_0 = \frac{w'}{z} = \frac{\sqrt{2}\,\lambda}{\pi\,a} \approx \frac{\lambda}{2a} \tag{8.57}$$

This is the classical formula.

For L small (horn sufficiently short)

$$\theta_0 = \frac{w'}{z} \approx \frac{a}{\sqrt{2L}} \qquad (8.58)$$

The angular aperture is given by the geometry of the horn. It is in principle independent of frequency, as can be expected from stationary phase condition. This property is particularly well satisfied in practical cases by corrugated horns. This is one of their advantages (Fig. 8.22).

Phase centre: The curvature of the wave surfaces is defined by the abscissa $z_c = z - R$ of the phase centre

$$z_c = -\frac{1 + \dfrac{z}{L}}{\dfrac{1}{L} + z\left(\dfrac{1}{L^2} + \dfrac{4\lambda^2}{\pi^2 a^2}\right)} \qquad (8.59)$$

- Near-field: z small: $z_c = -L$. C close to the cone apex (at infinity if the aperture is equiphase)

- Far-field: $z >> L$

  L large (small aperture horn): $z_c \approx 0$. C close to O, centre of curvature.

  L small (large aperture horn): $z_c \approx -L$. Close to the apex cone.

The conditions are those of stationary phase.

Optimal horn with minimum dispersion: The problem is as follows. Consider a horn of limited length for reasons of size. This horn is arranged to illuminate a zone in a plane normal to an axis located at a distance z. What aperture a of the horn gives the maximum field concentration (or the minimum dispersion) in this plane ? This problem occurs for instance in the design of a Cassegrain system. The existence of a minimum is evident from the expression of the dispersion itself (equation (8.56)). This is obtained by differentiating w'^2/z^2 with respect to a^2.

$$a_{\text{opt.}} = \sqrt{\frac{2}{\pi}} \sqrt{\frac{\lambda L}{1 + \dfrac{L}{z}}} \qquad (8.60)$$

In the far field $(z = \infty)$, we obtain the optimal angular aperture

$$\theta_{0_{\text{opt}}} = \sqrt{\frac{2}{\pi}} \sqrt{\frac{\lambda}{L}} \quad \text{and} \quad a_{\text{opt}} = \sqrt{\frac{2\lambda L}{\pi}} \tag{8.61}$$

All these results can be compared with those of rectangular horns (§5.1.2) and flared horns (§8.2.3).

FURTHER READING

1. Aubry, C. and Bitter, D., 'Radiation pattern of a corrugated conical horn in terms of Laguerre-Gaussian functions', *Electronics Letters*, Vol. 11, No. 7, pp154-156, April 1975.
2. Clarricoats, P.J.B. and Olver, A.D., *Corrugated Horns for Microwave Antennas*, Peter Peregrinus, Stevenage, 1984.
3. Drabowitch, S. and Ancona, C., *Antennes II: Applications*, Masson, Paris, 1978; translated into English as *Antennas II*, North Oxford Academic Press, 1986.
4. Elliot, R.S., 'On the theory of corrugated plane surfaces', *IRE Trans. Antennas & Propagation*, Vol. AP-2, No. 2, pp71-81, April 1954.
5. Jeuken, M.F.J., 'Corrugated conical horn antenna with small flare angle', *De Ingenieur*, Jrg.84/No. 34, August 1972.
6. Love, A.W., 'The diagonal horn antenna', *Microwave Journal*, Vol. 5, pp117-122, March 1962.
7. Love, A.W., *Electromagnetic Horn Antennas*, IEEE Press, New York, 1976.
8. Olver, A.D., Clarricoats, P.J.B., Kishk, A.A. and Shafai, L., *Microwave Horns and Feeds*, Peter Peregrinus, Stevenage, UK, 1994; co-published with IEEE Press.

EXERCISES

(The reader can make use of the data given in §8.1.3, equation (8.22) and Fig. 8.9).

8.1 *Circular horn*

A circular horn, excited with vertical polarization, has an approximately Gaussian main lobe whose beamwidth is different in the E-plane (vertical) from that in the H-plane. With the direction **u** specified in spherical coordinates (θ, φ), the characteristic radiation function can be modelled as

H-plane:
$$E(\theta,0) = \exp-\left(\frac{\theta^2}{\theta_1^2}\right)$$

E-plane:
$$E\left(\theta,\frac{\pi}{2}\right) = \exp-\left(\frac{\theta^2}{\theta_2^2}\right)$$

In what follows we can set $\dfrac{1}{\theta_1^2} = a+b$; $\dfrac{1}{\theta_2^2} = a-b$

Q1 What is the expression for the characteristic function?

(a) in the nominal (vertical) polarization (e_α): $E_\alpha(\theta,\varphi)$

(b) in the cross-polar polarization (e_β): $E_\beta(\theta,\varphi)$

Q2 What is the cross-polar ratio $m(\theta, \varphi)$, where

$$m(\theta,\varphi) = \left|\frac{E_\beta(\theta,\varphi)}{E_\alpha(0,0)}\right|$$

For what values of φ and θ is this ratio maximum?

Q3 For small deviations from boresight, we set

$$\Delta\theta = \theta_2 - \theta_1 \qquad\qquad \theta = (\theta_2 + \theta_1)/2$$

Find a simplified expression for m_{max} as a function of $(\Delta\theta/\theta)$.

[Typical value: $\Delta\theta/\theta = 10\%$].

ANSWERS

Q1 $E_\alpha(\theta,\varphi) = (\exp{-a\theta^2})\left[\cosh b\theta^2 + \sinh b\theta^2 \cos 2\varphi\right]$

$E_\beta(\theta,\varphi) = (\exp{-a\theta^2})\left[\sinh b\theta^2 \sin 2\varphi\right]$

Q2 $m(\theta,\varphi) = (\exp{-a\theta^2})\sinh b\theta^2$

m is maximum for $\varphi = \dfrac{\pi}{4} + k\pi$, $\theta = \dfrac{1}{\sqrt{a}}$, where $k = 0, 1, 2, ...$

$m_{max} = \dfrac{1}{e}\left|\dfrac{b}{a}\right|$

Q3 For small deviations from boresight $\Delta\theta = \theta_2 - \theta_1$, we have approximately

$m_{max} = \dfrac{1}{e}\left|\dfrac{\Delta\theta}{\theta}\right|$

Example: $\Delta\theta/\theta = 10\%$ gives $m_{max} \approx \dfrac{1}{27}$, which is about 28.6 dB.

8.2 *Astigmatic horn*

Consider a circular horn excited with vertical polarization, whose main lobe is approximately Gaussian. Here the beamwidth is *the same* in the E-plane (vertical) and H-plane, but the phase centres O′ and O″ are different, and separated by a distance O′O″ = δ. Under these conditions it can be shown that the characteristic radiation function is modified by a quadratic phase factor such that

$$E(\theta,0) = \exp{-(a - jb)\theta^2}$$

$$E\left(\theta,\tfrac{\pi}{2}\right) = \exp{-(a + jb)\theta^2}$$

with

$b \cong \dfrac{\pi\,\delta}{2\,\lambda}$ and $a = \dfrac{1}{\sqrt{\theta_0}}$, where θ_0 is the beamwidth.

Q1 Calculate the characteristic function in the nominal $\mathbf{e}_\alpha = E_\alpha(\theta,\varphi)$ and cross-polar $E_\beta(\theta,\varphi)$ polarization components.

Q2 What is the cross-polar ratio $m(\theta,\varphi) = \dfrac{|E_\beta(\theta,\varphi)|}{|E_\alpha(0,0)|}$

For what values of φ and θ is this maximum ? What is the corresponding value m_{max} ? (Suppose that δ is small, so that $b/a \gg 1$)

Q3 From this relation between m_{max}, θ_0 and δ, deduce the maximum tolerable astigmatism (δ/λ) with a horn of beamwidth $\theta_0 = \pi/6$ and a cross-polar ratio of 20 dB ($m_{max} = 0.1$).

ANSWERS

Q1 $E_\alpha = \left(\exp-\alpha\theta^2\right)\left[\cos b\theta^2 + j \sin b\theta^2 \cos 2\varphi\right]$

$E_\beta = j\left(\exp-\alpha\theta^2\right)\left[\sin b\theta^2 \sin 2\varphi\right]$

Q2 $m(\theta,\varphi) = \left(\exp-\alpha\theta^2\right)\left[\sin b\theta^2 \sin 2\varphi\right]$

m_{max} for $\varphi = \dfrac{\pi}{4} + k\pi$, $\theta = \theta_1 \cong \dfrac{1}{\alpha} = \theta_0$

$m_{max} \cong \dfrac{1}{e}\dfrac{\pi}{2}\dfrac{\delta}{\lambda}\theta_0^2$

Q3 $\dfrac{\delta}{\lambda} \leq \dfrac{36e}{5\pi^3} \approx 0.63$

9

Axially-symmetric systems

3.7m diameter C-band reflector antenna for Doppler weather radar
Photo: AMS (formerly Siemens ATM).

9.1 INTRODUCTION

By analogy with optics, the term 'axially-symmetric systems' is used here to denote radiating structures whose properties are invariant under a rotation about an axis ZZ', in other words, structures of revolution. They may contain reflecting or refracting surfaces with rotational symmetry, as well as multimode or monomode metallic or dielectric circular waveguides.

The main attraction of axially-symmetric systems is a consequence of their axis of symmetry, which makes them very suitable as tracking antennas (tracking radars, antennas for satellite communications, etc.) and makes them easy to manufacture (mass production) and to control mechanically.

We have already indicated (§8.1.3) in considering primary feeds, various properties of these types of structures, which we shall make use of in this chapter.

In general, the axially-symmetric system is a focusing system. We shall assume that except in the vicinity of the focal points, the fields may be treated as locally plane, which allows us to use ray theory as a first approximation (§7.3.4). A common type of focusing axially-symmetric system is stigmatic for a point F and a point at infinity on the FZ axis. But we may also have stigmatic systems for conjugate points at a finite distance (e.g. energy transmission systems using beams) or with both conjugate points located at infinity (non focal systems).

The strict stigmatism condition is not essential; in fact, it is sufficient that the caustic surface of the system is matched to that of the primary feed to ensure good efficiency (§8.1.4). Two important examples of axially-symmetric systems are the centre-fed paraboloid, and Cassegrain systems and their derivatives. Apart from these important applications, certain properties of the axisymmetrical systems may be generalized to non-axisymmetrical systems, such as rotationally-symmetric reflectors with non-circular section, or offset reflectors (§10.4).

9.2 SYMMETRY PROPERTIES - PROPAGATION OF POLARIZATION, RADIATION PATTERNS

Any meridian plane is a plane of symmetry of the system. This means that if an incident ray lies in such a plane it cuts the axis, and *vice versa*. If the polarization of this ray lies on the meridian plane ('radial' polarization), it is conserved at all points along the propagation path. Likewise, if the incident polarization is perpendicular to the meridian plane ('tangential' polarization), the polarization remains tangential in the system; these obvious properties result from the general theory of rays.

The conservation of the radial and tangential polarization components leads to an important consequence for the concepts of co-polarization and cross-polarization defined in connection with the primary radiation (§8.3.6).

Consider an axially symmetric system, stigmatic for a focus F and a point at infinity on the axis FZ. A primary spherical wave from a point F corresponds to a plane wave propagating in the axis direction, and vice versa.

The coordinates used result from the symmetry of the structure; if a meridian plane is chosen ('horizontal'), the coordinates associated with any direction of the vector $\mathbf{u}$ of origin F are spherical coordinates (ϕ, θ) with the corresponding unit vectors ($\mathbf{e}_\phi$, $\mathbf{e}_\theta$). Likewise, polarization is also conserved if we choose a reference plane π perpendicular to the axis and crossing it at the point O, the location of an arbitrary point M' of this plane is characterized by its polar or Cartesian coordinates (Fig. 9.1).

Consider an incident wave from the direction of the axis with constant polarization

$$\mathbf{p} = \mathbf{e}_x$$

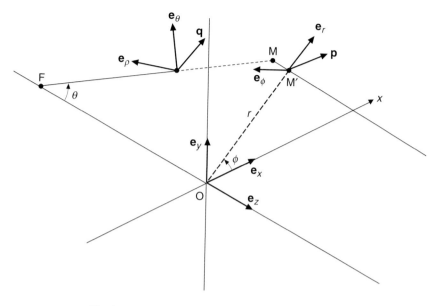

Fig. 9.1 Propagation of polarization towards the focus.

This can be expressed using the radial and tangential components in the following form

$$\mathbf{p} = \mathbf{e}_x = \mathbf{e}_r \cos\phi - \mathbf{e}_\phi \sin\phi \tag{9.1}$$

Conservation of the radial and tangential components gives rise on the spherical wave emerging towards F, to the component $\mathbf{e}_\theta$, $\mathbf{e}_\phi$ the combination of which gives the vector polarization

$$\mathbf{q} = \mathbf{e}_\theta \cos\phi - \mathbf{e}_\phi \sin\phi \tag{9.2}$$

$\mathbf{q}$ is recognized as the reference polarization vector denoted $\mathbf{e}_\alpha$ (§7.4.5, eqn. 7.63).
Reciprocally, if a primary feed of phase centre F radiates a wave polarized according to the vector $\mathbf{e}_\alpha$, that is to say a pure polarization (Huygens source, §8.1.2) we obtain, over the aperture π, a wave of constant polarization.
This important property of axially-symmetric systems emphasizes the interest in the concepts of polarization developed in Chapter 7.

In the general case, a primary pattern may be characterized by a vector function of the form

$$\mathbf{f}(\mathbf{u}) = \mathbf{p}(\mathbf{u}) f(\mathbf{u}) = \mathbf{e}_\alpha f_\alpha(\mathbf{u}) + \mathbf{e}_\beta f_\beta(\mathbf{u}) \tag{9.3}$$

where the scalar functions f_α, f_β are the radiated patterns in the co- and cross-polar components, $\mathbf{e}_\alpha$ and $\mathbf{e}_\beta$ respectively.

The primary gain function results from the primary pattern in the following form (equation (7.3))

$$g(\mathbf{u}) = A^2 |f(\mathbf{u})|^2 \tag{9.4a}$$

where

$$A^2 = \frac{4\pi K_1}{\iint_\Omega |f|^2 \, d\Omega} = \frac{4\pi K_1}{\iint_\Omega \left(|f_\alpha|^2 + |f_\beta|^2\right) d\Omega} \tag{9.4b}$$

and K_1 is a coefficient of efficiency due to the various losses in the feed (care has to be taken not to consider K_1 as the gain factor of the feed) and Ω is the solid angle. To separate the polarization component contributions, let us set

$$g(\mathbf{u}) = g_\alpha(\mathbf{u}) + g_\beta(\mathbf{u}) \tag{9.5}$$

$$g_\alpha(\mathbf{u}) = A^2 |f_\alpha(\mathbf{u})|^2 \; ; \quad g_\beta(\mathbf{u}) = A^2 |f_\beta(\mathbf{u})|^2 \tag{9.6}$$

9.3 PRINCIPAL SURFACE

9.3.1 Definition

The concept of a principal surface comes originally from the domain of optical instruments. In the typical case of a system that is stigmatic for the focus F and the point at infinity on the axis, this surface is defined as the locus of the intersection M of the rays from the focus F and the corresponding emerging rays parallel to the axis (Fig. 9.2).

This definition is easily generalized to other cases (astigmatic systems). For non-focal systems, the principal surface is at infinity.

The equation of the meridian of this axisymmetrical surface may be defined by the relation between the angle θ of the rays from the point F with the axis, and the distance to the axis r, of the emerging rays.

The radius vector may be used as well: $\rho = \mathbf{FM} = \mathbf{r}/\sin\theta$. The concept of a principal surface is important since its form has an effect upon the illumination function of the equivalent aperture of the system, and on its aberrations (§9.7.2).

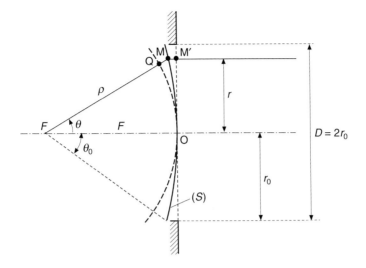

Fig. 9.2 Principal surface.

Example: axisymmetrical paraboloid

The principal surface is that of the reflector itself. Its equation is

$$r = 2F \tan \frac{\theta}{2} \tag{9.7}$$

where F is the focal length.

9.3.2 Pupil - aperture angle - focal length

Let r_0 be the distance to the axis of the emerging ray which is the furthest from the axis. This defines the diameter $D = 2r_0$ of the equivalent aperture or by analogy with optics, the system *pupil*. The corresponding angle $2\theta_0$ is the *angular aperture*. We shall also make use of the associated solid angle

$$\Omega_0 = 2\pi \left(1 - \cos \theta_0 \right) \tag{9.8}$$

The focal length is defined by analogy with that of the paraboloid

$$F = \lim_{\theta \to 0} \left(\frac{r}{\theta} \right) \tag{9.9}$$

9.3.3 Equivalent aperture of the system

Consider an axially-symmetric system illuminated by a primary feed. The plane π is normal to the axis at point O. The emergent parallel rays produce a zone of diameter D where there exists at each point M′, a certain electromagnetic field distribution which is, as we have already indicated, locally plane

$$\mathbf{E}(M') = E_x(M')\mathbf{e}_x + E_y(M')\mathbf{e}_y \tag{9.10}$$

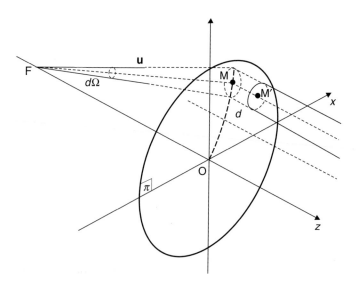

Fig. 9.3 Equivalent radiating aperture of the system.

This is the aperture illumination. It is obtained from the primary pattern $\mathbf{f}(\mathbf{u})$ (related to the point F) and by the principal surface. An elementary cone of rays from F, confined around the direction $\mathbf{u}$, of solid angle $d\Omega$, generates a cylinder of emerging rays surrounding the point M′ and cutting out an elementary area dS on the plane π. The power density dW/dS at point P is related to the radiated primary power density per steradian $dW/d\Omega$ by the relations giving the Poynting vector flux over the plane π

$$\frac{|E|^2}{2\eta} = \frac{dW}{dS} = \frac{dW}{d\Omega}\frac{d\Omega}{dS} \tag{9.11}$$

where $\eta = 120\pi$ ohms.

If dW designates the primary feed power, we know from the definition of gain in the absence of losses, or directivity (§3.3.2) that

$$dW/d\Omega = Wg(\mathbf{u})/4\pi$$

Then

$$|E(M')|^2 = 60Wg(\mathbf{u})\frac{d\Omega}{dS} \qquad (9.12)$$

Using spherical coordinates, we can write

$$|E(M')|^2 = 60Wg(\mathbf{u})\frac{\sin\theta}{r}\frac{d\theta}{dr} \qquad (9.13)$$

Vector expression for the illumination

Taking into account the properties of polarization propagation stated in §9.2, we may express the illumination vector field over the aperture in the following form:

$$E_x(M') = \sqrt{60W}\,Af_\alpha(\mathbf{u})\sqrt{\frac{d\Omega}{dS}}\,; \quad E_y(M') = \sqrt{60W}\,Af_\beta(\mathbf{u})\sqrt{\frac{d\Omega}{dS}} \quad (9.14)$$

where A is given by equation (9.4b). These expressions express the simultaneous influence of the primary feed and the principal surface of the system on the illumination function.

9.4 TRANSFER FUNCTION

To illustrate the contribution of the system itself to the illumination function, it is written in the following form (for a polarization component):

$$E_x(M') = E_0 Af_\alpha(\mathbf{u})T(\mathbf{u}) \qquad (9.15)$$

where $E_0 = \sqrt{60W}/F$.

$T(\mathbf{u})$ is the transfer function defined by

$$T = F\sqrt{\frac{d\Omega}{dS}} = F\sqrt{\frac{\sin\theta\,d\theta}{r\,dr}} \qquad (9.16)$$

This is a real function. Nevertheless, an aberration or a stigmatism defect (a non-plane emerging wave) may produce a complex transfer function (equation (9.5)).

T depends only on θ or on r. Sometimes this function will be denoted $T(\theta)$ or $T(r)$, when no ambiguity is possible. We verify that by definition (equation (9.9)) of the focal length, T is equal to unity along the axis $T(0) = 1$.

E_0 is interpreted as the field at the centre of aperture with a primary feed whose gain is unity and without cross polarization. Note that the transfer function defines the principal surface and *vice versa*.

Examples:

(1) *Paraboloid transfer function*: the equation of its surface is given by equation (9.7) from which we find (§7.4.2. eqn. 7.21)

$$T(r) = \cfrac{1}{\left[1 + \left(\cfrac{r}{2F}\right)^2\right]}$$ (9.17)

(2) *Principal surface of a system whose transfer function is constant*: the meridian equation is as follows:

$$r = 2F \sin\frac{\theta}{2}$$ (9.18)

(3) *Transfer function of a system which fulfils Abbe's aplanetism condition* (§9.7.2): the principal surface is a sphere centred at F whose equation is

$$r = F\sin\theta$$ (9.19)

We find

$$T(r) = \frac{1}{\sqrt{\cos\theta}} = \left[1 - \left(\frac{r}{2F}\right)^2\right]^{-\frac{1}{4}} \approx 1$$ (9.20)

The transfer function is almost constant if the focal distance is not too small.

(4) *Transfer function leading to uniform illumination function*: this condition leads to a gain factor maximum (superdirectivity effects not included (§13.4)). It will be used to optimize the gain factor of a Cassegrain System (§10.1.5). Assuming that the pattern is axisymmetrical and with zero cross polarization, the field on the aperture must be constant. Taking into account equation (9.15), we have

$$T(\theta) = \frac{f_\alpha(0)}{f_\alpha(\theta)}$$ (9.21)

9.5 SYSTEM GAIN

9.5.1 General expression

In order to calculate the axial gain of the system ($\theta = 0$), we start from the general expression for the aperture gain, with an illumination which is locally plane (§7.4.8, equation (7.74))

$$G_0 = 4\pi \frac{\left| \iint_\Sigma \mathbf{E}(\mathbf{v}) d\Sigma \right|^2}{\iint_\Sigma \left| \mathbf{E}(\mathbf{v}) \right|^2 d\Sigma} \qquad (9.22)$$

with

$$\mathbf{v} = \frac{\mathbf{OM'}}{\lambda}; \qquad d\Sigma = \frac{ds}{\lambda^2}$$

Let us separate the contributions of the co-polar and cross-polar components by setting

$$G_0 = G_{0\alpha} + G_{0\beta} \qquad (9.23)$$

where

$$G_{0\alpha} = 4\pi \frac{\left| \iint_\Sigma E_x(\mathbf{v}) d\Sigma \right|^2}{\iint_\Sigma \left(\left| E_x(\mathbf{v}) \right|^2 + \left| E_x(\mathbf{v}) \right|^2 \right) d\Sigma} \qquad (9.24)$$

with an analogous expression for $G_{0\beta}$.

Often, by symmetry (§8.1.3), the cross polarization component $E'_y(x,y)$ is an odd function of x and y and therefore $G_{0\beta} = 0$, i.e. the axial cross polarization is zero. Then we have

$$G_0 = G_{0\alpha} \qquad (9.25)$$

9.5.2 Expression obtained from the primary gain g' and the transfer function

In equation (9.24) the denominator represents the radiated primary power, apart from a constant coefficient

$$W = \frac{1}{2\eta} \iint_S |\mathbf{E}|^2 \, dS \tag{9.26}$$

Using equation (9.15), we obtain for one polarization component

$$G_{0\alpha} = \frac{1}{\lambda^2 F^2} \left| \iint_S A f_\alpha(\mathbf{u}) T(\mathbf{u}) \, dS \right|^2 \tag{9.27}$$

If $f_\alpha(\mathbf{u})$ is real (equiphase primary pattern) we may write, using equation (9.4a)

$$G_{0\alpha} = \frac{1}{\lambda^2 F^2} \left[\iint_S \sqrt{g'_\alpha(\mathbf{u})} T(\mathbf{u}) \, dS \right]^2 \tag{9.28}$$

9.5.3 Effect of various factors in the gain function

The gain function (9.27), using equation (9.4b) may be written as follows:

$$G_{0\alpha} = \frac{4\pi K_1}{\lambda^2 F^2} \frac{\left| \iint_S f_\alpha T \, dS \right|^2}{\iint_\Omega \left(|f_\alpha|^2 + |f_\beta|^2 \right) d\Omega} \tag{9.29}$$

Let us modify the form of this equation so as to emphasize the role of various factors of the gain function. Taking into account the expression of the transfer function, we can write

$$G_{0\alpha} = \frac{4\pi S}{\lambda^2} K_1 K_2 K_3 K_4 \tag{9.30}$$

In this expression

- K_1 is the ohmic loss coefficient

- K_2 is the cross polarization loss coefficient

$$K_2 = \frac{\iint_{4\pi} |f_\alpha|^2 \, d\Omega}{\iint_{4\pi} \left(|f_\alpha|^2 + |f_\beta|^2 \right) d\Omega} \tag{9.31}$$

$K_1 = 1$ for $f_\beta \equiv 0$, i.e. when the feed produces a pure polarization.

- K_3 is the spillover loss coefficient

This is related to wide-angle radiation from the feed, which is not intercepted by the system pupil. Diffuse primary radiation also affects the noise temperature of the antenna (§7.2.2: low-noise antennas, and §10.2.1: Cassegrain antennas).

$$K_3 = \frac{\iint_\Omega |f_\alpha|^2 \, d\Omega}{\iint_{4\pi} |f_\alpha|^2 \, d\Omega}$$

(9.32)

- K_4 is the illumination coefficient

$$K_4 = \frac{\left| \iint_S f_\alpha T d\Omega \right|^2}{\iint_S |f_\alpha T|^2 \, d\Omega}$$

(9.33)

Schwartz's inequality shows that this is maximum, and equal to unity for

$$f_\alpha T = \text{const}$$

(9.34)

This assumes that

(1) the characteristic surface of the primary radiation (co-polar) is rotationally symmetrical; this condition is related to the purity of the polarization. The primary characteristic function may then be written $f_\alpha(\theta)$.

(2) a transfer function T inversely proportional to this function

$$T(\theta) = \frac{f_\alpha(0)}{f_\alpha(\theta)}$$

(9.35)

In particular, the phase variations of the transfer function and of the primary characteristic function must be complementary.

Frequently certain other loss factors have to be taken into account, for instance:

- a factor K_5 related to fabrication and alignment tolerances,

- a factor K_6 related to blockage by structures placed in front of the aperture (primary feed or auxiliary reflector in the case of a Cassegrain - feed supports) (§10.1.7),

- a factor related to the mismatch at the feed terminal (§8.1.2) etc.

9.5.4 Concept of optimal primary directivity

Among the different factors that we have mentioned so far, two should be considered in particular since they vary inversely with the primary directivity:

- K_3: spillover loss coefficient;

- K_4: illumination coefficient.

If the primary pattern is represented by an axially symmetrical surface, by using equation (9.16) we can write

$$K_3 = \frac{\int_0^{\theta_0} |f_\alpha|^2 \sin\theta\, d\theta}{\int_0^\pi |f_\alpha|^2 \sin\theta\, d\theta}, \quad K_4 = \frac{\left| \int_0^{\theta_0} \dfrac{f_\alpha}{T} \sin\theta\, d\theta \right|^2}{\int_0^{\theta_0} \dfrac{\sin\theta}{|T|^2}\, d\theta \int_0^{\theta_0} |f_\alpha|^2 \sin\theta\, d\theta} \tag{9.36}$$

The maximum of their product allows us to define an optimal primary directivity. To illustrate this property, consider a constant transfer function (case of a paraboloid or a Cassegrain, with a long focal distance or a profile chosen to meet this criterion)

$$T = 1, \quad r = 2F \sin\frac{\theta}{2} \tag{9.37}$$

Consider a primary pattern class represented by the characteristic function

$$f_\alpha = \cos^{2n}\frac{\theta}{2} = \left(1 + \tan^2\frac{\theta}{2}\right)^{-n} \tag{9.38}$$

This form of pattern represents reasonably well the main lobe of primary feeds of good quality (corrugated horns, for instance), but it should be often corrected in order to take into account the first and subsequent sidelobes. The directivity parameter is the exponent n. Let us set

$$x = \log\left(1 + \tan^2\frac{\theta}{2}\right) \tag{9.39}$$

The characteristic function takes the form

$$f_\alpha = \exp(-nx) \tag{9.40}$$

For small values of θ, we have approximately

$$f_\alpha = \exp\left(n \tan^2 \frac{\theta}{2}\right) \tag{9.41}$$

The pattern has a Gaussian form.

The illumination taper at the edge, in dB, is given by

$$N = 20 \log f_\alpha (\theta_0) = -8.66 \, n x_0 \tag{9.42}$$

We find that the coefficients, K_3 and K_4, schematically represented in Fig. 9.3, vary in opposite senses as a function of n. The expression for their product shows clearly the influence of the primary directivity

$$K_3 K_4 = \frac{2n+1}{(n+1)^2} \frac{\left\{1 - \exp\left[-(n+1)x_0\right]\right\}^2}{1 - \exp(-x_0)} \tag{9.43}$$

This function is represented in Fig. 9.4. Its maximum is obtained for

$$(n+1)x_0 \approx 1.27 \tag{9.44}$$

This represents, if x_0 is small, an *optimal illumination taper* of the primary pattern close to

$$N_0 = -11 \text{ dB} \tag{9.45}$$

This a classical rule[1]. We often choose a lower illumination taper (for instance -12 dB or -13 dB): at the cost of a small loss in directivity. The level of the sidelobes is significantly reduced (§9.6.2).

---

[1] Silver, S. (ed), *Microwave Antenna Theory and Design*, MIT Radiation Lab. Series, Vol. 12, McGraw-Hill, New York, 1949; reprinted by Peter Peregrinus, Stevenage, 1984.

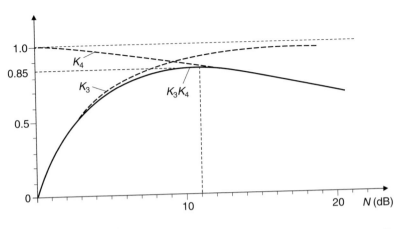

Fig. 9.4 Effect of the amplitude illumination taper (N dB) on the spillover coefficient K_3, the illumination coefficient K_4, and their product.

9.6 RADIATION PATTERNS

9.6.1 Equivalent aperture illumination

This illumination is defined in the plane π normal to the axis at point O (§9.3.3). Since it is locally plane wave, we have defined it from the primary pattern and the transfer function in terms of the Cartesian coordinates of the electric field (equation (9.14))

$$E_x(M') = E_0 A f_\alpha(\mathbf{u}) T$$

$$(9.46)$$

$$E_y(M') = E_0 A f_\beta(\mathbf{u}) T$$

The patterns radiated in the direction of the unit vector $\mathbf{u}'$ (spherical coordinates: ϕ', θ') can be obtained from the results given in Chapter 8 by Fourier transformation and are expressed in terms of the polarization components $\mathbf{e}_{\alpha'}, \mathbf{e}_{\beta'}$, in the forms (8.12), (8.13), (8.14).

9.6.2 Axisymmetric primary pattern with pure polarization

We have $f_\beta = 0$. The illumination may be written in the form

$$E_x(r) = E_0 A f_\alpha(\theta) T(r)$$

$$E_y(r) = 0 \tag{9.47}$$

The pattern at infinity can then be written in the form (8.28) with $B = 0$, where the 'obliquity factor' $(1+\cos\theta)/2$ has been omitted.

$$F_\alpha(\theta) = \frac{2j\pi}{\lambda^2} \int_0^D E_x(r) J_0\left(\frac{2\pi r}{\lambda}\sin\theta\right) r \, dr \tag{9.48}$$

$$= j\frac{\pi}{2}\frac{D^2}{\lambda^2} \int_0^1 E_x\left(\frac{D}{2}u\right) J_0(uv) u \, du \tag{9.49}$$

with

$$u = \frac{2r}{D}, \qquad v = \frac{\pi D}{\lambda}\sin\theta \tag{9.50}$$

The integral obtained is a finite Hankel transform. It can be calculated with the use of series expansion processes indicated in relation with diffraction patterns (§9.8.6).

This is a particularly suitable technique, to expand the electrical field E_x in a series as a function of u^2 in the vicinity of the edge of the pupil in the form

$$E_x = \sum_0^\infty a_n \left(1-u^2\right)^n \tag{9.51}$$

The integral (9.49) may then be expressed in the form

$$F_\alpha(\theta) = j\frac{\pi}{2}\frac{D^2}{\lambda^2} \sum_0^\infty \frac{a_n}{n+1} \Lambda_{n+1}(v) \tag{9.52}$$

The function Λ, often used in optics, is tabulated in standard tables of functions[2]. The form of the transfer function influences the illumination and the radiated pattern as shown from the following examples

---

[2] Abramowitz, M. and Stegun, I.A., *Handbook of Mathematical Functions*, Dover, 1965.

Uniform illumination

$$E_x = a_0$$

The pattern is given to within a multiplicative constant by the function

$$\Lambda_1(v) = \frac{2J_1(v)}{v} \tag{9.53}$$

This function is represented in Fig. 9.5. We observe the following characteristics

3 dB beamwidth: $\qquad \theta_{3\,dB} = 1.03\ \lambda/D$ (radians) $\tag{9.54}$

level of first sidelobe: $\qquad N_{dB} = -17.6\ dB$ $\tag{9.55}$

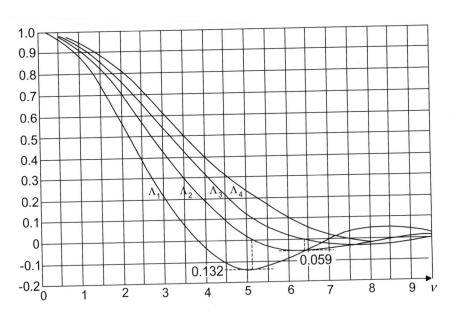

Fig. 9.5 Λ functions; characteristics of a circular aperture.

Tapered illumination

This type of illumination is obtained, for instance, taking $a_0 = 0$, $a_1 = 1$. The pattern then takes the form $\Lambda_2(v)$ represented in Fig. 9.5. We observe a significant lowering of the first sidelobe (−24.6 dB) and a small increase in the −3 dB beamwidth $\theta'_{3dB} = 1.27\,\lambda/D$ (radians).

The pattern obtained is a linear combination of the functions Λ_1 and Λ_2.

9.6.3 Effect of blockage

In many antennas, the aperture centre is blocked by the presence of an obstacle ('mask'): the primary feed in the case of a paraboloid, the auxiliary reflector in the case of a Cassegrain. It is easy to calculate the pattern by modifying the limits of the integral (equation (9.49)).

If D' is the diameter of the mask we have

$$F_\alpha = j\frac{\pi D^2}{2\lambda^2}\left[\int_0^1 E_x J_0\left(u\right)du - \int_0^{\frac{D'}{D}} E_x J_0\left(u\right)du\right]\tag{9.56}$$

Thus, the pattern appears as the difference between an unperturbed pattern and a parasitic pattern produced by an aperture D'/D times smaller.

Example:
Case of uniform illumination: This case leads to a pattern of the form

$$\Lambda_1\left(v\right) - \left(\frac{D'}{D}\right)^2 \Lambda_1\left(\frac{D'}{D}v\right)\tag{9.57}$$

Thus, superimposed on the unperturbed pattern is a pattern with the same form, but multiplied by a coefficient $\left(D'/D\right)^2$ and dilated by the ratio D/D'.

Fig. 9.6 shows the form of the final pattern. We observe a decrease in gain, a slight reduction in the main lobe, an increase of the first sidelobe and a lowering of the second.

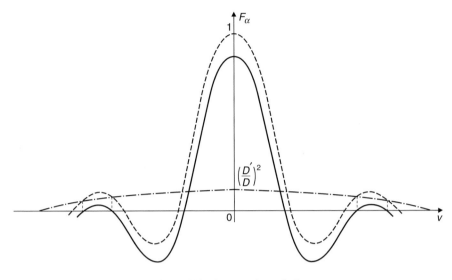

Fig. 9.6 Effect of blockage on the radiation pattern.

9.7 ABERRATIONS IN AXIALLY-SYMMETRIC SYSTEMS

9.7.1 Introduction

An axially-symmetric system is designed to operate with a primary feed whose phase centre coincides with the focus. It is interesting to look for a wider extension of the possibilities of the system trying to use at least a part of the focal plane around the focus, or by using several feeds as a kind of 'retina' (as in the case of 3-D radars), or by using a mechanical system to move a single feed (conical-scanning antennas, §10.2).

   If the feed undergoes a transverse displacement d, the main effect observed is a deviation of the radiated pattern by an angle close to $\alpha = d/F$. This effect, due to the tilt of the emerging wavefront, is accompanied by various distortions, known in optics as *aberrations*. These aberrations are quantified, as we shall see, by performing a limited expansion of the wavefront equation in terms of phase errors of quadratic, cubic and higher orders. The study of these aberrations is important since various different causes may produce the same effects: fabrication tolerances, deformations owing to weight, to wind and so on.

9.7.2 Main aberrations in the defocusing plane

Assume that the phase centre of the feed is located at F', at a distance d from the focal axis

$$F'OF = \alpha; \quad F'O = f' \tag{9.58}$$

The emerging wavefront has an equation of the form $Z_\alpha(x,y)$ (x, y are the Cartesian coordinates in the plane of the equivalent radiating aperture). First consider a section of the defocusing plane ($y = 0$). Note the following points.

- $Z_\alpha = 0$ for $\alpha = 0$ and for $x = 0$.

- The focus F is chosen so that the second order term in x^2 is zero. Thus, the only term of second order is αx.

- When the sign of α changes, the wavefront becomes symmetric with the preceding one, therefore

$$Z_{-\alpha}(-x,0) = Z_\alpha(x,0) \tag{9.59}$$

Thus, it is not possible to find terms of the third order such as $\alpha^2 x$ or αx^2.

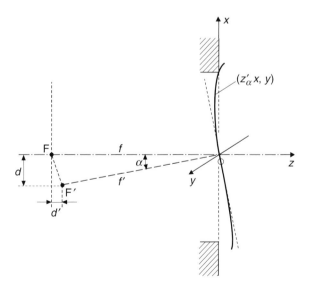

Fig. 9.7 Aberrations in the defocusing plane xOz.

If we limit the expansion to the fourth order, we obtain the following form:

$$Z_\alpha(x,0) = \alpha x + A\alpha^3 x + B\alpha^2 x^2 + C\alpha x^3 + Dx^4 \qquad (9.60)$$

The corresponding phase function over the aperture is

$$\varphi_\alpha(x,0) = \frac{2\pi}{\lambda} Z_\alpha = \varphi_1 + \varphi_2 + \varphi_3 + \varphi_4 \qquad (9.61)$$

Each term corresponds to a particular type of aberration producing a specific effect. These are examined in turn below.

Distortion

$$\varphi_1 = \frac{2\pi}{\lambda}\left(\alpha x + A\alpha^3 x\right) \qquad (9.62)$$

The phase function is linear with respect to x. This produces a deviation of the beam without distortion. The maximum radiation is, however, in the direction θ' which may be different to α

$$\frac{\theta'}{\alpha} = 1 + A\alpha^2 \qquad (9.63)$$

Astigmatism - Tangential focal line   (Fig. 9.8, Fig. 9.9)

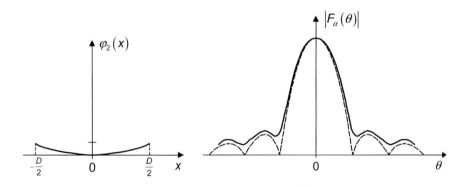

Fig. 9.8 Astigmatism: effect of a quadratic phase error.

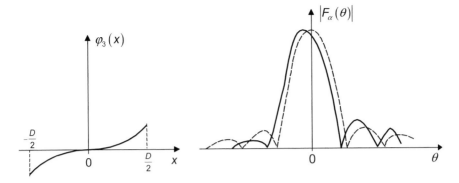

Fig. 9.9 Astigmatism: effect of a cubic phase error.

$$\varphi_2 = \frac{2\pi}{\lambda} B\alpha^2 x^2 \qquad (9.64)$$

The astigmatism does not shift the main lobe (by symmetry) but fills the nulls between the sidelobes without increasing their level. Consequently, the gain is reduced.

In general, to avoid this phenomenon the feed is shifted longitudinally (parallel to the focal axis) by a distance d'. For each defocusing, a position F' exists of coordinates d and d' for which the astigmatism is zero. The locus of the corresponding points is a curve tangential to the focal plane, termed the 'tangential focal line'.

Coma - aplanatism - Abbe condition (Fig. 9.10)

$$\varphi_3 = \frac{2\pi}{\lambda} C\alpha x^3 \qquad (9.65)$$

Even on the tangential focal line, a transverse defocusing produces, in general, a cubic phase error whose amplitude is proportional to the defocusing angle α

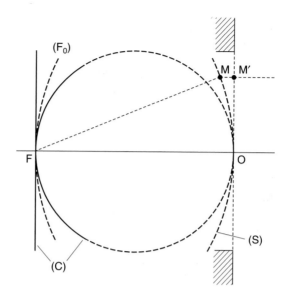

Fig. 9.10 Abbe condition: the principal surface (S) is a sphere centred on the focus F. (C) focal lines; (F$_0$) lines of best focusing.

This phenomenon appears even for small defocusing. This results in an asymmetry of the secondary pattern which produces a deviation of the direction of maximum radiation and an important increase of the level of the sidelobe towards the system axis. To prevent this phenomenon it can be shown that the system must satisfy the Abbe condition: the principal surface of the system must be a sphere of radius f centred on the focus. The system is then said to be *aplanatic*.

The transverse shift d of the feed introduces in the primary radiation pattern a phase function $\varphi = Kd\sin\theta$, where θ is the angle OFM. This phase function is linear with respect to x over the aperture if we have a relation of the form $x = F_0\sin\theta$ which expresses the Abbe condition. This condition is not satisfied by many antennas and in particular by parabolic reflectors (the principal surface is the paraboloid itself).

On the other hand, systems with not one, but at least two reflection or refraction surfaces, allow Abbe's condition of aplanatism to be added to the usual conditions for strict stigmatism at the focus. This is the case with certain dioptric systems, and also with Cassegrain-type systems with two reflectors (known as Schwartzschild optics, §10.1).

Spherical aberration

This is the only aberration of the fourth order ($\varphi_4 = Dx^4$), which does not depend on α. Thus, *a priori*, it may exist even for $\alpha = 0$, that is to say when the feed is at the focus. Nevertheless, if the system is stigmatic at the focus, the aberration is zero. In fact, the aberration may appear in certain non perfectly stigmatic systems such as the spherical reflector, which possesses other properties. The observed effect is of the same type as that of astigmatism, but it is often small.

Aberration in the plane perpendicular to the defocusing plane

In this case, $x = 0$. Consider the section $Z_\alpha(0, y)$ of the surface of the emerging wave. By symmetry, only even order aberrations are present: astigmatism and spherical aberration

$$Z_\alpha(0, y) = B'\alpha^2 y^2 + D'y^4 \tag{9.66}$$

$$\varphi = \varphi'_2 + \varphi'_4 \tag{9.67}$$

It is possible to cancel the astigmatism φ'_2, as was already done in the case of a longitudinal defocusing d''. In effect, we find for each defocusing angle α, an optimal focal line which cancels astigmatism on this plane. This line, known as the 'sagittal focal line' is in general distinct from the tangential focal line (Fig. 9.11). In practice, we choose a compromise which tolerates a certain 'astigmatism' and which locates the phase centre on an intermediary curve, known in optical terminology as the 'locus of circles of least diffusion' (F_0).

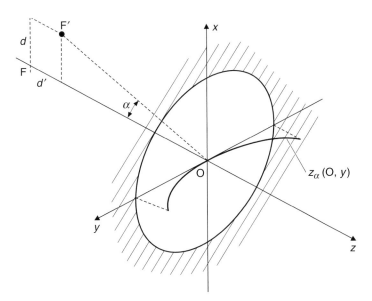

Fig. 9.11 Aberrations in the xOz plane, perpendicular to the defocusing plane.

9.8 AXIALLY SYMMETRIC SYSTEM CONSIDERED IN
RECEPTION: DIFFRACTION PATTERN

The determination of the field structure around the focus – the diffraction pattern – can be considered as a particular case of a general problem treated in Chapter 8: the radiation from a radially-symmetric structure. It can be deuced from the general results. But for a better understanding this problem will be treated here as a specific case.

9.8.1 Effect of transfer function

Consider a stigmatic axially-symmetric system such that an incident plane wave of axial direction, gives rise through the system to a spherical emerging wave which converges towards the focus F (Fig. 9.12).

Associated with the Cartesian coordinate system (e_x, e_y) of a plane P normal to the axis is the so-called 'polarization' coordinate system (e_α, e_β) on a sphere (S') of centre F and of radius FS $= f$.

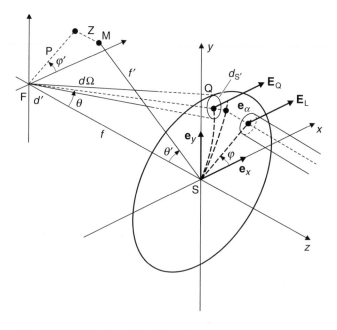

Fig. 9.12 Calculation of the diffraction pattern around the focus.

If the incident plane wave is of constant polarization, the field at each point L of the plane (P) is of the form

$$\mathbf{E}_L = E_0 \mathbf{e}_x \tag{9.68}$$

We saw that due to the conservation of the radial and tangential components of the field, the field converging towards the focus is polarized in the direction of vector $\mathbf{e}_\alpha$.

An elemental tube of rays surrounding L and which cuts on (P) an elemental area dS gives rise to a cone of rays converging to F which cuts on (S′) an area dS' surrounding a point Q. The conservation of energy flux is expressed in the form

$$\frac{dW}{dS'} = \frac{dW}{dS}\frac{dS}{d\Omega}\frac{d\Omega}{dS'} \tag{9.69}$$

or

$$\frac{|E_Q|^2}{2\eta} = \frac{|E_L|^2}{2\eta}\frac{F^2}{|T(r)|^2}\frac{1}{F^2} \tag{9.70}$$

Therefore

$$\mathbf{E}_Q = \frac{E_0}{T(r)}\mathbf{e}_\alpha \tag{9.71}$$

Thus the field on the emerging sphere is deduced from the field on the incident plane wave by *dividing* it by the transfer function $T(r)$.

9.8.2 Diffraction in the vicinity of the focus F of an element dS' of a spherical wave S'

The elemental field at M in the vicinity of F is deduced from the field $\mathbf{E}_Q$ by the general expressions of an elemental spherical wave from Q (equation (7.5))

$$d_2\mathbf{E}_M = \mathbf{e}_\alpha \frac{\exp\left(-j2\pi\dfrac{MQ}{\lambda}\right)}{\dfrac{MQ}{\lambda}}\frac{E_0}{T(r)}\frac{dS'}{\lambda^2} \tag{9.72}$$

This expression can be transformed. The denominator may be simplified using the approximation

$$MQ \cong FQ = f \tag{9.73}$$

In the exponential term, a better approximation to MQ should be made since a small variation of the parameter can produce a large variation of phase. We have **MQ** = **MS** + **SQ**. It can easily be verified that

$$|\mathbf{MQ}|^2 = |\mathbf{MS}|^2 - 2\mathbf{FM.SQ} \tag{9.74}$$

Let us set MS = f'; if MF/f is small

$$MQ \approx f + \frac{f'^2 - f^2}{2f} - \frac{\mathbf{FM.SQ}}{f} \tag{9.75}$$

The elemental field at M is of the form

$$d_2\mathbf{E}_M = \mathbf{e}_\alpha \frac{\exp(-jkf)}{f} \frac{E_0}{T(r)} \exp\left(-jk\frac{f'^2 - f^2}{2f}\right) \exp\left(2\pi j\frac{\mathbf{FM.SQ}}{f\lambda}\right) \frac{dS'}{\lambda^2} \tag{9.76}$$

If λ is taken as unity, we can set

$$\frac{f}{\lambda} = \phi, \qquad \frac{\mathbf{FM}}{f} = \boldsymbol{\tau}, \qquad \frac{\mathbf{SQ}}{\lambda} = \mathbf{v}, \qquad \frac{dS'}{\lambda^2} = d\Sigma', \qquad \frac{\mathbf{FQ}}{f} = \mathbf{e}_\alpha$$

and substituting in equation (9.76) we find the total field at M to be

$$\mathbf{E}_M(\boldsymbol{\tau}) = \frac{\exp(-j2\pi\phi)}{\phi} \left[\exp\left(-j2\pi\frac{\phi'^2 - \phi^2}{2\phi}\right)\right] \iint_\Sigma \mathbf{e}_\alpha(\mathbf{u}) \frac{E_0}{T(r)} \exp(j2\pi\boldsymbol{\tau}.\mathbf{v}) d\Sigma' \tag{9.77}$$

Remarks:

(1) The integral has the same form as the Fourier transform of a function proportional to the inverse of the transfer function. The vector $\mathbf{v}$ plays the role of a 'spatial frequency' (§13.2.1).

(2) The exponential term in brackets characterizes a phase which depends on ϕ', and hence on the position of the observation point M. This phase is constant if $\phi' = \phi$, that is to say *if the point* M *is on a sphere of centre* S *tangential to the focal plane at* F; this is the focal sphere of the system. In this case $|\boldsymbol{\tau}| \approx \theta'$, which is the angle that the direction SM makes with the axis.

(3) Case for which the focal distance becomes infinite: the field $\mathbf{E}_M$ represents the field at infinity diffracted by a planar equiphase aperture: this shows again the role of the Fourier transform.

9.8.3 Analysis of a diffraction pattern - contribution of an elementary crown of the spherical wave - hybrid waves

Let (ρ, φ', z) and (f', r, θ') be the cylindrical and spherical coordinates respectively of point M (Fig. 9.12).

(f, φ, θ) are the spherical coordinates of point Q of the spherical wave.

(φ, r) are the polar coordinates of L.

It can be shown that the elemental field at M, given by equation (9.72), may be written as the product of various factors, each of them having specific characteristics

$$d_2\mathbf{E}_M = \mathbf{e}_\alpha \exp\left[j2\pi \frac{\rho}{\lambda} \sin\theta \cos(\varphi - \varphi') d\varphi \right] dE'(\theta) \qquad (9.78)$$

where the scalar differential factor $dE'(\theta)$ is independent of φ

$$dE'(\theta) = h(\rho, z) \exp\left[-j2\pi \frac{z}{\lambda}(1 - \cos\theta) \right] \frac{E_0}{T(r)} \sin\theta \, d\theta \qquad (9.79)$$

$h(\rho, z)$ only depends on the position of the point M

$$h(\rho, z) = f \exp\left(-j\frac{2\pi f}{\lambda} \right) \exp\left[-j\frac{2\pi}{\lambda} \frac{\rho^2 + (f - z)^2 - f^2}{2f} \right] \qquad (9.80)$$

The elemental field created by a crown of spherical wave contained in a cone $(\theta, \theta + d\theta)$ is given by the integral

$$d\mathbf{E}_M = dE'(\theta) \int_0^{2\pi} \mathbf{e}_\alpha(\theta, \varphi) \exp\left[j2\pi \frac{\rho}{\lambda} \sin\theta \cos(\varphi - \varphi') \right] d\varphi \qquad (9.81)$$

We recognize the expression of a conical spectrum of plane waves, polarized as $\mathbf{e}_\alpha$, of incidence θ (§8.3.2). *Thus, the field created by this crown is a hybrid wave*[3].

9.8.4 Axial field

On the axis, $\rho = 0$, then taking into account equation (7.63) for $\mathbf{e}_\alpha$ (§7.4.6) and with

$$k = \frac{2\pi}{\lambda} \tag{9.82}$$

we find by integration of equation (9.81), within the factor of phase $\exp\left(-j2\pi\dfrac{\rho}{\lambda}\right)$, that the field is locally plane and the axial electric field is given by

$$\frac{E_x}{E_0} = 2\pi \frac{f}{\lambda} \exp\left(-jk\frac{z^2}{2f}\right) \exp\left(jkz\right) \int_0^{\theta_0} \frac{1}{T(r)} \exp\left[-jkz\left(1 - \cos\theta\right)\right] \frac{1 + \cos\theta}{2} \sin\theta \, d\theta \tag{9.83}$$

Example: paraboloid

The preceding integral is evaluated and we obtain (the phase term in z^2 is ignored)

$$\frac{E_x}{E_0} = 2\pi \frac{f}{z_0} \exp\left(jk'z\right) \frac{\sin\left(\pi \dfrac{z}{z_0}\right)}{\pi \dfrac{z}{z_0}} \tag{9.84}$$

with

$$z_0 = \frac{\lambda}{1 - \cos\theta_0}, \quad k' = k\frac{1 + \cos\theta_0}{2} \tag{9.85}$$

Axial amplitude distribution

The function is of the form

$$\frac{\sin\left(\pi z/z_0\right)}{\pi z/z_0}$$

The parameter z_0 represents the axial half-width of the diffraction pattern between two consecutive nulls. We verify that this axial depth z_0, which in optical

---

[3] Minett, H., Mac, B. and Thomas, A., 'Fields in the image space of symmetrical focus reflectors', *Proc. IEE*, Vol. 15, No. 10, October 1968.

terminology is called 'depth of field', increases when the aperture (D/f) of the instrument decreases.

$$\frac{2z_0}{\lambda} \approx \left(\frac{2f}{D}\right)^2 \tag{9.86}$$

It is obvious that the tolerance of longitudinal positioning of a primary feed increases with the 'depth of focus' z_0. An error z produces a loss of gain given by equation (9.84).

Maximum field at the focus

The same equation (9.84) gives the field at the focus

$$\frac{E_x}{E_0} = \frac{2\pi f}{\lambda}(1 - \cos\theta_0) = \frac{\pi D}{\lambda}\frac{D}{4f} \tag{9.87}$$

9.8.5 Transverse distribution of the diffracted field in the focal plane

Starting from equations (9.79) and (9.80) with $z = 0$, and neglecting a constant phase of $\exp(-j2\pi\varphi)$, we obtain

$$\frac{dE'_\theta}{E_0} = \frac{f}{\lambda}\exp\left(-jk\frac{\rho^2}{2f}\right)\frac{1}{T(r)}\sin\theta \, d\theta \tag{9.88}$$

Integrating equation (9.81) with respect to θ yields

$$\frac{\mathbf{E}(M)}{E_0} = \mathbf{F}(M)kf\exp\left(-jk\frac{\rho^2}{2f}\right); \quad \frac{\mathbf{H}^+(M)}{E_0} = \mathbf{G}(M)kf\exp\left(-jk\frac{\rho^2}{2f}\right) \tag{9.89}$$

The Cartesian components of the vector functions $\mathbf{F}(M)$ and $\mathbf{G}(M)$ are given by

$$\left.\begin{array}{ll} F_x = A(\rho) + B(\rho)\cos(2\varphi) & G_x = B(\rho)\sin(2\varphi) \\ F_y = B(\rho)\sin(2\varphi) & G_y = A(\rho) - B(\rho)\cos(2\varphi) \\ F_z = -jC(\rho)\cos\varphi & G_z = -jC(\rho)\sin\varphi \end{array}\right\} \tag{9.90}$$

$$A(\rho) = \int_0^{\theta_0} \frac{1+\cos\theta}{2} \frac{1}{T} J_0(k\rho\sin\theta)\sin\theta\,d\theta$$

$$B(\rho) = \int_0^{\theta_0} \frac{1-\cos\theta}{2} \frac{1}{T} J_2(k\rho\sin\theta)\sin\theta\,d\theta \qquad (9.91)$$

$$C(\rho) = \int_0^{\theta_0} 2\tan\frac{\theta}{2} \frac{1+\cos\theta}{2} \frac{1}{T} J_1(k\rho\sin\theta)\sin\theta\,d\theta$$

General properties

- Equations (9.89) show that on the focal plane, the phase of the diffraction pattern follows a quadratic function of radius. This is to be related to the remark made in §9.8.2 concerning the focal sphere of the system,

- *In the focal plane, the distribution of the electromagnetic field (equations (9.89) and (9.90)) is that of a balanced hybrid wave.* It has the symmetrical properties already mentioned (§8.1.3): magnetic field lines are deduced from electric field lines by a rotation of $\pi/2$ about the axis. Let us now study in greater detail the character of these fields.

9.8.6 Axially-symmetric systems with a small aperture θ_0

In this case, the functions B and C (equation (9.91)) are negligible compared with the function A. The field is locally plane. In addition, we can make the following approximation:

$$\sin\theta = \frac{r}{f}, \qquad \frac{1+\cos\theta}{2} \approx 1 \qquad (9.92)$$

Setting

$$u = \frac{2r}{D}, \qquad v = \frac{\pi D\rho}{\lambda f} \qquad (9.93)$$

The field is expressed in the form of *a finite Hankel transform*

$$F_x = \frac{D^2}{4f^2} \int_0^1 \frac{1}{T} J_0(uv)u\,du = G_y$$

$$\qquad (9.94)$$

$$F_y = G_x = 0$$

To calculate this integral, several types of series expansion of the function $1/T$ may be used. The most frequent consists of expanding $1/T$ in a power series with respect to u^2 in the vicinity of the edge of the pupil $(u \cong 1)$

$$\frac{1}{T} = \sum_0^\infty a_n \left(1 - u^2\right)^n \tag{9.95}$$

a_n is characterized by the successive derivatives of $1/T$ at the edge of the pupil. This gives

$$F_x(\rho) = \frac{D^2}{4f^2} \sum_0^\infty \frac{a_n}{2(n+1)} \Lambda_{n+1}(v) \tag{9.96}$$

The function Λ, which has already been shown for a circular aperture, is defined by

$$\Lambda_m = 2^m m! \frac{J_m(v)}{v^m} \tag{9.97}$$

The function is represented in Fig. 9.5 for $m = 1, 2, 3, 4$. Often two terms are sufficient, and if the transfer function is approximately constant, a single term is enough and gives simple and standard results which at least have a qualitative value for more complex cases.

For these reasons we shall analyse this special case in more detail below.

9.8.7 Constant transfer function

We find

$$\frac{E(\rho)}{E_0} = \frac{\pi D^2}{4\lambda f} \exp\left(-jk\frac{\rho^2}{2f}\right) \Lambda_1(v) \tag{9.98}$$

with

$$\Lambda_1 = \frac{2J_1(v)}{v} \tag{9.99}$$

Properties of the diffraction pattern

• At the focus, we find the field already calculated above (equation (9.87)),

• The transverse distribution function of the field is as follows:

$$\psi(v) = \frac{E(\rho)}{E(0)} = \Lambda_1(v) \tag{9.100}$$

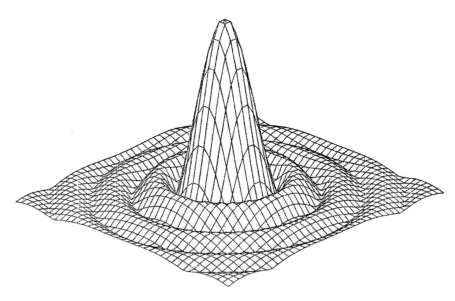

Fig. 9.13 Airy diffraction pattern.

This function is represented by the classical 'Airy diffraction function pattern' well known in optics. It comprises a main lobe surrounded by a set of ring-shaped sidelobes ('bright rings') separated by circles for which the field is zero ('dark rings'), a perspective view is given in Fig. 9.13. The level of the first ring corresponds to −17.6 dB with respect to the axis level.

- The diameter of the main spot is given by the first null of the Bessel function. We obtain the classical result

$$\delta = \frac{1.22\lambda}{\sin\theta_0} \approx \frac{2.44\lambda f}{D} \qquad (9.101)$$

This diameter is in general of the same order of magnitude as the wavelength. It is inversely proportional to the aperture of the system. Knowing the value of δ, we can easily calculate the effective dimension of the aperture of the primary feed matched to the diffraction pattern of the system (§9.3.2).

Distribution of the power flow over the diffraction pattern

It is interesting to know the percentage of the total power contained in the main lobe and, more generally, in a circle of arbitrary radius a. This is given by

$$\eta(a) = \frac{\int_0^a |\psi|^2 \rho \, d\rho}{\int_0^\infty |\psi|^2 \rho \, d\rho} = 1 - J_0^2 \left(\frac{2\pi a}{\lambda} \sin \theta_0 \right) - J_1^2 \left(\frac{2\pi a}{\lambda} \sin \theta_0 \right) \quad (9.102)$$

If we consider only the successive dark rings ($J_1 = 0$), we obtain the following results (Fig. 9.14):

No. of dark ring	% of enclosed power
first	83.8
second	91
third	94

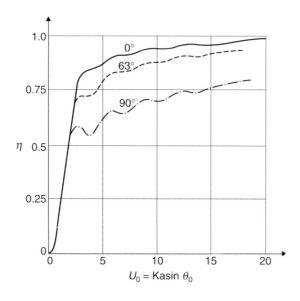

Fig. 9.14 Power flux distribution in the diffraction pattern.

9.8.8 Non-constant transfer function

Decreasing transfer function (tapered)

This is a frequent case (example: paraboloid). If we consider the pattern at infinity, it has already been seen (§9.6.2) that tapering of the transfer function allows the sidelobe level to be reduced. On the other hand, this raises the level of the bright rings of the diffraction pattern and decreases the percentage of energy contained in the main lobe. For example, if we assume a function of the form

$$ T = \frac{1}{1+u^2} \qquad \frac{1}{T} = 2 - \left(1 - u^2\right) \tag{9.103} $$

The diffraction pattern is then defined by the function

$$ \psi(v) = \Lambda_1(v) - \frac{\Lambda_2(v)}{4} \tag{9.104} $$

whose plot shows an significant increase of the level of the first bright ring.

Increasing transfer function

In this case, the function $1/T$ is decreasing. The opposite result to the above is obtained. The level of the first bright ring decreases and the percentage of the energy contained in the central pattern increases. Such a structure is particularly well suited to the use of a primary feed comprising a circular aperture whose illumination amplitude decreases towards its edges - a corrugated horn, for example (§10.1.5).

9.8.9 General case: system with a very large aperture

In the case of a large angular aperture θ_0, simplified expressions are no longer valid and we must use equations (9.90) and (9.91). We reproduce below the results for a paraboloid of aperture $\theta_0 = 63°$ given in the reference below[4].

---

[4] Minett, H., Mac, B. and Thomas, A., 'Fields in the image space of symmetrical focus reflectors', *Proc. IEE*, Vol. 15, No. 10, October 1968.

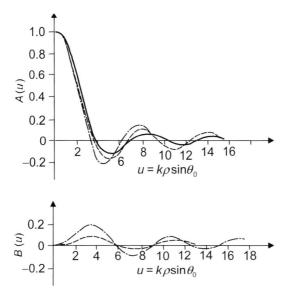

Fig. 9.15 The functions A and B.

The functions A and B are illustrated in Fig. 9.15, and the field lines and level lines in Fig. 9.16. The character of balanced hybrid wave was discussed in §9.8.3. So far as the function B is not negligible, the polarization of the field is not perfectly parallel to that of the incident wave except on the circles where $B = 0$ and over the principal planes $(\varphi' = 0, \varphi' = \pi/2)$.

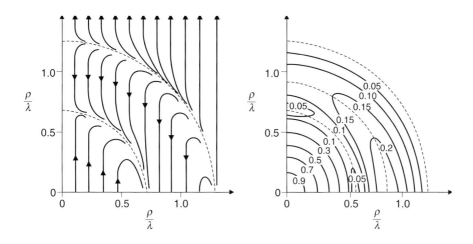

Fig. 9.16 Electric field lines: contours of constant field.

We observe a different amplitude distribution on these two planes. In these planes, the transverse components vary as the functions $A(\rho) - B(\rho)$ and $A(\rho) + B(\rho)$. Figure 9.14 gives the percentage of enclosed power for $\theta_0 = 0$, $63°$ and $90°$.

9.9 SYSTEM CONSIDERED IN RECEPTION: TRANSFER OF THE ENERGY CONTAINED IN THE DIFFRACTION PATTERN TO THE PRIMARY APERTURE

9.9.1 Diffraction pattern produced around the focus by an incident non-axial plane wave

This problem is obviously more complex than that of an axial wave. A simplified solution will be given in the case of an aplanatic axially-symmetric system of small angular aperture. Consider a plane wave originating from the direction of the unit vector **u**, different but close to the direction of the unit normal vector **n** (Fig. 9.17). Let us set

$$\mathbf{u} = \mathbf{n} - \boldsymbol{\tau}_1 \qquad\qquad (9.105)$$

where $\boldsymbol{\tau}_1$ is a 'small' vector.

With the assumptions we have made, it can be shown that the diffraction pattern keeps *the same form* as at normal incidence but undergoes a shift of vector $f\boldsymbol{\tau}_1$. The centre of the diffraction pattern goes to F_1

The shift of the diffraction pattern is then

$$\mathbf{FF}_1 = f\boldsymbol{\tau}_1 \qquad\qquad (9.106)$$

Setting, for any point M of the surface $\mathbf{FM}/f = \boldsymbol{\tau}$, the diffraction pattern is then written in the form

$$\psi\left(\frac{\mathbf{F}_1\mathbf{M}}{f}\right) = \psi(\boldsymbol{\tau} - \boldsymbol{\tau}_1) \qquad\qquad (9.107)$$

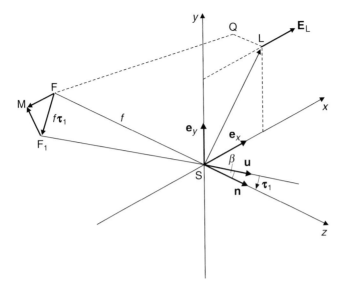

Fig. 9.17 Diffraction pattern: case of non-axial incident plane wave.

9.9.2 Radiation pattern of the system associated with a given primary aperture

Consider a primary aperture at the centre of a transmission of an illumination which is locally plane and which is represented by its electric field $\mathbf{E}(\mathrm{M})$ or $\mathbf{E}(\tau)$. We have seen that the transfer coefficient in reception of the diffraction pattern towards the primary aperture is given by equation (8.31). This is the gain factor of the system

$$K = \frac{\left|\iint_S \mathbf{E}(\mathrm{M}).\psi(\mathrm{M})\,dS\right|^2}{\iint_S |\mathbf{E}(\mathrm{M})|^2\,dS \iint_S |\psi(\mathrm{M})|^2\,dS} \qquad (9.108)$$

The integral of the numerator, which gives to within a constant the signal received by the feed as a function of the direction $\mathbf{u} = \mathbf{n} - \tau$ of the incident wave, represents the radiation pattern of the system in the polarization of the incident wave. From equation (9.107) we obtain

$$F(\tau_1) = \iint_S \mathbf{E}(\tau_1).\psi(\tau - \tau_1)\,dS \qquad (9.109)$$

If the diffraction pattern is of constant polarization, for example $\psi = \mathbf{e}_x \psi$ and if we set

$$\mathbf{E}(\tau) = \mathbf{e}_x E_x(\tau) + \mathbf{e}_y E_y(\tau) \tag{9.110}$$

we have

$$F(\tau_1) = \iint_S E_x(\tau_1) \cdot \psi(\tau - \tau_1) \, dS \tag{9.111}$$

or in Cartesian coordinates

$$F(\xi, \eta) = \int_{-\infty}^{+\infty} \int_{-\infty}^{+\infty} E_x(x, y) \psi(x - \xi, y - \eta) \, dx dy \tag{9.112}$$

(ξ, η) being the direction cosines of the direction of incidence $\mathbf{u}$ (or of the vector $\mathbf{n} - \mathbf{u}$).

The integrals may be formally extended to infinity, since E is zero outside the primary aperture. Thus the radiation pattern appears in the form of the *convolution of the diffraction pattern with the illumination function of the aperture*. This result is used in the theory of antennas considered as spatial frequency filters (§13.7). It allows us to show the identity between the radiation patterns considered either from the point of view of reception, or of transmission. We can say that the *secondary pattern F is the image of the primary illumination function* seen through the filter constituted by the antenna.

9.9.3 Examples of applications

Patterns of a system operating in a planar space

Consider an axially-symmetric system with a small angular aperture which operates in a plane space (example: optics between parallel planes). Since its transfer function is unity, the diffraction pattern produced by an incident wave of direction $\mathbf{u}$ making an angle β with the axis is of the form

$$\psi(x - \xi) = \frac{\sin\left(\dfrac{\pi D}{\lambda} \dfrac{x - \xi}{f}\right)}{\dfrac{\pi D}{\lambda} \dfrac{x - \xi}{f}} \tag{9.113}$$

with $\xi = f \sin \beta$.

The width between nulls of the diffraction main lobe is

$$\delta = 2\lambda\frac{f}{D} \tag{9.114}$$

If the primary aperture whose width is a is located at the focus, its illumination function, assumed uniform, is of the form[5]

$$E(x) = \text{rect}\left(\frac{2x}{a}\right) \tag{9.115}$$

The radiation pattern is obtained from the convolution integral (9.112) which yields

$$F_\alpha(\xi) = \frac{\lambda f}{\pi D}\left[\text{Si}\frac{\pi D}{\lambda f}\left(\xi + \frac{a}{2}\right) - \text{Si}\frac{\pi D}{\lambda f}\left(\xi - \frac{a}{2}\right)\right] \tag{9.116}$$

where Si denotes the sine-integral function.

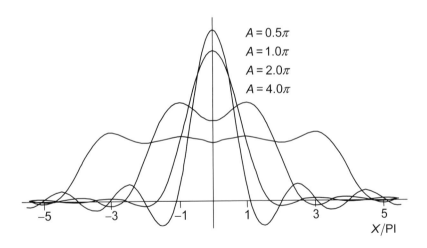

$A = 0.5\pi$
$A = 1.0\pi$
$A = 2.0\pi$
$A = 4.0\pi$

Fig. 9.18 Radiation pattern as images of the primary feed contour.

---

[5] By definition $\begin{cases} \text{rect } X = 1 \text{ if } |X| \le 1 \\ \text{rect } X = 0 \text{ if } |X| > 1 \end{cases}$

The pattern has the general aspect indicated in Fig. 9.18, where the parameter A is

$$A = \frac{\pi D a}{2 \lambda f}$$

This is the more or less distorted shape of the rectangular illumination function of the primary aperture. If the width a of the aperture is less than that of the diffraction pattern δ, the secondary pattern has the familiar form with a single maximum on the axis. When a increases, the near sidelobes begin to decrease and the main lobe is broadened a little. This is the phenomenon of 'tapering' due to the primary directivity. When a continues to increase the pattern becomes more and more square, forming a more and more accurate 'image' of the primary illumination (§13.2.1) and (§13.2.5).

9.9.4 Axial gain of an axially-symmetric system - effect of the diameter of the primary aperture

We shall see, using equation (9.108), how the gain factor varies as a function of the diameter of the primary aperture (that is to say, its directivity) and we shall determine the existence of an optimum diameter. We shall find the same results of §9.5.4 but in a slightly different form.

Consider, in the focal plane of the system, a radiating circular aperture of diameter $2a$, whose illumination is assumed constant in order to simplify the calculations (a non constant illumination leads to similar results):

$$\mathbf{E}(M) = E_0 \mathbf{e}_x \, \text{rect} \left(\frac{r}{a} \right) \tag{9.117}$$

(r, θ are the coordinates of M in the focal plane).

The axial diffraction pattern of an axially-symmetric system is assumed to be of the following form (§9.8.7, long focal length or narrow aperture):

$$\mathbf{E}'(M) = E_0' \mathbf{e}_x \Lambda_1 (u \, r) \tag{9.118}$$

with $\Lambda_1 = 2 J_1 (ur)/(ur)$; $\quad u = 2\pi/\lambda \sin \varphi_0$

Equation (9.108) takes here the following form:

$$K = \frac{\left| \int_0^a \Lambda_1(u\,r)\,r\,dr \right|^2}{\dfrac{a^2}{2} \int_0^\infty |\Lambda_1|^2\, r\,dr} = \frac{\left| \int_0^{ua} \Lambda_1(x)\,x\,dx \right|^2}{\dfrac{a^2 u^2}{2} \int_0^\infty |\Lambda_1(x)|^2\, x\,dx} \qquad (9.119)$$

The calculation is carried out making use of the following identities.

$$\int_0^\infty |\Lambda_1(x)|^2\, x\,dx = 2 \qquad (9.120)$$

$$\frac{d[J_0(x)]}{dx} = -J_1(x) \qquad (9.121)$$

We obtain

$$K(ua) = \left[\frac{1 - J_0(ua)}{\dfrac{ua}{2}} \right]^2 \qquad (9.122)$$

This function is shown in Fig. 9.19. We observe that it passes through a maximum for $ua = 2.7$. With respect to the diameter δ of the central diffraction pattern, the optimum diameter is

$$2\,a_{\text{opt}} \approx 0.71\,\delta \qquad (9.123)$$

with

$$\delta = \frac{1.22\lambda}{\sin\varphi_0} \qquad (9.124)$$

For this diameter, the theoretical gain factor is

$$\eta_{\text{max}} \approx 0.71 \qquad (9.125)$$

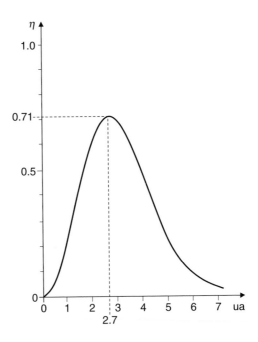

Fig. 9.19 Axial gain factor: effect of the diameter of the primary aperture.

9.10 RADIATION IN THE FRESNEL ZONE OF A GAUSSIAN ILLUMINATION - APPLICATION TO THE TRANSPORT OF ENERGY BY RADIATION (GOUBEAU'S WAVES)[6]

In §7.3.8, we gave the expression for the radiated field for an aperture using the Kirchhoff–Fresnel approximation: it appears as the Fourier transform of the illumination function with a quadratic phase factor. As an eigenfunction of the Fourier transform is a Gaussian function, this means that the radiated field at any point by a Gaussian illumination takes a Gaussian form. This is a basic result for the transport of energy by means of multiple focusing systems. In effect, if it is possible to find a plane in front of the aperture where the field distribution in amplitude possesses the *same variance* as over the aperture, then it is sufficient to place a focusing system on this plane which restores the quadratic phase function to meet the initial conditions by recurrence. We also can establish a chain of focusing

---

[6] Goubeau, G., 'Beam waveguides', *Advances in Microwaves*, Vol. 126, No. 3, p67, 1968.

systems which transport the energy, in principle, over an unlimited distance with negligible losses.

If we consider a Gaussian illumination on the plane $Z = 0$, centred about OZ, of variance W, focusing at a distance $OF = F$ (Fig. 9.20), the illumination is of the form

$$f(r) = \exp\left(-A\rho^2\right), \quad A = \frac{1}{W^2} + j\frac{k}{2F}, \quad k = \frac{2\pi}{\lambda} \tag{9.126}$$

We then find a plane where the field has a minimum dispersion W_m', not at the focus, but at a distance:

$$Z_m = \frac{F}{1 + \left(\dfrac{2F}{kW^2}\right)^2} \tag{9.127}$$

with

$$\left(\frac{W_m'}{W}\right)^2 = \frac{\left(\dfrac{2F}{kW^2}\right)^2}{1 + \left(\dfrac{2F}{kW^2}\right)^2} \tag{9.128}$$

On this plane the wave is plane; beyond this it is divergent. The initial dispersion W is found at a distance $Z_0 = 2Z_m$. It is in this plane that a focusing system must be placed to re-establish the initial conditions. These properties are summarized in Fig. 9.20.

Application: In many Earth Station antennas of the Cassegrain type, a periscope system using moveable plane and elliptic reflectors allows the main reflector to be illuminated, whatever its orientation, from a fixed primary feed (generally a corrugated horn (§8.3).

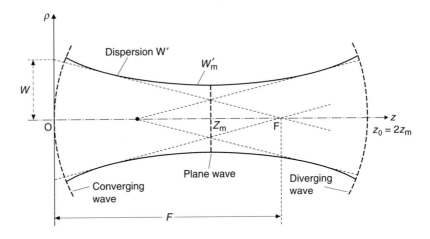

Fig. 9.20 Radiation of a Gaussian beam focused in the Fresnel region.

FURTHER READING

1. Debye, P., 'Das Verhalten von Lichtwellen in der Nähe eines Brempunktes oder einer Brennlinie', *Ann. Phys.*, Vol. 30, pp755-776, 1909.
2. Jull, E.V., *Aperture Antennas and Diffraction Theory*, Peter Peregrinus, Stevenage, 1981.
3. Love, A.W. (ed), *Reflector Antennas*, IEEE Press, New York, 1978.
4. Minett, H., Mac, B. and Thomas, A., 'Fields in the image space of symmetrical focus reflectors', *Proc. IEE*, Vol. 15, No. 10, October 1968.
5. Rusch, W.V.T. and Potter, P.D., *Analysis of Reflector Antennas*, Academic Press, New York, 1970.

EXERCISES

9.1 Axial gain of a focusing system operating in a planar space

In the focusing system of §9.9.3, calculate the gain factor as a function of the aperture length a of the primary feed, assumed to be uniformly illuminated (see §9.9.2, eqn. 9.108). What is its maximum value ?

ANSWER

$$\eta = \frac{2}{\pi X} \operatorname{si}^2 X \quad \left(X = \frac{\pi Da}{2\lambda f} \right)$$

is maximum and close to 0.7 for $a \approx 2\lambda F/D$

where si X is the integral sine function defined by $\operatorname{si} X = \int_0^X \sin x \, dx$.

9.2 Characteristic radiation function of an annular feed with radial polarization

In the design of a tracking antenna, a Cassegrain antenna (§10.2) is to be used, of long equivalent focal length (§10.2.4), with a primary feed consisting of two elements (Fig. 9.21).

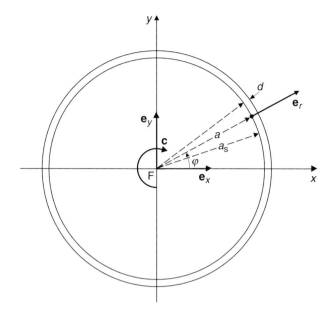

Fig. 9.21 Primary feed aperture **c**, circularly-polarized, associated with a circular slot of radial polarization **e**$_r$.

The first is an aperture of radius a_s, illuminated by a field with the characteristics of a plane wave of circular polarization (amplitude, phase and polarization constant). The signal received by this aperture is the 'reference signal' S.

The gain of this forms the subject of §9.9.4. It is given by the expression (9.122)

$$\eta_s\left(ua\right) = \left(\frac{1-J_0\left(ua_s\right)}{a_s/2}\right)$$

with

$$u = \frac{2\pi}{\lambda}\sin\alpha_0$$

where α_0 is half the angle subtended by the aperture. As we have seen (§9.9.4) the axial gain is maximum for a diameter $2a_s = 0.71\delta$, where δ is the diameter of the diffraction spot.

The second element is formed by an annular radiating slot, of width d, radial polarization $\mathbf{e}_r$ and radius a, surrounding the first element.

Q1 Calculate the characteristic radiation function of this aperture. Note that the radial polarization can be decomposed into a horizontal component e_x and a vertical component e_y, in the form

$$e_r = e_x \cos \varphi + e_y \sin \varphi$$

where (r, φ) are polar coordinates in the plane of the aperture.

ANSWER

Q1 Here the results cannot be deduced from §8.1.3 (radially symmetric structures); the polarization in the aperture is not constant, but radial. We start from equation 7.31 (§7.4.3). The characteristic function is given by equation 7.25 (§7.4.3). If we set

$$\mathbf{f}_T(M) = \mathbf{e}_r f(r)$$

we obtain, to within a multiplicative constant

$$\mathbf{E}(\theta) = \int_0^\infty \int_0^{2\pi} (\mathbf{n} \times \mathbf{e}_x \times \mathbf{u}) f(r) \exp\left(j \frac{2\pi r}{\lambda} \sin\theta \cos(\varphi - \phi) \right) d\varphi \, r dr$$

where (r, φ) are polar coordinates in the plane of the aperture, and (θ, ϕ) are spherical coordinates of the direction in space of the unit vector $\mathbf{u}$.

Since the illumination is circularly-symmetric, the characteristic function is as well. We can therefore consider just the plane $(\mathbf{n}, \mathbf{e}_x)$ defined by $\phi = 0$. Thus $\mathbf{u} = \mathbf{n} \cos\theta + \mathbf{e}_x \sin\theta$.

Setting $\mathbf{e}_r = \mathbf{e}_x \cos\varphi + \mathbf{e}_y \sin\varphi$. Only the component $\mathbf{e}_x$ contributes to the field at P. We have therefore

$$\mathbf{E}(\theta) = \int_0^\infty \int_0^{2\pi} \left[(\mathbf{n} \times \mathbf{e}_x) \times \mathbf{u} \right] \cos\varphi \exp(jz\cos\varphi) \, d\varphi \, r dr$$

where we have set $z = \dfrac{2\pi r}{\lambda} \sin\theta$.

Since the slot is narrow, the integration with respect to r can be carried out immediately. Thus

$$\mathbf{E}(\theta) = ad \int_0^{2\pi} \left[(\mathbf{n} \times \mathbf{e}_x) \times \mathbf{u} \right] \cos\varphi \exp(jz\cos\varphi) \, d\varphi$$

where now $z = \dfrac{2\pi a}{\lambda} \sin\theta$.

The polarization is necessarily radial; this can be verified by evaluating the factor in square brackets

$$\mathbf{n} \times \mathbf{e}_x = \mathbf{e}_y \qquad \mathbf{e}_y \times \mathbf{u} = \mathbf{e}_x \cos\theta - \mathbf{n}\sin\theta = \mathbf{e}_\theta$$

We have therefore

$$\mathbf{E}(\theta) = \mathbf{e}_\theta ad \int_0^{2\pi} \cos\varphi \exp(jz\cos\varphi) \, d\varphi$$

$$= \mathbf{e}_\theta ad \; 2\pi j \; \mathrm{J}_1(z)$$

$$= 2\pi jad \; \mathbf{e}_\theta \; 2\pi j \; \mathrm{J}_1\left(\frac{2\pi a}{\lambda}\sin\theta \right)$$

We therefore have a characteristic radiation surface which is toroidal in form, with a null on-axis (Fig. 9.22). These results will be useful in a further exercise (10.2) at the end of the next Chapter.

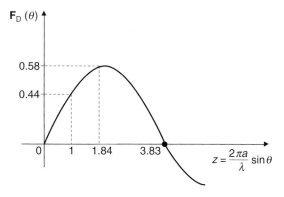

Fig. 9.22 Radial section of characteristic radiation surface of the circular slot.

9.3 *Radiation in the Fresnel zone of a Gaussian illumination*

Q1   Show that there exists an 'optimal' focal length $OF = F_{opt}$, for which the distance z_m of minimum dispersion is maximum.

Q2   What is this distance $z_m = z_m^{max}$?

Q3   What is the value of the dispersion $W'_m = W'^{opt}_m$ at this distance?

ANSWERS

Q1 $\dfrac{F_{opt}}{W} = \pi \dfrac{W}{\lambda}$

Q2 $z_m^{max} = \dfrac{1}{2} F_{opt}$

Q3 $W'^{opt}_m = \dfrac{W}{\sqrt{2}}$

10

Focused systems

Elements of a multi-mode monopulse primary feed. Photo: THALES (formerly Thomson-CSF).

10.1 INTRODUCTION

In this chapter we consider some examples of focused systems. Of these, some are axially-symmetric and have already been mentioned. In Chapter 7, the parabolic reflector was introduced as an application of the optical path law (§7.3.4). Dielectric lenses were mentioned following the definition of the refractive index of a dielectric medium (see equation (7.13)) - Luneburg, Eaton lenses and geodesic lenses.

In this chapter we consider in particular the axially-symmetric dual-reflector system known as the Cassegrain antenna, already introduced in Chapter 7 in the context of low noise antennas. Axially-symmetric focused antennas are frequently used in tracking systems, whose techniques we also consider (conical scanning and monopulse). We shall see that the angular accuracy of these systems is limited, not only by the antenna angular beamwidth or the signal-to-noise ratio (§10.3), but also by the depolarizing properties of the radar target (Chapter 12). Finally, the chapter ends with a study of offset-fed systems, designed to overcome the blockage effect of the primary feed, and with the synthesis of radiation patterns.

10.2 THE CASSEGRAIN ANTENNA

10.2.1 Introduction[1]

In 1663, Gregory, a Scottish mathematician, proposed (without actually building it), a dual-reflector telescope with concave mirrors, with one of them (the subreflector), giving an uninverted image of that reflected from the other mirror.

Guillaume Cassegrain was a sculptor and foundryman at Chartres, in France. In 1672 he proposed a modified version of the dual-reflector telescope in which the subreflector was convex, reflecting the image back through a hole in the centre of the main reflector.

In terms of geometrical optics, the meridians of the reflector should be calculated so that the image of a point at infinity on the axis is stigmatic. A classical and rigorous solution was proposed by Newton: if an intermediate stigmatic image is imposed, the optical path law shows that the main reflector is parabolic and the subreflector is ellipsoidal in the Gregorian system and hyperbolic in the case of the Cassegrain. Instead of forcing the stigmatism condition for the intermediate image (which *a priori* is unimportant), it is better to choose another supplementary property as was proposed by Schwarzschild who added the aplanatism condition of Abbe that leads to a spherical principal surface (§9.7).

In the case of large microwave antennas, the choice of the first Cassegrain antennas was motivated by exactly the same advantages as those of the optical telescopes: the presence of an image close to the centre of the main reflector allows the primary feed to be located at an easily accessible point. Also, the system may be *aplanetic* allowing multiple-feed antennas to be realized. Various other advantages have subsequently become apparent:

- The first concerns the *noise temperature* of the antenna. When a conventional parabolic reflector is pointed at the sky, its primary feed points towards the ground and therefore picks up ground noise. On the other hand, when a Cassegrain antenna is pointed at the sky, its primary feed also points at the sky, so it only receives sky noise. It is therefore far less sensitive to ground noise.

---

[1] Hannan, P.W., 'Microwave antennas derived from the Cassegrain telescope', *IRE Trans. Antennas and Propagation*, March 1961.

- Another advantage lies in the ability to obtain *high gain*. The use of two reflectors allows us to choose the form (within certain limits) of the *transfer function*: as we know, if the latter varies as the inverse of the primary pattern amplitude, the illumination of the equivalent aperture is constant (§9.4) and yields maximum gain. This is the principle of the so-called 'shaped' Cassegrain (§10.2.5).

10.2.2 Geometry

The Cassegrain antenna consists of two axially-symmetric reflectors. The primary feed is placed at the apex of the main reflector and illuminates the latter by means of a subreflector placed in front of it. The shapes of the two radially-symmetric surfaces are calculated so that, after the two reflections, the primary wave is planar. The classical solution consists of choosing a homofocal hyperboloid and paraboloid, so that the primary phase centre C' coincides with one of the foci F' of the hyperboloid (Fig. 10.1) (see §10.2.7). Fig. 10.2 depicts the case where one of the reflectors is planar.

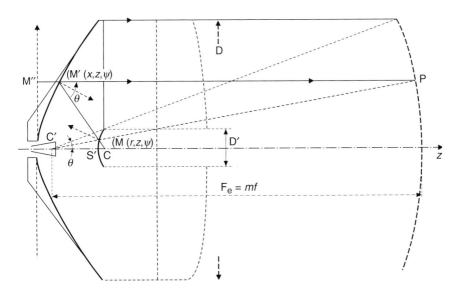

Fig. 10.1 Cassegrain antenna geometry.

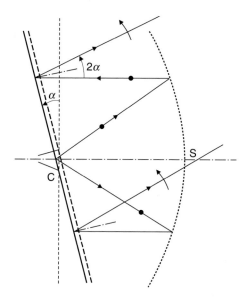

Fig. 10.2 Cassegrain with scanned planar main reflector for mechanical scanning of the beam. A polarization rotating system is necessary (§10.2.7).

Consider a plane containing the axis of revolution SS′ , and let us use the following notations.

- focal length of the main reflector: CC′ = F;

- useful diameters of the reflectors: D, D′ ;

- path of a ray from the primary feed: C′MM′P ;

- lengths of elements of rays: C′M = ρ, MM′ = ρ′;

- angles between the elements of rays and the axis: θ, θ′ ;

- cylindrical coordinates of the point M: (r, z, ψ);

- cylindrical coordinates of the point M′: (X, z, ψ);

- hyperboloid magnification: $m = \dfrac{\overline{C'S'}}{\overline{S'C}}$ (see below);

The reflector surfaces are defined, for example, by the following equations.

- paraboloid
$$\frac{X}{2F} = \tan\frac{\theta'}{2} \qquad (10.1)$$

- hyperboloid
$$m = \frac{\tan\theta'/2}{\tan\theta/2} \qquad (10.2)$$

The following approximate relation can be demonstrated: $D/D' \approx m+1$ (see exercise 10.1). If the subreflector is plane, for example, then $m = 1$, and this relation shows that its useful diameter is half the diameter of the main reflector.

10.2.3 Equivalent primary feed

The primary feed placed at C' and the subreflector (R') are equivalent to a single primary feed placed close to C, whose pattern produces the same illumination of the main reflector.

Let $f(\theta)$ be the primary pattern, and $h(\theta')$ the corresponding pattern of the equivalent source. The elementary solid angle between two cones of apex angle θ and $\theta + d\theta$ is equal to
$$d\Omega = 2\pi\sin\theta\,d\theta$$

The radiated primary power in this angle is proportional to
$$dW = |f(\theta)|^2\,2\pi\sin\theta\,d\theta \qquad (10.3)$$

After reflection, this power is contained in an elementary solid angle $d\Omega'$ comprised between two cones of apex angle θ' and $\theta' + d\theta'$, equal to
$$d\Omega' = 2\pi\sin\theta'd\theta'$$

and produces a pattern $h(\theta')$ so that
$$dW = |h(\theta')|^2\,2\pi\sin\theta'd\theta' \qquad (10.4)$$

From equations (10.3) and (10.4) we obtain the relationship between the primary pattern and that of the equivalent feed
$$|h(\theta')|^2\sin\theta'd\theta' = |f(\theta)|^2\sin\theta\,d\theta \qquad (10.5)$$

The relation between θ and θ' depends on the shape of the reflector. If the latter is a hyperboloid, it is defined by equation (10.2). For small angles, we have

$$m \approx \frac{\theta'}{\theta}$$

and

$$h(\theta') \approx \frac{1}{m} f\left(\frac{\theta'}{m}\right)$$

This result justifies the term 'magnification' given to the parameter m; the pattern undergoes a transformation which divides the angles by m. This corresponds to a virtual primary feed m times smaller than the real feed.

10.2.4 Principal surface (Fig. 10.1)

The principal surface of the system is by definition the locus of the intersection points of the rays C′M from the primary feed with the emerging rays M′P parallel to the axis. In the classical case of the paraboloidal-hyperboloidal combination we find, from equations (10.1) and (10.2), that the locus is a paraboloid whose focal length (equivalent focal length) is $F_e = mF$ (F being the focal length of the paraboloid). The electrical characteristics of the Cassegrain, as with any axially-symmetric system, are defined by its principal surface or by its transfer function. In particular, a primary feed whose pattern is $f(\theta)$ results in an illumination $k(X)$ which may be obtained by consideration of conservation of energy, as in the preceding paragraph. We obtain the following relation.

$$\left|k(X)\right|^2 XdX = \left|f(\theta)\right|^2 \sin\theta \, d\theta \tag{10.6}$$

Since the transfer function is (§9.4.15)

$$T = F_e \sqrt{\frac{\sin\theta \, d\theta}{XdX}} \tag{10.7}$$

then

$$k(X) = \frac{1}{F_e} Tf(\theta)$$

In the case of the paraboloidal-hyperboloidal combination, the transfer function is that of the equivalent paraboloid

$$T = \frac{1}{1 + \left(\dfrac{X}{2F_e}\right)^2} \tag{10.8}$$

10.2.5 Cassegrain with shaped reflectors[2]

We saw that the subreflector associated with the primary feed is equivalent to a feed with a pattern $h(\theta')$ related to the primary pattern $f(\theta)$ by equation (10.5) which depends on the shape of the subreflector, that is to say, the relation between θ and θ'. We may use this property so that the illumination should be theoretically constant and without spillover loss, which is desirable for both gain and noise temperature. To reach this result, we use the following steps.

(1) Find a subreflector of shape such that its equivalent primary pattern $h(\theta')$ produces an illumination of theoretically constant amplitude,

(2) In such a case, in general, the pattern $h(\theta')$ is no longer equiphase, or in other words, the reflected waves are no longer spherical. To correct the phase errors the shape of the reflector is modified.

Subreflector

From the energy conservation equations (10.5) and (10.6), we can relate the illumination function assumed constant $k(X) = k_0$ to the equivalent feed pattern $h(\theta')$

$$k_0^2 \, X dX = \left|h(\theta')\right|^2 \sin\theta' d\theta' \tag{10.9}$$

Since the main reflector is close to a paraboloid, X and θ' are approximately related by equation (10.1). From this, the shape of the pattern of the equivalent feed is therefore (Fig. 10.3) given by its characteristic function

$$\left|h(\theta')\right| = h_0 \left(1 + \tan^2 \frac{\theta'}{2}\right) = \frac{h_0}{\cos^2 \dfrac{\theta'}{2}} \tag{10.10}$$

Since the equivalent pattern is determined from equation (10.5) it is possible to define the profile of the subreflector. We choose a primary feed with a sufficiently

---

[2] Williams, W.F., 'High-efficiency antenna reflectors', *Microwave Journal*, July 1965.

directive pattern $f(\theta)$ to avoid spillover losses[3] (§9.5.3). Equation (10.5) is a differential equation defining the desired profile via a relation between θ and θ'. The polar equation may be obtained from the known differential equation relating the vector radius and its angle with the normal. This is the incident angle, whose value is $V = (\theta+\theta')/2$ (Snell's law). Thus we have

$$\frac{1}{\rho}\frac{d\rho}{d\theta} = \tan\left(\frac{\theta+\theta'}{2}\right) \qquad (10.11)$$

and

$$\rho = \exp\left[\int \tan\left(\frac{\theta+\theta'}{2}\right)d\theta\right]$$

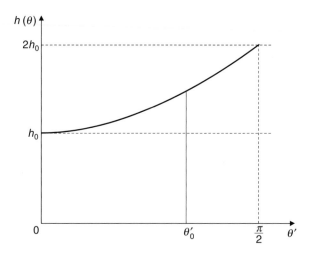

Fig. 10.3 Pattern of the equivalent feed source of a Cassegrain for maximizing the gain.

If we compare (Fig. 10.4) the meridian obtained with that of the corresponding hyperboloid, we observe at the apex a more pronounced curvature, and less towards the edges. This can be explained by the fact that the reflector produces a redistribution of paraxial primary energy towards the reflector edge.

---

[3] A corrugated horn or dual-mode horn is particularly suitable.

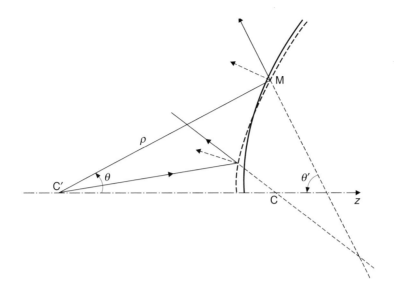

Fig. 10.4 Meridian of the subreflector for maximising the gain.

Main reflector

The profile of the main reflector may be determined by geometrical optics. The coordinate Z is obtained from the optical path applied to the stigmatism condition: the optical path measured along a ray from a point C and ending at an arbitrary plane perpendicular to the axis is constant. If the plane $Z = 0$ is chosen, the virtual path $(-Z)$ must be considered. We obtain

$$\rho + \frac{X - \rho \sin\theta}{\sin\theta'} - Z = C \qquad (10.12)$$

from which $Z(\theta), X(\theta)$ can be obtained with the help of equations (10.1) and (10.5). The profile obtained can be compared with that of the original paraboloid and, occasionally, the calculation iterated. In fact, to determine the main reflector profile, it is better to start with the calculation of the diffraction pattern of the subreflector and to deduce $Z(\theta)$ by correcting the phase variations of this pattern (§10.2.6).

Generalization

Within the framework of geometrical optics, it is not necessary to assume that the reflector has a profile close to a paraboloid. In effect, since the principal surface (or the transfer function) is defined by an arbitrary condition (constant illumination, aplanatism, etc.), we obtain a first relation $X(\theta)$. Snell's laws applied at the points M and M′ give equation (10.11) and the equation

$$\frac{dZ}{dX} = \tan\frac{\theta'}{2} \qquad (10.13)$$

Finally, the optical path law gives equation (10.12), then we have a system of four differential equations relating five unknowns (X, Z, θ, θ', ρ), which may be solved to express each one as a function of one variable. We thus obtain the profile of the two reflectors as a function of the angle θ, for example.

10.2.6 Diffraction pattern of the subreflector

Calculation assumptions

A simplified solution to this problem will be presented here by using a method developed by Daveau[4] which is based on two approximations:

• geometrical optics is used to calculate the incident field on the subreflector,

• physical optics is used to calculate the reflected field and the field re-radiated by the subreflector.

These approximations are valid if:

• in the vicinity of the subreflector the primary wave is locally plane,

• the curvature of the subreflector is sufficiently small compared with the wavelength so that the reflection laws on a plane are applicable at each point M of the reflector.

Under these conditions it is possible to consider each element dS of the reflector surface as an aperture to which the Kottler formula (§7.4.5, eqn. 7.49) can be applied; the integral is calculated on the surface of the reflector. This formula also results from the treatment of the diffraction problem by physical optics

$$\mathbf{E}_p(\mathbf{u}) = -j\frac{\exp(-jkR)}{2\lambda R}\mathbf{u}\times\int_S\left[\mathbf{n}\times\mathbf{E}+\eta(\mathbf{n}\times\mathbf{H})\times\mathbf{u}\right]\exp(jk\mathbf{CM}.\mathbf{u})\,ds \qquad (10.14)$$

---

[4] Daveau, D. 'Synthèse et optimisation des réflecteurs de forme spéciale pour antennes', *Revue Technique Thomson–CSF*, Vol. 2, No. 2, March 1970.

where $\eta = 120\pi \; \Omega$

   n = unit vector normal to the reflector at M
   C = reference point

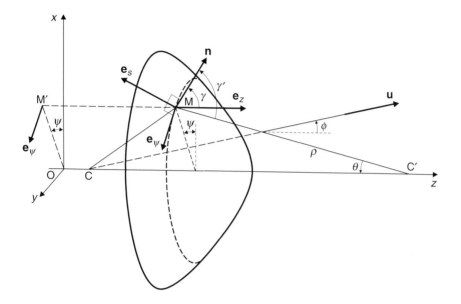

Fig. 10.5 Diffraction from the subreflector.

In fact, the calculation of the integral can be simplified if we include the assumptions:

- the primary feed is an axially-symmetric 'Huygens source' whose polarization is linear and horizontal; it radiates an axially-symmetric pattern $f(\theta)$;
- the subreflector is axially-symmetric, and it follows from the general properties of axially-symmetric structures that the diffracted pattern itself is then axially-symmetric with no cross polarization, i.e. that of a Huygens source (§8.1.2).

Notations used

In addition to the notations defined above we use the notations:

   $\mathbf{e}_\xi$ = unit vector associated with the coordinate ξ,
   φ = angle between the direction **u** and the axis,
   γ = angle between the vector **n** and the axis,

ς = incidence angle with the normal ($\gamma' = \gamma + \theta$),
s = curvilinear abscissa of point M in a meridian plane,
$(z, r, \psi), (\rho, \theta)$ = cylindrical and polar coordinates of M,
C′ C = phase centre of the primary feed. Reference points.
$\mathbf{H}^{+} = \eta\mathbf{H}$
$\eta = 120\pi\,\Omega$

Since the pattern is axially-symmetric and of known polarization, it is sufficient to calculate on the plane *y*0*z* where its polarization is 'horizontal'. Therefore, we have

$$\mathbf{u} = \mathbf{u}_y \sin\varphi + \mathbf{e}_z \cos\varphi \tag{10.15}$$

Incident field at M

Since the primary feed is an axially-symmetric Huygens source, its pattern $f(\theta)$ is axially-symmetric. The incident electric field at M is of the form

$$\mathbf{E}' = a(\theta)\mathbf{e}_\alpha \tag{10.16}$$

with (§8.1.2)

$$\mathbf{e}_\alpha = \mathbf{e}_\theta \cos\psi - \mathbf{e}_\psi \sin\psi$$

and

$$a(\theta) = f(\theta)\frac{\exp(-jk\rho)}{\rho} \tag{10.17}$$

in the same way, the incident magnetic field at M is

$$\mathbf{H}'^{+} = a(\theta)\left(\mathbf{e}_\theta \sin\psi + \mathbf{e}_\psi \cos\psi\right) \tag{10.18}$$

Reflected field at M

The form of the reflected field is deduced from the boundary conditions of a perfectly-conducting surface. The resulting electric field ($\mathbf{E} + \mathbf{E}'$) must be normal to the surface

$$\mathbf{E} + \mathbf{E}' = 2(\mathbf{n}.\mathbf{E}')\mathbf{n}$$

So, we have

$$\mathbf{E} = 2(\mathbf{n}.\mathbf{E})\mathbf{n} - \mathbf{E}'$$

Taking into account the following equation:

$$\mathbf{e}_\theta = \mathbf{n}\sin\gamma' + \mathbf{e}_s \cos\gamma' \tag{10.19}$$

We have

$$\left.\begin{array}{l} \dfrac{\mathbf{E}}{a} = \mathbf{e}_s \cos\gamma' \cos\psi + \mathbf{e}_\psi \sin\psi - \mathbf{n}\sin\gamma'\cos\psi \\[2em] \dfrac{\mathbf{H}^+}{a} = -\mathbf{e}_s \cos\gamma' \sin\psi + \mathbf{e}_\psi \cos\psi + \mathbf{n}\sin\gamma'\sin\psi \end{array}\right\} \quad (10.20)$$

We can verify that this field is locally plane.

Calculation of integral [5]

Let us calculate

$$\left.\begin{array}{l} \dfrac{\mathbf{E}\times\mathbf{n}}{a} = -\cos\gamma' \cos\psi\,\mathbf{e}_\psi + \sin\psi\,\mathbf{e}_s \\[2em] \dfrac{\mathbf{H}^+\times\mathbf{n}}{a} = \cos\gamma' \sin\psi\,\mathbf{e}_\psi + \cos\psi\,\mathbf{e}_s \end{array}\right\} \quad (10.21)$$

Taking into account the following equations:

$$\left.\begin{array}{l} \mathbf{e}_\psi = -\sin\psi\,\mathbf{e}_x + \cos\psi\,\mathbf{e}_y \\[1.5em] \mathbf{e}_s = \cos\gamma\cos\psi\,\mathbf{e}_x + \cos\gamma\sin\psi\,\mathbf{e}_y - \sin\gamma\,\mathbf{e}_z \end{array}\right\} \quad (10.22)$$

we obtain

$$\left.\begin{array}{l} \dfrac{\mathbf{E}\times\mathbf{n}}{a} = A\sin(2\psi)\mathbf{e}_x - \big[B + A\cos(2\psi)\big]\mathbf{e}_y - \sin\gamma\cos\psi\,\mathbf{e}_z \\[2em] \dfrac{\mathbf{H}^+\times\mathbf{n}}{a} = -\big[B - A\cos(2\psi)\big]\mathbf{e}_x + A\sin(2\psi)\mathbf{e}_y - \sin\gamma\cos\psi\,\mathbf{e}_z \end{array}\right\} \quad (10.23)$$

with

$$A = \cos\gamma' - \cos\gamma, \qquad B = \cos\gamma' + \cos\gamma \quad (10.24)$$

---

[5] Aubry, C. and Bitter, D, 'Radiation patterns of a corrugated conical horn in terms of Laguerre-Gaussian functions', *Electronics Letters*, Vol. 11, No. 7, pp154-156, April 1975.

On the other hand, we have

$$\mathbf{CM.u} = z\cos\varphi + r\sin\varphi\sin\psi \tag{10.25}$$

and

$$dS = rdr\frac{d\psi}{\cos\gamma} \tag{10.26}$$

The integral (10.14) is written in the form

$$\mathbf{E}_p(\mathbf{u}) = \frac{j}{2\lambda}\frac{\exp(-jkR)}{R}\int_0^{r_0} a\exp(jkz\cos\varphi)\mathbf{V}(r,\varphi)rdr \tag{10.27}$$

with

$$\mathbf{V}(r,\varphi) = \left\{\left[\int_0^{2\pi}\frac{\mathbf{H}^+\times\mathbf{n}}{a}\exp(jur\sin\psi)d\psi\right]\times\mathbf{u} + \int_0^{2\pi}\frac{\mathbf{E}\times\mathbf{n}}{a}\exp(jur\sin\psi)d\psi\right\}\times\mathbf{u} \tag{10.28}$$

and $u = k\sin\varphi.$

Taking into account the properties of the integral representations of the Bessel functions, we have

$$\int_0^{2\pi}\exp(jur\sin\psi)\ d\psi = 2\pi J_0(ur)$$

$$\int_0^{2\pi}\sin\psi\exp(jur\sin\psi)\ d\psi = 2\pi j J_1(ur)$$

$$\int_0^{2\pi}\cos(2\psi)\exp(jur\sin\psi)\ d\psi = 2\pi J_2(ur)$$

$$\int_0^{2\pi}\sin(2\psi)\exp(jur\sin\psi)\ d\psi = 0 \tag{10.29}$$

Finally, we obtain

$$\mathbf{E}_p(\varphi) = \frac{2\pi j}{2\lambda}\frac{\exp(-jkr)}{R}F(\varphi)\mathbf{e}_\alpha \tag{10.30}$$

with

$$F(\varphi) = \frac{1+\cos\varphi}{2}I_0 + \frac{1-\cos\varphi}{2}I_2 + j\sin\varphi I_1 \tag{10.31}$$

and

$$I_0 = \int_0^{r_0} \left(1 + \frac{\cos \gamma'}{\cos \gamma}\right) J_0\left(ur\right)\frac{f(\theta)}{\rho}\exp\left[-jk\left(\rho - z\cos\varphi\right)\right]r\,dr$$

$$I_2 = \int_0^{r_0} \left(1 - \frac{\cos \gamma'}{\cos \gamma}\right) J_2\left(ur\right)\frac{f(\theta)}{\rho}\exp\left[-jk\left(\rho - z\cos\varphi\right)\right]r\,dr \qquad (10.32)$$

$$I_1 = \int_0^{r_0} \tan \gamma \, J_1\left(ur\right)\frac{f(\theta)}{\rho}\exp\left[-jk\left(\rho - z\cos\varphi\right)\right]r\,dr$$

The first term is dominant. We verify that the diffracted field appears as a superposition of hybrid modes, i.e. as a conical spectrum of plane waves. This result generalizes the calculation of the diffracted field on the focal plane of an axially-symmetric system (§9.8.5).

Numerical results

Fig. 10.6 illustrates an example of the diffraction pattern of a Cassegrain antenna subreflector. Compared to the dashed line which corresponds to the geometrical optics solution, we observe that the general increasing form is maintained, but oscillations occur in the useful part of the pattern, and the slope at the edges is finite. These two phenomena are less important when the diameter of the reflector increases, that is to say, when we are close to the geometrical optics condition.

The meridian of the main reflector can then be calculated so that the phase errors cancel, and hence produce an equiphase illumination of the equivalent aperture.

In reference[5] an iterative method is proposed that leads to optimum meridians.

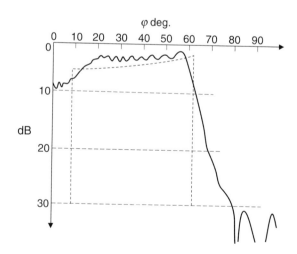

Fig. 10.6 Diffraction pattern of the subreflector (or pattern of the equivalent source).

10.2.7 Blockage by the subreflector

Blockage effect

The illumination of the system extends over the region of diameter D. This illumination has a central 'hole' of diameter D' due to the blockage by the subreflector. This illumination may be considered as the sum of the unperturbed illumination and that of a perturbation illumination whose amplitude is equal to the maximum amplitude of the preceding one but of opposite phase. The overall radiation pattern results from the superposition of these illuminations (§9.6.3). In particular, the parasitic illumination generates a pattern m times larger than the main lobe in the form of sidelobes which decrease the axial gain.

Let A be the *effective area* (§4.2) of the antenna ($A = KS$, K being the gain factor and S the area of the axial projection of the reflector), and S' the area of the projection of the subreflector. The central perturbation illumination is fed with a fraction S'/A of the total power and radiates with a gain also proportional to the fraction S'/A. The relative level of perturbation radiation is then, in power

$$N^2 = \left(\frac{S'}{A}\right)^2 = \frac{1}{K^2}\left(\frac{D'}{D}\right)^4 = \frac{1}{K^2 m^4} \qquad (10.33)$$

Example: for $K \approx 1$ and $D'/D = 0.1$, we have $N^2 = 10^{-4}$, that is to say -40dB. This is a typical figure which is often acceptable. The loss of gain is given by the relative decrease of the effective area, i.e.

$$\frac{G'}{G} = \frac{A - S'}{A} = 1 - \frac{1}{K}\frac{S'}{S} \qquad (10.34)$$

In the preceding example this gives 0.04 dB[6]

Remark: It should be noted that the theoretical pattern in the absence of blockage has sidelobes of a given level N_0 (for instance -17.6 dB for a theoretical uniform illumination). The overall level N' results from the combination of N_0 and N. We find in the most unfavourable phase combination

---

[6] Note that this result is true only if the radiation pattern of the equivalent source has a 'hole' on the axis. Use of shaped Cassegrain reflectors make this possible (Fig. 10.6). If it is not the case, the radiated power intercepted and diffracted by the subreflector is lost.

$$N' = \frac{N_0 + N}{1 - N} \qquad (10.35)$$

The better the theoretical level N_0, the greater the relative importance of the disturbance due to the subreflector. If N_0 is of the same order of magnitude as N, the sidelobe level increases by about 6 dB.

Suppression of blockage effect by rotation of the plane of polarization

Principle: This system is convenient for applications where a single polarization is used. The principle of the process is indicated in Fig. 10.7. The primary feed radiates a horizontal polarized wave. The subreflector is composed of horizontal metallic wires that reflect this wave towards the main reflector. The latter is covered with a layer having the property of rotating the plane of polarization by 90°.

  The structure of this coating is described in the following section. The polarization of the plane wave reflected by the paraboloid is then vertical, and this wave passes through the horizontal wires of the subreflector without reflection. The blockage effect is thus suppressed.

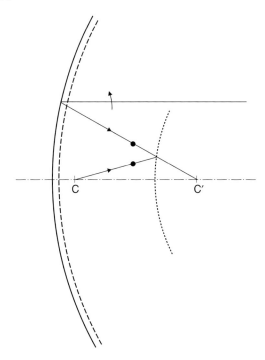

Fig. 10.7 Cassegrain antenna with rotation of polarization.

Structure of the coating: In its simplest form this is composed of a sheet of parallel wires, tilted at ±45° with respect to the vertical, located on a surface parallel to that of the reflector, and at a distance of a quarter of wavelength from it.

Consider (Fig. 10.8) a normally incident wave $\mathbf{P}_1$, (horizontally polarized). This wave may be considered as the superposition of two composite equiphase waves, whose polarizations are tilted at 45°: one of them being parallel to the wires $\mathbf{P}_1'$, and the other perpendicular $\mathbf{P}_1''$. The first component is reflected by the wires, the second passes through them and is reflected by the main reflector. It travels a further half wavelength, and after reflection it has a phase difference of π with respect to the first wave $\mathbf{P}_2''$. The combination of the two reflected components creates a vertically polarized wave: $\mathbf{P}_2$. This mode of operation assumes that the sheet of tilted wires is perfectly reflecting for the component whose polarization is parallel to the wires, and perfectly transparent (with no phase difference) for the orthogonal component. This condition is more exactly met as the wires are thinner and more closely spaced. The residual phase differences can be corrected by adjusting the distance separating the sheet of wires and the reflector. This system only functions in principle at a single frequency. In practice, we cannot obtain a relative bandwidth of more than a few percent. There is a larger bandwidth system that uses more widely spaced wires in which there is a trade-off between the selectivity, the reflection and transmission characteristics of the sheet of wires and that of the optical path separating the wires and the reflector.

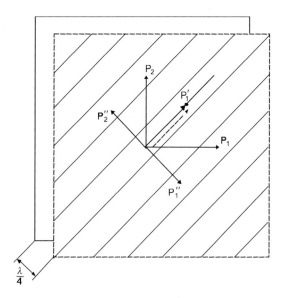

Fig. 10.8 Rotation of polarization.

At oblique incidence making an angle $\theta/2$ with normal, the distance should be $\lambda/\left[4\cos\left(\theta/2\right)\right]$. This means that in the case of a paraboloidal reflector the sheet of wires should form a confocal paraboloid whose focal length differs by a quarter of wavelength from that of the reflector.

Realization: The metallic wires are, in general, held in place by means of transparent plastic material.

Note that this process allows the arbitrary choice of the diameter of the subreflector. The main reflector may be planar. In this case we have $F = \infty$ and $m = 0$ with $F_e = mF$ remaining finite; we obtain the 'reversed Cassegrain'. A tilt through an angle α of the planar reflector produces a scanning through an angle of 2α without aberration (Fig. 10.2).

10.2.8 Schwarzschild aplanatic reflector

We have seen (§10.1.5) that if an additional condition is added to that of stigmatism it is possible to completely define, by means of geometrical optics, the reflector meridians. This condition may be that of aplanatism; the principal surface being then the Abbe sphere. Such a system was first designed for a telescope by Schwarzschild, and was studied for antenna applications by White and DeSize[7]. We observe that this process requires quite long focal lengths associated with moderate magnitudes m (2 to 5) which usually require some technique to prevent blockage (§10.1.7). Under these conditions, the process is excellent for moderate scanning angles (less than 10°). For larger scan angles, some of the rays reflected by one of the mirrors are no longer intercepted by the other mirror. Toroidal reflectors are then preferred.

10.3 TRACKING SYSTEMS

10.3.1 Introduction

Tracking systems are used to maintain a narrow radiation beam in the direction of a distant moving object (for example a radar target or communications satellite).

The antenna is in general a focused system whose angular movement is obtained by means of a rotating mount usually possessing a vertical axis (azimuth axis) and a horizontal axis (elevation axis). This mount is operated by servomechanisms

---

[7] White, W.D. and DeSize, L.K., 'Scanning characteristics of two-reflector antenna systems', *IRE Int. Conv. Rec.*, Part I, 1962.

incorporating feedback loops for precise pointing. The azimuth and elevation error signals that drive them are obtained from signals received by the antenna, by various interpolation methods. One such method uses conical scanning of the beam: the signal error results from the modulation of the received signal. Another method (monopulse) consists of comparing the signals received by pairs of adjacent beams. In fact, it can be shown that these two methods consist of minimizing the phase error of the incident wave over the antenna aperture, so that the tracking antennas will tend to orientate themselves orthogonally to the incident wavefront.

The angular accuracy obtained (measured by the standard deviation σ of the tracking error) mainly depends on the sharpness of the beam (θ_{3dB} aperture), the signal-to-noise ratio integrated over the duration of the measurement of R (equations (10.78) - (10.120))

$$\sigma \approx \frac{\theta_{3dB}}{\sqrt{R}}$$

(10.36)

The antenna axis (Fig. 10.9) is assumed to be oriented in a direction parallel to the unit vector $\mathbf{u}$ of spherical coordinates (e, a) where e and a denote elevation and azimuth angles respectively. The unit vectors $(\mathbf{e_e}, \mathbf{e_a}, \mathbf{u})$ define a base related to the antenna. The direction $\mathbf{u}'$ of the tracked object, close to $\mathbf{u}$, is expressed in this base in Cartesian or polar coordinates, (α, β) or (θ, ψ) respectively, in the form

$$\mathbf{u}' = \alpha \mathbf{e_a} + \beta \mathbf{e_e} + \gamma \mathbf{u}$$

(10.37)

with

$$\alpha = \sin \theta \cos \psi, \quad \beta = \sin \theta \sin \psi, \quad \gamma = \cos \theta$$

(10.38)

Often we shall set

$$\tau = \sin \theta$$

(10.39)

It can be verified that the pointing errors (α, β) are related to the elevation and azimuth errors of $\mathbf{u}'$ with respect to $\mathbf{u}$ by the relations

$$\Delta a = \frac{\alpha}{\cos e}, \quad \Delta e = \beta$$

(10.40)

which show that in the vicinity of the zenith ($e = \pi/2$) in automatic tracking the azimuth angle of the steerable mount should vary very quickly, otherwise significant errors may occur. This is a severe performance constraint for this type of mount.

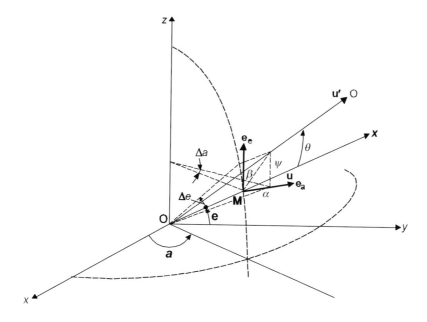

Fig. 10.9 Coordinate system used for analysis of tracking antennas.

In general, we should distinguish the problem of radar target tracking from that of beacon tracking. In effect, radar echoes are generally more or less depolarized, whereas it is possible to use the polarization characteristics of beacons to simplify their tracking systems (§10.3.5).

Angular and angle-error tracking

We should distinguish these two functions: the 'angle-error tracking' function consists of the processing of signals giving the angular differences (α, β) between the direction of the target and an axis related to the antenna. If the antenna has a mount of the azimuth-elevation type, equations (10.80) allow the determination of the absolute coordinates of the target, with respect to the ground.

In the 'angular tracking' function these differences (α, β), expressed as electrical signals, act as 'error signals' that control the feedback servomechanisms so that the steerable mount forces these errors to zero.

Quite often these two functions are associated. Imperfections in the electromechanical system of the servomechanisms (time constants, natural frequencies ....) and fluctuations due to targets (amplitude fluctuation) or to polarization (§10.3.3) will not always allow to force errors (α, β) to zero. The residual deviations may allow us to correct the data coming directly from the steerable mount. Equally, abnormal fluctuations of the error signals are sometimes usefully filtered by the time constant of the mount.

10.3.2 General characteristics of radar echoes

Scattering pattern of a target

The ideal radar target is point-like and reflects a spherical wave at the same polarization as that from the radar transmitter. Under these conditions, the orientation of the tracking antenna is such that the incident wave over its aperture is equiphase, so that its azimuth and elevation direction is the same as that of the target.

In fact, practical targets in general have a complex structure[8,9]. The waves reflected by the different parts of the target interfere, and the scattering pattern of the target (analysed, for instance, in the incident polarization) shows an irregular lobed appearance (Fig. 10.10). If L is the length of the target projected onto the radar-target axis, the angular width is of the order of λ/L radians. As the attitude of the target evolves as a function of time, the echo signal fluctuates according to a probability function which may be closely approximated by the Rayleigh function[10]. On the other hand, the phases of two adjacent lobes are often opposite. Assume that two lobes scan the aperture of the tracking antenna; the servomechanisms will steer the antenna so that the phase differences between the various points of the antenna are zero, which results in a pointing error whose amplitude may reach λ/D radians (D is the diameter of the aperture). Since the received signal under these conditions is weak, this effect is not really troublesome except for large targets at close range.

Nevertheless, in the case of an aircraft target flying at low altitude above a flat surface, such as the sea for instance, its specular image reflected by the ground interferes with the target and appears to the radar as a dual-point target. As a result, an *angular pointing noise* is observed whose probability function and frequency spectrum depend on the target, its motion, and on the servomechanism bandwidth. This phenomenon is often called 'specular point fluctuation' or 'glint' noise.

---

[8] Maffet, A.L., 'Scattering matrices', in *Methods of Radar Cross Section Analysis*, Academic Press, New York, 1968.

[9] Copeland, 'Radar target classification by polarization properties', *Proc. IRE*, July 1960.

[10] Various probability functions were studied by Swerling. The so-called Rayleigh function is the simplest of them (Swerling type 1).

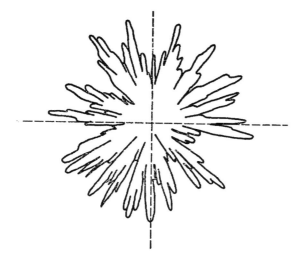

Fig. 10.10 Amplitude diffraction pattern of a radar target (observed at a single polarization).

Polarization of the echo - the polarization matrix

The polarimetric aspects of radar targets will be treated in Chapter 12 (§12.4). We shall limit ourselves in this section to the most important concepts relevant to angular tracking systems.

The polarization of the wave reflected by a target is almost always of a different polarization to that of the transmitted wave; the echo is often partially or totally depolarized. This plays an important role in relation with the cross polarization patterns of tracking antennas (§10.2.4). Let **u**, **v** be the two unit vectors of horizontal and vertical polarization, forming a right-handed set with the direction of propagation **U**. The incident polarization **p** in this base, is represented either by the column matrix $\begin{bmatrix} a \\ b \end{bmatrix}$ or the vector

$$\mathbf{p} = a\mathbf{u} + b\mathbf{v}$$

$$(10.41)$$

The polarization of the echo wave **p′** propagating in the direction **U′** = −**U**, will be represented in the vector bases (**u′**, **v′**, **U′**) forming a right-handed set, thus we have

$$\mathbf{p}' = a'\mathbf{u}' + b'\mathbf{v}'$$

with **u′** = **u**, **v′** = **v**, **U′** = −**U** and

$$\begin{bmatrix} a' \\ b' \end{bmatrix} = [\mathbf{P}] \begin{bmatrix} a \\ b \end{bmatrix} \tag{10.42}$$

[P] being the polarization matrix of the target or Sinclair matrix (this is a function of the attitude of the target with respect to the radar)

$$[\mathbf{P}] = \begin{bmatrix} \alpha & \beta \\ \gamma & \delta \end{bmatrix} \tag{10.43}$$

In general a target has two characteristic polarizations defined by the eigenvectors of the matrix **P** (§12.4.2).

Remark: The elements of the matrix [P] depend on the coordinate system used.

Exercise: calculate the elements of the matrix obtained by taking as base vectors the orthogonal circular components defined by

$$\mathbf{c}_1 = \frac{\mathbf{u} + j\mathbf{v}}{\sqrt{2}}, \quad \mathbf{c}_2 = \frac{\mathbf{u} - j\mathbf{v}}{\sqrt{2}}, \quad \mathbf{c}_1' = \frac{\mathbf{u}' + j\mathbf{v}'}{\sqrt{2}}, \quad \mathbf{c}_1' = \frac{\mathbf{u}' - j\mathbf{v}'}{\sqrt{2}}$$

Examples

- Dipole normal to the axis of the antenna, tilted at an angle θ with respect to the horizontal. Its characteristic polarizations are the linear polarizations parallel and perpendicular to the dipole. Then

$$[\mathbf{P}] = \begin{bmatrix} \cos^2 \theta & \sin \theta \cos \theta \\ \sin \theta \cos \theta & \sin^2 \theta \end{bmatrix} \tag{10.44}$$

- Sphere (and more generally rotationally-symmetric structures). This type of target is characterized by the fact that any linear polarization is a characteristic polarization and that its reflection response to a right-hand circular polarization is a left-hand circular polarization. We find

$$[\mathbf{P}] = \begin{bmatrix} 1 & 0 \\ 0 & -1 \end{bmatrix} \tag{10.45}$$

- Dihedral corner reflector (edge normal to the axis, tilted at an angle θ, dual-sector plane passing by the axis). Right-hand and left-hand circular polarizations are the characteristic polarizations. An incident vertical polarization gives a reflected polarization that undergoes a rotation through an angle of 2θ, thus

$$[\mathbf{P}] = \begin{bmatrix} \cos(2\theta) & \sin(2\theta) \\ -\sin(2\theta) & \cos(2\theta) \end{bmatrix} \tag{10.46}$$

In a general way, the coefficients of the matrix will be found by searching for the characteristic polarizations which result from symmetries of the target.

Target identification (see also §12.4)

Conversely, calculation of the characteristic polarizations of a target from the measurement of its polarization matrix may give interesting indications about its symmetry properties.

Case of a complex target: a complex target (an aircraft or a ship, for instance) may often be considered as an assembly of corners, of dipoles whose apparent relative positions may fluctuate. To characterize the polarization we define new matrices whose elements result from second-order averages of the preceding matrix elements. In Chapter 12 we shall define the coherence matrix, the Stokes parameters and the Mueller matrix.

Let us recall here some results. A complex target (an aircraft, for example) may be considered as an incoherent set of spheres, dihedrals and dipoles which have very different polarization properties. In fact we see that the properties of its matrix are often close to those of a cloud of dipoles whose orientations θ are random variables with a uniform probability density. We can then calculate the following mean values (coherence matrix, see §12.3.2).

$$\overline{|\alpha|}^2 = \overline{|\delta|}^2 = \frac{1}{2\pi} \int_0^{2\pi} \cos^4\theta \, d\theta = \frac{3}{8}; \quad \overline{|\beta|}^2 = \frac{1}{8} \tag{10.47}$$

$$\overline{\alpha\beta^*} = \overline{\beta\delta^*} = 0 \tag{10.48}$$

We deduce the mean powers reflected in each polarization component and the correlation coefficients of the fields received in these polarizations (§12.5)

$$\overline{|a'|}^2 = \frac{3|a|^2 + |b|^2}{8}; \quad \overline{|b'|}^2 = \frac{|a|^2 + 3|b|^2}{8} \tag{10.49}$$

$$\overline{|a'b'^*|} = 2\operatorname{Re}(ab^*)$$

Applications:

- Vertical linear incident polarization: $a = 1, b = 0$

  The depolarization of the received wave is measured by the ratio

$$\frac{\overline{|b'|}^2}{\overline{|a'|}^2} = \frac{1}{3} \tag{10.50}$$

  The correlation between orthogonal components is zero: $\overline{a'b'} = 0$

- Circular incident polarization: $a = 1, b = j$

  The depolarization is total and the correlation is zero

$$\frac{\overline{|b'|}^2}{\overline{|a'|}^2} = 1, \qquad \overline{a'b'*} = 0 \tag{10.51}$$

Comment: the loss of echo in circular polarization is small except if the target has a simple form. This explains why circular polarization is preferred for surveillance radars. It allows cloud echoes, composed of nearly spherical water droplets, to be suppressed on radar displays. On the other hand, aircraft of complex form clearly appear on the displays.

10.3.3 Conical scanning

Principle of operation - receiver block diagram

A conical scanning antenna in general comprises a rotationally-symmetric focused system (paraboloid, lenses, etc.), and a primary feed whose phase centre traces a circle of radius r centred on the focus in the focal plane. This means that the axis of the radiation beam follows in space a cone (Fig. 10.11). If the focal length f of the system is large enough, the half-angle at the apex of this cone (or defocalization angle) is approximately given by

$$\theta_1 = \frac{r}{f} \tag{10.52}$$

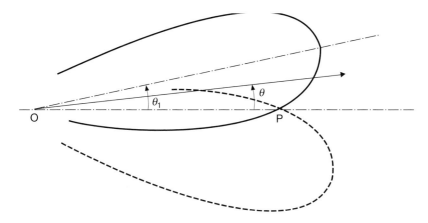

Fig. 10.11 Conical scanning.

The radiation beam cuts the axis of symmetry at a point P corresponding to a gain that remains constant in the course of the rotation. If the tracked target is assumed to be a point, and rigorously in the direction of this axis of symmetry, the amplitude of the received signal, from the primary feed, is theoretically constant. If the target is not exactly in the axis direction, there will be a modulation of the received signal at the rotation frequency of the primary feed. The amplitude and phase of this modulation characterize the spherical coordinates (θ, ψ) of the tracked target with respect to the antenna.

In order to show how we can deduce the Cartesian coordinates α and β from the modulated received signal, we shall summarize the operation of a scanning radar receiver. After downconversion, amplification and detection, an automatic gain control system (AGC) allows us to obtain a low frequency modulated signal whose mean level is independent of that of the received signal. Its expression is of the form

$$h(t) = 1 + p\tau \cos(2\pi Ft + \psi) \qquad (10.53)$$

where F is the scanning frequency and p a factor (the relative slope) that depends on the antenna pattern.

This signal is demodulated by means of a reference sinusoidal signal synchronized with the motion of rotation of the primary feed. In fact, two reference voltages shifted by $\pi/2$ are applied to the inputs of two demodulators whose outputs provide two d.c. voltages such that

$$x = Kp\tau\cos\psi \Bigg\}$$
$$y = Kp\tau\sin\psi$$
(10.54)

which combined with equations (10.38), yield

$$x = Kp\alpha \Bigg\}$$
$$y = Kp\beta$$
(10.55)

These signals are functions of the elevation and azimuth deviations of the target with respect to the axis direction of the antenna (equation 10.40). They are applied to the servomechanisms in elevation and in azimuth respectively.

Main parameters

(a) General expressions of the slope of 'angle-error tracking' and of the axial gain

First assume that the beam is pointing in the direction of the antenna axis of symmetry. Its pattern (in a nominal polarization) may be expressed as a function of coordinate pairs (α, β) or (τ, ψ) in the form

$$f_0(\alpha,\beta) = h_0(\tau,\psi) = h_0(\tau)$$
(10.56)

Now assume that the beam is pointing in the direction (α_1, β_1) which traces an axially-symmetric cone of semi-angle θ_1 about the axis as a function of time, then we have

$$\alpha_1 = \tau_1\cos\psi_1 \Bigg\}$$
$$\beta_1 = \tau_1\sin\psi_1$$
(10.57)

with

$$\tau_1 = \sin\theta_1 \text{ and } \psi_1 = 2\pi Ft$$
(10.58)

The pattern takes the form

$$h_1(\tau,\psi) = f_1(\alpha,\beta) = f_0(\alpha_1 - \alpha, \beta_1 - \beta)$$
(10.59)

Since in the vicinity of the axis, α and β are small, we can limit the expansion to first order

$$h_1(\tau,\psi) = h_0(\tau_1)\left[1 + p\tau\cos(2\pi Ft - \psi)\right] \qquad (10.60)$$

where

$$p = \frac{1}{h_0(\tau_1)}\frac{\partial h_0}{\partial \tau_1} \qquad (10.61)$$

The slope of the angle-error tracking p is the relative derivative (or logarithmic) on the axis of the unfocused pattern. The factor $h_0(\tau_1)$ represents the relative axial gain.

(b) Expression of the basic parameters in a Gaussian model

We can often take a Gaussian approximation to the main lobe of the radiated pattern in the following form:

$$h_0(\tau) = \exp\left(-\frac{\tau^2}{\tau_0^2}\right)$$

where τ_0 defines the 3 dB aperture of the pattern

$$\tau_0 \approx 0.85\,\theta_{3\mathrm{dB}}$$

The expressions for the relative axial gain $h_0(\tau_1)$ and of the slope (10.61) then take the form

$$h_0(\tau_1) = \exp\left(-\frac{\tau_1^2}{\tau_0^2}\right) \qquad (10.62)$$

$$p = \frac{2\tau_1}{\tau_0^2} \qquad (10.63)$$

The relative loss of axial gain is given in dB by the following equation:

$$\Delta G = 20\log\frac{1}{h_0(\tau_1)} \approx 8.6\left(\frac{\tau_1}{\tau_0}\right)^2 \qquad (10.64)$$

If ΔG is fixed, eliminating τ_1 between equations (10.64) and (10.63) shows that

$$p = \frac{1}{\tau_0}\sqrt{\frac{\Delta G}{2.15}} \qquad (10.65)$$

This relation shows that it is often necessary to trade-off the maximum range of the radar and its angular accuracy.

Remark: if the antenna is a circular aperture of diameter D, we have approximately $\tau_0 \approx \lambda/D$ radians. We obtain the following practical formula:

$$p = \frac{D}{\lambda} \sqrt{\frac{\Delta G}{2}} \qquad\qquad (10.66)$$

If we allow a loss of gain of 2 dB, we obtain the typical value used in practical systems

$$p = \frac{D}{\lambda} \qquad\qquad (10.67)$$

Angular accuracy of a conical scanning radar antenna related to thermal noise

(a) Generalities

The angular accuracy of a tracking radar depends on a large number of factors.

Factors related to the characteristics of the target: We have already noted the phenomena of angular pointing noise and the influence of echo fluctuations related to the scanning frequency. Equally, the velocity and the angular acceleration of the target, related to the cut-off frequency of the servomechanisms, influence the angular accuracy. In §10.2.4 we shall see the effect of depolarization.

Factors related to the characteristics of the rotating mount: Reflector rigidity, natural frequency and inertial momentum of the moving equipment, gearing performance and transmission tolerances, etc.

Factors related to the radar electrical characteristics: The study of these factors will be treated in the following sections. The knowledge of the role of the antenna characteristics will allow us to define the optimum characteristics.

(b) Expression of the standard deviation as a function of the signal-to-noise ratio

We will assume for simplicity that the radar is of the 'continuous' type. We will also assume that the antenna is pointing at the target, and we will calculate the standard deviation of the signal due to thermal noise, at the demodulator output.

At the mixer output, the signal has a power W_S and a constant amplitude measured by $\sqrt{W_S}$. In other respects, the noise is a fluctuating signal of power W_B whose amplitude variations may always be put, in the vicinity of the scanning frequency F, in the form

$$n(t) = \sqrt{W_B} \left[a(t)\cos(2\pi Ft) + b(t)\sin(2\pi Ft) \right] \qquad (10.68)$$

where $a(t)$ and $b(t)$ are two independent zero-mean random functions of the same variance such that

$$\overline{|a(t)|^2} + \overline{|b(t)|^2} = 1 \qquad (10.69)$$

and whose spectrum is assumed to be limited to the bandwidth $(2\Delta f)$ of the intermediate frequency (IF) circuits. The gain Γ of the IF amplifier is such that the amplitude of the output signal is constant

$$\Gamma \sqrt{W_S} = K \qquad (10.70)$$

The noise signal amplitude at the amplifier output is then

$$\Gamma n(t) = \Gamma \sqrt{W_B} \left[a(t)\cos(2\pi Ft) + b(t)\sin(2\pi Ft) \right] \qquad (10.71)$$

$$= K \sqrt{\frac{W_B}{W_S}} \left[a(t)\cos(2\pi Ft) + b(t)\sin(2\pi Ft) \right] \qquad (10.72)$$

We obtain at the demodulator outputs signals of zero mean value

$$x(t) = Ka(t)\sqrt{\frac{W_B}{W_S}}, \qquad y(t) = Kb(t)\sqrt{\frac{W_B}{W_S}} \qquad (10.73)$$

which may be interpreted as pointing errors $\alpha(t), \beta(t)$, such that

$$x(t) = Kp\alpha(t), \qquad y(t) = Kp\beta(t) \qquad (10.74)$$

whose mean square standard deviation is

$$\overline{\alpha^2} = \overline{\beta^2} = \frac{1}{p^2}\frac{W_B}{2W_S} \qquad (10.75)$$

Since the pointing errors α and β are independent, zero mean Gaussian, the mispointing θ follows a Rayleigh distribution with a mean-square standard deviation

$$\sigma_\theta^2 = \overline{\alpha^2} + \overline{\beta^2} = \frac{1}{p^2} \frac{W_B}{W_S} \qquad (10.76)$$

The servomechanism, of bandwidth $\Delta f'$, performs an integration of the signals such that the signal-to-noise ratio takes the form

$$R = \frac{W_S}{W_B} \frac{\Delta f'}{\Delta f} \qquad (10.77)$$

The angular fluctuation of the mount is finally given by

$$\sigma_\theta = \frac{1}{p\sqrt{R}} \qquad (10.78)$$

With a Rayleigh distribution we recall that the probability for $\theta \le \sigma_\theta$ is $1-1/e$, that is to say about 63%.

(c) Optimum defocalization

The radar equation (4.23) shows that the integrated signal-to-noise ratio R is proportional to the square of the on-axis gain, that is to say

$$R = B^2 G(\tau_1)^2 = B^2 \eta^2 \left(\pi \frac{D}{\lambda} \right)^4 |h_0(\tau_1)|^4 \qquad (10.79)$$

η being the gain factor of the aperture and B a coefficient which does not depend on the antenna. By combining equations (10.64) and (10.78) the fluctuation takes the form (h_0 is assumed real)

$$\sigma_\theta = \frac{2}{B\eta \left(\dfrac{\pi D}{\lambda} \right)^2 \dfrac{d\left[h_0(\tau_1) \right]^2}{d\tau_1}} \qquad (10.80)$$

Conclusion: The fluctuation is minimum when the defocusing is such that the power pattern (or gain) cuts the axis at a point of maximum slope, in general at a point of inflexion.

$$\frac{d^2\left(h_0^2\right)}{d\tau_1^2} = 0$$

(10.81)

Application to the Gaussian approximation to the pattern. From equation (10.55) we obtain the optimum defocusing, $\tau_1/\tau_0 = 1/2$ and the level of optimum overlapping

$$\Delta G_{dB} = 2.15$$

(10.82)

We then have from equation (10.60), $p = 1/\tau_0 = D/\lambda$ per radian and we have

$$\sigma_{\theta_{\min}} = \frac{\tau_0}{\sqrt{R}} = \frac{\lambda/D}{\sqrt{R}}$$

(10.83)

or, more explicitly

$$\sigma_{\theta_{\min}} = \frac{\sqrt{e}}{B\eta\pi^2}\left(\frac{\lambda}{D}\right)^3$$

(10.84)

Numerical example: consider a radar tracking a target with a signal-to-noise ratio of 10 dB per pulse. If the pulse repetition frequency is 1000 Hz and if the servomechanism time constant is one second and the signal-to-noise ratio 40 dB, if the diameter of the antenna is 3 metres and the operation frequency 5 GHz, then equation (10.84) shows that the tracking error is $\sigma_\theta = 0.2$ mrad and that an increase of 25% in the diameter doubles the accuracy.

Examples of conical scanning antennas

An important condition to fulfil is that of polarization stability as the beam rotates. In effect, a variation of the polarization may lead to a parasitic modulation of the echo (§10.2.2) which can give false measurements. That is why we should give to the primary feed not a rotation motion about the system focus but a 'nutation' in which the feed remains parallel to itself. This condition has the disadvantage of complicating the mechanics, particularly if the desired scanning frequency is high (> 50 Hz). For that reason the following system is sometimes used: a fixed circular cross-section guide carries the polarization to be radiated (linear or circular). This guide is connected by means of a rotating joint to a moving guide having a 'crank' shaped form which rotates, and at its end is fixed the primary feed. Despite this motion, the radiated polarization is in principle constant.

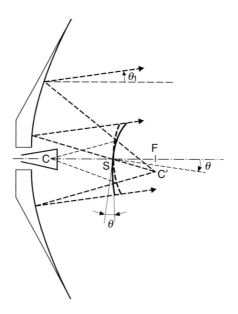

Fig. 10.12 Cassegrain tracking antenna. Beam scanning is obtained by conical rotation of the tilted subreflector.

In the case of large antennas ($D > 30\lambda$ or 50λ), it is often advantageous to use a Cassegrain system (Fig. 10.12). In this case we use an axially-symmetric fixed primary feed (dual-mode horn or corrugated for instance) and a subreflector tilted at an angle θ' and rotating about the focal axis CF. Geometrical optics shows that the image of the phase centre C of the feed moves to C' so that $FC' = 2SF$. This yields a defocusing angle

$$\theta_1 = \frac{FC'}{CF} = \theta' \frac{SF}{CS + SF}$$

The subreflector magnification is $m = CS/SF$. Hence

$$\theta_1 = \frac{2\theta'}{m+1} \tag{10.85}$$

If θ_1 is small and m not too large, the resulting aberration is acceptable. We may consider the calculation of the reflector profiles in order to minimise this aberration.

10.3.4 'Monopulse' antennas

Principle of radar monopulse antennas

(a) Diffraction pattern analysis (Fig. 10.13) Consider a focused axially-symmetric system (for instance a reflector), approximately oriented in the direction of the target. This target produces in the antenna focal plane an image located in the vicinity of the focus. This image is not a point. It is a diffraction pattern whose diameter δ of the main spot is a function of the relative aperture of the antenna ($\sin\theta_0 = D/2F$) with D as diameter, F as focal distance), and of the wavelength according to the classical formula (§9.8, eqn. 9.101).

The problem of the determination of the target direction with respect to the axis of symmetry of the antenna consists of determining the exact position of the diffraction pattern centre with respect to the focus. To solve this problem, we compare the level of the diffraction pattern at different points located at a circumference centred on the focus; if these levels are equal, this shows that the pattern centre coincides with the focus. We can say that the principle of the conical scanning antenna consists of *successively scanning these points by means of a unique primary aperture whose motion is circular*, and also consist in analysing the modulation of the received signal.

The monopulse antenna, on the other hand, compares the received amplitudes *simultaneously* by means of several fixed primary apertures (in general four) symmetrically situated about the focus: the angular errors of the target direction, with respect to the antenna axis, are measured by the received level differences. The differences are compared to the sums in order to obtain tracking signals which are independent of the overall level of the received signal. Tracking is then independent of signal amplitude fluctuations; this is the main advantage of the monopulse technique – *only one pulse is necessary* !. Note that this method of measurement assumes that the diffraction pattern is symmetric. This is the case for a true point target. On the other hand, in the case of a complex target, the diffraction pattern may be distorted and the measurements false. This results in *angular pointing noise* (§10.3.2 and §14.3.2).

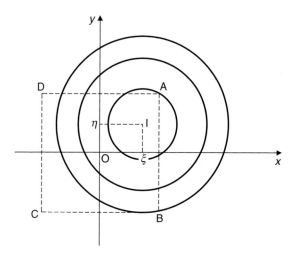

Fig. 10.13 Diffraction pattern analysis in the focal plane.

(b) Sum and Difference patterns We saw that we should process the sums and the differences of the received amplitudes in the primary horns placed about the focus. The amplitude variations of sum and difference as a function of the angular errors of the target constitute the sum and difference patterns. First let us assume that the whole radar operates in a plane space; the target is characterized by only one coordinate α (Fig. 10.14). The focal plane is reduced to a straight line where the coordinates are referenced with one coordinate x. The image of the target is a diffraction pattern centred on a point of the abscissa

$$\xi = f\alpha$$

and whose amplitude variation is characterized by an even function (§9.9.1)

$$\psi(x - \xi)$$

Consider two primary horns whose phase centres are symmetrically located at a distance $a/2$ either side of the focus F. At these points, the amplitudes of the diffraction function are

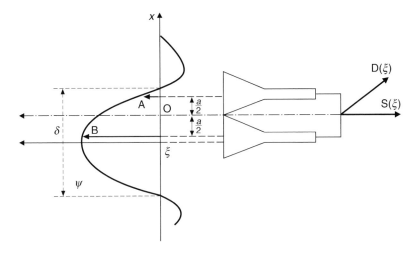

Fig. 10.14 Monopulse: interaction between the diffraction pattern and two symmetrical primary horns.

$$A(\xi) = \psi\left(\frac{a}{2} - \xi\right)$$

$$B(\xi) = \psi\left(-\frac{a}{2} - \xi\right) = \psi\left(\frac{a}{2} + \xi\right)$$

(10.86)

Let us consider, as a first approximation, that the received signal levels at each horn are proportional to A and B. These signals are applied to a coupler of the 'magic tee' type that provides their sum and difference. The corresponding patterns are expressed by the following functions

$$S(\xi) = A + B = \psi\left(\frac{a}{2} - \xi\right) + \psi\left(\frac{a}{2} - \xi\right)$$

$$D(\xi) = A - B = \psi\left(\frac{a}{2} - \xi\right) - \psi\left(\frac{a}{2} + \xi\right)$$

(10.87)

We verify that the sum is an even function and the difference an odd function. These functions are schematically represented in Fig. 10.15. Close to the axis, we can replace the patterns by their main parts

$$S = S(0) \approx 2\psi\left(\frac{a}{2}\right)\Bigg|$$

$$D = D'_\alpha(0) \qquad\qquad (10.88)$$

where $D'_\alpha(0)$ denotes the derivative of $D(\alpha)$ at $\alpha = 0$: $D'_\alpha(0) = 2f\psi'\left(\frac{a}{2}\right)$.

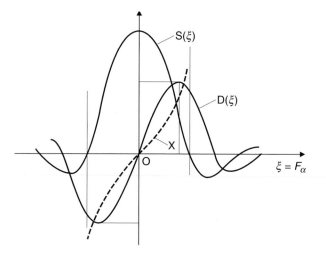

Fig. 10.15 Sum and Difference patterns of a monopulse antenna.

The corresponding gains close to the axis are of the form

$$G_S(0) = \eta\left(\frac{\pi D}{\lambda}\right)^2, \qquad G_D(\alpha) \approx \frac{1}{2}G''_D(0)\alpha^2 \qquad (10.89)$$

where η is the gain factor of the sum channel, G_S and G_D are the gains of the sum and difference channels and $G''_D(0)$ the second derivative of $G_D(\alpha)$ on the axis.

The performances of the antenna depend thus on two main parameters: *the axial gain* $G_S(0)$ which characterizes the range of the radar and *the axial second derivative of the difference gain* $G''_D(0)$ which characterizes the sensitivity of the system.

(c) Utilization of the sum and difference signals - Block diagram of a monopulse receiver

The main mode of utilization of the monopulse technique is called 'amplitude comparison' since it is based on a comparison of the amplitudes of the sum and difference signals. The block diagram is shown in Fig. 10.16.

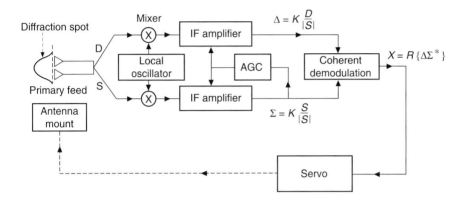

Fig. 10.16 Block diagram of a monopulse receiver.

To make use of the two signals S and D from the magic tee, it is necessary to use two receiver channels, each consisting of a down conversion mixer and IF amplifier. To obtain the characteristic information of the angular errors, and independent of the amplitude variations of the received signal, an automatic gain control (AGC) is inserted in the sum channel. The amplifier output signal has thus a constant amplitude which may be symbolized by the following expression

$$\Sigma = gS = \frac{K}{|S|} S$$

(10.90)

where $g = K/|S|$ is the gain of the amplifier and K a constant. The AGC control voltage is also applied to the difference amplifier. The output signal from this is then of the form

$$\Delta = gD = \frac{K}{|S|} D$$

(10.91)

The two signals are then applied to a coherent detector. This delivers an output signal X which is the scalar product of the vectors associated with the signals Σ and Δ. Using complex notation, we have

$$x = \mathrm{Re}\{\Sigma^*\Delta\} = K^2\,\mathrm{Re}\left\{\frac{S^*D}{|S|^2}\right\} \tag{10.92}$$

If S is assumed *real*, we have

$$x = K^2\,\mathrm{Re}\left\{\frac{D}{S}\right\} \tag{10.93}$$

If we replace S and D by their main parts, we obtain

$$x = K^2 p\alpha \tag{10.94}$$

from which and from equation (10.89), we have

$$p = \sqrt{\frac{G_D''(0)}{2G_S(0)}} \tag{10.95}$$

Main characteristics of a classical monopulse antenna

(a) Gain associated with the sum channel - slope and gain associated with the difference channel - tradeoffs between Sum and Difference characteristics in conventional monopulse antennas
   The two main characteristics are the gain of the sum channel $G_S(0)$ which determines the signal-to-noise ratio, and the slope associated with the difference channel (10.95) which determines the system sensitivity. Let us see how these two factors are related.
   Consider a one-dimensional scheme (Fig. 10.14) and two adjacent horns of aperture a. In the sum channel, the two feeds connected in phase operate as a single aperture $2a$. Under these conditions, it can be demonstrated that the sum gain is maximum, that is to say, that the received energy is maximum when the aperture $2a$ is of the order of magnitude of the diameter δ of the central lobe of the diffraction pattern (§9.9.3)

$$2a \cong \delta \tag{10.96}$$

In effect, if the aperture $(2a)$ is smaller than the central lobe, there will be some energy lost around it. If it is larger, it will not be properly excited since it will include the sidelobes that are out of phase with the main spot.

In the difference channel, due to an angular displacement of the target, the diffraction pattern successively scans the two apertures of diameter a. The main lobes are generated when the diffraction pattern successively passes over the two apertures. The lobes will have a maximum level that will lead to the maximum slope when each aperture will have a diameter close to δ, that is to say

$$a \cong \delta \tag{10.97}$$

Conditions (10.96) and (10.97) are incompatible; there is a kind of conflict between the sum gain and the difference slope, as that which opposes gain and slope on the axis of a scanning antenna. To optimize both the sum and the difference, it would be necessary to realize an odd illumination function of total width 2δ double that of the even illumination. Therefore, the dimensions of a *conventional* primary feed should result from a certain trade-off depending on the sum gain and on the difference slope of the overall characteristics of the antenna.

(b) Optimum primary illumination functions. Multimode horns
The 'gain factor' function of an antenna in terms of angular errors in the sum and difference channels can be expressed in the form of a convolution product of the diffraction pattern $\psi(x, y)$ with the corresponding illumination function: $s(x, y)$ for the sum; $d(x, y)$ for the difference (§8.1.4), (9.92).

$$G_S(\xi,\eta) = \left(\frac{\pi D}{\lambda}\right)^2 \frac{\left|\iint_S s(x,y)\psi(x-\xi,y-\eta)\,dxdy\right|^2}{\iint_S |s|^2\,dxdy \iint_S |\psi|^2\,dxdy} \tag{10.98}$$

$$G_D(\xi,\eta) = \left(\frac{\pi D}{\lambda}\right)^2 \frac{\left|\iint_S d(x,y)\psi(x-\xi,y-\eta)\,dxdy\right|^2}{\iint_S |d|^2\,dxdy \iint_S |\psi|^2\,dxdy} \tag{10.99}$$

It can be demonstrated using Schwartz's inequality that the maximum gain factor in the sum channel is obtained when the sum illumination function reproduces the diffraction pattern (or at least its complex conjugate ψ^*), whereas the maximum slopes of the difference patterns are obtained when the difference primary illumination functions are equal to the *partial derivatives* of the diffraction pattern

$$
\left.\begin{array}{ll}
\text{Sum} & s(x,y)=\psi^{*}(x,y) \\[4pt]
\text{Elevation difference} & d_{e}(x,y)=\dfrac{\partial}{\partial x}\psi^{*}(x,y) \\[4pt]
\text{Azimuth difference} & d_{a}(x,y)=\dfrac{\partial}{\partial y}\psi^{*}(x,y)
\end{array}\right\}
\tag{10.100}
$$

The aspect of the corresponding illumination functions is illustrated in Fig. (10.17). If only the main lobes are considered, it is clearly shown that the width of the odd illumination should be larger than that of the even illumination. We have seen in §8.2.4 how these conditions are fulfilled by the use of *multimode horns*. Under these conditions, the main part of the difference gain factor close to the axis is of the form

$$
G_{D}(\alpha)=\left(\frac{\pi D}{\lambda}\right)^{2}\left(\frac{\pi D}{4\lambda}\right)^{2}\alpha^{2}
\tag{10.101}
$$

Then

$$
G_{D}''(0)=2\left(\frac{\pi D}{\lambda}\right)^{2}\left(\frac{\pi D}{4\lambda}\right)^{2}
\tag{10.102}
$$

If the illumination is not an ideal function $\psi_{x}'^{*}$, we have in general

$$
G_{D}''(0)=2\eta'\left(\frac{\pi D}{\lambda}\right)^{2}\left(\frac{\pi D}{4\lambda}\right)^{2}
\tag{10.103}
$$

with $\eta' \le 1$. η' is a slope factor. Under these conditions, the 'angle-error tracking' slope becomes

$$
p=\sqrt{\frac{G_{D}''(0)}{2G_{S}(0)}}=\sqrt{\frac{\eta'}{\eta}}\,\frac{\pi}{4}\frac{D}{\lambda}\ \left(\text{rad}^{-1}\right)
\tag{10.104}
$$

If the gain factors (η) and of slope (η') are of the same order of magnitude, we have

$$
p\approx\frac{\pi}{4}\frac{D}{\lambda}\ \left(\text{rad}^{-1}\right)
\tag{10.105}
$$

If we compare this result to that of the slope (equation (10.6)) of a conical scanning antenna, we observe a loss of $\pi/4$ due to the fact that the axial gain $G_{S}(0)$ does not undergo the defocusing loss which, as we know, is of the order of 2 dB, representing approximately a loss of $4/\pi$.

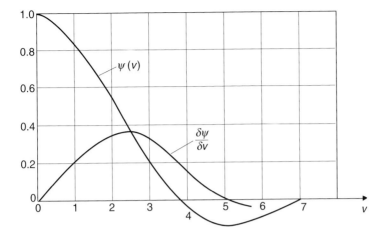

Fig. 10.17 Amplitude diffraction pattern at the focus and its derivative.

(c) Sidelobe effects

The aspect of the sum and difference secondary patterns is represented in Fig. 10.15. The difference sidelobes play an important role in the tracking characteristics. Indeed, if a small parasitic signal of amplitude ε emanating from, for instance, a ground echo, is picked up in the sum channel by a sidelobe, it does not generate any angle-error, and it does not modify the relative slope p. On the other hand, if it is picked up in the difference channel by a sidelobe of relative amplitude L (with respect to the sum gain), it can generate an angle-error α such that we have

$$p\alpha = L\varepsilon \qquad (10.106)$$

Therefore, it can be important to reduce the difference sidelobe level L and increase the slope p, and hence the diameter of the antenna. Moreover, if the diameter increases, the sidelobes of same level are closer to the axis, therefore less troublesome.

(d) Null depth

In practice, because of mechanical tolerances, the difference pattern is never exactly zero on the axis. With respect to the axial sum signal, a parasitic signal q *in phase quadrature* measures the 'null depth' (a *cophasal* parasitic signal would produce an angular shift). This is an important property of the monopulse antenna. Levels lower than -30 dB or -40 dB are often required. In the vicinity of the axis, defined by the minimum of D, the sum and difference patterns with respect to the sum signal may be expressed in the form

$$D(\alpha) = p\alpha + jq \left.\vphantom{\begin{array}{c}\\\\\end{array}}\right\}$$
$$S(\alpha) = 1$$

$$(10.107)$$

(e) Influence of both the null depth and a parasitic differential phase difference between the two channels sum and difference at the demodulator input
With this assumption, the demodulator input signals take the form

$$D(\alpha) = (p\alpha + jq)\exp(j\varphi) \left.\vphantom{\begin{array}{c}\\\\\end{array}}\right\}$$
$$S(\alpha) = 1$$

$$(10.108)$$

The 'angle-error tracking' signal is then

$$x = \mathrm{Re}\left\{\frac{D}{S}\right\} = p\alpha\cos\varphi - q\sin\varphi$$

$$(10.109)$$

Forcing it to zero leads to a pointing error

$$\alpha = \frac{q}{p}\tan\varphi \quad \text{or} \quad \frac{\alpha}{\theta_{3dB}} \approx q\tan\varphi$$

$$(10.110)$$

Numerical example: a null depth of -30 dB (that is $q \approx 1/30$) associated with a phase error of $20°$ leads to a pointing error of $1/100$ of the beamwidth. Thus, significant phase errors can be tolerated.

Angular accuracy of monopulse radar - Influence of antenna characteristics

(a) Angular fluctuation expression as a function of the signal-to-noise ratio
Assume for simplicity that the radar is of the CW type. Consider the sum and difference signals at the mixer outputs. The signal power in the sum channel is

$$|S|^2 = W_S$$

$$(10.111)$$

If the signal level is strong enough, we may neglect the influence of thermal noise on the sum channel. In the difference channel, noise may represented by an axially-symmetric stationary random complex function

$$U(t) = u(t) + jv(t) \tag{10.112}$$

which is superimposed on the difference signal D.

The spectrum of this function is assumed to be limited to the IF band ($2f$).

The corresponding noise power is the mean square value

$$W_B = \overline{|U|^2} = 2\overline{|u|^2} = 2\overline{|v|^2} \tag{10.113}$$

The components $u(t)$ and $v(t)$ are independent Gaussian functions. The output demodulator signal is given by equation (10.87) (§10.3.4). Hence

$$x(t) = K^2 \mathrm{Re}\left\{ \frac{S^*[D + U(t)]}{|S|^2} \right\} = K^2 \mathrm{Re}\left\{ \frac{S^* D}{|S|^2} \right\} + K^2 \mathrm{Re}\left\{ \frac{S^* U(t)}{|S|^2} \right\} \tag{10.114}$$

The deviation of x with respect to its mean value is, assuming that S is real

$$\Delta x = K^2 \mathrm{Re}\left\{ \frac{S^* U}{|S|^2} \right\} = K^2 \frac{u(t)}{S} \tag{10.115}$$

This deviation corresponds to an angular error $\Delta\alpha$ such that

$$\Delta x = K^2 p \Delta\alpha(t) \tag{10.116}$$

therefore

$$\Delta\alpha(t) = \frac{1}{p} \frac{u(t)}{S} \tag{10.117}$$

The pointing fluctuation is measured by

$$\sigma_\alpha^2 = \overline{|\Delta\alpha|^2} = \frac{1}{p^2} \frac{\overline{|u|^2}}{|S|^2} = \frac{1}{p^2} \frac{W_B}{2W_S} \tag{10.118}$$

and for the total standard deviation σ_θ^2

$$\sigma_\theta^2 = \sigma_\alpha^2 + \sigma_\beta^2 = \frac{W_B}{p^2 W_S} \tag{10.119}$$

In fact, the servomechanisms, with time constant T, are equivalent to a filter that reduces the noise power to the frequency band $1/T$.

Therefore, we have

$$\sigma_\theta = \frac{1/p}{\sqrt{R}}$$

$$(10.120)$$

where R is the sum signal to difference noise ratio integrated over a time equal to the time constant of the servomechanisms. This expression is identical to that which we found for conical scanning antennas. With expressions (10.104) for p, we find

$$\sigma_\theta = \sqrt{\frac{\eta}{\eta'}} \frac{\left(\dfrac{4\,\lambda}{\pi\,D}\right)}{\sqrt{R}}$$

$$(10.121)$$

(b) Influence of antenna characteristics

Here W_S is proportional to the square of the sum gain

$$R = B^2 G_S^2(0) = B^2 \eta^2 \left(\frac{\pi D}{\lambda}\right)^4$$

$$(10.122)$$

Therefore, we have

$$\sigma_\theta = \frac{4}{B\pi^3\sqrt{\eta\eta'}} \left(\frac{\lambda}{D}\right)^3$$

$$(10.123)$$

In radar tracking, always with the assumption of an adequate signal-to-noise ratio in the sum channel, the fluctuation depends upon the product of the difference slope factor η' with the sum gain factor η. In addition, the fluctuation is inversely proportional to the cube of the diameter. It has been shown that conventional monopulse primary feeds (§10.4.2) do not allow us to optimize the sum gain and the difference slope simultaneously. In such a case, we should design the primary feed so that the product $\eta\eta'$ is maximum.

Various types of classical monopulse antennas

Independently of the type of focused system used (paraboloidal reflector, lens, Cassegrain system, etc.), monopulse antennas differ in the structure of their primary feed. We saw (§10.4.2) that for a 2-D system, the simplest primary feed is composed of two horns symmetrically located about the focus and connected to a magic tee that provides the sum and difference patterns (Fig. 10.14). In the case of an actual system operating in a 3-D space, we generally use four horns whose phase centres are placed at the corners of a square centred on the focus. Two configurations are then possible: the diamond (Fig. 10.18) and the square (Fig. 10.19).

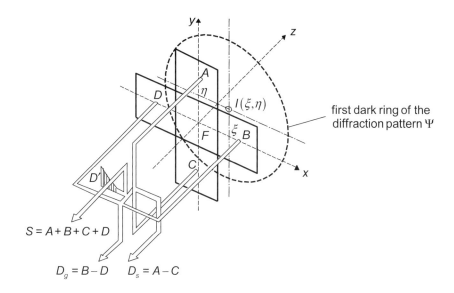

$$S = A + B + C + D$$

$$D_g = B - D \qquad D_s = A - C$$

Fig. 10.18 Monopulse primary feed - diamond configuration (3 magic tees).

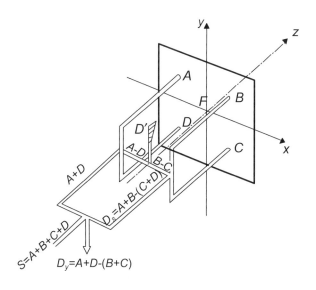

Fig. 10.19 Monopulse primary feed - square configuration (4 magic tees).

Primary feeds allowing the simultaneous optimization of both sum and difference patterns

It has been shown (§10.3.4) that the ideal primary illumination functions related to the sum and the difference are respectively represented by the diffraction spot and its derivative. In practice, this reduces to the realization of an odd illumination of width double of that of the even illumination. To reach this goal, several techniques have been used.

It seems that the most efficient method to provide illumination functions as close as possible to the ideal ones, and to independently control the difference and sum patterns, consists of using the *multimode-antenna technique* (§8.2.4, Fig.8.13).

Influence of cross polarization on the tracking accuracy

(a) Case of an axially-symmetric system
Using the unit vectors of co-polar and cross-polar, respectively **u** and **v**, the vector form of the sum pattern of the axially-symmetric system illuminated by an imperfect primary feed is deduced from the general results (8.1.3) and equation (8.26). If we assume a Gaussian approximation of the pattern, we obtain in the vicinity of the axis

$$\mathbf{S}(\alpha,\beta) = \exp\left(-p^2\tau^2\right)\left(\mathbf{u} + q^2\alpha\beta\,\mathbf{v}\right) \qquad (10.124)$$

where q is a cross-polar component and

$$\alpha = \tau\cos\psi, \quad \beta = \tau\sin\psi, \quad p = D/\lambda.$$

We notice that in these expressions the cross polarization is zero on the principal planes and maximum on the diagonal planes ($\psi = \pm\pi/4$).

The maximum level of relative cross polarization on a diagonal plane is given by

$$N = \frac{q^2}{2ep^2} \qquad (10.125)$$

where e is the base of natural logarithms.

The difference patterns $\left(\mathbf{D}_\alpha, \mathbf{D}_\beta\right)$ in the vicinity of the axis are deduced (if the antenna is optimized) from the partial derivatives of $\mathbf{S}$ ($\partial\mathbf{S}/\partial\alpha$, $\partial\mathbf{S}/\partial\beta$) by normalizing so that in D_α for instance, the coefficient of α is the angular tracking slope p

$$\mathbf{D}_\alpha(\alpha,\beta) = \exp\left(-p^2\tau^2\right)\left(\mathbf{u}p\alpha - \mathbf{v}eNp\beta\right) \qquad (10.126)$$

If an incident wave is characterized by a polarization

$$\mathbf{P} = a\mathbf{u} + b\mathbf{v} \tag{10.127}$$

where b/a indicates the degree of cross-polarization (b/a is *a priori* complex), the received signals in the sum and difference channels are given by the following scalar products.

$$s = \mathbf{S}^*.\mathbf{P} = \exp\left(-p^2\tau^2\right)\left(a + 2ep^2Nb\right) \left.\begin{array}{c}\\\\\\\\\end{array}\right\}$$

$$d = \mathbf{D}_\alpha^*.\mathbf{P} = \exp\left(-p^2\tau^2\right)\left(ap\alpha - epNb\beta\right) \tag{10.128}$$

In the case where b/a is real (polarization rotation) the 'angle-error tracking' signal given by equation (10.88) (§10.2.4) is zero, not for $\alpha = 0$, but for

$$\alpha = eN\frac{b}{a}\beta \tag{10.129}$$

This produces an *intermodulation* between elevation and azimuth mispointings measured by the product of the cross-polarization level in the sum channel by the cross-polarization ratio of the received wave.

Example: for a cross-polarization level of 20 dB, we have $N = 1/10$ if the incident wave cross-polarization is $b/a = 1/3$, we obtain an intermodulation of $\alpha/\beta \cong 10\%$.

Conclusion: in the case of an axially-symmetric system, the cross polarization does not produce a pointing error but an elevation-azimuth intermodulation that may produce angle-errors and servomechanism instabilities.

(b) Case of a non axially-symmetric system
In this case it will be shown (§10.4.1) that the cross polarization pattern associated with the sum pattern is odd on the plane orthogonal to the offset plane, and in phase quadrature with the direct polarization component. The vector pattern is then of the form (Gaussian model)

$$\mathbf{S}(\alpha,\beta) = \exp\left(-p^2\tau^2\right)\left(\mathbf{u} + jq'\beta\mathbf{v}\right) \tag{10.130}$$

where q' is a cross-polarization coefficient. The maximum of the cross polarization relative to the maximum level in the plane ($\alpha = 0$) is given by

$$N' = \frac{q'}{p\sqrt{2e}} \qquad (10.131)$$

The difference pattern **D**, as can be derived from **S**, is then associated with a non-zero cross polarization on the axis

$$\mathbf{D}_\beta(\alpha,\beta) = \exp\left(-p^2\tau^2\right)\left(p\beta\mathbf{u} - jN'\sqrt{\frac{e}{2}}\mathbf{v} \right) \qquad (10.132)$$

The tracking signal is given by equations (10.92) and (10.93). If the incident polarization is elliptical, there is a pointing error on the axis. For example, if $b = -ja$ (circular polarization) then

$$p\beta = \sqrt{\frac{e}{2}}N' \qquad (10.133)$$

This means that expressed as a fraction of 3 dB beamwidth ($1/p$) the pointing error is approximately equal to the cross polarization level N'. With the same numerical example, in the symmetrical case we would have an error of 10% of the beamwidth.

Comparison of conical scanning and monopulse

(a) Influence of thermal noise on pointing accuracy
Equations (10.79) and (10.118) allow the comparison of the standard deviations σ and σ' respectively obtained with a conical scanning and a monopulse system. By assuming that gain factors (η) and slope (η') are of the same order of magnitude, we obtain

$$\frac{\sigma'}{\sigma} = \frac{4}{\pi\sqrt{e}} \approx 0.78 \qquad (10.134)$$

which is an improvement of 1 dB in favour of the monopulse system.

This result can be explained by the fact that the only additional loss to be taken into account in the case of a conical scanning is a loss of axial gain, only in transmission, of about 2 dB. This corresponds, according to the radar equation, to 1dB of pointing fluctuation. In effect, in reception we get the product of the square root of the gain with the relative slope, that is to say what can be called the absolute slope of the system. If equation (10.100) seems to show a loss of relative slope of the monopulse system (about 1 dB), this is only due to the use of the axial sum gain as a reference signal.

It can be concluded from these considerations that a conical scanning using an unfocused beam in transmission should in principle have the same qualities as a monopulse with respect to thermal noise. In both cases, the 'angle-error tracking' signal is coherently demodulated by a signal of very high signal-to-noise ratio: the sinusoidal reference in the case of a conical scanning, the sum signal in the case of a monopulse system.

(b) Other comparison points

In fact, conical scanning exhibits other disadvantages. The rapid rotation of the primary system sometimes creates some difficult mechanical problems. On the other hand, echo fluctuations in amplitude may produce pointing noise to which monopulse is insensitive (§10.3.4).

At the price of an increase of the receiver complexity (three receiving channels instead of one), the monopulse system allows us to independently process the angular error and reference signals. In particular, it is possible (using multimode feeds) to obtain optimum even and odd patterns. The result is an increase of range and accuracy. Finally, since the antenna is completely static, all risks of vibration are eliminated.

10.3.5 Beacon tracking

The tracking processes described in the preceding sections are in general also applicable to the tracking of a beacon on board a missile, an aircraft or a satellite. Nevertheless, the optimization conditions are not in general the same; the signal-to-noise ratio is proportional to the antenna gain and not to its square. In space telecommunications, in particular in the case of large antennas, with a high gain slope, it is often not necessary to extract the entire available odd mode in the primary feed where it is generated by the offset of the diffraction spot. This often allows better optimization of the sum channel transmitting the telecommunication signals.

If the beacon is circularly polarized, it can be shown that the use of only the TM_{01} mode in a circular guide, demodulated in phase and in quadrature by the sum signal is sufficient to provide the elevation and azimuth errors. If the polarization is linear, two modes are necessary; TM_{01} and TE_{01}, for example.

10.4 NON AXIALLY-SYMMETRIC SYSTEMS

10.4.1 Offset reflector

Illumination

To avoid the blockage effect produced by the primary feed when it is placed in front of a reflector, a common solution is to use a device made up of a reflector whose aperture is offset with respect to the axis, with the primary feed being tilted at an angle θ_0 to point approximately at the reflector centre (Fig. 10.20).

The properties of such a system are similar to those of an axially-symmetric system. In particular, the illumination function can be considered as the product of the primary pattern with the transfer function (at least in the nominal polarization (§10.3.1)). Consider the common case of a parabolic reflector. The focal length from the focus F to the reflector centre O plays the role of an 'equivalent focal length' f'

$$f' = \frac{2f}{1 + \cos\theta_0} \qquad (10.135)$$

The offset is measured by the distance r_0 from O to the axis of revolution

$$r_0 = f'\sin\theta_0 = 2f\tan\frac{\theta_0}{2} \qquad (10.136)$$

In a base Oxyz, the transfer function is expressed by equation (9.16)

$$T(x,y) = \frac{f'}{\rho} = \frac{f'}{f}\frac{f}{\rho} \approx 1 - \frac{y}{2f}\sin\theta_0 + ... = 1 - \frac{r_0}{2ff'}y + ...\,(10.137)$$

which has an asymmetrical trapezoidal form. This may produce distortions of the radiated patterns (in particular, by reduction of the null depth in a difference monopulse pattern).

An important disadvantage of offset axially-symmetric systems is due to their polarization properties. We saw (§9.2) that an axially-symmetric system shows a pure polarization if it is illuminated by a Huygens source having the same axis. If such a source is oriented in the direction θ_0 with respect to the axis, the field lines of the aperture undergo a distortion called 'scallop-shell' (Fig. 10.20) which we shall now evaluate.

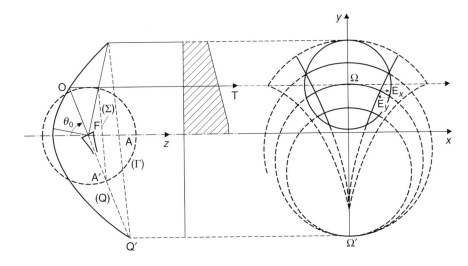

Fig. 10.20 Field lines on the aperture of an offset reflector fed by a Huygens source.

Cross polarization[11]

Consider a horizontally-polarized primary feed. The plane Q (normal to the plane of symmetry) which contains the pointing axis of the primary feed, also contains the radiated electric field. The corresponding field line over the aperture is the projection of the intersection of this plane Q with the paraboloid (P). The equations of the paraboloid and the plane are the following.

$$\left. \begin{array}{c} x^2 + y^2 = 4fz \\[2mm] y = -(z-f)\tan\theta_0 \end{array} \right\} \qquad (10.138)$$

Eliminating z from the above equations we find the projection we were looking for. A circle with the following equation is found

[11] Dijk, J., 'The polarization losses of offset paraboloid antennas', *IEEE Trans. Antennas and Propagation*, Vol. AP-22, July 1974.

$$x^2 + \left(y - \frac{2f}{\tan\theta_0}\right)^2 - \frac{4f^2}{\sin^2\theta_0} = 0$$

(10.139)

The slope of the field lines in the vicinity of the symmetry plane, that is to say the cross polarization rate, is obtained by differentiating

$$\frac{E_y}{E_x} = \frac{dy}{dx} = -\frac{x}{y - \dfrac{2f}{\tan\theta_0}} \approx -\frac{x}{2f}\sin\theta$$

(10.140)

$$\frac{E_y}{E_x} = -\frac{x}{f'}\tan\frac{\theta_0}{2}$$

This equation shows that the cross polarization is zero for an offset $\theta_0 = 0$, and maximum for $\theta_0 = \pi/2$ (case of a reflector horn).

By generalizing, we find that the electric and magnetic field lines over the aperture plane are the projections of the intersections of the paraboloid with the cones of apex F that are supported by the lines of polarization of the feed Σ. These are formed, in the case of a Huygens source, of two tangential beams of circles that are orthogonal at the point Ω' with coordinates

$$x_{\Omega'} = 0, \qquad y_{\Omega'} = -\frac{2f}{\tan\dfrac{\theta_0}{2}}$$

(10.141)

Equation (1.136) is thus a first order approximation of the cross polarization over the whole aperture. This function is odd with respect to x and in phase with the co-polar illumination. The patterns are obtained by Fourier transformation; the result is that the odd pattern (cross polarization) is in phase quadrature with the even pattern in direct polarization. In a Gaussian approximation we shall be able to use the form of §10.3.4 (equation (10.130))

$$\mathbf{S}(\alpha,\beta) = \exp\left(-p^2\tau^2\right)\left(\mathbf{u} + jq'\beta\mathbf{v}\right)$$

(10.130)

Peak level of cross polarization with a Gaussian model of the illumination function: Assume the following illumination in the principal polarization.

$$E_x(x,y) = \exp\left(-\frac{x^2 + y^2}{x_0^2}\right)$$

(10.142)

The relative level of the taper at the edge of the reflector is (in decibels)

$$N_{dB} \approx 2.15 \frac{D^2}{x_0^2}$$

By taking, for example, $N_{dB} = 8.6$ dB, we obtain $x_0 = D/2$.

The cross polarization illumination is given by

$$E_y(x, y) = -\frac{1}{f'} \tan\frac{\theta_0}{2} x \exp\left(-\frac{x^2 + y^2}{x_0^2}\right) \qquad (10.143)$$

which is an odd function of x.

We also encounter the same form in tracking antennas (§10.2.4). Its maximum is obtained on the plane $y = 0$ for $x = x_0/\sqrt{2}$.
We obtain the relative maximum level

$$N' = \left|\frac{E_{y_{max}}}{E_x(0,0)}\right| = \frac{1}{\sqrt{2e}} \frac{x_0}{F} \tan\frac{\theta_0}{2} \qquad (10.144)$$

and for $x_0 = D/2$, we have

$$N' = \frac{1}{\sqrt{2e}} \frac{D}{2f'} \tan\frac{\theta_0}{2} \qquad (10.145)$$

Numerical example: Consider a reflector, illuminated by an ideal corrugated horn whose illumination is approximated by a Gaussian function. Suppose $f' = 3D$. We have $\theta_0 = \pi/2$, then $N' = 23$ dB. Note that if the horn were smooth, the cross polarization would be far larger due to the use of the TE_{11} mode. Note that it is possible to avoid the cross-polarization of an offset reflector by use of several techniques; reflector construction with parallel metallic plates is equivalent to a polarization filter. Use of a multimode feed with odd modes then cancels the cross polarization.

10.4.2 Shaped reflectors - pattern synthesis

Introduction

We saw in Chapter 7 how the operational characteristics of a radar often require antenna patterns of a particular form (particularly a cosec^2 pattern in elevation). The synthesis of such patterns may be obtained in several different ways. One consists of considering the desired pattern as the image at infinity of a primary illumination composed of a set of feeds whose excitation amplitudes are weighted according to the desired function (§9.9.2), seen through a stigmatic system (paraboloid). This process is usable as long as the defocusing of the feeds the farthest from the focus does not lead to prohibitive aberrations (§9.7), that will limit the total angular aperture of the synthesized pattern to moderate values (<30° for instance). In this case, this process leads to best results for a given dimension of the reflector. For larger angular apertures (sometimes larger than 70°), it is better to use a single primary feed associated with a reflector of particular shape, termed 'dual curvature'. The synthesis of the reflector is carried out in a simple manner using the techniques of geometrical optics (Dunbar's method[12]). This is akin to pattern synthesis by the method of 'stationary phase' (§13.2).

Definition of the reflector shape from that of the 'main curve'

The aim is to obtain a characteristic radiating surface whose cross-section in a vertical plane of symmetry (π) should have a given shape ('elevation' pattern) and whose cross-section in any plane (P) orthogonal to (π) should have the maximum directivity ('azimuth' pattern). Such a target will be reached if the emerging wave is a cylinder whose generatrix is horizontal, and whose directrix line is defined by the elevation pattern to be realized (Fig. 10.21). In fact, the field at infinity in a direction **u** on the plane P normal to the plane of symmetry π and to the surface of the emerging wave (S) of an angle θ with respect to the horizontal, this plane, depends, as a first approximation (according to the stationary-phase principle), only on the field on the generatrix S∩P and on the curvature of this generatrix. Since the field is equiphase, the directivity is maximum in the plane of symmetry π (§13.2.4).

---

[12] Dunbar, A., 'Calculation of doubly curved reflectors for shaped beams', *Proc. IRE*, Vol. 36, Oct. 1948.

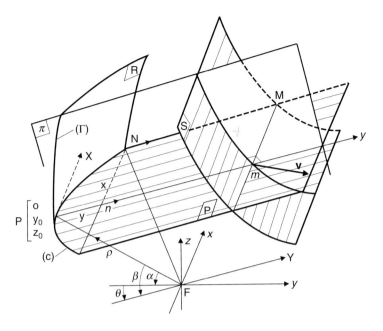

Fig. 10.21 Pattern synthesis with a shaped reflector.

Consider the cross-sections (Γ) and (C) of the reflector (R) by the planes π and P and the primary 'rays' FP (on π) and FN. These rays emerge parallel to OY on Pm and NM. It results that P and N belong to a paraboloid of axis FY parallel to the plane P, tangential to (R) along the curve (C); only this surface is stigmatic for F and the direction FY. The focal length of this paraboloid is deduced from the vector radius ρ = FP and the angle β between FP and FY according to the polar equation of the paraboloid (§7.2.4)

$$f_\theta' = \rho \cos^2 \frac{\beta}{2} = \rho \cos^2 \frac{\theta + \alpha}{2} \qquad (10.146)$$

where α is the angle (PF, FY).

The cross-section (C) of (R) in the plane (P) is then a parabola of the same focal length f_θ'. In the base (PXY) of P, the equation of this parabola giving the coordinates of the point N, is

$$X^2 = 4 f_\theta' Y \qquad (10.147)$$

On the other hand, from the base (F, x, y, z), the coordinates (O, y_0, z_0) of the point P of Γ and since the corresponding direction θ is assumed to be known (§10.3.2.3), we may deduce the coordinates of the point M by projecting the relation

$$\mathbf{FN} = \mathbf{FP} + \mathbf{PN} \qquad (10.148)$$

so that

$$
\left.\begin{array}{l}
x = X \\
y = y_0 + Y \cos \theta \\
z = z_0 + Y \sin \theta
\end{array}\right\}
\tag{10.149}
$$

Determination of the central curve Γ *(Fig. 10.22)*

A pyramid of rays, centred on F, of elementary aperture $d\alpha dy$ gives, after reflection, an elementary prism of width $\rho d\alpha$ and aperture $d\theta$. Energy conservation implies a relation of the form

$$
g(\alpha) d\alpha dy = KG(\theta) \rho \, dy d\theta
$$

or

$$
\frac{1}{\rho(\alpha)} g(\alpha) d\alpha = KG(\theta) d\theta
\tag{10.150}
$$

$g(\alpha)$ being the primary gain function and $G(\theta)$ the desired secondary gain. The proportionality coefficient K is defined by the conservation of the total primary and secondary energy. If (α_1, α_2) are the limits of the primary beam, and (θ_1, θ_2) those of the secondary pattern, we have

$$
\int_{\alpha_1}^{\alpha_2} \frac{1}{\rho(\alpha)} g(\alpha) d\alpha = K \int_{\theta_1}^{\theta_2} G(\theta) d\theta
\tag{10.151}
$$

Numerical integration of equation (10.150) gives the variation of θ as a function of α. The profile (Γ) is obtained from the known relation between the polar coordinates (ρ, α) of P and the angle $\beta/2$ of the normal at P and with the vector radius

$$
\frac{1}{\rho} \frac{d\rho}{d\alpha} = \tan \frac{\theta + \alpha}{2}
\tag{10.152}
$$

From this can be deduced $\rho(\alpha)$, the coordinates (y_0, z_0) of P, and hence those of an arbitrary point N of the reflector from equations (10.149).

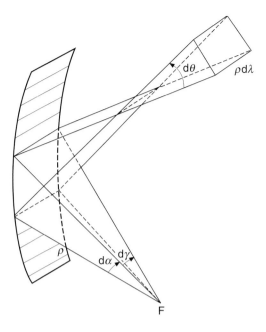

Fig. 10.22 Determination of the central curve Γ.

Diffraction pattern

Daveau's paper, already referred to[4], describes a method allowing the optimization of the pattern by successive approximations starting from the above calculated profile (Γ). Since the elevation pattern is principally defined by the central curve (Γ), we shall calculate the diffraction pattern of a narrow cylindrical vertical strip, of width ℓ, sustaining the curve (Γ). Let us start from the formula (§10.2.6) for the Cassegrain antenna that we shall apply to the locally reflected field at P on (Γ) (Fig. 10.23). We use, in addition to the already used base (F,$\mathbf{e}_x$,$\mathbf{e}_y$,$\mathbf{e}_z$), a local base at P (**n**, $\mathbf{e}_s$, $\mathbf{e}_x$) where the unit vectors **n** and $\mathbf{e}_s$ are respectively normal and tangential to (Γ). The first base is deduced from the second by a rotation about $\mathbf{e}_x$ through an angle $\gamma = \alpha - \beta/2$. If the primary incident field ($\mathbf{E}_i$, $\mathbf{H}_i$) shows, for instance, a vertical polarization with amplitude E₀, the reflected field is deduced taking into account the boundary conditions (§10.2.6) and takes the following form

$$\frac{\mathbf{E}}{\mathbf{E}_0} = \mathbf{n}\sin\frac{\beta}{2} - \mathbf{e}_s\cos\frac{\beta}{2} \qquad (10.153)$$

$$\frac{\mathbf{H}^+}{\mathbf{E}_0} = -\mathbf{e}_x$$

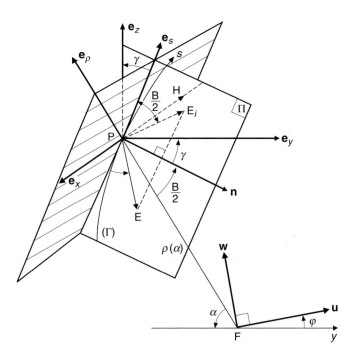

Fig. 10.23 Diffraction pattern.

If $\mathbf{w}$ is the unit vector of the plane (π) orthogonal to the considered direction $\mathbf{u}$ we calculate successively

$$\left(\frac{\mathbf{E} \times \mathbf{n}}{\mathbf{E}_0}\right) \times \mathbf{u} = \mathbf{w}\cos\frac{\beta}{2} \qquad (10.154)$$

$$\left[\frac{\left(\mathbf{H}^+ \times \mathbf{n}\right) \times \mathbf{u}}{\mathbf{E}_0}\right] \times \mathbf{u} = \mathbf{w}\cos\left(\gamma + \varphi\right) = \mathbf{w}\cos\left(\gamma + \varphi - \frac{\beta}{2}\right) \qquad (10.155)$$

In addition

$$E_0 = E_1 \frac{\sqrt{g(\alpha)}}{\rho(\alpha)} \tag{10.156}$$

With $E_1 = \sqrt{60W}$, W being the feed power. (This formula results from the power density radiated by a primary feed having a gain $g(\alpha)$).

On the other hand, the elementary curvilinear abscissa ds on (Γ), is

$$ds = \frac{\rho \, d\alpha}{\cos \frac{\beta}{2}} \tag{10.157}$$

Finally, the radiated electric field $\mathbf{E}(\varphi)$ at the distance R is given by

$$\mathbf{E}(\varphi) = E_1 \mathbf{w} \frac{\rho}{2\lambda R} F(\varphi) \tag{10.158}$$

where

$$F(\varphi) = \int_{\alpha_1}^{\alpha_2} \left[1 + \cos(\varphi + \alpha) + \tan \frac{\beta}{2} \sin(\varphi + \alpha) \right] \sqrt{g(\alpha)} \exp\left\{ jK\rho \left[1 + \cos(\varphi + \alpha) \right] \right\} d\alpha \tag{10.159}$$

Remark: owing to stationary phase conditions, the field contribution at P is principally dominant in the regions where the reflected ray is precisely in the considered direction φ. We then have $\varphi + \alpha \approx \beta$ and the quantity between brackets in the above integral is close to 2.

Numerical example

Fig. 10.24 depicts an example of a diffracted pattern compared with an ideal cosecant pattern before and after optimization.

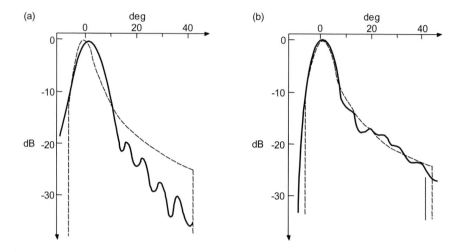

Fig. 10.24 Diffraction pattern of a shaped reflector: (a) before optimization, (b) after optimization.

FURTHER READING

1. Aubry, C. and Bitter, D., 'Sum and Difference radiation patterns of a corrugated conical horn by means of Laguerre-Gaussian functions', *Proc. V European Microwave Conference*, Hamburg, September 1975.
2. Dijk, J., 'The polarisation losses of offset paraboloid antennas', *IEEE Trans. Antennas & Propagation*, Vol. AP-22, July 1974.
3. Drabowitch, S., 'A new tracking mode coupler using a corrugated feed for satellite communication Earth-Station antenna', *Proc. VI European Microwave Conference*, Rome, pp165-168, September 1976.
4. Hannan, P.W., 'Microwave antennas derived from the Cassegrain telescope', *IRE Trans. Antennas & Propagation*, March 1961.
5. Silver, S. (ed), *Microwave Antenna Theory and Design*, MIT Radiation Lab. Series, Vol. 12, McGraw-Hill, New York, 1949; reprinted by Peter Peregrinus, Stevenage, 1984.
6. Williams, W.F., 'High-efficiency antenna reflectors', *Microwave Journal*, July 1965.

EXERCISES

10.1 *Cassegrain antenna*

Demonstrate the following relationship between the diameters of the main reflector (D) and subreflector (D'), and the magnification of the hyperboloid

$$\frac{D}{D'} = (m+1)\left[1 - \frac{1}{m}\left(\frac{D}{4F}\right)^2\right]$$

Observe that for $D/4F \ll 1$ or $m \gg 1$ this expression reduces to

$$\frac{D}{D'} = m+1$$

ANSWER

It can be shown from equations (10.1) and (10.2) that the polar equation of the equivalent parabola is

$$\rho' = \frac{2Fm}{1+\cos\theta}$$

Then $x = 2Fmt$ with $t = \tan\theta/2$

On the other hand, the polar equation of the hyperboloid can be written

$$\rho = \frac{c^2 - a^2}{c\cos\theta - a} \qquad \text{with} \quad 2c = C'C = F$$

Since the eccentricity is $e = c/a$, we have

$$m = \frac{C'S'}{S'C} = \frac{c+a}{c-a} = \frac{e+1}{e-1} \qquad \text{then} \quad e = \frac{m+1}{m-1}$$

We deduce

$$r = \rho \sin\theta = \rho \frac{2t}{1+t^2}$$

Then

$$r = \frac{c+a}{1 - \frac{c+a}{c-a}t^2}$$

$$\frac{x}{r} = (m+1)(1-mt^2)$$

For $r = \frac{D'}{2}$, $x = \frac{D}{2}$ and

$$\frac{D'}{D} = (m+1)(1-mt_0^2) \quad \text{with} \quad t_0 = \frac{D}{4Fm}$$

Finally

$$\frac{D}{D'} = (m+1)\left[1 - \frac{1}{m}\left(\frac{D}{4F}\right)^2\right]$$

10.2 *Monopulse null depth*

A double horn monopulse feed (§10.3.4, fig.10.14) is slightly tilted with an error angle ε. What is the resulting null depth?

ANSWER

Each horn receives the incident signal with a positive and negative phase error $\pi a \varepsilon / \lambda$. The individual signals take the form

$$A' = \psi\left(\frac{a}{2} + \xi\right)\exp\left(j\pi\frac{a}{\lambda}\varepsilon\right) \qquad B' = \psi\left(\frac{a}{2} - \xi\right)\exp\left(-j\pi\frac{a}{\lambda}\varepsilon\right)$$

Then we have

$$S(0) \cong 2\psi\left(\frac{a}{2}\right) \qquad D(0) = 2j\psi\left(\frac{a}{2}\right)\pi\frac{a}{\lambda}\varepsilon$$

The relative null depth is then:

$$q = \left|\frac{D(0)}{S(0)}\right| = \pi \frac{a}{\lambda}\varepsilon$$

In a focused system, we optimize the 'sum' channel by choosing $2a = \delta = 2\lambda\dfrac{f}{D}$ (eqn. 9.114). Then we have

$$q = \pi \frac{f}{D}\varepsilon$$

Numerical example: For a parabolic reflector with relative aperture $D/f = 3$, we have $q \cong \varepsilon$. With a $2°$ tilt, $\varepsilon = 1/30$. In decibels, $Q = -20 \log q = -30\text{dB}$.

10.3   *Monopulse feed with radial polarization*

In an exercise (9.1) in the preceding Chapter, we have seen that an annular aperture with radial polarization gives a symmetric primary pattern with radial polarization. From this, the corresponding *secondary* pattern has the same symmetries. In the vicinity of the axis it is therefore of the form

$$\mathbf{F}_{\mathrm{D}}(\theta) = \mathbf{e}_\theta q \theta$$

where q is the 'slope' of the pattern in the vicinity of the axis.

Suppose that there is an incident wave of circular polarization **c**, incident from a direction of scanning $\mathbf{u}(\theta, \varphi)$. The signal received in the corresponding channel is of the form

$$D(\theta,\varphi) = \mathbf{F}_{\mathrm{D}}(\theta).\mathbf{c} = q\theta\,\mathbf{e}_\theta.\mathbf{c}$$

The complex unit vector **c** is related to the linear components by the relation (§12.2.4):

$$\mathbf{c} = \frac{\mathbf{e}_\alpha + j\mathbf{e}_\beta}{\sqrt{2}}$$

where the vectors ($\mathbf{e}_\alpha$, $\mathbf{e}_\beta$) define the Huygens coordinates (equation (7.63))

$$\begin{cases} \mathbf{e}_\alpha = \mathbf{e}_\theta \cos\varphi - \mathbf{e}_\phi \sin\phi \\ \mathbf{e}_\beta = \mathbf{e}_\theta \sin\varphi + \mathbf{e}_\phi \cos\phi \end{cases}$$

We have therefore

$$\mathbf{c} = \frac{1}{\sqrt{2}} \left(\mathbf{e}_\theta + j\mathbf{e}_\varphi \right) \exp j\varphi$$

From which it follows that

$$D(\theta, \varphi) = \frac{1}{\sqrt{2}} q\theta \exp j\varphi$$

The factor $1/\sqrt{2}$ expresses the fact that only half of the available power (in circular polarization) is received by the slot.

Coherent demodulation of the signal D by the reference signal S (§10.3.5) allows us, in a similar way to monopulse, to find the coordinates (θ, φ) of the direction $\mathbf{u}$ of the circularly-polarized incident signal.

The following exercise consists of evaluating the sensitivity of the system, by calculating the slope q, close to the axis. In order to solve the problem, a first method would be to explicitly calculate the characteristic secondary function $D(\mathbf{u})$. Another, more simple method is to consider the received signal by calculating the *convolution* of the diffraction spot ψ with the vectorial illumination of the annular slot, according to equation (9.109), which gives the *difference* gain factor

ANSWER

Q2   A wave of circular polarization $\mathbf{c}$ incident in the plane $(\mathbf{n}, \mathbf{e}_x)$ for example, and whose direction, given by the vector $\mathbf{u}$, makes an angle θ with $\mathbf{n}$, gives rise in the focal plane to a *vectorial* diffraction spot ψ which is eccentric with respect to a focus at a distance $\xi = F\theta$, where F is the equivalent focal length of the Cassegrain.

Assuming geometric aberration to be negligible, this takes the form

$$\psi_\theta(x, y) = \psi(x - \xi, y) \approx \psi(x, y) - \xi \frac{\partial \psi}{\partial x} = \mathbf{c} \left(\psi - \xi \frac{\partial \psi}{\partial x} \right)$$

$\mathbf{c}$ being the unit vector of the circular polarization

$$\mathbf{c} = \frac{1}{\sqrt{2}} \left(\mathbf{e}_x + j\mathbf{e}_y \right)$$

and ψ the *scalar* diffraction spot.

On the other hand, the illumination of the circular slot with radial polarization, and of radius a and width d, can be written as

$$\mathbf{f}(r,\varphi) = \mathbf{e}_r \, \text{rect}\left(2\frac{r-a}{d}\right)$$

The integral in the numerator of equation (9.109) for the gain factor is written as

$$I = \int_0^\infty \int_0^{2\pi} \mathbf{f} \cdot \psi_\theta \, ds = \int_0^\infty \text{rect}\left(2\frac{r-a}{d}\right)\left[\int_0^{2\pi} \mathbf{e}_r \cdot \mathbf{c}\left(\psi - \xi\frac{\partial\psi}{\partial x}\right)d\varphi\right] r dr$$

Since the slot is narrow we can set

$$I = ad \left[\int_0^{2\pi} \mathbf{e}_r \cdot \mathbf{c} \, \psi \, d\varphi - \xi \int_0^{2\pi} \mathbf{e}_r \cdot \mathbf{c} \, \psi \, \frac{\partial\psi}{\partial x} d\varphi\right]$$

Then, since $\mathbf{e}_r = \mathbf{e}_x \cos\varphi + \mathbf{e}_y \sin\varphi$, we have $\mathbf{e}_r \cdot \mathbf{c} = \exp j\varphi$. The first integral is therefore zero.

Further
$$\frac{\partial\psi}{\partial x} = \frac{\partial\psi}{\partial r}\frac{\partial r}{\partial x}$$

with
$$r^2 = x^2 + y^2$$

i.e.
$$\frac{\partial r}{\partial x} = \frac{x}{r} = \cos\varphi$$

The integral is therefore written

$$I = -\xi ad \int_0^{2\pi} \exp j\varphi \, \cos\varphi \left(\frac{\partial\psi}{\partial r}\right)_{r=a} d\varphi$$

$$= -\xi ad \left(\frac{\partial\psi}{\partial r}\right)_{r=a} \int_0^{2\pi} \cos^2\varphi \, d\varphi$$

It follows that

$$I^2 = \pi^2 a^2 d^2 \left(\frac{\partial \psi}{\partial r}\right)^2 \xi^2$$

On the other hand

$$\int_0^\infty \int_0^{2\pi} |f(r)|^2 r\, dr\, d\varphi = 2\pi a d$$

The gain factor that we seek is therefore

$$\eta_D = \frac{ad}{4} \frac{\left(\dfrac{\partial \psi}{\partial r}\right)^2_{r=a}}{\displaystyle\int_0^\infty |\psi|^2\, r\, dr} \xi^2$$

In the case where the diffraction spot has the classical form

$$\psi(r) = \Lambda_1(ur)$$

we have

$$\int_0^\infty |\psi|^2\, r\, dr = \frac{2}{u^2}$$

and

$$\left(\frac{\partial \psi}{\partial r}\right)_{r=a} = u\Lambda_1'(ua)$$

where Λ_1' is the first derivative of the Λ_1 function (Fig. 9.5).

So finally

$$\eta_D = \frac{ad}{8} u^2 \left[\Lambda_1'(ua)\right]^2 u^2 \xi^2$$

where

$$u = \frac{2\pi}{\lambda} \sin \alpha_0$$

The received signal, in amplitude, is proportional to the square root

$$D \cong \sqrt{\eta_D} = \frac{1}{\sqrt{2}} q\theta$$

where q is the angular slope

$$q = F\sqrt{ad}u^2 \Lambda_1'(ua)$$

Search for maximum sensitivity
This is obtained when the coefficient q is maximum, i.e. when the slope of the diffraction spot ψ is maximum, i.e. when the radius a of the slot corresponds to the *point of inflexion*. Fig. 9.5 shows that this is approximately achieved for $ua = 2.5$, which corresponds to a slot diameter of

$$2a_D \cong 0.66\,\delta$$

where δ is the diameter of the diffraction spot.

If this result is compared with that of §9.3.2, recalled at the beginning of exercise 9.1, the latter gives maximum gain for the reference channel, for a diameter

$$2a_s \cong 0.71\delta$$

We see that the two values are similar, but that the optimum sensitivity is obtained at the cost of a small loss in axial gain.

11

Arrays

The two array faces of the SAMPSON naval multifunction phased array radar, developed by AMS (photo: AMS).

11.1 INTRODUCTION

Array antennas are used in a large number of applications, such as radioastronomy, space and terrestrial telecommunications - but undoubtedly it is radar that has played the most important role in their development.

A first aspect of interest is electronic scanning. Due to their large inertia, mechanically-steered radar antennas have time constants of the order of a second. This means a waste of valuable time when pointing at multiple successive targets. It is obvious that such an inertia is incompatible with the time constants of the operating electronic circuits which are often of the order of a microsecond or less.

In contrast, a system which can steer the beam without inertia allows the radar to exploit its full possibilities. So the beam of a surveillance radar scanning the horizon can pause to check an occasional detection. Equally, an electronically-scanned tracking radar can track multiple targets, looking successively at each with no waste of time. Finally, a single radar may simultaneously carry out both surveillance and tracking functions by properly scheduling the various tasks according to the operational scenario.

A second operational aspect is that of 'electromagnetic imaging'. The signal environment of antennas, both in radar and other applications, is generally very complex. This complexity is steadily increasing, so that in general there will be a multiplicity of useful signals coming from various directions with various frequencies and polarizations, as well as natural or deliberate unwanted signals (jammers), multipath, and so on.

The identification, selection and representation of these signals in space and time are the problem of imaging.

Array antennas with appropriate processing allow this to be done. In effect, the elements which make up the array provide spatial samples of the incident field. Thus, it is possible to analyse the field in amplitude, phase and polarization by temporal sampling and spectral analysis, frequently using digital processing techniques. All this relates to the domain of *signal processing antennas* which will be further considered in Chapter 14.

Firstly, let us consider the more classical aspects of array antennas in terms of electronic scanning.

11.1.1 Phased arrays

A process, known for many years, allows electronic scanning to be obtained with phased arrays. This consists of feeding a set of radiating elements regularly disposed on a plane (planar array) by means of phase shifters, in such a way that the phase variations along the array follow an arithmetical progression whose increment is the phase shift between two adjacent elements. Thus, this array generates a plane wave whose direction depends on this phase difference. During the early 1960s, reliable phase shifters were developed which allowed these antennas to be realized in practice.

At the same time, the digital computer became the ideal control and programming tool for these phase shifters. Present-day phase shifters are almost all of the digital type, but the quantized nature of the phase differences imposes certain limitations on the array performances, which will be examined later (§11.5).

11.1.2 Bandwidth - use of delay lines - subarrays

Strictly speaking, delay lines should be used rather than phase shifters. Consider a set of N antenna elements whose phase centres are located at different points $A_0,...A_n,...A_{N-1}$. Consider a wavefront incident from the direction of the unit vector **u**

(Fig. 11.1). This reaches the element A_n with a time advance τ_n with respect to the element A_0, measured by the distance $A_0 A_n$ projected onto **u**

$$\tau_n = \frac{\mathbf{A_0 A_n \cdot u}}{c} \tag{11.1}$$

where c is the velocity of propagation.

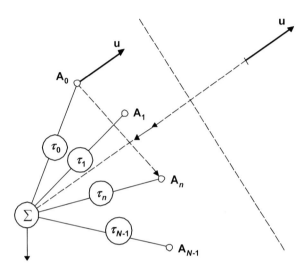

Fig. 11.1 Delay-line array.

If these elements are connected to a common receiver by lines of equal length, and if each line includes a delay which exactly compensates the time advance τ_n, the individual signals add *coherently*. The resulting signal is then a maximum. Correspondingly, on transmit the elements generate a wavefront in the direction **u**.

In which cases can we replace these delay lines by phase shifters? When the frequency bandwidth is sufficiently narrow. The signal delayed by τ_n may be written in complex form

$$\exp\left[j2\pi f \left(t - \tau_n \right) \right] = \exp\left[j \left(2\pi f t - \varphi_n \right) \right] \tag{11.2}$$

with

$$\varphi_n = 2\pi f \tau_n = \frac{2\tau}{\lambda} \mathbf{A_0 A_n \cdot u} \tag{11.3}$$

If the phase remains constant over the frequency range of operation, distortion of the radiated wave may result, and in particular, a pointing error (§11.3.4).

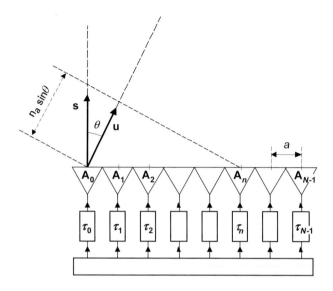

Fig. 11.2 Linear delay-line array.

Comment: In the case of large arrays of large bandwidth, to avoid having to provide each element with a delay line (too expensive), we can divide the array into *phased* subarrays interconnected with delay lines. The pointing error is then avoided, however the sidelobes may be adversely affected (Exercise 11.2).

11.1.3 Active arrays

The phase shifters or delay lines mentioned so far are assumed to be passive. They are thus perfectly linear but some losses inevitably result. Increasingly these components are combined with amplifiers in so-called 'active array modules'. The properties of such modules and arrays are treated in §11.8. We give here only some general characteristics.

Active arrays in transmission

When a phased array is fed by a single transmitter, the connections between transmitter and radiating elements result in significant losses. The feeder lines, the rotating joint (if used) and the phase shifters alone often dissipate more than half of the power of the transmitter.

The development of solid state transmitters now allows us to shift the 'transmission' function into modules which feed the radiating elements directly with no transmission loss. Another advantage is that these modules generally use transistors, which do not require high voltages and stable regulated power supplies, such as would be the case with tube transmitters.

Also, in the case of failure of some modules, the system can continue to operate with slightly reduced performance; the operating reliability is thus improved. This property is sometimes known as *graceful degradation*.

Finally, the phase shifters associated with each radiating element may operate at *low power level* if they are placed at the *input* of the amplifier. The low power phase shifters are often more accurate, faster and more reliable than the high power phase shifters used with conventional techniques.

All these advantages explain the increasing development of active arrays in transmission.

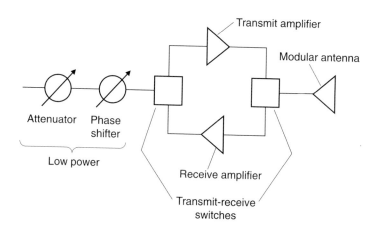

Fig. 11.3 Example of an active array module.

Active arrays on receive

In a 'passive' phased array on receive, the transmission losses that we have mentioned are still present, reducing the signal-to-noise ratio of the received signals. Nevertheless, with certain radar equipment the operational range is limited by the external noise or clutter, and so the external signal-to-noise ratio is not affected by this attenuation.

However, the main benefit of active modules on receive lies in the possibility to control the amplitude and phase weighting (and eventually, the polarization) of the signals received by each element of the array. In fact, passive phased arrays only allow us to control the phase of such signals. Use of variable attenuation would increase losses in an undesirable way. On the other hand, with active modules, it is possible to control the amplitude without affecting the internal signal-to-noise ratio, provided that the attenuator is placed after the amplification stage.

This amplitude control allows us to modify arbitrarily the form of the receiving pattern, in particular the sidelobe level and position of nulls. This possibility is exploited in adaptive antennas where the weights are adaptively controlled. This is the domain of *signal processing antennas* (§14.5).

Transmitting-receiving active arrays

Of course, the transmitting and receiving functions may be combined in the same module, but then it becomes relatively complex. The development of active arrays is limited by the cost of the antenna which may involve several thousands of modules. The significant advance in the development of monolithic microwave integrated circuits (MMICs) using gallium arsenide (GaAs) has allowed the realization of operational equipment using active arrays.

11.2 GENERAL STRUCTURE OF A PHASED ARRAY (EXAMPLES)

11.2.1 General structure

A 'passive' phased array may take on very different forms, but always includes all of the parts described below (Fig. 11.4). Considering the antenna on transmit, we find successively:

- a feed network or power splitter distributing energy to the different elements of the array by means of phase shifters according to a desired amplitude function. This amplitude function does not vary with the scan angle (except for adaptive arrays and spatial filtering (§14.7));

- the set of phase shifters;

- a computer which calculates the phases for the required pointing direction and a driver that controls the phase shifters (providing control voltages/currents to diode or ferrite phase shifters);

- the set of array elements;

- in some cases, a complementary focusing system (reflector or lens).

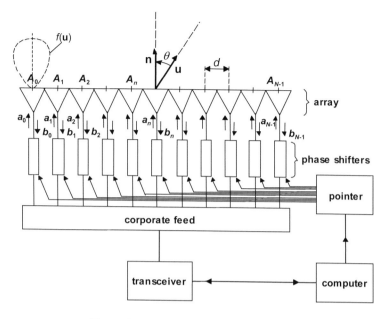

Fig. 11.4 General structure of a phased array.

Feed network - comment

We shall see that due to mutual coupling, the elements have a reflection coefficient which varies with the beam pointing angle ('active' reflection coefficient (§11.4.3)). Therefore, they reflect back a set of signals to the splitter outputs. If these elements are not *individually impedance-matched and isolated*, these signals are reflected once more toward the elements with uncontrolled phases and amplitudes that result in a parasitic radiation pattern which is superimposed on the required pattern. This condition, as we shall see later, plays an important role in the design of feed networks where it is often necessary to use directional couplers and matched loads.

Phase shifter structures

There are several different types of phase shifter that may be used.

Switched line phase shifter: This consists of a number of sections employing two-way switching between different line lengths to achieve a digitally controllable phase shift, as shown in Fig. 11.5.

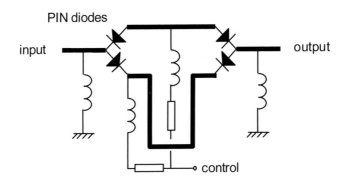

PIN diodes

input

output

control

Fig. 11.5 Switched-line phase shifter.

The switching elements are usually PIN diodes, although FETs may also be used. PIN diodes behave as resistors to RF signals, where the resistance is controlled by the dc current. Typically, a forward-biased diode would have a resistance of a few ohms and a reverse-biased diode would have a resistance of 10 - 100 kΩ. The dc bias may be applied through an inductor - with a high reactance compared to 50 Ω at the operating frequency - to avoid any effect on the ac performance.

This type of phase shifter is widely used because it allows digital control of the phase shift and is easy to incorporate in a microprocessor-controlled system. However, it can tend to be lossy - the insertion loss of each phase shifting section being around 0.5 dB - and relatively expensive, because of the number of PIN diodes that are required.

Phase shifter resolution is dependent on the number of sections, or bits, n, according to $\Delta\phi = 2\pi/2^n$. The accuracy of a phase shifter using line lengths is limited by the signal bandwidth, Δf, since the phase shift is only precise at one frequency. In the worst case, when a phase shift of 2π is introduced, its accuracy is $\Delta\phi_e = 2\pi\Delta f / f_c$. An expression for the usable bandwidth of an n-bit phase shifter can be obtained by equating $\Delta\phi_e$ to the resolution $\Delta\phi$, giving a usable fractional bandwidth of

$$\frac{\Delta f}{f_c} = \frac{1}{2^n} \qquad (11.4)$$

It is clear that this type of phase shifter is only suitable for relatively narrow-band applications. (Note: although setting the *delays* appropriately would give very broadband performance, the phase shifter sets the phase shifts modulo-2π, thus limiting the bandwidth as discussed above). A more broadband phase shifter could be produced by using all-pass filter networks in place of the line lengths. These filters are designed to produce broadband phase shifts with little attenuation.

Phase shifter performance is specified in terms of:

• accuracy (number of bits, accuracy of each phase increment);

• frequency stability (this type of phase shifter is relatively insensitive to changes in temperature);

• insertion loss;

• impedance match, reproducibility, cost.

Usually, phase shifters have 2 to 5 bits. Their accuracy is of the order of half a bit and their insertion losses range typically from 1 to 2 dB.

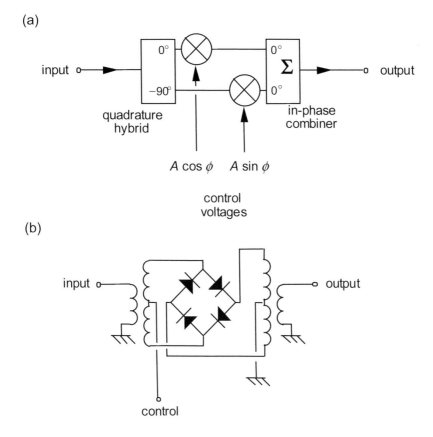

Fig. 11.6 (a) Vector modulator phase shifter; (b) Quad PIN diode attenuator.

Vector modulator: The vector modulator (Fig. 11.6a) is able to control both the phase and amplitude of a signal. It uses the weighted combination of an in-phase (I) and quadrature (Q) phase shifted signal to introduce a phase shift variable over a $\pm\pi$ range. To enable the device to work in all four quadrants the attenuator elements must be bipolar, that is they pass the signal either uninverted or inverted depending on the polarity of the control voltage. Quad PIN diode attenuators are available commercially for this purpose (Fig. 11.6b), though their attenuation-vs-current characteristic is non-linear, and in some applications may need to be linearized by (for example) a digital lookup table and digital-to-analogue converter.

Denoting the input signal by $V_{in}\cos\omega t$, the signals emerging from the 0° and –90° ports of the quadrature hybrid are $\left(V_{in}/\sqrt{2}\right)\cos\omega t$ and $\left(V_{in}/\sqrt{2}\right)\sin\omega t$, respectively. These signals are then weighted by factors of $A\cos\phi$ and $A\sin\phi$ respectively and are combined in a reactive combiner. The output signal is therefore

$$V_{out} = \frac{V_{in}}{\sqrt{2}}\left[\frac{\cos\omega t}{\sqrt{2}}A\cos\phi + \frac{\sin\omega t}{\sqrt{2}}A\sin\phi\right] = \frac{AV_{in}}{2}\cos(\omega t - \phi) \qquad (11.5)$$

If A is held constant so that the weighting factors are $\cos\phi$ and $\sin\phi$, the vector modulator acts as a constant-amplitude phase shifter. The theoretical insertion loss is 6 dB and in practice it is likely to be slightly higher (nearer 8 or 9 dB) because of the finite minimum attenuation of the attenuators. This may be quite a severe disadvantage in many phased array applications, often restricting the use of vector modulators to IF stages rather than at the antenna elements themselves. The bandwidth, however, is usually quite good, being limited to typically an octave by the quadrature hybrid. The ability of the vector modulator to control the signal amplitude as well as phase makes them particularly useful in adaptive arrays, as discussed in §14.7.

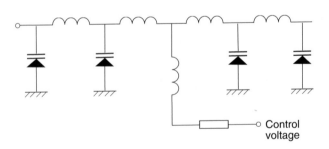

Fig. 11.7 Varactor phase shifter.

Varactor-controlled phase shifter: Fig. 11.7 shows another phase shifter configuration, consisting of a filter containing varactors in place of the usual capacitor elements. The filter is of the 'all-pass' type, producing a broadband phase shift with little attenuation, the phase shift being varied by changing the bias on the varactors. This type of phase shifter is capable of a reasonably low insertion loss, but is highly non-linear because of the response of the varactors and also because of the non-linear relationship between their capacitance and the insertion phase shift.

11.2.2 Examples of array structures

Corporate feed

This is composed of a set of successive power dividers. These are either directional couplers or magic tees whose unused outputs on transmit are terminated by matched loads. This precaution is necessary to satisfy the matching and isolation conditions defined in the preceding section (Fig. 11.8).

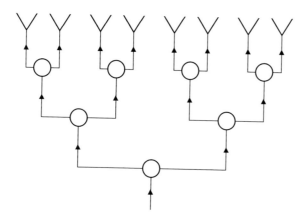

Fig. 11.8 Corporate feed.

Advantages:

• planar symmetric aperiodic structure.

Disadvantages:

• structural complexity;

- losses (this disadvantage does not occur with an active array (§11.8));

- difficulty in obtaining the optimum amplitude weighting function, particularly if the sum and the difference channels are desired simultaneously (see monopulse antennas §10.3.4).

Small planar arrays can easily be realized using printed circuit lines, incorporating the phase shifters, or in some cases using the radiation elements themselves (printed dipoles).

This is a good technique for realizing subarrays combined with intermediate active modules and/or delay lines.

Feed network with waveguides and directional couplers (Fig. 11.9)

The energy distribution is obtained from a central guide whose energy is tapped by means of scaled directional couplers. It is an example of a Blass Matrix (§11.7.4, Fig. 11.24).

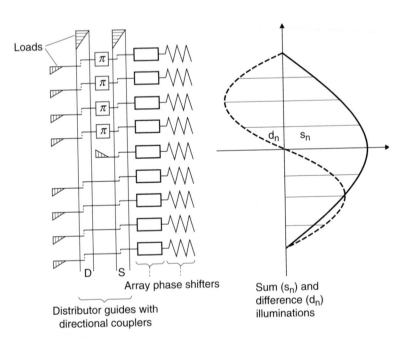

Fig. 11.9 Phased array feed with waveguides and directional couplers.

Advantages:

- Although the preceding corporate feed network distributes the signal in a widening surface, the present feed network has a quasi-linear structure and due to that it can be rather small (this is particularly attractive for high frequencies). In addition, the use of directional couplers ensures the absorption of reflected waves by the couplers (§11.2.2).

Disadvantages:

- frequency sensitivity of the phase function (this may sometimes be applied deliberately) (§11.6);

- many different coupling coefficients for the directional couplers;

- asymmetrical structure, which is generally not suitable for tracking antennas.

If a difference channel is desired, the structure is complicated since it requires its a separate feed guide and a second set of directional couplers giving the desired antisymmetric distribution. The orthogonality of the sum and difference distributions leads to isolation between the two guides, however interactions may occur, and the adjustment should be made with great care (see §11.7.4: Blass matrices).

Feed network using a lens between two parallel planes

A primary element radiates between two parallel planes (with occasionally a sum and a difference channel). An array of elements correctly arranged between two planes (preferably on the arc of a circle) collects the primary energy with amplitude distribution functions controlled by the primary patterns. The collected waves pass through the phase shifters before being radiated (see §11.7.5: Rotman lens, Fig. 11.26).

Advantages:

- relatively simple structure;

- easy adjustment by altering the directivity or the position of the primary element, which is not difficult to adjust;

- symmetrical structure with a large bandwidth.

Disadvantages:

• bulky.

'Disk' feed network

Consider two parallel planes limited by a circle of diameter D and centre O. A central primary element is able to transmit a radial wave propagating between these two planes (Fig. 11.10).

The energy available is collected by means of couplers (preferably directional couplers) placed on concentric circles. It is then phase shifted to compensate for the phase shift produced during the propagation of the radial wave (the phase shift is proportional to the radius), and, on the other hand, to define the correct phase gradient to produce the desired orientation of the beam.

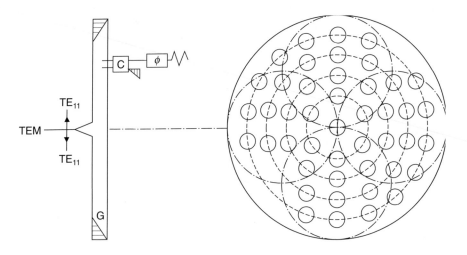

Fig. 11.10 Disk feed network for a planar array.

It is possible, by means of a multimode element, to generate illuminations corresponding to the sum and difference patterns of a monopulse antenna. A circular guide propagating the radial TM_{01} mode will give a TEM wave with an electric field perpendicular to the walls, suitable for an illumination of the 'sum' type. The same guide propagating the TE_{11} mode at vertical or horizontal polarization will generate antisymmetric illuminations in these same planes, and will produce patterns of the 'elevation difference' and 'azimuth difference' types.

Advantages:

- planar structure, longitudinally compact;

- its symmetry makes it suitable for tracking antennas.

Disadvantages:

- the array is fed by a travelling wave, so its phase is a function of frequency. The controlling computer should take into account the frequency to determine the correct phase.

Electronic phased lens

The lens is composed of two arrays (internal and external). The primary element radiates in free space and defines the illumination functions of the lens whose phase shifters are located between the input and output arrays. A spherical input face allows the Abbe condition to be satisfied (§9.7.2), and to use several primary elements simultaneously, permitting electronic scanning of several beams at the same time (§11.7.5: Rotman lens).

Advantages:

- simple feed;

- easy adjustment of the feed by altering only the primary elements;

- symmetrical structure.

Disadvantages:

- difficult access to phase shifters for assembly, maintenance and replacement;

- two arrays are necessary.

Phased reflector array

A 2-D array is illuminated by a primary element which may be non axially-symmetric. The collected waves pass through short-circuited phase shifters that reflect them toward the array with phase shifts $2\phi_k$, where ϕ_k is the differential phase shift corresponding to the transmission of the phase shifter number k. The phases ϕ_k transform the spherical wave from the primary element into a plane wave in the desired direction.

Advantages:

- simple feed, a single array;

- easy adjustment by means of the primary element;

- easy access to the phase shifters.

Disadvantages:

- The matching condition of the feed network of §11.2.1 is difficult to realize.

The interelement coupling between produces an 'active' reflection coefficient (§11.4.3) that reflects part of the primary wave without the correct phase shift. This energy produces a parasitic pattern called the 'primary pattern image' which degrades the angular accuracy and increases the sidelobes. All the advantages of this structure can only be obtained if the matching of the array to the incident and emergent spherical primary waves is carefully thought out. Note that the relative importance of these phenomena decreases as the array becomes larger.

11.3 LINEAR ARRAY THEORY[1]

11.3.1 Basic equation - array factor

The properties of linear arrays can simply be expressed and can easily be extrapolated to the cases of planar arrays (particularly when structures and illuminations are 'separable').

Consider a set of N identical elements whose phase centres A_0, A_1, ... A_{N-1} are aligned along an axis (Fig. 11.4). The spacing between two adjacent phase centres or 'interelement spacing' of the array, plays an important role with respect to the properties of the array. The elements are assumed to be fed by a matched feed network according to an illumination function defined in amplitude and phase by a set of complex numbers

$$a_0, a_1, \dots\dots a_n \dots\dots a_{N-1}$$

Suppose that each element, the others being present and connected to matched loads, possesses the same pattern $\mathbf{f(u)}$. Due to the principle of linear superposition and the

---

[1] Hansen, R.C. (ed), *Microwave Scanning Antennas*, Vol. II, Academic Press, New York, 1964, (reprinted by Peninsula Publishing, Los Altos, USA, 1985).

translation theorem (§3.1.2), the pattern of the overall array is obtained by summing the element patterns, taking into account their excitation amplitude and their position with respect to a chosen reference point, for instance the point A_0,

$$\mathbf{F}(\mathbf{u}) = \sum_{n=0}^{N-1} a_n \mathbf{f}(\mathbf{u}) \exp\left(j\frac{2\pi}{\lambda}\mathbf{A_0A_n}\cdot\mathbf{u}\right) \tag{11.6}$$

$$\mathbf{F}(\mathbf{u}) = \mathbf{f}(\mathbf{u})R(\mathbf{u}) \tag{11.7}$$

where $R(\mathbf{u})$ is the array factor, which depends only on the spacing and on the illumination function. If θ is the angle between the direction $\mathbf{u}$ and the normal to the array $\mathbf{n}$, we can write

$$\mathbf{A_0A_n}\cdot\mathbf{u} = nd\sin\theta \tag{11.8}$$

Setting

$$\tau = \sin\theta, \qquad \frac{d}{\lambda} = \Delta\nu$$

we find

$$R(\mathbf{u}) = \sum_{0}^{N-1} a_n \exp(j2\pi n\Delta\nu\tau) \tag{11.9}$$

11.3.2 Uniform illumination and constant phase gradient

In this case we can write

$$a_n = \exp(-jn\varphi) \tag{11.10}$$

The array factor then takes the form

$$R(\mathbf{u}) = \sum_{0}^{N-1} \exp\left[jn(2\pi\tau\Delta\nu - \varphi)\right] \tag{11.11}$$

Setting

$$x = \exp\left[j(2\pi\tau\Delta\nu - \varphi)\right]$$

we have

$$R(\mathbf{u}) = \sum_{0}^{N-1} x^n = \frac{x^N - 1}{x-1} = x^{(N-1)/2}\frac{\left(x^{N/2} - x^{-N/2}\right)}{\left(x^{1/2} - x^{-1/2}\right)}$$

Taking out the constant amplitude factor

$$Nx^{(N-1)/2}$$

we have

$$R(\mathbf{u}) = \frac{\sin\left[N\left(\pi\Delta v\tau - \frac{\varphi}{2}\right)\right]}{N\sin\left(\pi\Delta v\tau - \frac{\varphi}{2}\right)} \qquad (11.12)$$

This is a periodic function, and is plotted in Figs. 11.11 and 11.12. It is defined for any value of τ, even for $|\tau| > 1$, which can be interpreted as imaginary, or 'invisible' space (§13.2.2). It passes through a maximum ± 1 each time the denominator and numerator are simultaneously equal to zero, that is to say

$$\pi\Delta v\tau - \frac{\varphi}{2} = k\pi \qquad (k = \text{integer}) \qquad (11.13)$$

A maximum is obtained for the direction τ_0 such that

$$\pi\Delta v\tau_0 - \frac{\varphi}{2} = 0 \qquad (11.14)$$

Thus, a pointing direction τ_0 is obtained with the phase gradient

$$\varphi = 2\pi\Delta v\tau_0 = 2\pi\frac{d}{\lambda}\sin\theta_0 \qquad (11.15)$$

According to equation (11.13) two consecutive maxima are separated by an interval $\Delta\tau$ such that

$$\Delta v\Delta\tau = 1 \qquad (11.16)$$

or

$$\Delta\tau = \frac{\lambda}{d} \qquad (11.17)$$

These results show an important property of uniformly-spaced arrays, namely the existence of several 'grating lobes' periodically located at intervals inversely proportional to the element spacing. These lobes are in general undesirable because they are sources of losses and directional ambiguity (sometimes they are called ambiguity lobes by analogy with temporal signals). To avoid them, or to separate them sufficiently, we should use a small interelement spacing, for example, for $d = \lambda/2$, $\Delta\tau = 2$, and the grating lobes lie outside the domain $(-1, +1)$ of the real directions. The polarity of the grating lobes depends on the parity of N.

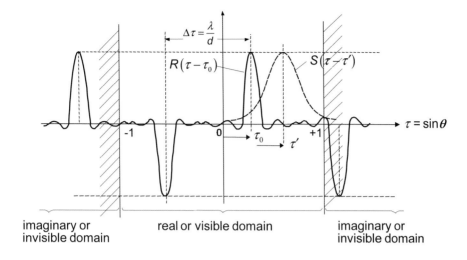

Fig. 11.11 Array factor, even illumination.

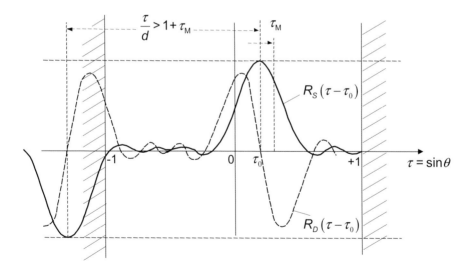

Fig. 11.12 Array factor, odd illumination.

In terms of the pointing direction τ_0 (equation (11.14)), the array factor can be written

$$R_{\tau_0}(\tau) = R(\tau - \tau_0) = \frac{\sin\left[\pi N \Delta v (\tau - \tau_0)\right]}{N \sin\left[\pi \Delta v (\tau - \tau_0)\right]}$$

(11.18)

The phase difference produces a shift of the array factor in τ space. This result corresponds to that known in the time domain, according to which a linear phase slope as a function of frequency does not modify the signal shape, but produces an temporal advance or delay.

Comment: Dispersivity in terms of frequency of an array antenna. Equation (11.15) shows that if the phase gradient φ remains constant as the frequency (and therefore the wavelength) varies, this results in a variation of the pointing direction. A frequency deviation df produces an angular error which can be obtained by differentiating equation (11.15).

$$d\varphi = 0$$

then

$$d\theta = -\tan\theta \frac{df}{f}$$

(11.19)

11.3.3 Half-power beamwidth

If we take the angle θ as variable instead of $\sin\tau$, the array factor undergoes a distortion as a function of the pointing angle θ_0. In particular, the half-power beamwidth (3 dB width) of the main lobe is increased. This is defined by the condition

$$R = \frac{\sin\left[\pi N \Delta v (\sin\theta - \sin\theta_0)\right]}{N \sin\left[\pi \Delta v (\sin\theta - \sin\theta_0)\right]} = \frac{1}{\sqrt{2}}$$

(11.20)

where

$$\theta = \theta_0 + \frac{\theta_{3dB}}{2}$$

This is approximately obtained for

$$\pi N \Delta v \left[\sin\left(\theta_0 + \frac{\theta_{3\,dB}}{2}\right) - \sin\theta_0\right] = \frac{\pi}{2}$$

(11.21)

from which we obtain

$$\theta_{3\,dB} \approx \frac{1}{\pi \Delta \nu \cos \theta_0} = \frac{\lambda}{Nd \cos \theta_0} = \frac{\lambda/L}{\cos \theta_0} \tag{11.22}$$

where $L = Nd$ is the total length of the array.

Therefore we have, in the scanning plane, a broadening of the beam by a factor $\cos \theta_0$. Note that for θ close to $\pi/2$ (*endfire arrays*), another approximation may be made, which gives

$$\theta_{3\,dB} \approx 2 \sqrt{\frac{\lambda}{L}} \tag{11.23}$$

11.3.4 Spectral bandwidth available on a phased array

According to a general rule, the available frequency bandwidth is limited to the frequency shift which scans the pointing direction within the 3 dB beamwidth. Comparing equations 11.19 and 11.22 gives

$$\frac{df}{f} = \frac{1}{\sin \theta_0} \frac{\lambda}{L}$$

or

$$df = \frac{1}{\sin \theta_0} \frac{c}{L}$$

where c is the velocity of propagation. This result shows that the available bandwidth does not depend on the centre frequency, but does depend on the array length L. As a general rule, this bandwidth can be considered as the *inverse* of the time required by the radiation to traverse the aperture of the array.

As mentioned in §11.1.2, this frequency sensitivity can be avoided by dividing the array into *phased* subarrays interconnected with delay lines, but at the cost of grating sidelobes. This topic is the subject of exercise 11.2.

11.3.5 Condition to prevent grating lobes from occurring in the scanning region

Consider an array aimed at scanning over a range of angles $\pm \tau_M$ ($\tau_M = \sin \theta_M$). The lack of grating lobe implies that at the scanning end ($\tau = \tau_M$), the closest grating lobe is still in the 'imaginary' domain, in a direction $\tau < -1$, and that the spacing between the lobes $\Delta \tau = 1/\Delta \nu$ is such that

$$\frac{1}{\Delta \nu} > 1 + \tau_M \tag{11.24}$$

from which the element spacing condition is obtained

$$\frac{d}{\lambda} < \frac{1}{1+\sin\theta_M}$$

(11.25)

Example:

$$\theta_M = 30°; \qquad \sin\theta_M = \frac{1}{2}; \qquad \frac{d}{\lambda} < \frac{2}{3}$$

This factor is important in array design. The reduction of the element spacing leads to various drawbacks: increased mutual coupling, increase of the total number of elements (and of phase shifters) for a given aperture (dimension which determines the gain), thus increasing the cost. Therefore, a tradeoff has to be found to best meet the requirements.

11.3.6 Effect of weighting the array illumination function

The illuminations considered so far have been of constant amplitude. The corresponding patterns possess quite high close-in sidelobes, of the order of -13 dB. To reduce them, we introduce an amplitude taper by decreasing the relative level of excitation of the elements at the edges. This reduces to a multiplication of the preceding illuminations by a continuous function; let this be $s(v)$, whose Fourier transform is $S(\tau)$. According to the convolution theorem, the pattern obtained is the convolution of the preceding pattern with the pattern $S(\tau)$ (Figs. 11.11, 11.12).

This means that each lobe of the array resembles to a certain extent the pattern $S(\theta)$ with close sidelobes more or less improved. The pattern is defined by the function

$$R_s(\tau - \tau_0) = \int_{-\infty}^{\infty} R(\tau' - \tau_0)S(\tau' - \tau)d\tau'$$

(11.26)

Application: difference pattern

For tracking antennas, the Δ pattern, which has a null in the pointing direction and two main lobes of opposite phase on either side, is frequently used. Such a pattern is obtained with an antisymmetric weighing function of the array illumination, so that $d(v) = d(-v)$. Its Fourier transform $D(\tau)$ is also antisymmetric. The convolution of the latter with the pattern $R(\tau - \tau_0)$ leads to a periodic pattern $R_D(\tau - \tau_0)$ composed of a set of anti-symmetric lobe pairs resembling the pattern $D(\tau)$ (Figs. 11.11, 11.12). The particular form of the pattern, its slope in the vicinity of the nulls, and its close-in sidelobes, are determined by the form of the weighing function $d(v)$.

11.3.7 Effect of element directivity

Since the pattern of an array F is the product of the pattern $f(\mathbf{u})$ of an element in the presence of the others with the array factor $R(\mathbf{u})$, we see that a first effect of the element directivity is to weight the occasional 'grating lobe' level (Fig. 11.13). We have here

$$F(\tau) = f(\tau) R(\tau - \tau_0)$$

Another important effect is to control the gain variations of the main lobe during scanning. It is desired that $f(\mathbf{u})$ should vary by only a small amount in the useful scanning region, and that it should be minimum outside this region to minimize the grating lobes. If the element spacing is large, such lobes may exist. In other respects, it is possible to influence the element pattern by the choice of element, and by introducing, if necessary, mutual coupling with the neighbouring elements. If the element spacing is small, the choice of the element does not have a large influence on the element pattern, which is almost entirely controlled by the usual couplings between the elements.

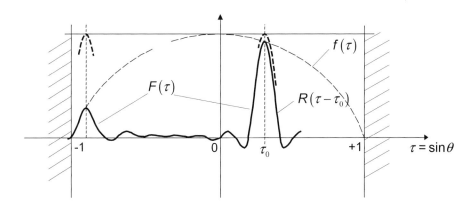

Fig. 11.13 Effect of the directivity of the element pattern.

11.4 VARIATION OF GAIN AS A FUNCTION OF POINTING DIRECTION[2, 3, 4]

11.4.1 Array operating on transmission

Consider a large planar array, of area S, fed by a matched feed network (§2.1.1) according to an illumination of uniform amplitude and a linear phase function which produces a beam pointing in the direction **u** (Fig. 11.14).

The element spacing of the array is assumed sufficiently small that only a single lobe is present in the angular range under consideration. Under these conditions it has been shown (§11.3.3; equation (11.22)) that the useful lobe width increases as $1/\cos\theta_0$. This means that the array directivity decreases as $\cos\theta_0$. This result can be confirmed by imagining the field transmitted by the array in the near field region. Due to the stationary phase conditions, the beam radiates a portion of plane wave tilted by an angle θ_0, of extent S', which is the projection of the array area S onto the pointing direction θ_0

$$S' = S\cos\theta_0 \tag{11.27}$$

The array directivity (or gain, when no loss is present) is of the form

$$D(\theta_0) = D_0\cos\theta_0 \tag{11.28}$$

where

$$D_0 = \frac{4\pi S}{\lambda^2} \tag{11.29}$$

is the array directivity when $\theta_0 = 0$. The concept of directivity is related to the 'shape' of the radiated pattern, particularly its 3 dB beamwidth. To evaluate the gain we should take into account the mismatch losses due to mutual coupling. If the array is assumed very large and the illumination uniform, the geometrical and electrical environment of an element does not depend on the particular element, to within the excitation phase gradient. Thus, in each of them we observe a reflected wave,

---

[2] Parad, L.I., 'Some mutual coupling effects in phased arrays', *Microwave Journal*, Vol. 5, No. 2, pp87-89, June 1962.

[3] Hannan, P.W., 'The element gain paradox for a phased array antenna', *IEEE Trans. Antennas & Propagation*, Vol. AP-12, pp423-433, July 1964.

[4] Allen, J.L., 'Gain and impedance variations in scanned dipole arrays', *IRE Trans. Antennas & Propagation*, Vol. AP-10, pp566-572, Sept 1962.

measured by an active reflection coefficient $\rho(\mathbf{u})$ which varies with the imposed phase gradient, but which is identical for all elements. Thus, the maximum array gain for the pointing angle θ_0 is deduced from the directivity (equation (11.28)) by multiplication by the transfer coefficient: $1 - |\rho(\mathbf{u})|^2$. This gives

$$G(\mathbf{u}) = D_0 \cos \theta_0 \left[1 - |\rho(\mathbf{u})|^2 \right]$$

(11.30)

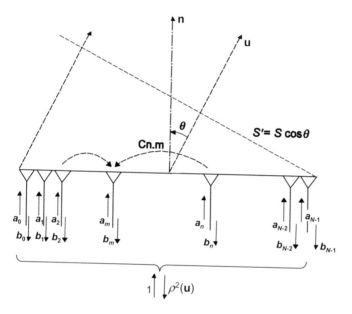

Fig. 11.14 Gain variations vs. pointing directions $\mathbf{u}$, in the presence of mutual coupling.

If now we consider the gain $g(\mathbf{u})$ of an element, in the presence of the other elements assumed to be terminated with matched loads, we can write, according to the principle of superposition

$$G(\mathbf{u}) = N g(\mathbf{u})$$

(11.31)

and

$$g(\mathbf{u}) = g_0 \cos \theta \left(1 - |\rho(\mathbf{u})|^2 \right)$$

(11.32)

where

$$g_0 = \frac{D_0}{N}$$

(11.33)

is the element directivity, in the normal direction. Let us point out that in these expressions, $\rho(\mathbf{u})$ is the active reflection coefficient observed when all elements are excited with the phase gradient in the direction $\mathbf{u}$. The modulus of the pattern $f(\mathbf{u})$ which is involved in equations (11.6) and (11.7) is proportional to the square root of the gain $g(\mathbf{u})$. It is thus strictly related to the active reflection coefficient $\rho(\mathbf{u})$.

11.4.2 Array on receive

Consider an incident plane wave, coming from the direction $\mathbf{u}$. We observe a specularly reflected wave, one or several evanescent waves, and a transmitted wave. In each element, we observe the *same* transmitted wave to within a constant phase gradient corresponding to the tilt of the incident wave.

The variations of power received at any element as a function of the incidence angle $\mathbf{u}$, allow us, in comparison with a standard element, to evaluate the gain $g(\mathbf{u})$ of an element in the presence of the others. Equations (11.31) and (11.32) lead then to the gain $G(\mathbf{u})$ and to the modulus $|\rho(\mathbf{u})|$ of the active reflection coefficient. This measurement allows us to use a simple prototype since *phase shifter and feed network are not involved*. On the other hand, $\rho(\mathbf{u})$ can be calculated from measurements of the coupling coefficients between elements, as we shall see later.

11.4.3 Array active reflection coefficient - mutual coupling[5]

Active reflection coefficient

If N elements of the array are excited in transmission by a set of emerging waves (Fig. 11.14), a_1, a_2, ..., a_n..., a_N, we observe, due to mutual coupling, N reflected waves $b_1, b_2, ... , ... b_N$, linearly related to the first ones, of the form

$$b_m = \sum_{n=1}^{N} C_{m,n} a_n \tag{11.34}$$

where the coefficients $C_{n,m}$ characterize the matrix of the coupling coefficients.

In the case of a uniform linear array excited with a constant phase gradient (see equations (11.14) and (11.15)) corresponding to the pointing angle $\tau_0 = \sin\theta$, we have

---

[5] Amitay, N., Galindo, V. and Wu, C.P., *Theory and Analysis of Phased Array Antennas*, Wiley-Interscience, London, 1972.

$$\varphi = 2\pi\Delta\nu\tau_0 = kd\tau_0, \quad \left(k = \frac{2\pi}{\lambda}\right) \tag{11.35}$$

and after equation (11.10)

$$a_n = a_0 \exp\left(-jknd\tau_c\right) \tag{11.36}$$

We observe at element m the reflected wave

$$b_m = a_c \sum_{n=1}^{N} C_{m,n} \exp\left(-jknd\tau_0\right) \tag{11.37}$$

which characterizes an 'active reflection coefficient'

$$\rho_m\left(\tau_0\right) = \frac{b_m}{a_m} = \sum_{n=1}^{N} C_{m,n} \exp\left[-jk\left(n-m\right)d\tau_0\right] \tag{11.38}$$

Form of the coupling coefficients

The coupling coefficients depend here only on the interelement spacing $d\,|n-m|$. They take the following form

$$C_{m,n} = \gamma\left(d\,|n-m|\right)\exp\left[-j\beta_S d\,|n-m|\right] \tag{11.39}$$

where γ is a decreasing positive function of the interelement spacing and β_S is close to k.

Under these conditions, the coefficient ρ takes the form

$$\rho_m\left(\tau_0\right) = \sum_{n=1}^{N} \gamma\left(d\,|n-m|\right)\exp\left\{-jd\left[\beta_S\,|n-m| + k\tau_0\left(n-m\right)\right]\right\} \tag{11.40}$$

The measurement of the coupling coefficients $C_{n,m}$ allows us to derive the active reflection coefficient of the array, and hence its radiation characteristics.

11.4.4 Blind angle phenomenon[5, 6, 7]

Mutual coupling

In some cases, we observe that for certain pointing directions τ_0 close to those corresponding to the appearance of the grating lobe in the visible domain, the active reflection coefficient becomes close to unity: almost all the transmitted power is reflected. This causes the appearance of a null in the radiation pattern of an element in the presence of the others and then the array gain becomes zero in the corresponding direction. Several theories have been proposed to explain this phenomenon. All concern the existence of a surface wave whose coupling phenomena are the effects and whose propagation constant is precisely the coefficient β_S.

This phenomenon may be explained by considering the form of the active reflection coefficient (equation (11.40)). If we apply the method of stationary phase to this expression, we see that $|\rho_m(\tau_0)|$ is maximum for pointing directions τ_0 such that

$$d\left[\beta_S |n-m| + k\tau_0 (n-m)\right] = 2\pi K \qquad (K = 1, 2, ...) \qquad (11.41)$$

For the elements of rank $n > m$, this is obtained for

$$d\left(\beta_S + k\tau_0\right) = 2\pi \qquad (11.42)$$

and for the elements of rank $n < m$, for

$$d\left(-\beta_S + k\tau_0'\right) = -2\pi \qquad (11.43)$$

Consequently

$$\tau_0 = \frac{\lambda}{d} - 1 - \left(\frac{\beta_S}{k} - 1\right), \quad \tau_0' = -\tau_0 \qquad (11.44)$$

We obtain a pair of symmetrical directions $\pm \tau_0$ for which the reflection coefficient is maximum, thus the minimum gain. Since β_S is close to k or slightly greater (slow wave) equation (11.44) shows that the pointing angle τ_0 giving the maximum for ∥,

---

[6] Knittel, H., Hessel, A. and Oliner, A., 'Element pattern nulls in phased arrays and their relations to guided waves', *Proc. IEEE*, Vol. 56, No. 11, pp1822-1836, December 1968.

[7] Lee, S.W., 'On the suppression of radiation nulls and broadband impedance matching of rectangular waveguide phased arrays', *IEEE Trans. Antennas & Propagation*, Vol. AP-19, No. 1, pp 41-51, January 1971.

is close or slightly less than τ_M for which the first grating lobe of the array appears (§11.3.4, equation (11.24)).

Remark: In equation (11.41) we have not chosen other values for K besides $K = 1$. This is because in the rather frequent case of a slow wave the corresponding directions are not in the real domain.

Expression for $\rho(\tau)$ in a simple theoretical model

In the case of a uniform infinite array the coefficient $\rho(\tau)$ is the same for all elements. If we number them from $-\infty$ to $+\infty$, we have for the elements of rank zero

$$\rho(\tau) = \sum_{-\infty}^{+\infty} \gamma(d|n|) \exp\left[-jd\left(\beta_s |n| + k\tau n\right)\right] \tag{11.45}$$

Since the blind angle direction τ_0 is given by equation (11.41), in the vicinity of $\tau = \tau_0$ only the elements of rank $n > 0$ play a significant role and we can write

$$\rho(\tau) = \sum_{0}^{+\infty} \gamma(dn) \exp\left[-jdnk\left(\tau - \tau_0\right)\right] \tag{11.46}$$

Let us assume for the mutual coupling coefficients an exponentially-decreasing function such as

$$\gamma(dn) = \gamma_0 \exp(-\alpha dn) \tag{11.47}$$

where α characterizes the reduction of the coupling from one element to the next. Let us set

$$y = \exp\left\{-d\left[\alpha + jk\left(\tau - \tau_0\right)\right]\right\} \tag{11.48}$$

then

$$\rho(\tau) = \gamma_0 \sum_{0}^{\infty} y^n = \frac{\gamma_0}{1-y} \tag{11.49}$$

and

$$\left|\rho(\tau)\right|^2 = \frac{\gamma_0^2}{1 + \exp(-2\alpha d) - 2\exp(-\alpha d)\cos\left[kd\left(\tau - \tau_0\right)\right]} \tag{11.50}$$

We see that for $\tau = \tau_0$ $\left|\rho(\tau)\right|$ takes a maximum value

$$\rho_M = \frac{\gamma_0}{1 - \exp(-\alpha d)} \leq 1 \tag{11.51}$$

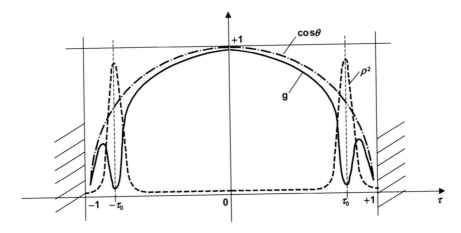

Fig. 11.15 Active reflection coefficient; blind angle phenomenon.

With this notation, we obtain

$$|\rho(\tau)|^2 = \frac{\rho_M^2}{1 + \left[\dfrac{\sin\left(kd\dfrac{\tau - \tau_0}{2}\right)}{\sinh\left(\dfrac{\alpha d}{2}\right)}\right]^2}$$

(11.52)

If the interelement spacing *d* is small, we have

$$|\rho(\tau)|^2 = \frac{\rho_M^2}{1 + \dfrac{k^2}{\alpha^2}(\tau - \tau_0)^2}$$

(11.53)

We obtain an analogous form in the vicinity of the symmetrical direction $\tau_0' = \tau_0$.

These expressions show that the maximum for $|\rho|^2$ in the direction τ_0 (minimum of radiation) is greater as α becomes smaller, that is to say as the fall-off of interelement coupling is reduced. The variations of $|\rho(\tau)|^2$ and of the corresponding element gain $g(\tau)$ given by equation (11.30) are represented in Fig. 11.15.

In the case of a small array, we obtain a result analogous to that of the case of a rapid decrease of α, namely the minimum of radiation in the blind angle directions becomes blunter.

11.4.5 Case where the element spacing is relatively large

In this case, mutual coupling is small and the element pattern is principally defined by the type of element chosen. Nevertheless, the pattern of such an element may be particularly distorted by the presence of neighbouring elements without the appearance of significant coupling coefficients $C_{m,n}$. This is due to the fact that the neighbouring elements can be excited by coupling in a high-order mode that does not propagate at the input but radiates and modifies the pattern of the element which is fed. This phenomenon is observed in the case of large horns whose aperture may locally propagate an antisymmetrical mode, for instance, of the $TE_{2,0}$ type. This is also observed in the case of longitudinal radiation elements.

11.4.6 Study of an array of open-ended guides considered as a periodic structure

An array can be considered as a periodic structure separating two media: the external medium ($z > 0$) is the semi-infinite space of free propagation (with its visible and invisible domains - see §11.3.1), and the internal medium ($z < 0$) composed of the guides feeding the array.

Consider a plane wave incident upon a receiving array. The main problem is to determine the transfer coefficient of the array, which is generally measured by the amplitude of the principal mode excited in the guides. We can deduce by reciprocity other characteristics corresponding to various conditions of excitation of the array, for example, active reflection coefficients, element pattern, etc.

To solve this problem, a method consists of expressing the transverse electromagnetic field in the plane of discontinuity $z = 0$ in two forms both as external field and internal field.

The external field comprises the incident wave and the field diffracted by the array. The latter, due to the periodic structure of the array, can be expressed in the form of a spectrum of a set of visible and invisible plane waves (that is to say, propagating and evanescent). The internal field, in an arbitrary guide, consists of the set of propagating and evanescent modes which propagate in this guide (towards the negative z). This gives a doubly-infinite set of linear equations whose coefficients are the transfer or reflection coefficients. They can be solved by limiting the number of modes to a certain rank[8]. Sometimes an angle of incidence is found for which the transfer coefficient is zero, which corresponds to a blind angle.

---

[8] Roederer, M.A., 'Etudes des reseaux finis de guides rectangulaires à parois épaisses', *L'Onde Electrique*, November 1971.

11.5 EFFECTS OF PHASE QUANTIZATION

11.5.1 Case where all phase shifters are fed in phase

To understand the effect of the phase quantization, we shall assume that the element spacing is small so that the array can be treated as a continuous aperture. If such an array points in the direction $\tau = \sin\theta$, we know that the phase at a point M has a constant gradient (§11.3.2), and thus varies linearly as a function of the abscissa x of M. In fact, this follows from equation (11.15) by substituting the element spacing d by the abscissa x, as follows.

$$\varphi(x) = 2\pi \frac{x}{\lambda} \tau \qquad (11.54)$$

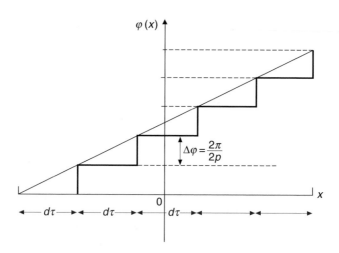

Fig. 11.16 Phase quantization (constant phase origin).

As the phase is quantized in increments $\Delta\varphi = 2\pi/2^p$, in fact we apply a scale variation to the array whose steps have a height of $\Delta\varphi$ and a length of d_τ such that the mean phase slope is maintained (Fig. 11.16)

$$\Delta\varphi = \frac{2\pi}{2^p} = 2\pi\frac{d_\tau}{\lambda}\tau \tag{11.55}$$

then

$$d_\tau = \frac{\lambda}{\tau}\frac{1}{2^p} \tag{11.56}$$

The illumination obtained has a constant phase in the intervals d_τ. This corresponds to an array of element spacing d_τ pointing in the direction τ. This array has a grating lobe in the direction τ_1 such that (eqn. 11.17)

$$\tau - \tau_1 = \frac{\lambda}{d_\tau} = \tau^p \tag{11.57}$$

The angular deviation of the quantization lobe is thus proportional to the pointing error τ. The array lobes are weighted by the pattern $f'(\tau)$ of the element of aperture d_τ, that is to say

$$f'(\tau) = \frac{\sin\left(\pi\frac{d_\tau}{\lambda}\tau\right)}{\pi\frac{d_\tau}{\lambda}\tau} \tag{11.58}$$

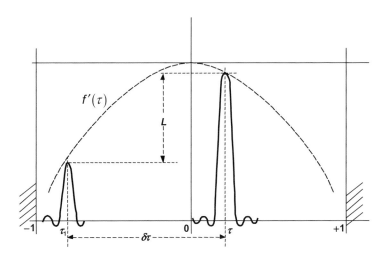

Fig. 11.17 Effect of phase quantization (constant phase origin).

The relative level of the useful lobe of direction τ and of the quantization lobe of direction τ_1 is then (Fig. 11.17)

$$L = \frac{f'(\tau_1)}{f'(\tau)} = \frac{1}{2^p - 1} \qquad (11.59)$$

This level is independent of the pointing angle. It only depends on the accuracy of the phase shifters, for example, $p = 3$ bits, $L = 1/7$, that is to say approximately $-17\,\mathrm{dB}$.

11.5.2 Effects of quantization when the phase origin varies from one phase shifter to another

From the preceding, the stepwise phase functions give a periodic character to the phase errors, which lead to localized parasitic lobes of relatively high levels, particularly for phase shifters of just a few bits. If we now assume that the phase displayed for each phase shifter involves a constant of 'origin' depending on the number of bits of the phase shifter (Fig. 11.18), this phase is of the form

$$\Phi_n = \varphi_n + k\Delta\varphi \qquad (k = 1, 2, \ldots) \qquad (11.60)$$

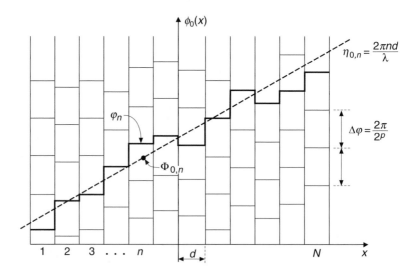

Fig. 11.18 Phase quantization (random phase origin).

Under these conditions, the periodic character of the phase errors along the array disappears. If the theoretical phase of the phase shifter n is $\varphi_{0,n}$, the phase error is

$$\Psi_n = \Phi_{0,n} - \left(\varphi_n + k\Delta\varphi\right) \tag{11.61}$$

k is chosen so that this error is minimum. It never exceeds $\Delta\phi/2$ but its exact value depends on the difference $\phi_{0,n} - \varphi_n$. If the origin phases φ_n are randomly distributed, we see that there is no correlation between the errors of the different elements. The probability density of the phase error Ψ_n of a element is uniform over $-\Delta\phi/2$ to $+\Delta\phi/2$. The illumination function of an uniformly illuminated array is thus of the form (to first order)

$$a_n = \exp\left[j\left(\Phi_{0,n} - \Psi_n\right)\right] \approx \exp\left(j\Phi_{0,n}\right)\left(1 - j\Psi_n\right) \tag{11.62}$$

The phase error introduces a parasitic illumination

$$\Delta a_n = j\Psi_n \exp\left(j\varphi_{0,n}\right) \tag{11.63}$$

and a parasitic pattern

$$\Delta R(\tau) = \sum_0^{N-1} \Psi_n \exp\left(j\Phi_{0,n}\right)\exp\left(j2\pi n\Delta v\tau\right) \tag{11.64}$$

which is superimposed on the desired pattern

$$R(\tau) = \sum_0^{N-1} \exp\left(j\Phi_{0,n}\right)\exp\left(j2\pi n\Delta v\tau\right) \tag{11.65}$$

In the direction of maximum radiation θ_0, we have

$$R(\tau_0) = N \tag{11.66}$$

The relative mean energy of the parasitic radiation (Fig. 11.19) is the mean square value

$$\overline{L} = \overline{\left|\frac{\Delta R(\tau)}{R(\tau)}\right|^2} = \frac{1}{N^2}\overline{\Delta R(\tau)\Delta R(\tau)^*}$$

$$\overline{L} = \overline{\left|\frac{\Delta R(\tau)}{R(\tau)}\right|^2} = \frac{1}{N^2}\sum_0^{N-1}\sum_0^{N-1}\overline{\Psi_n\Psi_m}\exp\left[j\left(\Phi_{0,n} - \Phi_{0,m}\right)\right]\exp\left[j2\pi\Delta v\tau\left(n - m\right)\right]$$

$$\tag{11.67}$$

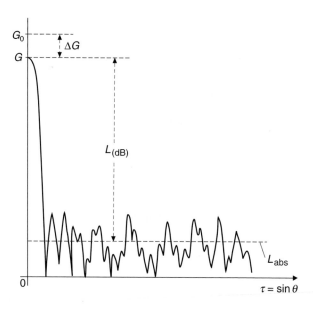

Fig. 11.19 Mean phase quantization sidelobe energy level.

By hypothesis

$$\overline{\Psi_n \Psi_m} = 0$$

except for $n = m$, in which case

$$\overline{\left|\Psi_n\right|^2} = \frac{1}{3}\left(\frac{\Delta\varphi}{2}\right)^2 \tag{11.68}$$

Then

$$\overline{L} = \frac{\left(\dfrac{\Delta\varphi}{2}\right)^2}{3N} = \frac{\pi^2}{3N2^{2p}} \tag{11.69}$$

or in decibels

$$\overline{L_{\mathrm{dB}}} = -\left(10\log N + 6p - 5\right) \tag{11.70}$$

This is plotted in Fig. 11.20. The rms quantization sidelobe level decreases by 6 dB for each additional bit and by 10 dB for each ten-fold increase in the number of antenna elements.

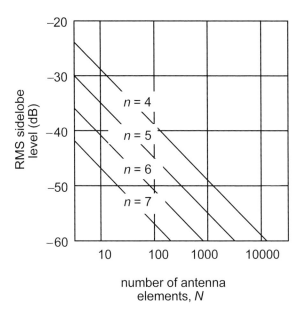

Fig. 11.20 Quantization sidelobe level as a function of number of bits *n* and number of elements *N*.

Influence on accuracy on-axis

The mean level of diffused radiation is found again in the null depth of the difference channel. If the slope of the null is $1/\theta_{3\,dB}$ (see §10.4.2), the standard deviation of the position of the axis in a plane, taking into account the fact that only the component in phase with the difference of this radiation produces a deviation, is given by

$$\frac{\sigma_e}{\theta_{3\,dB}} = \frac{\pi}{2^p \sqrt{6N}}$$

$$(11.71)$$

Numerical example

With a relative level of −40 dB the standard deviation is of the order of $\theta_{3\,dB}/100$.

Influence of the gain

The energy lost in the parasitic lobes involves a relative loss of gain

$$\frac{\Delta G}{G_0} = \frac{\sum_0^{N-1} \overline{|\Delta a_n|^2}}{\sum_0^{N-1} \overline{|a_n|^2}} = \overline{|\Psi_n|^2} = \frac{1}{3}\left(\frac{\Delta\varphi}{2}\right)^2$$

$$(11.72)$$

and in decibels

$$10\log\frac{G}{G_0} \approx -\frac{14}{2^{2p}}$$ (11.73)

Remark: The mean 'absolute' level of the quantization lobes is obtained by multiplication of the relative level by the principal gain

$$G = Ng_0(\mathbf{u})$$

where $g_0(\mathbf{u})$ is the element gain in the presence of the other elements. Thus we have

$$\overline{L_{\text{ABS}}} = \overline{GL} = \frac{\pi^2}{3.2^{2p}}g_0(\mathbf{u})$$

and in decibels

$$\overline{L}_{\text{ABS}}(\text{dB}) = g_0(\mathbf{u})_{\text{dB}} - 6p + 5$$ (11.74)

Thus, the *absolute* level *depends only on the accuracy of the phase shifters* (number of bits) and not on the number of elements.

Numerical example
We desire an absolute level of less than −7 dB with respect to isotropic. Assuming a typical element gain of 6 dB, the number of bits should be greater than 3.

Note: If phase shifters with random phase origins φ_n are used, then their values must be stored in a memory and used in the calculation of the displayed phases.

11.6 FREQUENCY-SCANNED ARRAYS

An electronically scanned antenna can be realized in a simple way by using a frequency-scanned array. Scanning is then obtained by varying the frequency. A simple means to obtain such an array is to use a length of transmission line to feed the array by means of directional couplers. Two adjacent elements of the array, spaced by *a*, are fed by two couplers separated in the transmission line by a length *b*. The array and transmission line properties allow us to show that a frequency deviation *df* about the central frequency f_0 produces an angular scanning $d\tau$ given in radians, to first order, by the following equation:

$$d\tau = \frac{b}{a}\frac{\lambda_g}{\lambda}\frac{df}{f_0}$$

where λ_g is the guide wavelength. We know that the phase shift produced by a length b of the transmission line is given by

$$\Delta\varphi = \frac{2\pi b}{\lambda_g} \tag{11.75}$$

This causes a deviation θ of the beam with respect to the normal, such that (equation (11.15))

$$2\pi\frac{a}{\lambda}\sin\theta = \Delta\varphi - 2K\pi \tag{11.76}$$

the integer K being chosen so that $|\sin\theta| \leq 1$.

Setting, as usual, $\tau = \sin\theta$, and taking the differential of this equation

$$ad\left(\frac{\tau}{\lambda}\right) = d\left(\frac{b}{\lambda_g}\right)$$

or

$$a\frac{\lambda d\tau - \tau d\lambda}{\lambda^2} = -b\frac{d\lambda_g}{\lambda_g^2} \tag{11.77}$$

In the vicinity of the normal ($\tau \approx 0$)

$$d\tau = -\frac{b}{a}\frac{\lambda}{\lambda_g}\frac{d\lambda_g}{\lambda_g} \tag{11.78}$$

If the transmission line has no frequency cut-off (coaxial line filled with a dielectric of dielectric constant ε), we have

$$\lambda_g = \frac{\lambda}{\sqrt{\varepsilon}}$$

Then

$$d\tau = -\frac{b}{a}\sqrt{\varepsilon}\frac{d\lambda}{\lambda} = +\frac{b}{a}\sqrt{\varepsilon}\frac{df}{f} \tag{11.79}$$

If the transmission line (waveguide) has a cut-off wavelength λ_c, we have the classical relationship

$$\frac{1}{\lambda_g^2} + \frac{1}{\lambda_c^2} = \frac{1}{\lambda^2} \qquad (11.80)$$

which gives, upon differentiation

$$-\frac{d\lambda_g}{\lambda_g^3} = -\frac{d\lambda}{\lambda^3}$$

and finally we obtain

$$d\tau = \frac{b}{a} \frac{\lambda_g}{\lambda} \frac{df}{f} \qquad (11.81)$$

The deviation in radians is proportional to the relative length (b/a) of the waveguide, to the ratio of guide and free-space wavelengths, and to the relative frequency deviation.

Exercise: Repeat the calculation of the two preceding cases by generalizing when the pointing direction is not close to the array normal ($\tau_0 \neq 0$).

11.7 ANALOGUE BEAMFORMING MATRICES

11.7.1 Introduction

We have seen (§7.2.2, §9.7.1) that multiple-beam antennas are sometimes used in radar and sonar applications, as an alternative to electronic scanning, to improve the range and the angular resolution of equipment. On the other hand, multiple beams and electronic scanning can be combined, since the angular scanning of a group of several beams allows us to increase the duration of the illumination of targets (this improves velocity measurements by the Doppler effect). In addition, it allows us to perform interpolations and combinations between signals received from multiple beams (this improves the angular accuracy and allows techniques to reduce the effect of jamming (§14.3, §14.4)).

Such antennas can be realized by using a kind of *retina* of multiple primary elements on the focal surface of a reflector or a lens. If we desire to combine multiple beams with electronic scanning we might use a reflectarray or an active lens (§11.2.2).

This technique is valid as long as the angular domain covered by the multiple beams is not too large (about 5 to 6 degrees). Beyond that, even using a long focal length and aplanatic systems, we are limited by geometrical aberrations which distort the beams (§9.7)

The use of array antennas combined with beamforming matrices allows us to avoid these limitations.

In general, these matrices are microwave structures known as 'reciprocal passive multipoles', in principle without losses, since their properties rule those of the multiple-beams that they fed.

Other types of matrices are used, e.g. analogue and digital matrices (§11.8) which frequently require a frequency shift of the received signals.

11.7.2 General properties of multi-port networks

'S' matrix

A multi-port network is composed of N ports in the form of waveguides, numbered from 1 to N. On each of them we define a reference plane π_n where we observe:

- an incident wave of complex amplitude a_n;

- a reflected or emerging wave of amplitude b_n;

The coefficients a_n and b_n are assumed to be normalized so that the transmitted powers are measured (in watts) by $|a_n|^2$ and $|b_n|^2$.

Assume that an single incident wave (a_p) is sent to the port (p), the other ports being terminated in matched loads. We observe emerging waves of the form

$$b_1 = S_{1,p}a_p, \quad ... \quad b_n = S_{n,p}a_p, \quad ... \quad b_p = S_{p,p}a_p, \quad ... \qquad (11.82)$$

Thus the elements $S_{n,p}$ are *transfer coefficients*. They can be arranged in the form of a matrix which is *symmetric* if the network is reciprocal. The elements of the principal diagonal ($S_{p,p}$) are the reflection coefficients

$$[\mathbf{S}] = \begin{bmatrix} S_{11} & S_{12} & .... & S_{1N} \\ S_{21} & S_{22} & .... & S_{2N} \\ & & & \\ S_{N1} & S_{N2} & .... & S_{NN} \end{bmatrix} \qquad (11.83)$$

In general, the set of incident and emerging waves may be represented by column matrices

$$[\mathbf{A}] = \begin{bmatrix} a_1 \\ a_2 \\ \vdots \\ a_N \end{bmatrix} \text{ and } [\mathbf{B}] = \begin{bmatrix} b_1 \\ b_2 \\ \vdots \\ b_N \end{bmatrix} \tag{11.84}$$

So we have the matrix relation

$$[\mathbf{B}] = [\mathbf{S}][\mathbf{A}] \tag{11.85}$$

Case of a lossless matrix

Consider again the case where we inject a wave (a_p) at the single port (p). The net emerging power must be equal to the incident power

$$|b_1|^2 + \dots + |b_N|^2 = |a_p|^2$$

then

$$|S_{1,p}|^2 + |S_{2,p}|^2 + \dots + |S_{N,p}|^2 = 1 \tag{11.86}$$

Thus, the sum of the squares of the transfer coefficient magnitudes of a row or a column is equal to unity.

Orthogonality of a lossless matrix

Consider again the case of an excitation of the single port (p) by an incident wave (a_p). We have seen that the emerging waves are expressed by

$$b_1 = S_{1,p}a_p, \quad \dots \quad b_n = S_{n,p}a_p, \quad \dots \quad b_N = S_{N,p}a_p$$

Now reverse the sense of propagation of the emerging waves; they become incident waves of same amplitude but of *conjugate phases* in order to take into account the reversal of the direction of propagation

$$a_1' = b_1^*, \quad a_2' = b_2^*, \quad \dots \ a_N' = b_N^* \tag{11.87}$$

By reciprocity all the power contained in these incident waves returns to the *single port* (p). Thus, the power transmitted to any other arbitrary port (n) is zero. Let us calculate this power. From the foregoing, the emerging signal at (n) is

$$b_n = S_{n,1}a'_1 + S_{n,2}a'_2 + \ldots + S_{n,p}a'_p + \ldots + S_{n,N}a'_N = 0$$

Taking into account relations (11.87), we obtain

$$S_{n,1}S^*_{1,p} + S_{n,2}S^*_{2,p} + \ldots + S_{n,N}S^*_{N,p} = 0$$

or still, by symmetry

$$S_{n,1}S^*_{p,1} + S_{n,2}S^*_{p,2} + \ldots + S_{n,N}S^*_{p,N} = 0 \qquad (11.88)$$

This relation yields the property of orthogonality: the hermitian product of two single rows or two single columns is zero.

11.7.3 Beamforming applications

Orthogonality of the illuminations of an array (Fig. 11.21)

An array of N radiating elements is fed by a multi-port network comprising $N+M$ ports: N 'outputs' feeding the radiating elements of the array and M 'inputs' among which we consider two denoted A and B. The latter are assumed matched and independent, that is to say isolated.

$$S_{AA} = S_{BB} = S_{AB} = S_{BA} = 0$$

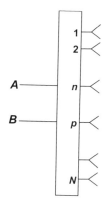

Fig. 11.21 Orthogonality of the array illuminations.

This gives the following results:

Conservation of energy:

$$\left|S_{A,1}\right|^2 + \left|S_{A,2}\right|^2 + \ldots + \left|S_{A,N}\right|^2 = 1$$

$$(11.89)$$

$$\left|S_{B,1}\right|^2 + \left|S_{B,2}\right|^2 + \ldots + \left|S_{B,N}\right|^2 = 1$$

Orthogonality:

$$S_{A,1}S^*_{B,1} + S_{A,2}S^*_{B,2} + \ldots + S_{A,N}S^*_{B,N} = 0 \qquad (11.90)$$

Interpretation: The illuminations of the array produced by excitation of the inputs *A* and *B* are respectively

$$S_{A,1}, S_{A,2}, \ldots S_{A,N}$$

$$S_{B,1}, S_{B,2}, \ldots S_{B,N}$$

from which we can conclude

Theorem: *In a lossless distribution the illuminations produced from two arbitrary isolated inputs are orthogonal.*

Orthogonality of the characteristic functions

We know that the characteristic functions of radiation are obtained by Fourier transformation of illuminations. The conservation of the hermitian product (Parseval's theorem) leads to the orthogonality of the characteristic functions. If (α, β, γ) are the direction cosines of the unit vector **u** $\left(|\alpha|^2 + |\beta|^2 + |\gamma|^2 = 1\right)$, we have

$$\int_{-\infty}^{\infty}\int_{-\infty}^{\infty} \mathbf{F}_A(\alpha,\beta)\mathbf{F}^*_B(\alpha,\beta)\,d\alpha d\beta = 0 \qquad (11.91)$$

In the case of a rectilinear array, if θ is the angle of **u** to the normal of the array (see notations of §11.3.1), we have

$$\int_{-\infty}^{\infty} \mathbf{F}_A(\tau)\mathbf{F}^*_B(\tau)\,d\tau = 0 \quad (\tau = \sin\theta) \qquad (11.92)$$

In practice, orthogonality is often verified in the real domain ('visible') that is to say for $-1 \leq \tau \leq 1$. In particular, this is the case of point elements of one half-wavelength spacing; the characteristic functions are periodic with period $\lambda/d = 2$. In this case it is sufficient to limit the integral to this domain

$$\int_{-1}^{1} F_A(\tau) \, F_B^* d\tau = 0 \tag{11.93}$$

Directional beams

Consider a regular rectilinear array of an element spacing d and length $D = Nd$ with uniform amplitude illumination. A matrix with M inputs $(A_0, \ldots A_m, \ldots A_{M-1})$ can create M orthogonal directional beams if the illuminations corresponding to each of these inputs $(S_{m,0}, S_{m,1}, \ldots S_{m,N-1})$ possess a phase illumination $(\varphi_{m,1}, \varphi_{m,2}, \ldots \varphi_{m,N})$ which varies linearly along the array, and if the phase shift relative to two decoupled inputs varies by an integer multiple of 2π along the array.

These conditions are satisfied if the transfer coefficient from an input port (m) to an output (n) is of the form

$$S_{m,n} = \frac{1}{\sqrt{N}} \exp\left(-j2\pi \frac{nm}{N}\right) \tag{11.94}$$

Comment: Discrete Fourier transform
If the input ports are excited with complex amplitudes $(a_1, \ldots a_m, \ldots a_M)$, the resulting output illumination is of the form

$$b_n = \frac{1}{\sqrt{N}} \sum_{m=0}^{N-1} a_m \exp\left(-j2\pi \frac{nm}{N}\right) \tag{11.95}$$

Conversely, on receive, if the signals collected by the N elements of the array are defined by the distribution b_n $(0 \leq n \leq N-1)$, the signals received at the M output ports are given by

$$a_m = \frac{1}{\sqrt{N}} \sum_{n=0}^{N-1} b_n \exp\left(j2\pi \frac{nm}{N}\right) \tag{11.96}$$

We observe that the two distributions (a_m) and (b_n) are related by a discrete Fourier transform.

Radiated patterns: These follow from the radiation characteristic functions which are the Fourier transforms of the illuminations. The array factors, which are the characteristics functions assuming that the elements are isotropic, are obtained from the transfer coefficients $S_{m,n}$ in the form of a Fourier series

$$R_m(\tau) = \frac{1}{N}\sum_{n=0}^{N-1} S_{m,n} \exp(j2\pi\Delta v\tau) = \frac{\sin\left[N\pi\Delta v(\tau-\tau_m)\right]}{N\sin\left[\pi\Delta v(\tau-\tau_m)\right]} \tag{11.97}$$

The τ_m define the sampled pointing directions

$$\tau_m = m\Delta\tau \tag{11.98}$$

$\Delta\tau$ is the pointing deviation of two adjacent beams

$$\Delta\tau = \frac{1}{2v_0} = \frac{1}{N\Delta v} = \frac{\lambda}{Nd} = \frac{\lambda}{D} \tag{11.99}$$

The patterns $R_m(\tau)$ are periodic orthogonal functions of period $1/\Delta v = \lambda/d$. The maximum of each pattern coincides with the nulls of the others. If the element spacing is small, of the order of $\lambda/2$, we can neglect the periodic character and represent the array factors in the form

$$R_m(\tau) = \frac{\sin\left[N\pi\Delta v(\tau-\tau_m)\right]}{N\pi\Delta v(\tau-\tau_m)} = \frac{\sin(N\pi\Delta v\tau - m\pi)}{N\pi\Delta v\tau - m\pi} \tag{11.100}$$

We shall encounter these functions again in Chapter 12.

11.7.4 Examples of matrices

Butler matrix[9]

This matrix is based on the use of 3 dB directional couplers.

---

[9] Butler, J. and Lowe, R., 'Beam forming matrix simplifies design of electronically scanned antennas', *Electronic Design*, Vol. 9, April 1961, pp170-173.

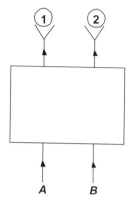

Fig. 11.22 Two-element array fed by a 3 dB directional coupler.

Two-element array: A 3 dB coupler can feed a two-element array and form two decoupled orthogonal beams (Fig. 11.22). If we consider the two ports A and B and the outputs 1 and 2 which excite the two elements of the array, the transfer coefficients characterizing the two illuminations are

$$S_{A,1} = \frac{1}{\sqrt{2}}, \quad S_{A,2} = \frac{1}{\sqrt{2}}\exp\left[-\left(j\frac{\pi}{2}\right)\right]$$

$$ \tag{11.101}$$

$$S_{B,1} = \frac{1}{\sqrt{2}}\exp\left[-\left(j\frac{\pi}{2}\right)\right], \quad S_{B,2} = \frac{1}{\sqrt{2}}$$

We verify the conservation of energy and the orthogonality of the matrix.
The array factors obtained (in a planar space) are of the form

$$F_A(\tau) = \cos\left(\pi\tau + \frac{\pi}{4}\right), \quad F_B(\tau) = \cos\left(\pi\tau - \frac{\pi}{4}\right) \tag{11.102}$$

Four-element arrays: The matrix has four inputs and can generate four orthogonal beams. It is composed of four 3 dB couplers connected through two $\pi/4$ phase shifters (Fig 11.23). The transfer coefficients obtained are represented below on a trigonometric circle.

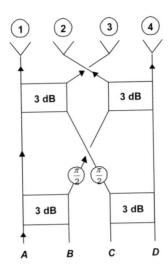

Fig. 11.23 Four-element array fed by a 4×4 Butler matrix.

The four patterns have the form indicated above, with an angular spacing between the beams (for a spacing $d = \lambda/2$): $\Delta\tau = 1/2$.

Generalization: To form N beams, the number C of couplers required is given by the function

$$C = \frac{N}{2}\log N \qquad (11.103)$$

The drawback of Butler matrices is in the complexity of their interconnection, particularly for large matrices. For large arrays it is better to use other types such as Blass matrices or quasi-optical matrices.

Comment: The Butler matrix is of didactic interest, since it is the circuit equivalent of the FFT (Fast Fourier Transform) where the 'Butterflies' are replaced by 3 dB couplers[10]. An application of the Butler matrix to circular arrays is given in §11.8.4.

---

[10] Ueno, M., 'A systematic design formulation for Butler matrix applied FFT algorithm', *IEEE Trans. Antennas & Propagation*, Vol.29, No.3, pp496-501, May 1981; see also correction in *IEEE Trans. Antennas & Propagation*, Vol.29, No.5, p825, September 1981.

Blass matrix

The M excitation guides $(1, 2, ... m, ... M)$ are connected to the N guides feeding the elements of the array $(1, ... n, ... N)$ through directional couplers. The coupling coefficients are chosen so that the transfer coefficients $S_{m,n}$ follow the desired illumination functions (Fig. 11.24). The guides are terminated with matched loads. An application of the Blass matrix is given in §11.2.2, Fig. 11.9.

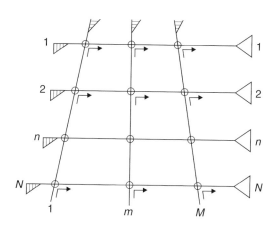

Fig. 11.24 Blass matrix.

Orthogonality: Since the couplers are directional, and the elements are assumed to be matched, the inputs are isolated, and the corresponding patterns are orthogonal.

Phase functions: The phases vary linearly along the array with an increment which deviates from the tilt angle of the excitation guide with respect to the feeding guides. This increment can be corrected by introducing a constant phase shift in front of each element. With circular polarization an easy solution consists of using 'helical' antennas: an axial rotation of the helix allows to create the necessary constant phase shift.

Electronic scanning: By inserting programmable phase shifters in front of the antenna elements it is possible to combine multiple beams and electronic scanning.

Advantage of the Blass matrix: The number C of couplers necessary to form M beams is

$$C = MN \qquad (11.104)$$

Thus, this system is particularly attractive for large arrays with a small number of beams. The number of couplers may be rather large, but the interconnection layout is simple and does not require overlapping or multilayer technologies.

Disadvantage: The phase increment depends on the wavelength and equally on the direction of the beams. Thus, the system is not well suited to large arrays if the bandwidth of the transmitted signals is very wide.

Application exercise: We wish to synthesize a passive antenna possessing a null in a fixed direction independent of the form or direction of the useful beam.

Such a characteristic is useful in radar applications; a null in the direction of the horizon can suppress clutter and occasional jamming.

To solve this problem, we can use a two port Blass matrix. One port (A) is associated with a uniformly illuminated beam with maximum gain in the horizontal direction. The other port (B) creates a steerable beam by means of phase shifters (Fig. 11.25).

Consider an incident plane wave from the horizontal direction. By reciprocity it can be shown that all the power is absorbed at A in the load. Therefore, no energy is sent to the port B; its pattern has a null in the horizontal direction.

More precisely, an incident wave of direction θ (with respect to the normal to the array) generates a signal in the channel n

$$a_n = \exp\left(jn\pi\tau\right) \quad \text{where } \tau = \sin\theta, \quad d = \frac{\lambda}{a} \qquad (11.105)$$

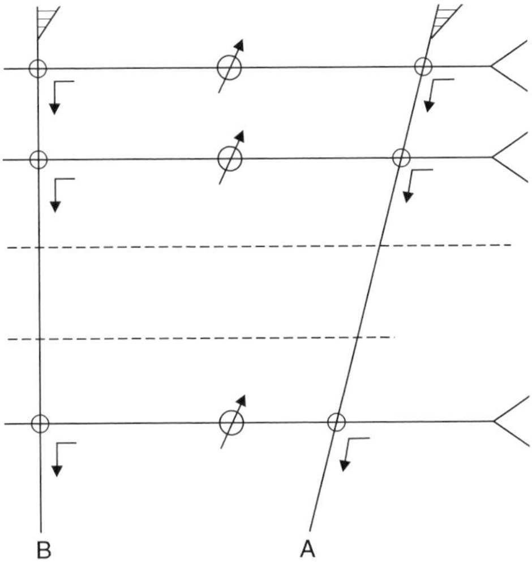

Fig. 11.25 Phased array radiating a pattern with a fixed null.

This can be separated into

- a constant illumination $\alpha_n = \alpha_0$ (which provides a signal at A);

- and a residual illumination β_n orthogonal to α_n

$$a_n = \exp\left(jn\pi\tau \right) = \alpha_0 + \beta_n$$

From which we obtain the residual incident wave

$$a'_n = \beta_n = \exp\left(jn\pi\tau \right) - \alpha_0$$

Since the latter is orthogonal to the distribution α_0, we have

$$\sum_{n=0}^{N-1} \alpha_0 \left[\left(\exp\left(jn\pi\tau \right) - \alpha_0 \right) \right] = 0$$

from which we obtain

$$\alpha_n = \mathrm{sinc}\left(\frac{N\pi\tau}{2} \right) \exp\left(j\pi \frac{N-1}{2} \right)$$

Then

$$\alpha'_0 = \exp\left(jn\pi\tau\right) - \exp\left(j\pi\frac{N-1}{2}\right)\mathrm{sinc}\left(\frac{N\pi\tau}{2}\right)$$ (11.106)

Let us form a pattern pointing in a direction τ_0. We sum these signals (a'_n) in a direction τ_0 with a phase shift $(-n\tau_0)$, i.e.

$$F_{\tau_0}\left(\tau\right) = \sum_0^{N-1} a'_n \exp\left(-jn\pi\pi_0\right)$$

$$= \left\{\mathrm{sinc}\left[\frac{N\pi\left(\tau\text{-}\tau_0\right)}{2}\right] - \mathrm{sinc}\left(\frac{N\pi\tau}{2}\right)\mathrm{sinc}\left(\frac{N\pi\tau_0}{2}\right)\right\}\exp\left(j\pi\frac{N-1}{2}\right)$$ (11.107)

↑	↑	↑
pointing in direction τ_0	pointing in direction $\tau_0 = 0$ (horizon)	weighted by the sidelobe of the preceding pattern

Thus, we obtain a pattern pointing in direction τ_0 but slightly distorted in order to generate a null in the direction ($\tau = 0$).

11.7.5 Non-orthogonal directional beams

Problem of multiple-beam angular coverage

We have seen that orthogonal adjacent directional beams of the form sinc(x) crossover at a relative level of $2/\pi$, i.e. approximately −4 dB. If an illumination taper is used in order to reduce the sidelobes of the orthogonal beams, the crossovers occur at an even lower level. This means a hole in the radar coverage may occur in the corresponding direction, and a lower detection probability will result.

A solution to this problem is to bring the beams closer to one another. If the chosen angular spacing $\Delta\tau'$ is lower than the spacing resulting from the orthogonality condition $(\Delta\tau = 1/2v_0 = \lambda/Nd)$, the relative overlap level becomes

$$L' = \mathrm{sinc}\left(\frac{\pi}{2}\frac{\Delta\tau'}{\Delta\tau}\right) \approx 1 - \frac{\pi^2}{24}\left(\frac{\Delta\tau'}{\Delta\tau}\right)^2$$ (11.108)

and in dB

$$L_{\mathrm{dB}} \approx 4\left(\frac{\Delta\tau'}{\Delta\tau}\right)^2$$ (11.109)

If for instance the angular spacing is reduced by half, the loss is approximately one dB.

However, the problem still remains of decoupling the corresponding ports. This problem may be solved, in the case of passive matrices, at the expense of some losses. This reduces the real gain but the improvement of the relative gain in the critical direction allows us to obtain an optimum solution.

Quasi optical matrix: Rotman lens

Description (§11.2.2: electronic phased lens): The Rotman lens is enclosed between two parallel conducting plates (Fig. 11.26). Its external face is a regular linear array, its internal face an array arranged along a curve constituting the 'principal surface' (§9.3) of the system. These two arrays are connected by two fixed-phase shifters ensuring focusing, and occasionally variable phase shifters if we want to combine electronic scanning with beam forming.

Multiple beams are created by means of a 'retina' of primary elements placed on the focal surface of the lens, which approximates to the arc of a circle of radius equal to the focal length F.

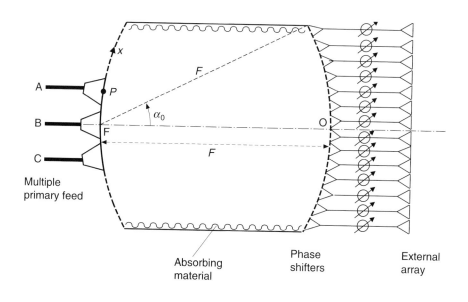

Fig. 11.26 Array with Rotman lens.

The shape of the principal surface (the internal face) is chosen in order to minimize the geometrical aberrations. One of the most troublesome is the *coma* aberration (§9.7.3) which creates a cubic distortion of the wavefront. To cancel it the Abbe condition means that the principal surface should be a sphere (or a circle in this case) centred at the focus. The inputs of the matrix are the primary sources, which in general have some small degree of mutual coupling. The radiated beams are not in general perfectly orthogonal, since part of the primary pattern is not intercepted by the lens and the undesirable spillover is generally absorbed by an appropriate material.

Diffraction spot: An axial incident wave produces, through the lens, a circular wave converging at the focus. At a point P in the vicinity of the focus F, the diffraction spot is obtained from the Huygens-Fresnel formula. If D is the lens length (array length), x the abscissa along the focal curve and α_0 the semi-angle subtended by the lens, seen from the focus, we obtain (§9.9.3, eqn. 9.113)

$$\Psi(x) = \mathrm{sinc}\left(\frac{2\pi x}{\lambda}\sin\alpha_0\right) \approx \mathrm{sinc}\left(\frac{\pi D x}{\lambda F}\right) \qquad (11.110)$$

The distance δ between nulls of this spot is given by (eqn. 9.114)

$$\delta = \frac{\lambda}{\sin\alpha_0} \approx \frac{2\lambda F}{D} \qquad (11.111)$$

Imaging and multiple beams: A multiple point 'object' configuration of radiating elements, at infinity, may be defined by a distribution $A(x)$, each point M of abscissa x defining a direction MO. These elements produce an image $B(x)$ in the focal plane through the system, which is the convolution of the distribution $A(x)$ with the diffraction spot $\Psi(x)$ (§13.5). Assuming negligible distortions, we find

$$B(x) = A(x) \otimes \Psi(x)$$

This image can be sampled with the maximum spacing Δx by applying the sampling theorem.

$$\Delta x = \frac{\lambda}{2\sin\alpha_0} \approx \frac{\lambda F}{D} \qquad (11.112)$$

that is to say $\Delta x = \delta/2$.

   This is half the size of the diffraction spot. There is therefore a tradeoff between the antenna gain that we can obtain from a primary element and the spacing sampling: the first is optimized for a primary aperture close to δ, the second for an aperture smaller than $\lambda/2$ (§9.9.3).

It is theoretically possible to reconcile these two requirements. In effect, we know how to realize primary elements whose illuminations are functions of the orthogonal sinc(x) type. These would produce sector primary patterns without spillover, and would lead to a uniform illumination of the lens, thus the maximum gain. Nevertheless, this solution gives high sidelobes. That is why, in practice, we can fit in the primary elements by means of loaded magic-T at the expense of some losses. On the other hand, we observe that these losses are only effective on receive. On transmit, if all the elements are excited in-phase, there are no losses. On receive we obtain an overlap rate of 50% between the elements. The patterns are optimized, but there is a loss of gain (Fig. 11.27).

In the operational systems where *external* noise is significant (this is frequently the case when low noise temperature receivers are used) this drawback is acceptable. It disappears to a large extent if each primary element comprises an *active module* that amplifies the received signal before it reaches the loaded magic-T.

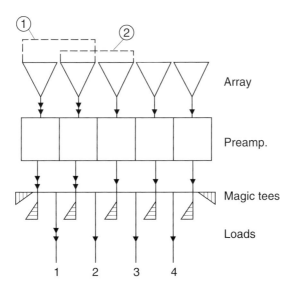

Fig. 11.27 Active multiple-channel primary feed with overlapping illuminations.

11.8 FURTHER TOPICS

11.8.1 Active modules

Passive and active phased arrays - requirements for modules

One of the principal applications of phased array antennas is in radar, offering advantages in respect of beam agility, adaptive pattern control and (ultimately) multiple simultaneous surveillance and tracking functions.

Phased array radars can be divided into two categories: passive phased arrays (Fig. 11.28a) use a single high-power transmitter, distributing the power to the individual array elements via a feed network incorporating phase shifters, and active phased arrays (Fig. 11.28b), in which each element or subarray has its own active module, comprising transmit power amplifier, phase shifter, receive low-noise amplifier, and possibly transmit upconverter/receive downconverter as well. Active phased arrays avoid the losses that occur in the feed network of a passive phased array. They also have reliability advantages, in that a fault in a single module does not cause a catastrophic failure of the whole radar (*graceful degradation*) and in that a failed module can be replaced without taking the entire radar out of commission. The principal disadvantage is that phased array active modules have proved expensive to develop, which probably accounts for the relative slowness of these concepts to find widespread use in practical radars.

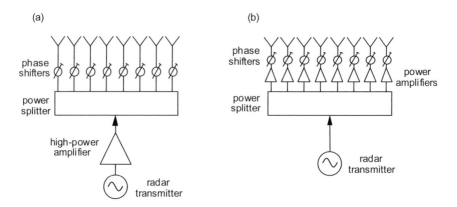

Fig. 11.28 (a) passive, and (b) active phased array antennas (shown in transmit mode).

Figure 11.29 shows a block diagram of a typical active phased array module. The interface and control signals that are needed to each module are: transmit and receive RF signals, phase shifter control, and dc supply. Alternatively, instead of the transmit and receive signals being at RF, these can be at IF, with a microwave local oscillator distributed to each module, or the local oscillator can even be in digital form. Clearly, the distribution of these signals to each module represents a technical challenge, and optical techniques represent an attractive option.

The entire module may be realized as a GaAs MMIC (or a small number of separate MMICs), with potential advantages of low production costs and high unit-to-unit reproducibility. Gallium Arsenide FET technology has advanced over the past decade or two to the point where power outputs of several watts at 10 GHz are achievable, and correspondingly more at lower frequencies. Figure 11.30 shows a typical module for an active phased array radar.

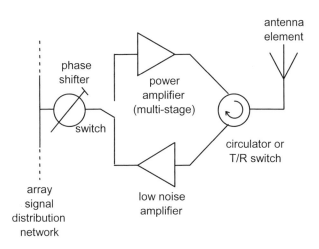

Fig. 11.29 Block diagram of a typical active phased array module.

Due to the finite efficiency of the power amplifiers (typically 40% at best), there will be a considerable amount of heat to be dissipated at the array face. Thus thermal design is also very important, particularly if the infra-red signature must also be minimized. It is also important that the phase and amplitude responses of the modules (for example, variations in power amplifier phase response due to thermal effects) track properly, particularly if low-sidelobe patterns are to be achieved.

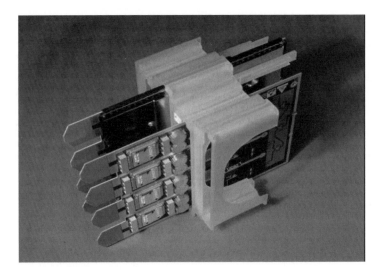

Fig. 11.30 Module for the SAMPSON active phased array radar (photo: AMS).

Optical control of phased arrays

As has already been remarked, the distribution of interface and control signals to the array modules represents a technical challenge, and some years ago[11] it was suggested that the use of optical fibres might provide a solution, avoiding the bulkiness, loss and weight of coaxial cables. In addition optical fibres are immune from electromagnetic interference.

The signals to be fed to each module are[12]:

• a microwave phase reference;
• a lower frequency transmit waveform;
• element control signals.

The corresponding output signals are:

• a received signal;
• element monitoring signals.

---

[11] Forrest, J.R., Richards, F.P., Salles, A.A. and Varnish, P., 'Optical fibre networks for signal distribution and control in phased array radars', *Proc. RADAR'82 Conference*, IEE Conf. Publ. No. 216, pp408-412, October 1982.

[12] Seeds, A.J., 'Optical technologies for phased array antennas'; *IECE Trans. Electronics*, Vol. E76-C, No. 2, pp198-206, February 1993.

Usually the transmit and received signals will be at IF, and the control and monitoring signals will be digital. Consequently, the most demanding signal distribution problem concerns the microwave phase reference.

Single-mode optical fibre is used, since this avoids the introduction of modal noise due to mechanical disturbance of the fibre. The losses of single-mode fibre are approximately 0.3 dB/km at 1.55 µm, and 3 dB/km at 850 nm. The microwave signal is intensity-modulated onto a semiconductor laser, and detected at the module, either using a photodiode, or directly controlling an oscillator device such as an IMPATT, bipolar transistor or MESFET. Modulation bandwidths of over 30 GHz have been reported. The advent of optical amplifiers means that optical transmission factors of greater than unity are possible, and this is important if the optical signal has to be split many ways to distribute the signal to a large number of modules.

It is also possible to use two optical sources phase locked together, such that their frequency difference forms the microwave signal (Fig. 11.31).

A more recent idea is to use direct digital synthesis within each module. This still requires the distribution of a high-frequency clock signal to each module, but gives enormous flexibility with respect to transmit signal generation, spatio-temporal coding (§14.2), and phase or true-delay beamforming.

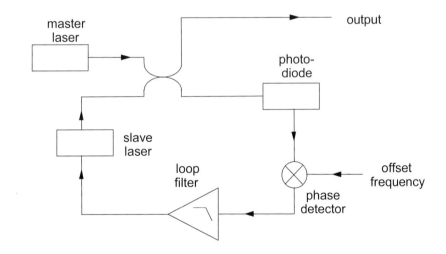

Fig. 11.31 Optical phase-locked loop, used to provide the microwave reference signal to a phased array module.

11.8.2 Digital beamforming

An important development over the past two decades or so has been the use of digital techniques for antenna beamforming, principally (though not exclusively) for receiving applications. Digital technology has obvious advantages in respect of flexibility, reliability and repeatability, and for these reasons there is a continuing trend to work with digital signals as early as possible in a receiving system. Essentially, the signal from each element of the array is digitized, and the phase and amplitude weighting performed by digital multiplication by a complex weight. The signals from all elements are then summed digitally, giving a number in digital form whose directional response is a function of the array geometry and the (digital) weights, in exactly the same way as with analogue beamforming.

The benefits offered by digital beamforming may be summarized as:

- the beam direction and shape may be varied very rapidly, in principle at the full digital clock rate, allowing digital implementations of the adaptive algorithms discussed in §14.4;
- it is possible to form multiple beams from a single array without the usual constraints on orthogonality (§11.7.3);
- the phase and amplitude errors of each receiver channel may be characterized and corrected, giving very accurate control of phase and amplitude, and hence of pattern shape and sidelobe level;
- in radar applications, the increased flexibility allows optimum use of available radar power and dwell time for surveillance and tracking functions, according to the particular scenario.

Implementation

A block diagram of a typical digital beamforming array is shown in Fig. 11.32. Each element (or subarray) feeds a dedicated receiver channel, providing amplification and selectivity. Note that the local oscillators must be coherent from one channel to another, so they are each derived from a single oscillator and distributed to each receive channel. The signal in each channel is digitized, usually in I/Q form at baseband. If the signal bandwidth is B, then the bandwidth of each of the I and Q channels is $B/2$. Hence the minimum (Nyquist) sample rate of the ADCs is B. The frequency and phase responses of the receiver channels must track closely, over the full dynamic range of signals to be handled, though if it can be arranged for calibration signals to be injected into each receive channel sequentially, any errors can be calibrated out. Also, errors in the complex downconverters must be kept to a low level - specifically the gains of the I and Q channels must be identical, they must be in exact phase quadrature, and there should be no dc offsets in the mixers.

Churchill et al.[13] have described a technique by which errors of these kinds may be characterized and corrected.

The desire to avoid such errors, and the advent of faster and faster ADCs, has led to the widespread use of direct digitization at IF (or even, ultimately at RF) with a single ADC per channel. This is helped by realizing that the Sampling Theorem (Nyquist) demands that the sample rate need only be at least twice the bandwidth of the signal to be sampled, rather than twice the highest frequency present. More specifically[14], the sample rate should respect

$$2B\left\{\frac{Q}{n}\right\} \le f_s \le 2B\left\{\frac{Q-1}{n-1}\right\} \tag{11.113}$$

where the signal bandwidth $B = f_{\mathrm{H}} - f_{\mathrm{L}}$, $Q = f_{\mathrm{H}}/B$, n is a positive integer and $n \le Q$. The form of the subsequent digital processing to extract the baseband I and Q signals is described in many books on digital signal processing[15].

---

[13] Churchill, F.E., Ogar, G.W. and Thompson, B.J., 'The correction of I and Q errors in a coherent processor', *IEEE Trans. Aerospace & Electronic Systems*, Vol. AES-17, No. 1, pp131-137, January 1981.

[14] Glover, I.A. and Grant, P.M., *Digital Communications* (second edition), pp173-175, Pearson, Harlow, 2004.

[15] Mulgrew, B., Grant, P.M. and Thompson, J., *Digital Signal Processing - Concepts and Applications* (second edition), Palgrave MacMillan, Basingstoke, 2003.

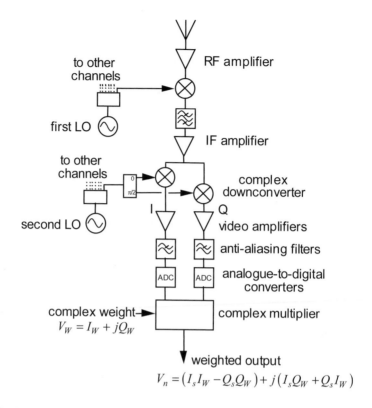

Fig. 11.32 Typical architecture of one channel of a digital beamforming receiver, with baseband digitization.

The weighting of the digitized signal, in amplitude and phase, is achieved as follows. Suppose that the baseband signal is represented by $V_s = I_s + jQ_s$ and the corresponding weight is $V_w = I_w + jQ_w$. The product of these is the weighted output from the nth channel

$$V_n = (I_s I_w - Q_s Q_w) + j(I_s Q_w + Q_s I_w) \qquad (11.114)$$

The weighted outputs are then summed digitally to form the beam

$$V_b = \sum_n V_n \qquad (11.115)$$

Of course, it is entirely possible to form more than one beam simultaneously from the same set of digitized signals V_s, in different directions and of different shapes.

The dynamic range of V_b depends on the number of receiver channels M of the array and the number of bits N of the ADCs. If the maximum voltage output per channel is A, then the maximum value of V_b is MA, corresponding to a power $(MA)^2/2$. The minimum power corresponds to the least significant bit of one ADC (allowing one bit for the sign), i.e. $(A/2^{N-1})^2$. The dynamic range is the ratio of these two power levels, which is $2^{2(N-1)}M$, or

$$\left[6(N-1) + 10\log_{10}M\right] \quad \text{dB} \qquad (11.116)$$

Array calibration

To obtain the full benefit of a digital beamforming array it is necessary to calibrate out all the errors of the analogue parts of the system. The calibration routine consists of injecting a test signal into each receiving channel in turn, and measuring the amplitude and phase of the received signal. Any departures from the desired amplitude and phase can then be compensated using corrected digital weights. In general, the larger the part of the system that lies within the calibration loop, the better. The interval between calibrations will, of course, depend on the timescale on which the channel responses may be expected to vary.

There are several options:

(i) the calibration signals may be injected into each of the receiver channels, immediately behind the antenna elements. This is relatively simple to implement, but care must be taken in the design of the network that distributes the calibration signals;

(ii) the calibration signals may be obtained by one or more sources in the near field of the array. The different path lengths to the array elements and the difference in their amplitude responses are assumed known. This method requires very precise knowledge of the position of the calibration source, and the calibration source may even cause blockage of the aperture;

(iii) the same technique can be used with a source in the far-field. This avoids the variation in path length and amplitude responses of the array elements.

11.8.3 MEMS technology in phased arrays

MicroElectroMechanical Systems (MEMS) technology is a relatively recent development that has the potential to make a great impact in phased arrays. MEMS components were originally developed in the 1970s for applications such as accelerometers and other sensors, but in the early 1990s the technology began to be applied to RF switches, and eventually to phase shifters and other RF components. Essentially they consist of miniature devices that use mechanical movement to realize a short circuit or open circuit in a transmission line.

MEMS switches

The key MEMS component for phased array applications is the RF switch. The actuation mechanism is usually electrostatic, although thermal, magnetostatic or piezoelectric activation is also possible. They have been demonstrated at frequencies from DC to 120 GHz. Table 11.1, taken from reference [11], summarizes the properties of electrostatic MEMS switches compared to those of FET and PIN diode switches.

From this it can be seen that RF MEMS switches have very low power consumption, very high isolation and low loss, and excellent intermodulation performance. Furthermore, they can be built on glass or low-cost silicon substrates. Switches capable of 1-10 billion switching cycles are already available, and this is likely to be improved upon by a factor of 10 in the near future. However, their switching speed and power handling capability are inferior to PIN and FET switch technology.

Parameter	RF MEMS	PIN	FET
Voltage (V)	20-80	±3-5	3-5
Current (mA)	0	3-20	0
Power consumption (mW)	0.05-0.1	5-100	0.05-0.1
Switching time	1-300 μs	1-100 ns	1-100 ns
C_{up} (series) (fF)	1-6	40-80	70-140
R_s (series) (Ω)	0.5-2	2-4	4-6
Capacitance ratio	40-500	10	n/a
Cutoff frequency (THz)	20-80	1-4	0.5-2
Isolation (1-10 GHz)	very high	high	medium
Isolation (10-40 GHz)	very high	medium	low
Loss (1-100 GHz) (dB)	0.05-0.2	0.3-1.2	0.4-2.5
Power handling (W)	< 1	< 10	< 10
Third-order intercept point (dBm)	+66-80	+27-45	+27-45

Table 11.1. Performance comparison of FET, PIN and RF MEMS electrostatic switches.

There are various configurations of switches. Figure 11.33 shows a photomicrograph of one example, which is a series switch developed by the Rockwell Scientific Company[16]. This has what has been termed a 'clam shell' configuration, and is fabricated on a GaAs substrate, with a 2 μm layer of silicon dioxide deposited using

---

[16] Mihailovich, R.E., Kim, M., Hacker, J.B., Sovero, E.A., Studer, J., Higgins, J.A. and DeNatale, J.F., 'MEM relay for reconfigurable RF circuits', *Wireless Comp. Letters*, Vol.11, No.2, pp53–55, February 2001.

a low-temperature PECVD process onto the substrate. The top electrodes are 75 × 75 μm and are fabricated using a 250 nm layer of gold over a 1.2 μm thick PECVD silicon dioxide membrane. Quoted performance figures include switching time: 8–10 μs, loss: 0.1 dB (0.1–50 GHz) and isolation: 50 dB (4 GHz), 30 dB (40 GHz) and 20 dB (90 GHz).

RF MEMS phase shifters

Phase shifters can be built using switches to switch lengths of transmission line, in the usual way (§11.2.1). Figure 11.34 shows an example of a 2-bit phase shifter realized by the University of Michigan and Rockwell Scientific. This occupies an area of 9.6 mm^2. Measured performance at 10 GHz gives a phase accuracy of ±2 degrees, an average insertion loss of 0.55 dB, and a return loss of better than 14 dB.

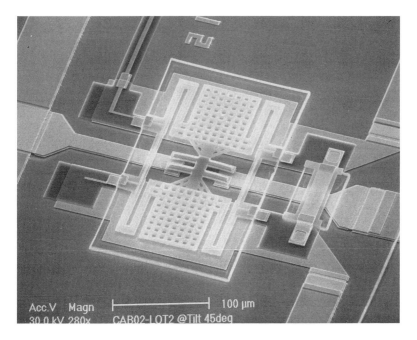

Fig. 11.33 Photomicrograph of a MEMS series switch (photograph: Rockwell Scientific Company).

Other MEMS components

It is also possible to realize other microwave components in MEMS technology, including tunable oscillators, tunable filters, physically movable antennas, and frequency selective surfaces.

There is no doubt that this technology will have a major impact on the development of future antenna array systems.

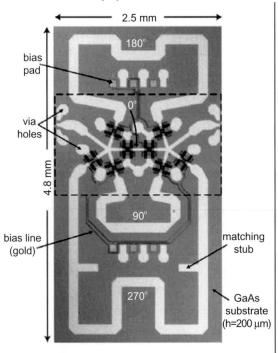

Fig. 11.34 RF MEMS phase shifter (photograph: Dr Gabriel Rebeiz).

11.8.4 Circular, cylindrical, spherical and conformal arrays

Circular and cylindrical arrays are used in applications that require coverage over the full 360° of azimuth. They also have some potentially-useful properties in respect of operation over wide instantaneous bandwidths. It is perhaps interesting to consider that the element density, when the array is viewed sideways-on, is greatest at the edges, effectively giving the array an inverse amplitude taper. Thus we might expect circular arrays to be good for direction-finding applications (where wide-spaced elements give a high sensitivity of differential phase to signal direction), but not so good for forming low-sidelobe beams.

Beam cophasal excitation

The simplest method of feeding a circular array is simply to arrange for the signals from all the elements to add coherently in the particular direction of interest (Fig. 11.33). For omnidirectional elements, and assuming that the number of elements is large, this gives an azimuth-plane beam of the form

$$D(\theta) = J_0\left(\frac{4\pi}{\lambda}\sin\frac{\theta}{2}\right) \tag{11.117}$$

where θ is the azimuth angle, J_0 is the Bessel function of order zero, r is the array radius and λ is the wavelength. This is plotted in Fig. 11.35 for the case of an array of diameter 5λ.

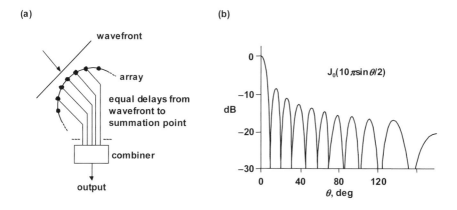

(a)

wavefront

array

equal delays from wavefront to summation point

combiner

output

(b)

$J_0(10\pi\sin\theta/2)$

dB

θ, deg

Fig. 11.35 (a) Beam cophasal excitation; (b) resulting radiation pattern (with no amplitude taper).

Whilst this form of excitation may be simple, steering of the beam (beyond limited scan) requires commutation of interlaced array sectors. In contrast with linear arrays, the pointing direction of the generated beam is *not* frequency dependent, but the use of phase shifters makes the beamwidth slightly frequency dependent, with an approximately quadratic variation. Moreover, the phase illumination of the array follows a rather complicated form.

A comparison between linear arrays and beam-cophasal excited circular arrays is given in Appendix A11.

Phase mode excitation

To overcome these limitations the idea of phase mode excitation was developed. Consider firstly a continuous circular array (i.e. one with an infinite number of elements and negligible interelement spacing), and suppose that the elements are omnidirectional. The excitation of the array F can be regarded as a periodic function of azimuth angle θ, of period 2π, and can therefore be expressed as a Fourier series

$$F(\theta) = \sum_{m=-N}^{N} C_m \exp(jm\theta) \tag{11.118}$$

Each term of the series is known as a *phase mode*, and the coefficient C_m (which is in general complex) is a *phase mode coefficient*. It can be seen that the zero-order phase mode ($m=0$) corresponds to an excitation of constant phase as a function of azimuth angle. The first-order phase mode ($m = 1$) corresponds to one cycle of phase over 360° of azimuth; the second-order phase mode ($m = 2$) corresponds to two cycles of phase over 360° of azimuth, and so on. The negative-order phase modes simply correspond to phase variation in the opposite sense.

When an array is excited by a single phase mode, the far-field pattern $D_m(\theta)$ has a similar form

$$D_m(\theta) = C_m j^m J_m(\beta r) \exp(jm\theta)$$

$$= K_m \exp(jm\theta) \tag{11.119}$$

where $J_m(.)$ is the Bessel function of order m, $\beta = 2\pi/\lambda$, and r is the array radius.

The far-field phase modes are omnidirectional in azimuth, but each with the same characteristic variation of phase with azimuth angle as the corresponding excitation phase mode. The frequency-dependence of this expression is contained in the Bessel function, so for omnidirectional elements the variation of the phase mode amplitudes with frequency is as shown in the dotted responses of Fig. 11.36. For directional elements the corresponding expression is no longer a single Bessel coefficient, but a series of terms, each corresponding to the Fourier coefficients of the element pattern[17]. For the particular case of an element pattern of the form $(1 + \cos\theta)$, which is a reasonable approximation to many practical elements, the equivalent expression to (11.119) is

$$D_m(\phi) = C_m j^m \left[J_m(\beta r) - jJ'_m(\beta r) \right] \exp(jm\theta) \tag{11.120}$$

---

[17] Rahim, T. and Davies, D.E.N., 'Effect of directional elements on the directional response of circular antenna arrays', *Proc. IEE*, Vol. 129, Pt. H., No. 1, pp18-22, 1982.

This gives a much flatter variation of phase mode amplitude with frequency, since the maxima of $J_m(\beta r)$ correspond to the zeroes of $J'_m(\beta r)$ (and vice versa), which is much more suitable for broadband operation (Fig. 11.36).

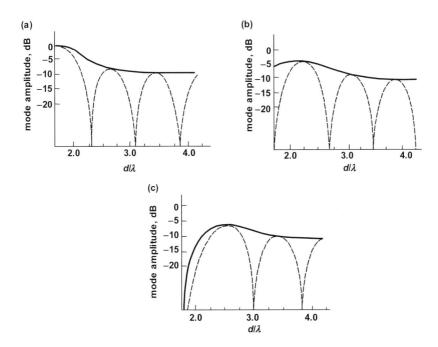

Fig. 11.36 Calculated variation of phase mode amplitude with frequency: (a) zero-order mode; (b) first-order mode; (c) second-order mode: ----- omnidirectional elements; —— $(1 + \cos\theta)$ directional element patterns (after ref. 14).

When a discrete array is used exactly the same theory applies, but the excitation function $F(\theta)$ is sampled at the element locations by a sampling function $S(\theta)$. For an array of n omnidirectional elements

$$S(\theta) = \sum_{q=-\infty}^{\infty} \exp(jnq\theta) = 1 + \sum_{q=1}^{\infty} \exp(jnq\theta) + \sum_{q=-\infty}^{-1} \exp(jnq\theta) \quad (11.121)$$

giving

$$D_m(\theta) = C_m j^m J_m(\beta r) \exp(jm\theta)$$

$$+\sum_{q=1}^{\infty}C_{m}j^{-g}J_{-g}\left(\beta r\right)\exp\left(-jg\theta\right)$$

$$+\sum_{q=-\infty}^{-1}C_{m}j^{h}J_{h}\left(\beta r\right)\exp\left(jh\theta\right) \qquad (11.122)$$

where $g = (nq-m)$ and $h = (nq+m)$.

The first part of this expression is identical to that for a continuous array (11.119). The two series represent ripple terms as a function of θ (spatial ripple). As a rule of thumb, to keep the spatial ripple acceptably low, the interelement spacing should be no greater than $\lambda/2$, which in practice provides an upper limit to the frequency of operation.

For an n-element array it is possible to generate $n+1$ phase modes, though the $+n/2$ and $-n/2$ phase modes are actually identical. The discrete Fourier Transform to generate the phase modes from the element signals is conveniently provided by a Butler Matrix. The bandwidth is usually limited to about one octave by that of the quadrature hybrid couplers that form part of the Butler Matrix, though with special-purpose components it is possible to improve upon this[18].

The behaviour of phase modes as a function of elevation angle can be understood by realizing that the zero-order mode is the co-phasal sum of all the element signals, so there is a maximum in the vertical direction, weighted by the element pattern in that direction. For all the other phase modes the element signals are summed with an integer number of cycles of phase, so there is a null in the vertical direction, irrespective of the element pattern.

As an illustration, Fig. 11.37 shows measured phase mode patterns of a 4-element circular array, at a frequency of 900 MHz. The array elements were $\lambda/4$ monopoles, and there was a single central monopole element which acted as a reflector, giving the other elements a more directional pattern and hence improving the bandwidth of operation. The characteristic variation of phase with direction is clearly evident, as well as the spatial ripple due to the discrete nature of the array.

---

[18] Withers, M.J., 'Frequency-insensitive phase-shift networks and their use in a wide-bandwidth Butler matrix', *Electronics Letters*, Vol. 5, No. 20, pp496-497, October 1969.

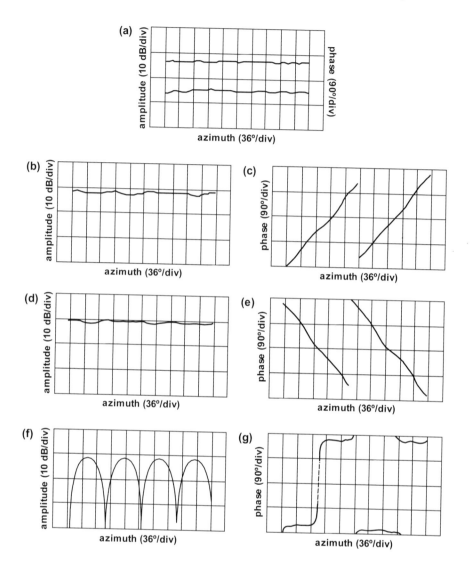

Fig. 11.37 Measured phase mode patterns of a 4-element circular array of monopole elements, at a frequency of 900 MHz, as a function of azimuth angle: (a) zero-order mode amplitude (upper) and phase (lower); (b) +1 order mode amplitude; (c) +1 order mode phase; (d) −1 order mode amplitude; (e) −1 order mode phase; (f) second-order mode amplitude; (g) second-order mode phase. Amplitude: 10 dB/div; phase 90°/div.; horizontal scale (azimuth angle) in each case 36°/div.

Null steering

A plane wave incident on the array will excite all phase modes simultaneously. Suppose that the zero-order phase mode and first-order phase mode are equalized in amplitude and added together (Fig. 11.38). It is easy to see that there is one direction in which they will cancel, giving a single null. The direction of this null can be steered by inserting a phase shifter in one of the paths, and the null direction is given directly by the setting of the phase shifter. Furthermore, if the phase mode coefficients K_m can be characterized and compensated by means of networks with appropriate transfer functions, then the null shape and direction can be maintained over the full instantaneous bandwidth of the array.

These results hold for any two phase modes whose orders differ by one.

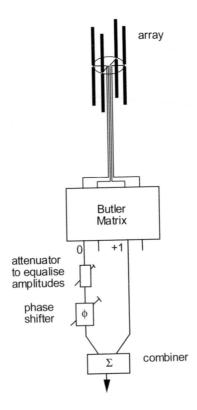

Fig. 11.38 Null-steering with a four-element circular array.

Direction finding

In a rather similar way, if the phase difference is measured between two adjacent phase modes when a plane wave is incident on the array, this phase difference provides a direct reading of the direction of arrival of the signal (Fig. 11.39), and the same comments about broadband operation apply. In fact, if phase modes whose orders differ by two are chosen, then the sensitivity is increased by a factor of two, but at the expense of introducing a 180° ambiguity (which can be resolved by the original measurement). As an example, an experimental system based on this principle has been constructed and demonstrated, operating over the band 2–30 MHz[19], and has been produced commercially.

It is also possible to use superresolution direction finding techniques with phase modes; this is discussed in §14.6.5.

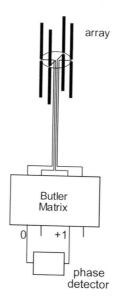

Fig. 11.39 Direction finding: a plane wave incident on the array will excite all phase modes simultaneously. The phase difference between two adjacent phase modes will provide a direct indication of the direction of arrival of the signal. Greater sensitivity (at the expense of ambiguity) can be obtained by using phase modes whose orders differ by more than one, and the ambiguities resolved by lower-order comparisons.

[19] Cvetkovic, M., Davies, D.E.N., Griffiths, H.D. and Collins, B.S., 'An HF direction-finding and null steering system employing a four-element circular array', *Proc. 4th IEE Intl. Conference on HF Radio Systems and Techniques*, London; IEE Conf. Publ. No. 284, pp221-225, April 1988.

Pattern synthesis

The results quoted above for null steering are an example of a much more general application of phase modes to synthesize radiation patterns. Davies[20] showed that phase modes may be treated in the same way as the elements of a uniformly-spaced linear array, and all the techniques developed for linear array pattern synthesis may be applied to circular arrays, subject to the following comments.

- Phase modes are omnidirectional in azimuth, so the 'elements' are also omnidirectional. The radiation pattern of the circular array corresponds to the array factor of the linear array.
- Phase modes are orthogonal (since they are the outputs of a discrete Fourier Transform), so there is no mutual coupling.
- The radiation patterns have the same shape (in θ space) as those formed by a linear array (in $kd\sin\theta$ space) with an interelement spacing of $\lambda/2$, independent of frequency.
- For a discrete array, exciting a single mode port excites a periodic sequence of modes. For example, if the $+1$ mode is excited on a four element array, the -7, -3, $+5$, and $+9$ modes are also excited. These additional harmonics correspond to additional linear array 'elements'.

As an example, Fig. 11.40 shows an example of an instantaneously broadband beam pattern with <-20 dB sidelobes, formed in this way from a four-element array of monopole elements, measured over the frequency band 8–12 GHz.

Isolated omnidirectional patterns

A further property of circular arrays excited by means of phase modes is that there is high isolation between the individual phase mode ports of the Butler Matrix (§11.7.3). This arises because of the orthogonality property of the Discrete Fourier Transform. The phase modes therefore act as isolated omnidirectional patterns, potentially over a broad bandwidth. In practice, imperfections in the Butler Matrix and variations in the impedances presented by the array elements limit the isolation, but it is easy to obtain 20 or 30 dB of isolation without taking any special precautions. The bandwidth is usually limited by that of the Butler Matrix to an octave or so.

A particular application for this idea lies in radiocommunications base stations, where it is desired to multiplex several transmitters to the same antenna, with each

---

[20] Davies, D.E.N., 'A transformation between the phasing techniques required for linear and circular aerial arrays', *Proc. IEE*, 1965, Vol. 112, No. 11, pp2041-2045.

having an omnidirectional pattern[21]. Each transmitter is fed to one phase mode port of the Butler Matrix feeding the array. This considerably relaxes the specification on any multiplexing filters that are used.

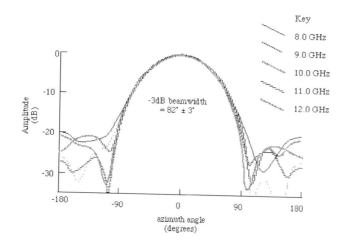

Fig. 11.40 Example of instantaneously broadband beam pattern synthesized over the band 8–12 GHz, using a 4-element array of monopoles.

Sectoral phase modes[22]

Whilst the omnidirectionality property of phase modes is in most cases an attraction, there may be instances where it is a disadvantage, for example in DF applications where there is more than one co-channel signal present. In fact, in the phase-comparison DF scheme of Fig. 11.39, when a 15 dB weaker co-channel signal is also received by the array, the worst-case DOA error is $2\arcsin 10^{-15/20} = 20.5°$.

There is therefore some attraction in the ability to form two sets of multibeam patterns, such that each angular sector is covered by a pair of low-sidelobe beams with linear and opposite phase slopes. Amplitude comparison may then be used to unambiguously determine the angular sector facing the incident signal, while a more accurate bearing is obtained by detecting and comparing the phases of the two directional beams covering that sector, just as with ordinary phase modes, but with a

---

[21] Guy, J.R.F. and Davies, D.E.N., 'Novel method of multiplexing radiocommunication antennas using circular array configuration', *Proc. IEE*, 1983, Vol. 130, Pt. H, No. 6, pp410-414.

[22] Griffiths, H.D. and Eiges, R., 'Sectoral phase modes from circular antenna arrays'; *Electronics Letters*, Vol.28, No.17, pp1581–1582, August 1992.

much-reduced susceptibility to out-of-sector interference. Such patterns are known as sectoral phase modes.

Excitation scheme: A pair of beams in direction $\phi_m = (2\pi/M)m$, $m = 0,..., M-1$ sharing the same directional amplitude pattern and characterized by equal but opposite phase slopes may be approximately synthesized by linearly combining a set of symmetrically-weighted but asymmetrically-numbered phase modes

$$F_{m1}(\phi) = \sum_{\mu=-L}^{L-L_0} a_{l^+} e^{j(2\pi/M)m\mu} \, \Phi_\mu(\phi,\omega)/K_{\mu 0}(\omega) \tag{11.123}$$
$$\Lambda A(\phi-2\pi m/M)e^{(L_0/2)(\phi-2\pi m/M)}$$

$$F_{m2}(\phi) = \sum_{\mu=-L+L_0}^{L} a_{l^-} e^{j(2\pi/M)m\mu} \, \Phi_\mu(\phi,\omega)/K_{\mu 0}(\omega) \tag{11.124}$$
$$\Lambda A(\phi-2\pi m/M)e^{-(L_0/2)(\phi-2\pi m/M)}$$

where

$$l^{\pm} = \text{Int}\left(\left|\mu \pm L_0/2\right| + 1/2\right) \tag{11.125}$$

$$A(\phi) = a_0 + 2\sum_{l=1}^{L-\text{Int}(L_0/2)} a_l \cos l\phi \tag{11.126}$$

$L < M/2$, $0 < L_0 < 2L$ and for an odd L_0: $a_0 = 0$. Under uniform excitation $A(\phi)$ becomes

$$A(\phi) = \sin\left[(2L-L_0)\phi/2\right]/\sin(\phi/2) \tag{11.127}$$

and its null-to-null beamwidth is $4\pi/(2L-L_0) < 4\pi/M$. A proper coverage of an angular sector of $2\pi/M$ radians is this ensured with a sidelobe level of approximately –13.5 dB. This may of course be improved, at the expense of an increased beamwidth, by applying an amplitude taper to the combined phase modes. Because each pair of synthesized patterns cover one of M angular sectors but otherwise behave as omnidirectional phase modes we have named them 'sectoral phase modes' or SPM beams.

Both the amplitude pattern $A(\phi)$ and the phase slope of SPM beams are affected by the choice of L_0. The larger L_0, the less directional are the beams but the steeper are their phase slopes. At the extreme, $L_0 = 2L$ and the two beams are simply the two omnidirectional phase modes Φ_{-L} and Φ_L. Thus, whereas the highest DF accuracy but poorest immunity is achieved by comparing the phases of the highest-order modes of the set $\Phi_{-L},...,\Phi_0,...,\Phi_L$, best immunity but least accuracy results from the comparison of two SPM beams with $L_0 = 1$. In the latter case the respective phase slopes of the two compacted beams are $1/2$ and $-1/2$, and

$$\arg F_{m1}(\phi) - \arg F_{m2}(\phi) = \phi - 2\pi m/M \qquad (11.128)$$

Naturally, the viability of the proposed SPM scheme for a wideband DF system depends on wideband alignment of the omnidirectional phase modes. As shown above, this has been demonstrated for a circular array of directional elements with an analogue beamformer, and can also be implemented in a digital beamforming network by appropriate filtering.

Null-steering enhancement: The need for a radiation pattern with a steerable null arises in both communication and DF applications, where the receiving array has to reject unwanted interference or jamming signals. Classically, a circular array pattern null may be synthesized by the subtraction of adjacent phase modes (Fig. 11.38) and steered by an intermediate phase shift β

$$\Phi_\mu/K_{\mu 0} - e^{j\beta}\Phi_{\mu+1}/K_{(\mu+1)0} \; \Lambda 2 e^{-j\mu\phi} e^{-j(\phi-\beta-\pi)/2} \sin(\phi-\beta)/2 \quad (11.129)$$

The nulled phase modes can then be linearly combined by an inverse DFT to yield a multiple pattern of M directional beams. The pattern formed at $\phi = \beta$ is, however, shared by all the directional beams, which for some applications may be somewhat restrictive. Additional independently-steered nulls may be incorporated in the system (at the expense of wider beams) but they too will be shared by all the beams.

Here we consider an alternative multibeam nulling scheme. Instead of combining adjacent phase modes, the same nulling concept is applied to M pairs of SPM beams characterized by $L_0 = 1$

$$F_m = F_{m1} - e^{j\beta_m} F_{m2} \; \Lambda 2 e^{j(\beta_m+\pi)/2} A(\phi - 2\pi m/M) \sin(\phi - \beta_m - 2\pi m/M)/2 \qquad (11.130)$$

where $A(\phi)$, F_{m1} and F_{m2} $(0 \le m \le M-1)$ are given by eqns. (11.126), (11.123) and (11.124) respectively. By controlling the M phase shifts $\{\beta_m\}_{m=0}^{M-1}$ the proposed modified configuration allows each directional beam to independently steer its own null.

Simulated examples: Fig. 11.41 displays the two amplitude plots and the comparative phase response for a pair of SPM beams, formed from a 16-element circular array by combining two sets of 14 phase modes each with low-sidelobe weighting. The comparative phase response is nearly linear with a slope of approximately unity over an angular sector larger than $\pm 360°/32 = 11.25°$.

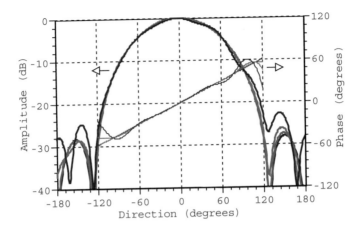

Fig. 11.41 Comparative phase plot and amplitude patterns for a pair of SPM beams formed by aligning and combining phase mode sets $\left\{\Phi_{-7}/K_{(-7)0},...,\Phi_{6}/K_{(6)0}\right\}$ and $\left\{\Phi_{-6}/K_{(-6)0},...,\Phi_{6}/K_{(7)0}\right\}$ in a 16-element array with interelement spacing of 0.4 wavelengths and element power pattern of $\cos^{2}\phi/2$. Mode weighting (in dB): { −15.5 −12.8 −8.3 −4.7 −2.3 −0.7 0 −0.7 −2.3 −4.7 −8.3 −12.8 −15.5}.

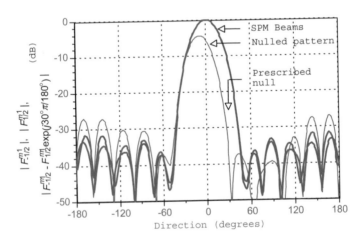

Fig. 11.42 Directional beam with prescribed pattern null at close to 30° formed by phased subtraction of two SPM beams.

The amplitude patterns, which are almost identical, both drop to a peak sidelobe level less than −31 dB outside an angular region of ±43°. A directional beam with a pattern null is shown in Fig. 11.42. This pattern is synthesized by applying a phase shift of 30° to one of the SPM beams and subtracting it from the other. The result is a $\sin\phi/2$-type null at $\phi = 32.3°$.

Conclusions: A simple circular array scheme for synthesizing broadband directional beams with phase-mode-like phase behaviour has been presented. Such beams may be formed by linearly combining a set of aligned phase modes, and it has been shown that they may be beneficially applied to modally-based DF and null-steering systems.

Spherical phase modes[23]

The preceding sections have demonstrated a wide range of applications of circular arrays (and by extension, of cylindrical arrays). It is interesting to consider whether the same properties may be shown by spherical arrays, considering their excitation in terms of spherical harmonics.

Consider an excitation function $E(\theta', \phi')$ on the surface of a sphere (Fig. 11.43), and consider one spherical harmonic term (spherical phase mode) $Y_{l'}^{m'}$ of this excitation, with $Y_l^m(\theta, \phi)$ defined as

$$Y_l^m(\theta, \phi) = \sqrt{\frac{(2l+1)(l-m)!}{4\pi(l+m)!}} P_l^m(\cos\theta) e^{jm\phi} \qquad (11.131)$$

where the associated Legendre functions P_l^m mean that the amplitude of the spherical harmonics are a function of direction. The directional dependence of phase is given by the exponential term, and the name 'spherical phase mode' is justified by analogy with their circular array counterparts.

The far-field radiation pattern is given by

$$D(\theta, \phi) = \frac{1}{4\pi} \iint_{s'} Y_{l'}^{m'}(\theta', \phi') e^{j\beta a \cos\psi} ds' \qquad (11.132)$$

where a is the radius of the sphere, $\beta = 2\pi/\lambda$ and λ is the wavelength. The plane waves in the far field can be expanded into a sum of partial waves using Bauer's formula:

$$e^{j\beta a \cos\psi} = \sum_{l=0}^{\infty} (2l+1) j^l \hat{J}_l(\beta a) P_l(\cos\psi) \qquad (11.133)$$

---

[23] De Witte, E. Griffiths, H.D. and Brennan, P.V., 'Phase mode processing for spherical antenna arrays', *Electronics Letters*, Vol.39, No.20, pp1430-1431, 2 October 2003.

where $\hat{J}_l$ is a spherical Bessel function of the first kind and P_l a Legendre polynomial, both of order l[24]. Also, using the addition theorem to expand the Legendre zonal harmonics into spherical harmonics

$$P_l\left(\cos\psi\right)=\frac{4\pi}{2l+1}\sum_{m=-l}^{l}Y_{l'}^{m'*}\left(\theta',\phi'\right)Y_l^m\left(\theta,\phi\right)\qquad(11.134)$$

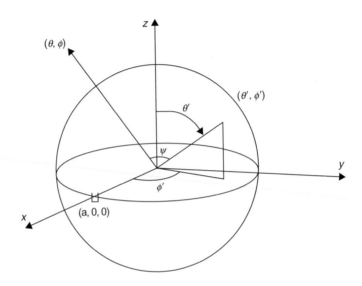

Fig. 11.43 Spherical array geometry.

The far field becomes:

$$D\left(\theta,\phi\right)=\sum_{l=0}^{m}\sum_{m=-l}^{l}j^l\hat{J}_l\left(\beta a\right)Y_l^m\left(\theta,\phi\right)\underbrace{\iint_{s'}Y_{l'}^{m'*}\left(\theta',\phi'\right)Y_l^m\left(\theta,\phi\right)ds'}_{\delta_{ll'}\delta_{mm'}}$$

$$(11.135)$$

where we have used the orthogonality property of spherical harmonics over the sphere. Finally, we obtain the far field

---

[24] Gradshteyn, I.S. and Ryzhik, I.M., *Tables of Integrals, Series and Products*, Academic Press, 1980.

$$D(\theta,\phi) = j^{l'}\hat{J}_{l'}(\beta a) Y_{l''}^{m'}(\theta,\phi) \qquad (11.136)$$

In this result we recognise the same spherical harmonic $Y_l^{m'}$ of the excitation, which shows that a spherical phase mode of the excitation function results in a far field radiation pattern with the same spherical phase mode form, just as with circular arrays.

Discrete spherical arrays: A practical array will consist of discrete elements whose effect can be represented as the product of the spherical harmonic Y_l^m with a sampling function $S(\theta,\phi)$. The sampling function can take various forms; one example is a weighted equiangular sampling grid of spatial bandwidth B, which has an azimuthal angle between elements of π/B and an elevation angle between elements of $\pi/2B$. For an equiangular topology of $2B \times 2B$ elements it has been shown[25] that any function of spatial bandwidth $l_{max} \le B$ can be perfectly reconstructed from its $4B^2$ samples. Furthermore, for this element distribution, the far field can be evaluated analytically. It can be shown that this element distribution, when excited according to a particular harmonic, radiates a far field that consists of the same harmonic, together with higher order harmonics (perturbation states). The latter are heavily attenuated by the spherical Bessel factor from equation (11.136), which in turn depends on the ratio of the wavelength to the radius of the sphere.

Fig. 11.44 shows an example to verify the preceding theory. Figure 11.44(a) shows a spherical harmonic of a continuous excitation, corresponding to $l = 3$, $m = 2$ in equation (11.136), with $a = 0.5$ m and $\lambda = 0.15$ m. Figure 11.44(b) shows the calculated far field pattern from a discrete set of elements on the same sphere. Each element radiates a weighted sample of the continuous function Y_3^2. The sample grid has a spatial bandwidth $B = 32$. The far field is calculated with the equivalent Riemann sum of equation (11.132):

$$D(\theta,\phi) = \frac{\sqrt{2}}{4B} \sum_{j=0}^{2B-1} \sum_{k=0}^{2B-1} a_j^{(B)} Y_l^m(\theta_j,\phi_k) \, e^{j\beta a \cos\psi} \qquad (11.137)$$

where the sample points are taken on the weighted equiangular grid $\theta_j = \pi j/2B$, $\phi_k = \pi k/B$, and the weights $a_j^{(B)}$ account for the $\sin\theta$ factor in the continuous integral formulation $(ds' = \sin\theta \, d\theta \, d\phi)$.

---

[25] Driscoll, J.R. and Healy, D.M. Jr., 'Computing Fourier transforms and convolutions on the 2-sphere', *Adv. Appl. Math.*, Vol.15, pp202-250, 1994.

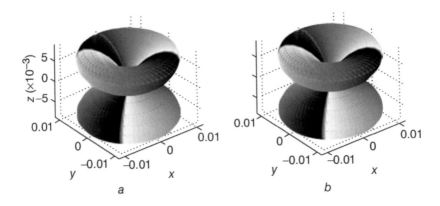

Fig. 11.44 Spherical harmonic Y_3^2 and corresponding far field pattern from sampled array. In both cases phase is represented on grey scale. (a) theoretical spherical harmonic; (b) calculated far field.

Practical implementations and applications: In a practical implementation the spherical phase modes may be generated by a stage of discrete Fourier transforms followed by a Legendre transform. The latter projects the sample vectors onto the associated Legendre functions. A bank of Butler matrices can implement the first stage. The output is a set of sample vectors for every value of m. The second stage can be implemented partially in the digital domain by first projecting the sample vector onto a cosine or sine base, depending on m, with Discrete Cosine and Discrete Sine transforms. The rest of the processing can be done digitally. As the number of nonzero terms in this projection is equal to the order l of the associated Legendre functions, this technique would minimise the number of A/D converters for a given spherical bandwidth.

In[17] it was shown how the use of directional element patterns improves the bandwidth of planar circular arrays. A similar observation is made for spherical arrays. Here the raised cosine patterns are rotated around the normal axis of the elements on the sphere. Since these patterns can be achieved by the mismatch with direction of circularly-polarised radiation and circularly-polarised elements, it can be anticipated that such elements will be optimum with spherical phase mode excitations.

Spherical phase modes generated in this way should be usable in all of the direction finding, null steering and pattern synthesis applications described earlier, in azimuth and elevation rather than just in azimuth. It is suggested that one of the more immediate applications might be to broadband beamforming and superresolution direction finding.

Conformal arrays

Conformal arrays are the most general type of phased array, and can be defined as an array whose elements are mounted flush on a non-planar surface. They have particular applications in arrays for missiles or high-speed aircraft, and also in systems that require wide-angle coverage. Some of their properties, advantages and disadvantages may be summarized as:

- because the elements are mounted flush on the surface, there is no aerodynamic drag;
- the error in scan angle that occurs with a radome in a linear or planar array is eliminated or reduced;
- since the array is three-dimensional it can be difficult to ensure that all elements radiate the same polarization. It is possible to use circular polarization, but it is not easy to obtain an element pattern that remains perfectly circularly polarized over a wide angle;
- it is difficult to etch elements on singly- or doubly-curved surfaces.

The techniques used in the design of conformal arrays are described in detail by Borgiotti[26]. A modern concept for the use of conformal arrays is the so-called 'smart skin', in which the surface of a platform is covered with antenna elements which form an array which can be used for a variety of applications in radar, communications, etc. The concept has some synergy with the waveform diversity techniques described in §14.7.11.

11.8.5 Sparse and random arrays

Introduction

The interelement spacing in a conventional array is constant. In these so-called 'regular' arrays the spacing should be of the order of $\lambda/2$, to avoid grating lobes (§11.3.4). This means that their various properties, such as the number N_0 of elements, the 3 dB beamwidth θ_{3dB}, the maximum gain G, are closely interrelated. For example, for a uniformly illuminated planar circular array with diameter D and N_0 elements each of gain g, we have the following approximate relationships.

---

[26] Borgiotti, G., 'Conformal arrays', Chapter 11 in *The Handbook of Antenna Design*, Vol.II, A.W. Rudge, K. Milne, A.D. Olver and P. Knight eds, Peter Peregrinus, Stevenage, 1983.

$$N_0 = \pi \frac{D^2}{\lambda^2} \ , \quad \theta_{3\text{dB}} = \frac{\lambda}{D} = \sqrt{\frac{\pi}{N_0}} \ , \quad G = N_0 g \qquad (11.138)$$

On the other hand, in order to lower the close-in sidelobes, it is necessary to apply an amplitude weighting function.

Several practical considerations lead us to seek a means of choosing these parameters more independently of each other. One reason results from the use of active modules on transmission. Such amplifiers are usually used in class C for maximum output power, so that amplitude weighting of the array is not practicable. To reduce the level of close-in sidelobes we can vary the *density* of elements, i.e. the number of elements per unit area, so that the density is reduced towards the periphery of the array.

A second reason is that of economy; the cost of an array depends on the number of radiating elements. This fixes the maximum gain G of the antenna. If the number of elements N is fixed, what is the optimum spatial arrangement of the modules ? By spreading the elements in this way the dimension D of the antenna is increased. This results in an increase in the main lobe directivity and hence in the resolving power (§13.2.2)

In some systems, such as bistatic radars or telecommunications systems, it is useful to optimize independently the transmitting and receiving antenna characteristics. In transmission, *gain* is an essential characteristic., On receive, an array is immersed in a complex environment of wanted signals and parasitic signals. The most important characteristic is then *the angular resolution*; this can be improved by use of *sparse arrays*.

In Chapter 14 (Signal Processing Antennas), we define the concept of *non-redundant* array which, although irregular, cannot be considered as 'sparse'. We will also see that, in certain cases, *multiplicative arrays* allow great savings to be made in the number of elements, whilst still giving good directivity.

All these examples show that the quality criteria for the transmit antenna pattern cannot simply be transposed to the case of the receiving antenna.

Synthesis techniques for irregular arrays

An irregular array is, by definition, sparse, because it is ineffective to use an interelement spacing of less than $\lambda/2$. The array is characterized by two factors: the dispersion rate Δ and the spatial distribution function of the elements.

The dispersion rate is the ratio $\Delta = N_0/N$, where N is the number of elements, and N_0 the number that the array would have if it were fully-populated (non-sparse), with an interelement spacing of $\lambda/2$. In the case of an array occupying an area S, we have

$$N_0 = \frac{4S}{\lambda^2} \quad \text{and} \quad \Delta = \frac{4S}{N\lambda^2} \qquad (11.139)$$

The spatial distribution function is chosen from an enormous number of possible functions, equal to the number of combinations of N elements in N_0 possible positions

$$C_{N_0}^N = \frac{N_0!}{N!\,(N_0 - N)!} \tag{11.140}$$

Example: $N_0 = 20$, $N = 10$, $\varDelta = 2$, $C_{N_0}^N = \dfrac{20!}{(10!)^2} = 184{,}756$

To help make this choice we can start from the idea explained in the introduction, i.e. to vary the element density according to a weighting function chosen at the outset. In this case the position of the elements can, in principle, be fixed deterministically. This method was originally reserved for 'small' arrays. We will now consider statistical (or probabilistic) methods.

Statistical methods

There exist several types of statistical method. We present here one inspired by that due to Skolnik, Sherman and Ogg[27].

Principle: Consider an array which, if it were fully populated, would have N elements. The thinned array is obtained by 'turning off' or 'leaving on' the elements (of equal amplitude), following a random process. We arrange that the average value of the illumination function follows a reference taper function chosen at the beginning.

Let a_n $(0 \le a_n \le 1)$ be the amplitude taper of the fully populated array. We choose a random binary amplitude excitation $\tilde{a}_n$ with the following probability distribution

$$\begin{aligned} \tilde{a}_n &= 1 \quad \text{with probability } a_n \\ \tilde{a}_n &= 0 \quad \text{with probability } 1 - a_n \end{aligned} \tag{11.141}$$

The random variable $\tilde{a}_n$ therefore has an average value

$$\langle \tilde{a}_n \rangle = 1 \times a_n + 0 \times (1 - a_n) = a_n \tag{11.142}$$

and second order moment

$$\langle \tilde{a}_n^2 \rangle = 1^2 \times a_n + 0^2 \times (1 - a_n) = a_n \tag{11.143}$$

---

[27] Skolnik, M.I., Nemhauser, G. and Shermann, J.W., 'Dynamic programming applied to unequally spaced arrays, *IEEE Trans. Antennas & Propagation*, Vol. AP-12, 1964.

Dispersion rate Δ: The number $\tilde{N}$ of elements turned on is also a random variable

$$\tilde{N} = \sum_{n=1}^{n=N_0} \tilde{a}_n \tag{11.144}$$

whose average value is

$$N = \left\langle \tilde{N} \right\rangle = \sum_{1}^{N_0} a_n \tag{11.145}$$

The dispersion rate is defined by $\Delta = N_0/N$.

Example: for uniform excitation $a_n = \text{const.} = a$; $\Delta = 1/a$

We see that the choice of a_n allows the illumination function and the dispersion rate to be chosen at the same time.

Average radiation pattern

The array factor can be written in the form:

$$\tilde{F} = \sum_{n=1}^{N_0} \tilde{a}_n \exp ju_n \tag{11.146}$$

where u_n contains the direction of radiation.

The average radiation pattern is identified with the pattern F_0 of the fully-populated array with a reference illumination function a_n

$$\left\langle \tilde{F} \right\rangle = \sum_{n=1}^{N_0} a_n \exp\left(ju_n \right) = F_0 \tag{11.147}$$

In general, the function a_n will be symmetric in the cardinal planes of the array, and the function F_0 is in general purely real in the direction of radiation.

Average gain (or average power pattern)

This is obtained by taking the average of the square modulus of $\tilde{F}$

$$\left\langle |F|^2 \right\rangle = \sum_n \sum_m \left\langle \tilde{a}_n \tilde{a}_m \right\rangle \exp j\left(u_n - u_m \right) \tag{11.148}$$

Separating out terms of the same order and of different order, the random variables $\tilde{a}_n$, $\tilde{a}_m$ being independent:

$$\left\langle |F|^2 \right\rangle = \sum_n \left\langle \tilde{a}_n^2 \right\rangle + \sum_{n \neq m} \sum \left\langle \tilde{a}_n \right\rangle \left\langle \tilde{a}_m \right\rangle \exp j\left(u_n - u_m\right) \qquad (11.149)$$

Taking into account equations (11.142) and (11.143)

$$\left\langle |F|^2 \right\rangle = \sum_n a_n + \sum_{n \neq m} \sum a_n a_m \exp j\left(u_n - u_m\right) \qquad (11.150)$$

which, by reintroducing the term a_n^2 into the double summation, can also be written

$$\left\langle |F|^2 \right\rangle = \sum_n a_n + \sum \sum a_n a_m \exp j\left(u_n - u_m\right) - \sum a_n^2 \qquad (11.151)$$

$$\left\langle |F|^2 \right\rangle = \left| \sum_{n=1}^{N_0} a_n \exp ju_n \right|^2 + \sum_1^{N_0} a_n \left(1 - a_n\right) \qquad (11.152)$$

or

$$\left\langle |F|^2 \right\rangle = F_0^2 + \sum_1^{N_0} a_n \left(1 - a_n\right) \qquad (11.153)$$

So the average power pattern appears as the superposition of

- the ideal reference pattern F_0^2, which is dominant in the main lobe region and in the close-in sidelobes, and
- an isotropic pattern, which is dominant in the far-out sidelobe region (diffuse sidelobes).

Relative level of diffuse sidelobes

On axis, $u_n = 0$, and $|F_0(0)|^2 = \left| \sum_1^{N_0} a_n \right|^2 = \left(\dfrac{N_0}{\Delta} \right)^2 = N^2$ $\qquad (11.154)$

The relative average level of the diffuse sidelobes is defined as

$$L_0 = \frac{\left\langle |F|^2 \right\rangle}{|F_0|^2} = \frac{1}{N^2} \sum_1^{N_0} a_n \left(1 - a_n\right) \qquad (11.155)$$

Example: The illumination function affects the diffuse sidelobes rather less than the dispersion rate Δ.

Consider a constant illumination

$$a_n = \text{const.} = a = \frac{1}{\Delta} = \frac{N}{N_0} \qquad (11.156)$$

We find that

$$L_0 = \frac{1-a}{N} = \frac{1-N/N_0}{N} = \frac{1-1/\Delta}{N} \qquad (11.157)$$

- If $N = N_0$: no dispersion, no diffuse sidelobes;

- If $N \ll N_0$: large dispersion, $L \approx 1/N$.

In moving from a dispersion rate of 2 to 4, with the same number of elements, the average sidelobe level is increased by a factor 3/2. This comes from the fact that with a constant gain (N), the main lobe narrows; part of the energy of the main lobe moves into the sidelobes.

Probability function of diffuse sidelobes

We can set $\tilde{F} = F_0 + x + jy$, where x and y are zero-mean Gaussian random variables with the same variance. This gives the result that far from the main lobe, in the diffuse sidelobe region, $\left|\tilde{F}\right|$ follows a Rayleigh function, and $\left(\dfrac{\left\langle\left|\tilde{F}\right|^2\right\rangle}{F_0^2}\right)$ follows an exponential function, with average value L_0. It is therefore completely defined. The probability that the relative level L exceeds kL_0 is given by

$$\text{Prob}\{L > kL_0\} = \exp(-k) \qquad (11.158)$$

11.8.6 Retrodirective and self-phasing arrays[28]

Retrodirective arrays

In some applications there is a requirement to reflect an incident signal back in the direction from which it has come, for example to enhance the radar detectability of a target. It may be desirable to do this over a broad range of angles and over a broad bandwidth, and in some cases it may be desired to impose modulation on the reflected signal.

The reflectivity of such an arrangement is characterized by the radar cross section, σ (§4.3.2), so the signal power P_r received by a radar at a range r is given by the radar equation

$$P_r = \frac{P_t G}{4\pi r^2}.\sigma.\frac{1}{4\pi r^2}.\frac{G\lambda^2}{4\pi} \tag{11.159}$$

There are several means of realizing a retroreflector. A conducting sphere of radius a much greater than the wavelength will provide an omnidirectional reflection corresponding to its physical cross-sectional area A_e, thus $\sigma = A_e = \pi a^2$. A physically-smaller retroreflector can be achieved by a Luneburg lens (§7.3.5) with a metallic coating on the rear face, which gives a radar cross section $\sigma = 4\pi A_e^2/\lambda^2$, and the same order of performance is achieved with a dihedral or trihedral corner reflector (§12.4).

Another approach is provided by the Van Atta array[29], which is shown in its basic linear array form in Figure 11.45(a). This may be regarded as the passive array equivalent of the corner reflector. Elements equally displaced from the centre of the array are connected by equal length transmission lines; thus the signal received at each element is reradiated from its counterpart with a phase such that a beam is formed in the direction of the incident signal. The retrodirective array will work over a range of directions determined by the beamwidths of the individual elements, which can in principle cover a broad range of angles. The basic idea can readily be extended to a planar array.

An n-element Van Atta array has a maximum RCS (in the absence of losses) of

$$\sigma = \frac{n^2 G_e^2 \lambda^2}{4\pi} \tag{11.160}$$

---

[28] Margerum, D.L., 'Self-phased arrays', Chapter 5 in Hansen, R.C. (ed), *Microwave Scanning Antennas*, Vol.III, pp341-407, Academic Press, New York, 1964, (reprinted in a single volume by Peninsula Publishing, Los Altos, USA, 1985).

[29] Van Atta, L.C., Electromagnetic Reflectors, US patent No. 2,908,002, October 6 1959.

where G_e is the gain of an individual element, which will be close to unity to give broad angular coverage. Since the retrodirective beam is formed by delays, the performance is broadband, and in practice is limited by the bandwidth of the elements.

The effective RCS can be increased by including bidirectional amplifiers in each path[30], as shown in Fig. 11.45(b); however, the gain of the amplifiers must not exceed the combined isolation of the circulators (or, indeed, the coupling between corresponding elements) or instability will result. For the same reason, it is important that the elements present a good impedance match. The maximum RCS of the arrangement of Fig. 11.45(b) is

$$\sigma = \frac{n^2 G_e^2 G_a \lambda^2}{4\pi} \tag{11.161}$$

where G_a is the amplifier gain, taking into account the losses of the circulators.

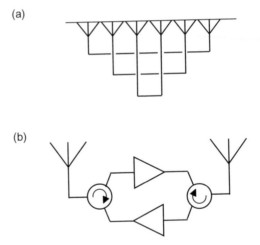

(a)

(b)

Fig. 11.45 (a) Basic passive form of Van Atta array; (b) the RCS can be increased significantly by using bidirectional amplifiers between pairs of elements.

It can be appreciated that Van Atta arrays are capable of giving very high RCS, particularly in their active form, over a broad range of angles and over a broad bandwidth, from a very compact configuration. It is also possible to modulate or

---

[30] Davies, D.E.N., 'Some properties of Van Atta arrays and the use of 2-way amplification in the delay paths', *Proc. IEEE*, Vol.110, p507, 1963.

switch the retrodirected signal by including modulators or switches in the signal paths of the arrangement of Fig. 11.45(b).

Self-phasing arrays

In other applications there may be a requirement to receive a signal with an antenna exhibiting gain, over a wide range of directions where the direction is not known *a priori*. This is sometimes known as 'omnidirectional gain'.

This effect can be achieved using a so-called self-phasing array[31]. It is necessary that the signal to be received contains a separate carrier, which may be in the centre of the signal band or at one edge. The basic arrangement is shown in Figure 11.46. At each element of the array the carrier is extracted, by a filter or a phase-locked loop, and used to downconvert the signal to baseband. The signals from all the array elements are then added together.

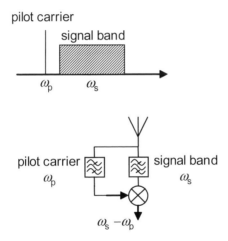

Fig. 11.46 Basic principle of self-phasing array. At each element the pilot carrier is extracted and used to downconvert the signal band.

The bandwidth is limited by the physical size of the array[32]. Considering a linear array of length L, the phases of the output signals at the extreme ends of the array are $(\omega_s - \omega_p)t$ and $(\omega_s - \omega_p)t + \psi_p - \psi_s$, where ψ_p and ψ_s are the phase shifts of the

[31] Cutler, C.C., Kompfner, R. and Tillotson, L.C., 'A self-steering array repeater', *Bell Syst. Tech. J.*, p2013, Sept. 1963.

[32] Brennan, P.V., 'An experimental and theoretical study of self-phased arrays in mobile satellite communications', *IEEE Trans. Antennas & Propagation*, Vol.37, No.11, pp1370-1376, November 1989.

pilot and signal wavefronts due to the difference in path lengths between the elements. If the array is small (or conversely, the overall bandwidth is small), the phase error $\psi_p - \psi_s$ is negligible. More precisely, the phase error is

$$\psi_s - \psi_p = \frac{L(\omega_s - \omega_p)\sin\theta}{c} = \frac{2\pi \sin\theta}{\lambda_{IF}} \qquad (11.161)$$

A practical implementation, in which the pilot carrier is extracted by means of a phase-locked loop, and the output is produced at a convenient IF rather than at baseband, is shown in Figure 11.47.

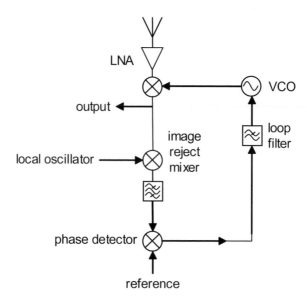

Fig. 11.47 Self-phasing array scheme using a phase-locked loop (after Brennan[32]).

Having extracted the pilot carrier at each element, it is then possible to phase the array on transmit by using the pilot carriers as local oscillators to upconvert the signal to be transmitted. The phases of the pilot carriers contain the phase information to form the transmit beam in the correct direction, without needing to know what that direction is. This may be useful for applications such as terrestrial mobile communications, or communications with satellites in low earth orbit.

Fusco and co-workers at Queen's University Belfast have developed integrated printed antennas with self-phasing receivers to use in these applications, and have demonstrated an elegant retrodirective duplex communication link based on these principles[33].

Random symmetrical pair arrays

A related scheme is the so-called random symmetrical pair array[34]. The array is formed of pairs of elements, each pair with a random spacing and random orientation.

Fig. 11.48 shows one such pair, of element spacing *d* and with a signal incident from an angle *θ* with respect to the normal to the axis of the pair. The signals from the two elements are combined in a hybrid coupler via equal length lines. We assume for the moment that the elements are omnidirectional, and that the signals received at each element are of unit amplitude.

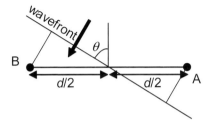

Fig. 11.48 One pair of a random symmetrical pair array.

The wavefront encounters the first element (A) with a phase lead of $\phi = \dfrac{\pi d}{\lambda} \cos\theta$, and the second element (B) with an equivalent phase lag.

The sum (S) and difference (D) of the two element signals are therefore

$$S = \frac{1}{\sqrt{2}}\left(e^{j\phi} + e^{-j\phi}\right) = \sqrt{2}\cos\phi \qquad D = \frac{1}{\sqrt{2}}\left(e^{j\phi} + e^{-j\phi}\right) = j\sqrt{2}\sin\phi \qquad (11.162)$$

It is evident that, for a signal of a given incidence angle, the sum signals from all such pairs will be purely real and either positive or negative, and the difference signals will be purely imaginary and either positive or negative.

---

[33] Karode, S. and Fusco, V.F., 'Self-tracking duplex communication link using planar retrodirective antennas', *Proc. 1998 Microwave Symposium*, Baltimore, USA, pp977-1000, June 1998.

[34] Benjamin, R., Titze, W.A.U., Brennan, P.V. and Griffiths, H.D., 'Symmetric-pair antennas for beam steering, direction finding or isotropic-reception gain'; *IEE Proc.*, Vol.138, Pt.H, No.4, pp368-374, August 1991.

This suggests a number of applications. Firstly, if the negative sum signals are inverted and all the sum signals added, the result is a beam in the direction of the incident signal. The inversion of polarity, where necessary, could automatically be achieved by squaring the signals, followed by filtering to remove unwanted products. It can be shown that the gain of such a scheme is only 0.91 dB less than that achieved with fully-cophasal addition.

Secondly, the scheme can be used in direction finding: the patterns of the polarities of the signals from each pair can be used to determine the direction of arrival of a signal. The polarities can be measured simply by one-bit phase comparators.

APPENDIX 11A: COMPARISON OF LINEAR AND CIRCULAR ARRAYS

11A.1 Gain of an arbitrary array

Condition for maximum gain

We know that the gain of a linear radiating aperture is maximum when the illumination function is constant, and the same result holds for a planar array. In the case of an *arbitrary* array, what is the condition for maximum gain ?

Consider an array made up of N arbitrary elements oriented in an arbitrary manner, and whose phase centres are located at the points A_n ($n = 0, 1, 2, ... N–1$)[35].

We suppose that their characteristic radiation functions $f_n(\mathbf{u})$, which we assume are real, are normalized such that their individual gains are

$$g_n(\mathbf{u}) = \left| f_n(\mathbf{u}) \right|^2$$

Each element is fed with a complex amplitude a_n, normalized in such a way that the feedpoint powers are

$$W_n = \left| a_n \right|^2$$

We can therefore write

$$a_n = \sqrt{W_n} \exp\left(j\varphi_n \right)$$

where φ_n are the feedpoint phases.

The total power is the sum of the individual feedpoint powers

$$W = \sum_{n=0}^{N-1} W_n$$

If O is taken as a reference point, the field radiated by the array is the sum

---

[35] If the elements do not possess a defined phase centre, the points A_n are just reference points and $f_n(\mathbf{u})$ are complex. However, the same conclusions hold.

$$F_R(\mathbf{u}) = \sum_{n=0}^{N-1} \sqrt{W_n} \exp(j\varphi_n) f_n(\mathbf{u}) \exp(-jk\mathbf{OA}_n.\mathbf{u})$$

A single omnidirectional element (of unity gain) fed with the total power W gives, under the same conditions, a field

$$F_0(\mathbf{u}) = \sqrt{W}$$

The gain is therefore

$$G_R(\mathbf{u}) = \left| \frac{F_R(\mathbf{u})}{F_0(\mathbf{u})} \right|^2$$

which can be written in the form

$$G_R(\mathbf{u}) = \frac{\left| \sum_{n=0}^{N-1} \left(\sqrt{W_n} \exp(j\varphi_n) \right) \left(f_n(\mathbf{u}) \exp(-jk\mathbf{OA}_n.\mathbf{u}) \right) \right|^2}{\sum_{n=0}^{N-1} \left| \sqrt{W_n} \exp(j\varphi_n) \right|^2 \sum_{n=0}^{N-1} \left| f_n(\mathbf{u}) \exp(-jk\mathbf{OA}_n.\mathbf{u}) \right|^2} \sum_{n=0}^{N-1} \left| f_n(\mathbf{u}) \right|^2$$

Making use of Schwartz's inequality, the gain is maximum if

$$\sqrt{W_n} \exp(j\varphi_n) = A f_n(\mathbf{u}) \exp(-jk\mathbf{OA}_n.\mathbf{u})$$

where A is an arbitrary scalar. We therefore have

(1) $\varphi_n = -k\mathbf{OA}_n.\mathbf{u}$

(2) $W_n = B g_n(\mathbf{u})$ (where B is an arbitrary positive scalar)

In this case the gain is maximum, and of value

$$G_R(\mathbf{u})_{max} = \sum_0^{N-1} g_n(\mathbf{u})$$

Conclusion
THEOREM: *If the feedpoint powers are proportional to the gains of the individual elements in the direction considered, the total gain is maximum and equal to the sum of the individual gains.*

It can be verified that if the elements are identical and oriented in the same direction, the illumination should be uniform and the resulting gain is N times the gain of an individual element. However, for an arbitrary array it is necessary to adjust both the phase *and the amplitude*. This problem can be solved by using *phase mode excitation* (§11.8.3)

Remark: Direct adjustment of the amplitude is difficult in the case of a passive array, since it results in losses. On the other hand, with an array composed of active modules this can be done without losses, by controlling each amplifier, at the same time as adjusting the phase.

11A.2 Gain of a beam cophasal circular array

It can be shown that the gain of an arbitrary circular array does not depend on the *directivity* of the elements, nor the array diameter, but only on the number N of elements and their directivity in the plane perpendicular to the array.

Suppose we have a circular array of diameter $D = 2R$, consisting of N elements with phase centres A_n (with polar coordinates R, α_n) spaced regularly around the circumference of the array. Thus

$$\alpha_n = n\Delta\alpha \quad \text{with} \quad \Delta\alpha = 2\pi/N$$

In an arbitrary direction θ in the plane of the array, the gain of an element is of the form

$$g_n(\theta) = g(\alpha_n - \theta)$$

The maximum resulting gain of the array is therefore

$$G_{\text{max}} = \sum_{n=0}^{N-1} g(\alpha_n - \theta) = N \frac{1}{2\pi} \sum_{0}^{N-1} g(\alpha_n - \theta)\Delta\alpha$$

If the element spacing is small, the summation can be replaced by an integral in terms of the average of the gains in the plane of the array

$$G_{\text{max}} = N\overline{g_n}$$

with

$$\overline{g_n} = \frac{1}{2\pi} \int_{0}^{2\pi} g_n(\alpha) \, d\alpha$$

This average, which is independent of n, only depends on the element directivity in the plane perpendicular to the array (equation (7.3)).

11A.3 Radiation pattern of a beam cophasal circular array

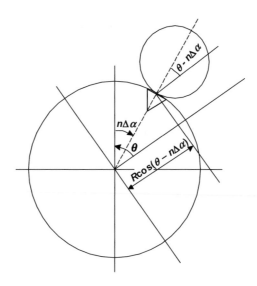

Fig. 11.49 Geometry of circular array.

Let $u_n = a_n \exp(j\varphi_n)$ be the illumination function of an antenna element of phase centre A_n, oriented radially outwards. Its characteristic function, with respect to this point, is real and of the form

$$f_n(\theta) = f(\theta - n\Delta\alpha) \qquad \text{with} \quad \Delta\alpha = d/R$$

The total field in the far field in direction θ is the sum

$$F(\theta) = \sum_{n=0}^{N-1} u_n f(\theta - n\Delta\alpha) \exp\left[jkR\cos(\theta - n\Delta\alpha) \right]$$

If we wish to point a beam in a direction θ_0, for example, $\theta_0 = 0$, we choose the feedpoint phases

$$\varphi_n = -kR\cos n\Delta\alpha$$

We have

$$F(\theta) = \sum_{0}^{N-1} a_n f(\theta - n\Delta\alpha) \exp\{jkR[\cos(\theta - n\Delta\alpha) - \cos n\Delta\alpha]\}$$

Condition for maximum gain

For this, the excitation amplitude a_n should be proportional to the amplitude of the element pattern in the desired direction θ_0, i.e.

$$a_n = f(-n\Delta\alpha)$$

We have therefore

$$F(\theta) = \sum_{0}^{N-1} f(\theta - n\Delta\alpha) f(-n\Delta\alpha) \exp\left[2jkR\sin\left(\frac{\theta}{2}\right)\sin\left(\frac{\theta}{2} - n\Delta\alpha\right)\right]$$

Array with a large number N of elements

In this case we can replace the above summation by an integral. Setting $\alpha = -n\Delta\alpha$, we have

$$F(\theta) = \int_{0}^{2\pi} f(\alpha + \theta) f(\alpha) \exp\left[2jkR\sin\left(\frac{\theta}{2}\right)\sin\left(\frac{\theta}{2} + \alpha\right)\right] d\alpha$$

11A.4 Example: $\cos\alpha$ element patterns

Consider the case of a directional element pattern of the form $f(\alpha) = \cos\alpha$. Setting

$$\beta = \frac{\theta}{2} + \alpha$$

We can separate the radiation pattern into two parts

$$F(\theta) = \int_{0}^{2\pi} \cos^2\beta \exp(jz\sin\beta) \, d\beta - \int_{0}^{2\pi} \sin^2\beta \exp(-jz\sin\beta) \, d\beta$$

where $z = 2kR\sin\left(\frac{\theta}{2}\right)$

This gives explicitly

$$\frac{F(\theta)}{F(0)} = \Lambda_1(z) - \sin^2\left(\frac{\theta}{2}\right) J_0(z)$$

Close to the axis the first term is dominant:

$$\frac{F(\theta)}{F(0)} \cong \Lambda_1(z) = \frac{2J_1(2kR\sin\theta/2)}{2kR\sin\theta/2}$$

The Λ_1 function is plotted in Fig. 9.14.

Note that:

- the 3 dB beamwidth $\theta_{3dB} \cong \dfrac{\lambda}{2R} = \dfrac{\lambda}{D}$;
- level of first sidelobe $L_1 = 0.13$, which is approximately -17 dB.

Further out, it is the second term which dominates. This is a consequence of radiation from the outermost elements, which point away from the axis.

Example: For $\theta = \pi/2$ the following relative levels L_2 are obtained:

- For $D/\lambda = 5$ $L_2 = 0.07$, i.e. approximately -23 dB
- For $D/\lambda = 30$ $L_2 = 0.03$, i.e. approximately -31 dB

11A.5 Comparison of linear and circular arrays

In beam-cophasal circular arrays, the *phase* of the excitation follows a sinusoidal function which is relatively complicated, and the amplitude has to be varied. However, this allows a pointing direction θ_0 which is independent of frequency, depending only on the symmetry of the excitation function with respect to the direction θ_0. This is not the case with linear arrays (§11.3.2, equation (11.19)). The -3 dB beamwidth of a circular array is independent of the pointing direction, but depends slightly on frequency (in a quadratic manner). This is the opposite of the case with linear arrays, where the -3 dB beamwidth does depend on the pointing direction (equation (11.22)), but very little on frequency.

Circular arrays do not suffer from localized grating lobes, such as can occur with linear arrays. They also do not suffer from the blindness phenomenon (§11.4.3). All these advantages should be taken into consideration when making tradeoffs.

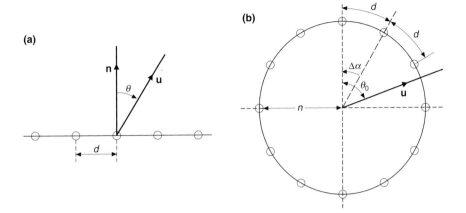

Fig. 11.50 (a) Linear array, and (b) circular array geometry.

FURTHER READING

1. Amitay, N., Galindo, V. and Wu, C.P., *Theory and Analysis of Phased Array Antennas*, Wiley-Interscience, London, 1972.
2. Barton, P., 'Digital beam forming for radar', *Proc. IEE*, Vol. 127, Pt. F, No. 4, pp266-277, August 1980.
3. Benjamin, R. and Seeds, A.J., 'Optical beam forming techniques for phased array antennas', *IEE Proc.*, Vol. 139, Pt. H, No. 6, pp526-534, December 1992.
4. Borgiotti, G., 'Conformal arrays', Chapter 11 in *The Handbook of Antenna Design*, Vol.II, A.W. Rudge, K. Milne, A.D. Olver and P. Knight (eds), Peter Peregrinus, Stevenage, 1983.
5. Brookner, E., 'Phased-array radars', *Scientific American*, pp94-102, February 1985.
6. Cohen, E.D., 'Trends in the development of MMICs and packages for active electronically scanned arrays (AESA)', 1996 IEEE International Symposium on Phased-Array Systems and Technology, Boston, MA, pp1-4, October 1996.
7. Davies, D.E.N., 'Circular arrays', Chapter 12 in *The Handbook of Antenna Design*, Vol.II, A.W. Rudge, K. Milne, A.D. Olver and P. Knight (eds), Peter Peregrinus, Stevenage, 1983.
8. Hansen, R.C. (ed), *Microwave Scanning Antennas*, Vols I-III, Academic Press, New York, 1964, (reprinted in a single volume by Peninsula Publishing, Los Altos, USA, 1985).
9. Mailloux, R., 'Phased array theory and technology', *Proc. IEEE*, Vol. 70, No. 3, March 1982.
10. Mailloux, R., *Phased Array Antenna Handbook* (second edition), Artech House, 2005.
11. Rebeiz, G.M., *RF MEMS: Theory, Design and Technology*, Wiley, 2003.
12. Schelkunoff, S.A., 'A mathematical theory of linear arrays', *Bell System Technical Journal*, Vol. 22, pp80-107, 1943.
13. Shenoy, R.P., 'Phased array antennas - Part 3: active aperture arrays', in *Advanced Radar Techniques and Systems*, G. Galati (ed), Peter Peregrinus, Stevenage, 1993.
14. Steinberg, B.D., *Principles of Aperture and Array Systems Design*, Wiley-Interscience, London, 1976.
15. Steyskal, H., 'Digital beamforming antennas: an introduction', *Microwave Journal*, pp107-124, January 1987.
16. Tsunoda, Y. and Goto, N., 'Sidelobe suppression of planar array antennas by the multistage decision method', *IEEE Trans. Antennas & Propagation*, September 1987.

EXERCISES

11.1 Design of an electronically-scanned array antenna

It is required to define the geometric and electric parameters of a planar array with the following external characteristics.

- wavelength: $\lambda = 10$ cm

- scan angle: in azimuth $\pm 45°$ either side of the normal to the array
in elevation $-5°$, $+55°$ with respect to the horizon

- sidelobes: close-in $L < -23$ dB
far-out (quantization lobes) - consider two cases $L < -43$ dB
$L < -49$ dB

- maximum on-axis gain: 37 dB
  taking into account ohmic losses K_1 of the phase shifters
    diode phase shifters   1.3 dB
    ferrite phase shifters   0.7 dB
  taking into account the loss K_2 due to the illumination taper
  taking into account the quantization loss K_3 of the phase shifters
    $p = 2$ bits
    $p = 3$ bits

- 3 dB elevation beamwidth θ_s equal to half the azimuth beamwidth θ_g:
  $\theta_s = \theta_g/2$

The goal of the exercise is to define the internal characteristics of the antenna, starting from the external characteristics stated above.

Q1   After a first-cut estimate of the illumination and quantization losses K_2 and K_3 (for example, $K_1 + K_2 + K_3 = 3$ dB), evaluate the area of the antenna, the horizontal and vertical dimensions H and V, the horizontal and vertical element spacing a_H and a_V (suppose that the array is inclined so as to equalize the phase shifts in elevation), and the number of elements N. Hence deduce the beamwidths θ_s and θ_g.

Q2   Evaluate the number of bits for the phase shifters, in order to meet the specification for the average far-out sidelobes. Evaluate the corresponding losses K_3, for the cases $p = 2$ and $p = 3$.

Q3 Evaluate the loss K_2 in gain due to the aperture weighting function, for the two cases:

  (a) a cosine weighting (without 'pedestal') both vertically and horizontally;
  (b) a cosine weighting both vertically and horizontally, with the largest pedestal compatible with the specified *close-in* sidelobe level (an attenuation at the edge of −12 dB is suitable).

ANSWERS

Q1   *Antenna area*: With $K_1 + K_2 + K_3 = 3$ dB, the required directivity is 40 dB. Thus:

$$\frac{4\pi S}{\lambda^2} = 10^4$$

which gives approximately $S = 8$ m^2.

Dimensions: We have $\theta_s \approx \dfrac{\lambda}{V}$; $\theta_g \approx \dfrac{\lambda}{H}$

It follows that $V = 2H$ and $S = VH = 8$
Thus: $H = 2$m, $V = 4$m

Horizontal and vertical element spacing: These result from equation (11.25). In elevation, a tilt of the plane of the array with respect to the vertical of 25° gives a maximum scan angle of ±30°. We therefore choose

$$a_H = \frac{2}{3}\lambda \approx 6.6 \text{ cm} \qquad a_V = \frac{\lambda}{1+\sqrt{2}/2} \approx 5.8 \text{ cm}$$

Number of elements:

$$N = \frac{H.V}{a_H a_V} = 2050$$

Beamwidths:

$$\theta_s \approx \frac{\lambda}{V} \approx 1.43° \qquad \theta_g \approx \frac{\lambda}{H} \approx 2.86°$$

Q2   *Number of bits for phase shifters*: We make use of equation (11.70). Since the number of bits must be an integer, we choose respectively the following values:

L_{max}	$P_{theoretical}$	P_{actual}	L_{actual}
−43 dB	2.5	3	−46 dB
−49 dB	3.5	4	−52 dB

Quantization loss: due to K_3.
Using equation (11.74), for $p = 3$ $K_3 = -0.22$ dB
$p = 4$ $K_3 = -0.055$ dB

Q3   *Loss due to aperture weighting function K_2:* We find (exercise 7.4)

$$K_2 = \left[\cfrac{1}{1 + \cfrac{A}{(1+\beta)^2}} \right]^2$$

with $A = \dfrac{\pi^2}{8} - 1$, $\beta = \dfrac{\pi}{2} \dfrac{a}{(1-a)}$

With a pedestal of −12 dB:

$a = 0.063$, $\beta = 0.104$, $K_2 \approx 0.70$, i.e −1.52 dB

11.2 *Grating lobes produced by use of phased subarrays fed by delay lines*

An array of this kind of length L can be considered as an array where the interelement spacing is equal to the length L' of a subarray.

Q1: For uniform illumination law and for a pointing direction θ_0, show that the array factor is of the form

$$R\left(\frac{\sin\theta}{\lambda} \right) = \frac{\sin\left[N\pi \dfrac{L'}{\lambda} (\sin\theta - \sin\theta_0) \right]}{N \sin\pi \dfrac{L'}{\lambda} (\sin\theta - \sin\theta_0)}$$

Obviously, the grating lobes are in the directions $\theta_{R,k}$:

$$\sin \theta_{R,k} - \sin \theta_0 = k \frac{\lambda}{L'}$$

Q2 Show that at a wavelength λ, the radiation function of a *phased* subarray is given by

$$F_{L'}\left(\frac{\sin \theta}{\lambda}\right) = \operatorname{sinc} \pi L' \left(\frac{\sin \theta}{\lambda} - \frac{\sin \theta_0}{\lambda_0}\right)$$

with

$$\operatorname{sinc} x = \frac{\sin x}{x}$$

Q3 With the overall radiation function of the array being given by the product

$$F\left(\frac{\sin \theta}{\lambda}\right) = RF_{L'}$$

show that the relative amplitude level of the grating lobes is given by:

$$M = \sin \theta_0 \frac{L'}{c} df$$

This is independent of the centre frequency; it depends only on the subarray dimension L' and the frequency shift df. As a general rule, it is given by the product of the frequency shift and the time required by the signal to traverse one subarray.

12

Fundamentals of polarimetry

Jules Henri Poincaré (1854–1912)

12.1 INTRODUCTION

12.1.1 Applications of polarimetry in radar and telecommunications

Polarization is a property specific to electromagnetic waves, which is not found in other types of radiation such as acoustic waves. At the beginning of the development of HF and VHF systems this property received little attention.

Nevertheless, very early radars used circular polarization to discriminate against echoes from rain (§7.1). Similarly, telecommunications systems took advantage of dual polarizations to provide isolation between transmitted and received signals.

The development of polarimetry is due, on one hand, to an increased demand from users who were aware of the technological advances in that field, and on the other hand to the progress made in the theoretical domain, allowing a better knowledge of the phenomena involved.

Thus, the depolarization of radar echoes allows the use of the 'characteristic' polarizations in order to reinforce wanted echoes and to minimize parasitic echoes, and at the same time to contribute to the identification of the nature and the attitude of radar targets.

Let us take as an example a vertically polarized radar illuminating a dipole (wire) tilted at a random angle α with respect to the vertical axis (the probability density of α is assumed uniform). Only the electric field component parallel to the target-dipole is relevant, and this varies as $\cos \alpha$. The field reflected by the target, in turn, is only dependent on the component parallel to the vertical polarization of the radar. Thus, the received echo amplitude s_{VV} (when vertical polarization is used) is of the following form.

$$s_{VV} = \cos^2 \alpha$$

Therefore, the average power loss is

$$\frac{1}{2\pi} \int_0^{2\pi} s_{VV}^2 \, d\alpha = \frac{3}{8}$$

which is approximately 4.3 dB.

If the radar uses both vertical and horizontal polarization, then the corresponding received amplitudes are of the form

$$s_{VV} = \cos^2 \alpha \qquad s_{HV} = \cos \alpha \sin \alpha$$

The total power is increased. The average power loss is 3 dB (only the component of the polarization which is transmitted orthogonal to the target is lost). Furthermore, the ratio of the two received signals indicates the attitude of the target with respect to the radar, defined in this case by the tilt angle α

$$\frac{s_{HV}}{s_{VV}} = \tan \alpha$$

Finally, if we have a 'polarimetric' radar with a feedback loop between the transmitted polarization and the polarization received from the target, then all the available power is used

$$s_H = \cos \alpha, \quad s_V = \sin \alpha$$

so that

$$|s_H|^2 + |s_V|^2 = 1$$

Ultimately, this example shows that the polarimetric radar allows us to increase the useful signal and contributes to defining the orientation and the characteristics of the target.

In telecommunications systems as well as in radars, the use of two independent orthogonal polarizations to transmit different pieces of information is quite common. So-called 'frequency re-use' techniques permit us to double the information carried in a given frequency bandwidth, provided that residual intermodulation is eliminated by means of a feedback loop between the polarized waves and a correlation criterion from the received signals.

12.1.2 Some historical references

The pioneers of optical polarimetry were mathematicians such as Stokes (1819–1903) and Poincaré (1854–1912). Among those involved in microwave polarimetry we should mention Roubine[1], Sinclair[2] and Deschamps, who introduced the complex vector notion. In the seventies important progress was made, in particular by Huynen[3] with his phenomenological theory of radar echoes. The main goal of this chapter is to present a summary of the results of all of this work.

12.1.3 Basics

Complex vectors

Since electromagnetic waves are transverse, the polarization of the electric field can be represented in a coordinate system by means of two orthogonal unit vectors $(\mathbf{u}, \mathbf{v})$ (for instance horizontal and vertical) in a plane orthogonal to the propagation direction. Two waves of the same frequency and polarized in the directions of the unit vectors are characterized by their amplitude and their relative phase as follows.

$$E_{\mathbf{u}}(t) = a\cos(2\pi ft)$$
$$E_{\mathbf{v}}(t) = b\cos(2\pi ft + \varphi)$$

The resulting field is

---

[1] Roubine, E. and Bolomey, J.C., *Antennas I*, p7, North Oxford Academic, London, 1987.
[2] Sinclair, G., 'The transmission and reflection of elliptically polarized waves', *Proc. IRE*, Vol. 38, pp148-151, February 1950.
[3] Huynen, J.R., *Phenomenological Theory of Radar Targets*, Ph.D Dissertation, Drukkerij Bronder-offset N.V., Rotterdam, 1970.

$$\mathbf{E}(t) = E_{\mathbf{u}}(t)\mathbf{u} + E_{\mathbf{v}}(t)\mathbf{v}$$

It is useful to consider the above expressions as the real parts of the following exponential forms

$$E_{\mathbf{u}}(t) = \mathrm{Re}\left[a \exp(j2\pi ft)\right]$$
$$E_{\mathbf{v}}(t) = \mathrm{Re}\left[b \exp(j\varphi)\exp(j2\pi ft)\right]$$

Then, we define a complex vector (where the time factor $\exp(j2\pi ft)$ has been omitted)

$$\mathbf{E} = a\mathbf{u} + b\mathbf{v}\exp(j\varphi)$$

from which we obtain the electric field

$$E(t) = \mathrm{Re}\left[\mathbf{E}\exp(j\pi ft)\right]$$

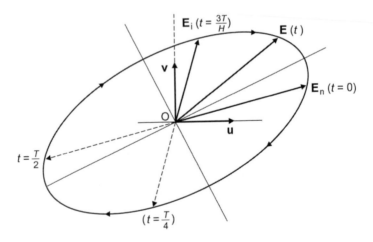

Fig. 12.1 Polarization ellipse.

Real and imaginary parts

Let us set

$$\mathbf{E} = \mathbf{E}_r + j\mathbf{E}_i$$

where

$$\mathbf{E}_r = a\,\mathbf{u} + b\cos\varphi\,\mathbf{v} \qquad\qquad \mathbf{E}_i = b\sin\varphi\,\mathbf{v}$$

Then

$$\mathbf{E}(t) = \mathbf{E}_r \cos(2\pi ft) - \mathbf{E}_i \sin(2\pi ft)$$

We observe that the tip of the vector traces an ellipse in a plane where the vectors $(\mathbf{E}_r, \mathbf{E}_i)$ represent the two conjugate axes of the polarization ellipse (Fig. 12.1). The tangents at the tip of the conjugate vectors are parallel to these axes.

Particular cases

Components in phase: If $\varphi = 0$, $\mathbf{E}_i = 0$, $\mathbf{E}(t) = \mathbf{E}_r \cos(2\pi ft)$ and the tip of vector $\mathbf{E}(t)$ traces a straight line segment. The wave is linearly polarized.

Components in quadrature: If $\varphi = \pm\pi/2$, then $\mathbf{E} = a\mathbf{u} \pm jb\mathbf{v}$. The axes of symmetry of the ellipse are in the direction of $(\mathbf{u}, \mathbf{v})$.

Furthermore, if the amplitudes are equal ($a = b$), then $\mathbf{E} = a(\mathbf{u} \pm j\mathbf{v})$ and the wave is circularly polarized, clockwise or counter-clockwise.

12.2 FULLY POLARIZED WAVES

12.2.1 Definition

A wave, at a given frequency, is said to be fully polarized if the tip of the rotating vector representing the electric field describes a well-defined and stable ellipse as a function of time. This definition will allow us, later on, to define a partially polarized wave.

12.2.2 Algebraic representation of elliptical polarization

The semi-major and semi-minor axes of the polarization ellipse (see Fig. 12.2) are chosen as a coordinate system where $\mathbf{u}_0$ and $\mathbf{v}_0$ are the unit vectors.

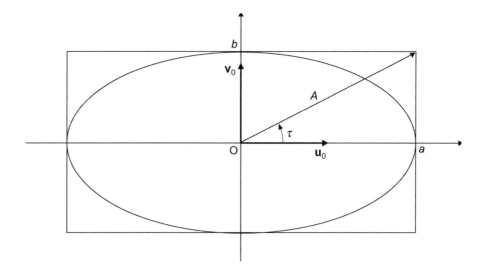

Fig. 12.2 Canonical coordinate system: algebraic ellipticity.

The real and imaginary parts of the electric field are

$$\mathbf{E}_r = a\mathbf{u}_0 \qquad \mathbf{E}_i = b\mathbf{v}_0 \tag{12.1}$$

therefore the resulting field is

$$\mathbf{E} = a\,\mathbf{u}_0 + b\,\mathbf{v}_0 \tag{12.2}$$

Algebraic ellipticity: this is the angle between the diagonal of the rectangle which circumscribes the ellipse and the semimajor axis, and is called the *ellipticity*

$$a = A\cos\tau \tag{12.3}$$

$$b = A\sin\tau \tag{12.4}$$

$$A^2 = a^2 + b^2 \tag{12.5}$$

then

$$\mathbf{E} = A\left(\mathbf{u}_0 \cos\tau + j\mathbf{v}_0 \sin\tau\right) \tag{12.6}$$

Note: the sign of τ allows us to define, as well as the ellipticity, the *sense* of the elliptical polarization.

τ	E	polarization type
0	$A\mathbf{u}_0$	linear parallel to $\mathbf{u}_0$
$+\dfrac{\pi}{4}$	$A\dfrac{\mathbf{u}_0 + j\mathbf{v}_0}{\sqrt{2}}$	right hand circular
$-\dfrac{\pi}{4}$	$A\dfrac{\mathbf{u}_0 - j\mathbf{v}_0}{\sqrt{2}}$	left hand circular
$\dfrac{\pi}{2}$	$jA\mathbf{v}_0$	linear, parallel to $\mathbf{v}_0$

12.2.3 Normalized Cartesian coordinate system

Such a coordinate system may be obtained by performing a rotation through an angle ψ about the origin of the system $\mathbf{u}_0, \mathbf{v}_0$. The previous unit vectors are related to the new ones by the following equations.

$$\mathbf{u} = (\mathbf{u}_0.\mathbf{u})\mathbf{u} + (\mathbf{u}_0.\mathbf{u})\mathbf{v} \tag{12.7}$$

$$\mathbf{v} = (\mathbf{v}_0.\mathbf{v})\mathbf{u} + (\mathbf{v}_0.\mathbf{v})\mathbf{v} \tag{12.8}$$

$$\begin{bmatrix} \mathbf{u}_0 \\ \mathbf{v}_0 \end{bmatrix} = \begin{bmatrix} \cos\psi & \sin\psi \\ -\sin\psi & \cos\psi \end{bmatrix} \begin{bmatrix} \mathbf{u} \\ \mathbf{v} \end{bmatrix} \tag{12.9}$$

From which we obtain a new expression for the field

$$\mathbf{E} = A(\mathbf{u}_0 \cos\tau + j\mathbf{v}_0 \sin\tau) \tag{12.10}$$

$$\mathbf{E} = \alpha\mathbf{u} + \beta\mathbf{v} \tag{12.11}$$

$$\begin{bmatrix} \alpha \\ \beta \end{bmatrix} = A \begin{bmatrix} \cos\psi & -\sin\psi \\ \sin\psi & \cos\psi \end{bmatrix} \begin{bmatrix} \cos\tau \\ j\sin\tau \end{bmatrix} \qquad (12.12)$$

A typical application is the measurement of the parameters of the elliptical polarization. A polarimetric receiver allows us to measure the amplitude and phase (α, β) in the coordinate system $(\mathbf{u}, \mathbf{v})$.

The question is: can we deduce the polarization ellipse parameters, that is to say the tilt angle ψ and the ellipticity τ? In order to solve this problem we use the base of circular polarizations.

12.2.4 Base of circular polarizations

It is useful to define a specific coordinate system to analyse the polarization.
Let us set

$$\mathbf{c} = \frac{\mathbf{u} + j\mathbf{v}}{\sqrt{2}}, \quad \mathbf{c}' = \frac{\mathbf{u} - j\mathbf{v}}{\sqrt{2}}$$

It can easily be verified that the vectors are orthonormal

$$|\mathbf{c}|^2 = |\mathbf{c}'|^2 = 1 \quad \text{and} \quad \mathbf{c}.\mathbf{c}'* = 0$$

Using the coordinate system defined by these vectors, we can express the electric field in the following form.

$$\mathbf{E} = \alpha\mathbf{u} + \beta\mathbf{v} = \alpha'\mathbf{c} + \beta'\mathbf{c}' \qquad (12.14)$$

where

$$\alpha' = \frac{\alpha + j\beta}{\sqrt{2}}, \quad \beta' = \frac{\alpha - j\beta}{\sqrt{2}} \qquad (12.15)$$

It can be seen (Fig. 12.3) that any elliptical polarization can be expressed as the sum of two opposite circular polarizations.

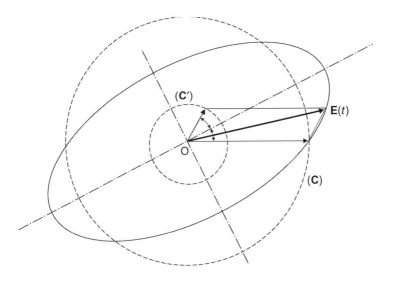

Fig. 12.3 Decomposition of elliptical polarization into two opposite circular polarizations.

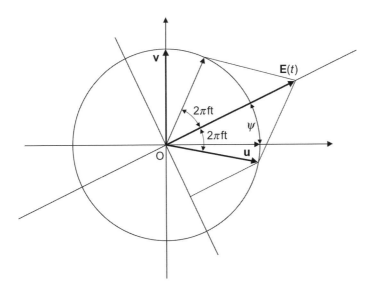

Fig. 12.4 Linear polarization as the sum of two opposite circular polarizations.

Example: linear polarization at angle ψ (Fig. 12.4). We have $\tau = 0$.
  Therefore,

$$\mathbf{E} = \cos\psi\ \mathbf{u} + \sin\psi\ \mathbf{v} \tag{12.16}$$

$$\alpha = \cos\psi, \quad \beta = \sin\psi$$

$$\alpha' = \frac{1}{\sqrt{2}}\exp(j\psi), \quad \beta' = \frac{1}{\sqrt{2}}\exp(-j\psi)$$

This example shows two circular polarizations of equal amplitude, with a phase difference of twice the tilt angle ψ of the linear polarization.

12.2.5 Polarization ratio

In any base the polarization ratio is defined as the ratio of the coordinates.
   Let us set:

- referred to a linear base: $R = \beta/\alpha$;

- referred to a circular base: $\rho = \beta'/\alpha'$.

From relation (12.15), we deduce

$$\rho = \frac{\alpha - j\beta}{\alpha + j\beta} = \frac{1 - jR}{1 + jR} \tag{12.17}$$

The linear and circular polarization ratios are related by a *homographic relationship*.

Expression of ρ as a function of ψ and τ
Equation (12.12) gives

$$\alpha = A(\cos\psi\cos\tau - j\sin\psi\sin\tau) \tag{12.18}$$

$$\beta = A(\sin\psi\cos\tau + j\cos\psi\sin\tau) \tag{12.19}$$

$$\alpha + j\beta = \frac{A}{\sqrt{2}}\exp(j\psi)\cos\left(\tau + \frac{\pi}{4}\right) \tag{12.20}$$

$$\alpha - j\beta = \frac{A}{\sqrt{2}}\exp(-j\psi)\cos\left(\tau - \frac{\pi}{4}\right) = \frac{A}{\sqrt{2}}\exp(-j\psi)\sin\left(\tau + \frac{\pi}{4}\right) \tag{12.21}$$

$$\rho = \tan\left(\tau + \frac{\pi}{4}\right)\exp(-j2\psi) \qquad (12.22)$$

Conclusions: The tilt angle ψ is just half the phase difference between the circularly polarized components.

The ellipticity τ is given by the expression

$$\tau = \tan^{-1}|\rho| - \frac{\pi}{4}$$

A typical application is in polarimetric measurements.

We now have the answer to the questions posed earlier. The measurement of the linear orthogonal (α, β) components of any electric field allows us to calculate the circular polarization ratio ρ from equations (12.20) and (12.21). The argument and magnitude of ρ given by equation (12.22) permits us to determine, respectively, the tilt angle ψ and the ellipticity $\tan\tau$.

12.2.6 Polarization diagram

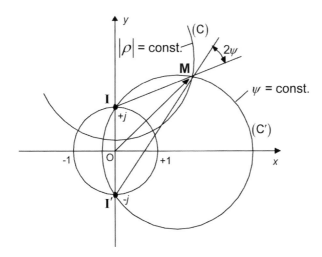

Fig. 12.5 Polarization diagram

In the complex plane (see Fig. 12.5) let us consider the points I $(0, j)$ and I' $(0, -j)$ and the vector OM image of the linear polarization of ratio $R = \beta/\alpha$. It can be verified that vector $\mathbf{I'M} = \mathbf{I'O} + \mathbf{OM} = \mathbf{OI} + \mathbf{OM}$ is the image of the complex number $j+R$ and $\mathbf{IM}$ is the image of $j-R$.

Let us calculate the ratio of distances. We have

$$\left|\frac{\mathbf{I'M}}{\mathbf{MI}}\right| = \left|\frac{j+R}{j-R}\right| = \left|\frac{1-jR}{1+jR}\right| = |\rho| \tag{12.23}$$

which is equal to the circular polarization ratio. We observe that the angle of the two vectors $\mathbf{I'M}$ and $\mathbf{IM}$ is the argument of their ratio

$$\arg\frac{R+j}{R-j} = \arg\frac{1-jR}{1+jR} = \arg\rho = -2\psi \tag{12.24}$$

Conclusion: the loci of the points M for which ψ is constant is a set of circles crossing the points I' and I. The locus of the points M where ρ is constant is a set of circles orthogonal to those for which ψ is constant (Fig. 12.5).

Poincaré sphere

Let us represent Fig. 12.5 in a vertical plane, namely Ox, Oy where I and I' are located on the Oy axis at a distance, respectively, of 1 and -1 from the origin O, and let Ox and Oy be orthogonal.

Let P be a point located on the Oz axis orthogonal to the plane Ox, Oy at distance unity from the centre O (see Fig. 12.6).

If we apply the following geometrical transformation known as inversion, that is to say $\mathbf{OM}.\mathbf{OM'} = 2$, it can be demonstrated that the plane Ox, Oy is transformed into a unit-radius sphere (Σ) whose poles are the points I I'. A very important property of this transformation is the invariance of the angles and the transformation of circles into circles when they do not cross the point P and into straight lines if they cross it. Since the points I and I' are invariant, the circles C' are transformed into meridians and the circles C being orthogonal to the circles C' are transformed into parallels of the sphere Σ. We observe that the equator is invariant and its points correspond to linear polarization. The upper and lower poles represent left and right circular polarizations.

The Cartesian coordinates of a point M' on the surface of the sphere (see Stokes parameters §12.3.7) are as follows (Fig. 12.6)

$$s_1 = \cos(2\tau)\cos(2\varphi) \tag{12.25}$$

$$s_2 = \sin(2\tau)\sin(2\varphi) \tag{12.26}$$

$$s_3 = \sin\ (2\tau) \hspace{3cm} (12.27)$$

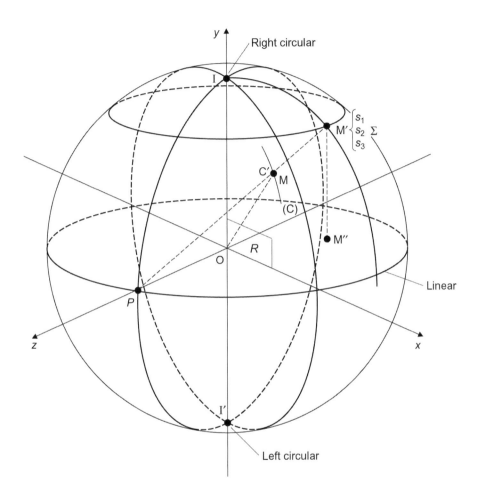

Fig. 12.6 The Poincaré sphere.

Polarization map

An orthogonal projection of the Poincaré sphere onto the equatorial plane (one map for each hemisphere) is quite a useful tool (Fig. 12.7). The vector radius of a general point M'' is related to the ellipticity $\cos(2\tau)$ and its polar angle represents twice the tilt angle 2ψ.

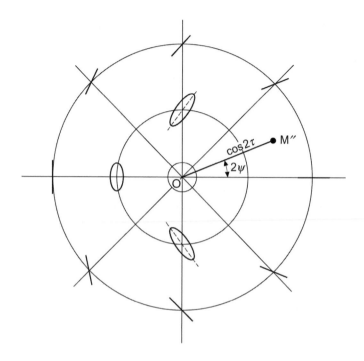

Fig. 12.7 Polarization map.

12.2.7 Polarization coupling to the receiving antenna

The concept of 'polarization factor' is described in Chapter 4.

Let us recall that for an antenna which transmits a wave whose polarization is defined by the complex vector $\mathbf{E}$ and receives a polarized wave defined by the complex vector $\mathbf{E}'$, the power input to the antenna is proportional to

$$\gamma = \frac{|\mathbf{E}.\mathbf{E}'|^2}{|\mathbf{E}|^2 \, |\mathbf{E}'|^2} \qquad (12.28)$$

where γ is known as the 'polarization factor'. Its value lies between zero and unity. This value of unity is obtained when the incoming vector is proportional to the complex conjugate of $\mathbf{E}$

$$\mathbf{E}' = k\mathbf{E}^* \tag{12.29}$$

where k is a scalar. The polarizations are then said to be matched.

Circular polarization is a good example to show how the conjugate vector operates.

Example 1: *Right handed circular polarization*
The complex vector transmitted by the antenna is of the form

$$\mathbf{E} = \mathbf{u} + j\mathbf{v} = \mathbf{u} + \mathbf{v}\exp\left(j\frac{\pi}{2}\right) \tag{12.30}$$

Since the incident polarization (travelling in the opposite direction to the transmitted wave) is also right-handed circularly polarized, the sense of rotation of the electric field must be reversed. The electric field, then, should be written

$$\mathbf{E}' = \mathbf{u} + \mathbf{v}\exp\left(-j\frac{\pi}{2}\right) = \mathbf{u} - j\mathbf{v} = \mathbf{E}^* \tag{12.31}$$

which is the complex conjugate of the transmitted vector.

Example 2: *Cross-polarized input signal*
In this case the power input to the antenna is zero.

Let us consider the case where a canonical coordinate system is used, so that we have

$$\mathbf{E} = a\mathbf{u} + jb\mathbf{v} \tag{12.32}$$

The scalar product is zero if

$$\mathbf{E}' = k\left(b\mathbf{u} + ja\mathbf{v}\right) \tag{12.33}$$

which means that for both waves the ellipticity is the same, that their major axes are orthogonal and that the their senses of rotation are opposite, if the waves are travelling in the same direction (Fig. 12.8). On the other hand, their senses of rotation are the same if the waves propagate in opposite directions (transmission and reception). Both cases are representative of right and left circularly polarized waves as well as linearly orthogonal polarized waves.

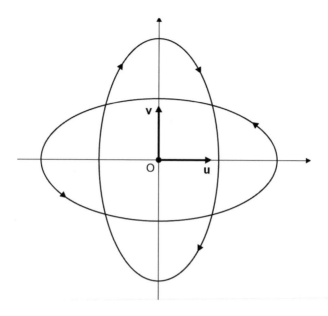

Fig. 12.8 Orthogonal polarization ellipses (same direction of propagation).

Example 3: *Linear polarizations*

Let us consider two linear polarizations making, respectively, angles θ and θ' with the unit vector **u** (see Fig. 12.9). We have

$$\mathbf{E} = \mathbf{u}\cos\theta + \mathbf{v}\sin\theta$$

$$\mathbf{E}' = \mathbf{u}\cos\theta' + \mathbf{v}\sin\theta'$$

From which we obtain

$$\mathbf{E}.\mathbf{E}' = \cos(\theta - \theta')$$

and the following expression for the polarization factor

$$\gamma = \cos^2(\theta - \theta') \tag{12.34}$$

It can be seen that the polarizations are matched if they are parallel, and crossed if they are orthogonal.

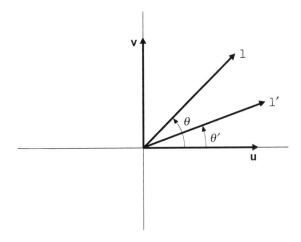

Fig. 12.9 Polarization factor for linear polarizations.

12.3 PARTIALLY POLARISED WAVES

12.3.1 Definition and physical origin

Several phenomena may be responsible for partial polarization of electromagnetic waves.

Let us consider two incoherent noise sources having the same average power, and let us observe them through filters whose bandwidth ΔF is centred on a carrier frequency f_0. These noises are represented according to the notation of Rice[4] by uncorrelated random functions $\alpha(t)$ and $\beta(t)$ with a time constant of $1/\Delta F$.

If these signals are received by two antennas, polarized horizontally (of unit vector **u**) and vertically (of unit vector **v**) respectively, the resulting field may be represented by the following field vector

$$\mathbf{E} = \alpha(t)\mathbf{u} + \beta(t)\mathbf{v} \tag{12.35}$$

---

[4]Rice, S.O., 'Mathematical analysis of random noise', *Bell Syst. Tech. J.*, Vol. 23, pp282-332 & Vol. 24, pp46-156.

Thus, we obtain a polarization ellipse changing quickly with respect to the measurement interval with time constant $1/\Delta F$, but slowly with respect to the carrier frequency ($\Delta F/f_o \ll 1$). Since they are uncorrelated their hermitian mean product is zero

$$\overline{\alpha(t)\beta^*(t)} = 0 \tag{12.36}$$

The corresponding powers are

$$\left. \begin{array}{l} A^2 = \overline{|\alpha(t)|^2} \\[2mm] B^2 = \overline{|\beta(t)|^2} \end{array} \right\} \tag{12.37}$$

If they are equal, the wave is said to be completely depolarized. If one of them largely predominates with respect to the other, then the wave is said to be partially polarized. For instance, if $A^2 \gg B^2$, the wave is largely horizontal.

The preceding example is referred to a coordinate system (**u**, **v**) of linear polarization. In fact, a coordinate system (**c**, **c′**) referred to circularly polarized waves, or any kind of orthogonal polarization coordinate system, may be used as well, provided that the appropriate transformation formulas are used.

Physical examples

It is known that the radiation from the Sun can be observed at different wavelengths, in the visible and infrared spectrum (optics) and at radio wavelengths (HF, VHF and microwave). It is generally observed that the waves are depolarized or partially polarized (owing to the action of the solar magnetic field on charged particles). The same observations can be carried out on galactic or extra galactic radio sources.

Another example can be taken from radar. Radar echoes are quite often reflected from objects having very complex forms such as aircraft. Convex surfaces, solid angles and lines reflect partial polarized echoes more or less delayed in time according to their longitudinal position with respect to the radar wave direction. The resulting field fluctuates in amplitude, phase and polarization, and as a function of time, owing to the evolution of the form and the attitude of the aircraft with respect to the radar. The resultant wave, which is the sum of the partial waves reflected by the different parts of the aircraft (Fig. 12.10), takes the following form

$$\begin{bmatrix} \alpha(t) \\ \beta(t) \end{bmatrix} = \sum_n \begin{bmatrix} \alpha_n(t) \\ \beta_n(t) \end{bmatrix} \exp\left[j2\pi f_0 \left(t - \frac{r_n(t)}{c} \right) \right] \tag{12.38}$$

In order to study the properties of partially polarized waves two particular analysis tools have been developed:

- the Coherence matrix (Wolf, 1959);
- the Stokes parameters (Stokes, 1852).

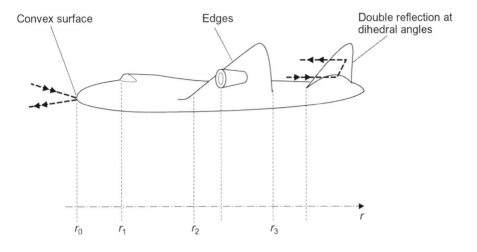

Fig. 12.10 Polarized radar echoes from an aircraft.

12.3.2 Coherence matrix

The physical origin of the coherence matrix concept (Fig. 12.11) emerges from the following problem. Consider an antenna with polarization unit vector **h**, oriented toward an incident polarized wave **E**. What is the average power $\overline{w}$ received by this antenna ?

The vectors **h** and **E** may be represented by their column matrices

$$\mathbf{h} = a\,\mathbf{u} + b\,\mathbf{v} \ \text{ or } \ \mathbf{h} = \begin{bmatrix} a \\ b \end{bmatrix} \quad \text{and} \quad a^2 + b^2 = 1$$

$$\mathbf{E}(t) = \alpha(t)\mathbf{u} + \beta(t)\mathbf{v} \ \text{ or } \ \mathbf{E} = \begin{bmatrix} \alpha(t) \\ \beta(t) \end{bmatrix}$$

The received amplitude is proportional to the scalar product of the two vectors, that is to say, to the product of the transpose of one of the matrices with the other.

$$A(h,t) = \mathbf{h.E} = \mathbf{h^t E} = \mathbf{E^t h}$$

The received power is then proportional to

$$w = \left| \mathbf{h^t E} \right|^2 = (\mathbf{h^t E})(\mathbf{h^t E})^* = \mathbf{h^t E}(\mathbf{E^t h})^*$$

$$w = \mathbf{h^t E}(\mathbf{E^t h}^*) = \mathbf{h^t E E^t h}^*$$

and its mean value:

$$\overline{w} = \mathbf{h^t J h}^* \tag{12.39}$$

where **J** is the coherence matrix

$$\mathbf{J} = \overline{\mathbf{E E}}^t = \overline{\begin{bmatrix} \alpha \\ \beta \end{bmatrix} (\alpha^* \beta^*)} = \begin{bmatrix} \overline{|\alpha|^2} & \overline{\alpha \beta^*} \\ \overline{\alpha^* \beta} & \overline{|\beta|^2} \end{bmatrix} \tag{12.40}$$

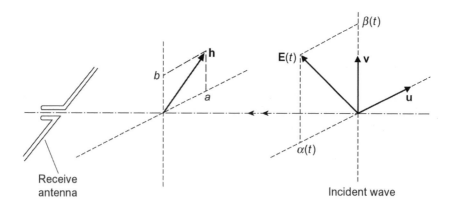

Fig. 12.11 Mean power received from a partially polarized wave: the coherence matrix.

The coherence matrix is then a hermitian matrix whose principal diagonal terms are the average power of the polarization components. The other diagonal terms represent the correlation coefficients between components. We can see that the determinant of this matrix is always positive or zero, $\text{Det } \mathbf{J} \ge 0$.

Available power

Horizontal polarization:

$$\mathbf{h}_{\mathrm{H}} = \begin{bmatrix} 1 \\ 0 \end{bmatrix}, \qquad \omega_{\mathrm{H}} = \overline{|\alpha|^2}$$

Vertical polarization:

$$\mathbf{h}_{\mathrm{V}} = \begin{bmatrix} 0 \\ 1 \end{bmatrix}, \qquad \omega_{\mathrm{V}} = \overline{|\beta|^2}$$

Total available power

Assuming reception of both polarizations

$$w_{\mathrm{TOT}} = \overline{|\alpha|^2} + \overline{|\beta|^2} = J \ = \text{trace of the matrix} \tag{12.41}$$

Degree of coherence

This is the normalized correlation coefficient between components

$$\mu_{\alpha\beta} = \frac{\overline{\alpha\beta^*}}{\sqrt{\overline{|\alpha|^2 + |\beta|^2}}} \tag{12.42}$$

Its value is dependent on the coordinate system chosen (**u**, **v**).

12.3.3 Completely unpolarized wave

This is a quite common case in radiometry and in radioastronomy. There is no correlation between the components, whose average powers are equal

$$\overline{\alpha\beta^*} = 0, \qquad \overline{|\alpha|^2} = \overline{|\beta|^2} = \frac{w}{2} \tag{12.43}$$

The average power received by any antenna is obtained from equation (12.39)

$$w(\mathbf{h}) = \begin{bmatrix} a & b \end{bmatrix} \begin{bmatrix} \overline{|\alpha|^2} & 0 \\ 0 & \overline{|\beta|^2} \end{bmatrix} \begin{bmatrix} a^* \\ b^* \end{bmatrix} = \left(|a|^2 + |b|^2 \right) \overline{|\alpha|^2} = \frac{w}{2} \tag{12.44}$$

This is independent of the polarization and equal to half the available total power.

In order to obtain the total power it is necessary to receive both orthogonal polarizations.

12.3.4 Completely polarized wave

In this case the components are perfectly correlated

$$\frac{\beta(t)}{\alpha(t)} = q = \text{const.} \quad \text{and} \quad \mu_{\alpha\beta} = 1$$

$$\mathbf{J} = \overline{|\alpha|^2} \begin{bmatrix} 1 & q^* \\ q & |q|^2 \end{bmatrix} \Rightarrow \det \mathbf{J} = 0 \tag{12.45}$$

The determinant of the coherence matrix is zero.

Examples

Linear polarization making an angle θ with the horizontal axis:

$$\mathbf{E} = \mathbf{u}\cos\theta + \mathbf{v}\sin\theta = \alpha\,\mathbf{u} + \beta\,\mathbf{v}$$

$$\mathbf{J}_\theta = \begin{bmatrix} \cos^2\theta & \sin\theta\cos\theta \\ \sin\theta\cos\theta & \sin^2\theta \end{bmatrix} \tag{12.46}$$

Horizontal polarization:

$$\mathbf{J}_H = \begin{bmatrix} 1 & 0 \\ 0 & 0 \end{bmatrix}$$

Vertical polarization:

$$\mathbf{J}_V = \begin{bmatrix} 0 & 0 \\ 0 & 1 \end{bmatrix}$$

Circular polarizations: $\alpha = 1, \quad \beta = \pm j$

$$\mathbf{J}_C = \begin{bmatrix} 1 & j \\ -j & 1 \end{bmatrix} \qquad \mathbf{J}_{C'} = \begin{bmatrix} 1 & -j \\ j & 1 \end{bmatrix} \qquad (12.47)$$

Remark: Any completely unpolarized wave may be considered as the sum of two cross polarized waves or as the sum of two inverse circularly polarized waves:

$$\mathbf{J} = \begin{bmatrix} \alpha_0^2 & 0 \\ 0 & \alpha_0^2 \end{bmatrix} = \alpha_0^2 \left(\begin{bmatrix} 1 & 0 \\ 0 & 0 \end{bmatrix} + \begin{bmatrix} 0 & 0 \\ 0 & 1 \end{bmatrix} \right)$$

$$= \frac{\alpha_0^2}{2} \left(\begin{bmatrix} 1 & j \\ -j & 1 \end{bmatrix} + \begin{bmatrix} 1 & -j \\ j & -1 \end{bmatrix} \right)$$

12.3.5 Stokes parameters (Stokes, 1819–1903)

These parameters are deduced from the coherence matrix $\mathbf{J}$ by decomposing it into its symmetric and antisymmetric parts (s_0, s_1 parameters), on one hand, and into its real and imaginary parts (s_2, s_3 parameters), on the other hand, as follows

$$\mathbf{J} = \begin{bmatrix} \overline{|\alpha|^2} & \overline{\alpha\beta^*} \\ \overline{\alpha^*\beta} & \overline{|\beta|^2} \end{bmatrix} = \frac{1}{2} \begin{bmatrix} s_0 + s_1 & s_2 + js_3 \\ s_2 - js_3 & s_0 - s_1 \end{bmatrix} a \qquad (12.48)$$

A Stokes vector, denoted $\mathbf{S}$ is also defined; its components are $\left(s_0, s_1, s_2, s_3 \right)$

$$\left. \begin{aligned} s_0 &= |\alpha|^2 + |\beta|^2 \quad \text{(total available power)} \\ s_1 &= |\alpha|^2 - |\beta|^2 \\ s_2 &= \overline{\alpha\beta^*} + \overline{\alpha^*\beta} = 2\mathrm{Re}\left[\overline{\alpha\beta^*} \right] \\ js_3 &= \overline{\alpha\beta^*} - \overline{\alpha^*\beta} = 2j\,\mathrm{Im}\left[\overline{\alpha\beta^*} \right] \end{aligned} \right\} \qquad (12.49)$$

12.3.6 Decomposition of a partially polarized wave

The introduction of these parameters allows us to show that any wave may be considered as the sum (in terms of power) of a depolarized wave and a polarized wave.

Let us set

$$\bar{\omega} = \frac{\sqrt{s_1^2 + s_2^2 + s_3^2}}{s_0} \tag{12.50}$$

Then, we can write

$$\mathbf{J} = \mathbf{J}_D + \mathbf{J}_P$$

with

$$\left. \begin{array}{c} \mathbf{J}_D = \dfrac{s_0}{2} \begin{bmatrix} 1 - \bar{\omega} & 0 \\ 0 & 1 - \bar{\omega} \end{bmatrix} \\[2em] \mathbf{J}_P = \dfrac{1}{2} \begin{bmatrix} \bar{\omega} s_0 + s_1 & s_2 + j s_3 \\ s_2 - j s_3 & \bar{\omega} s_0 - s_1 \end{bmatrix} \end{array} \right\} \tag{12.51}$$

It can be verified that the determinant of the matrix $\mathbf{J}_P$ is zero, as it must be since the wave is polarized. The power it transports is given by its trace, $\bar{\omega} s_0$. The first matrix $\mathbf{J}_D$ characterizes a depolarized wave; the power it transports is $s_0 (1 - \bar{\omega})$.

We are now in a position to define, in a intrinsic way (that is to say independently of the reference system) the *degree of polarization* as being the ratio of the polarized available power to the total available power

$$\frac{w_{\text{polarised}}}{w_{\text{total}}} = \frac{\mathbf{J}_P(\text{trace})}{\mathbf{J}(\text{trace})} = \frac{\bar{\omega} s_0}{s_0} = \bar{\omega} = \frac{\sqrt{s_1^2 + s_2^2 + s_3^2}}{s_0} \leq 1 \tag{12.52}$$

12.3.7 Geometrical interpretation of the preceding results: Stokes parameters and Poincaré sphere

Stokes parameters of a monochromatic polarized wave

Let us consider a polarized wave whose parameters are τ (ellipticity) and ψ (tilt angle) in an orthogonal Cartesian coordinate system (u, v) (Fig. 12.12).

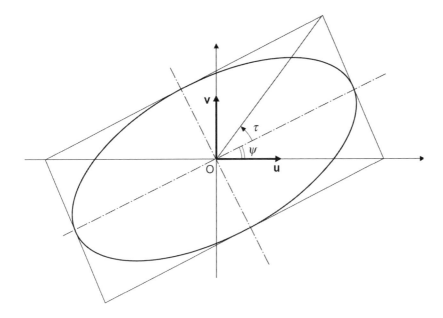

Fig. 12.12 Polarization ellipse and Stokes parameters.

Its components (α, β), (equations (12.18) and (12.19)) allow us to obtain

$$\left.\begin{array}{l} \alpha = \cos\psi\cos\tau + j\sin\psi\sin\tau \\ \beta = \sin\psi\cos\tau - j\cos\psi\sin\tau \end{array}\right\} \qquad (12.53)$$

From equation (12.49), we deduce the parameters related to the Stokes vector

$$\left.\begin{array}{l} s_0 = |\alpha|^2 + |\beta|^2 = 1 \\[2mm] s_1 = |\alpha|^2 - |\beta|^2 = \cos 2\psi \cos 2\tau \\[2mm] s_2 = \alpha\beta^* + \alpha^*\beta = \cos 2\psi \sin 2\tau \\[2mm] js_3 = \alpha\beta^* - \alpha^*\beta = j\sin 2\tau \end{array}\right\} \qquad (12.54)$$

It is easy to verify that (s_1, s_2, s_3) are the coordinates of a point E representing the polarization on the surface of the unit radius ($s_0 = 1$) Poincaré sphere (§12.2.6, Fig. 12.13). This allows us to find an interpretation for the points E′ inside the sphere

$$\mathrm{OE}' = \frac{r}{s_0} = \frac{\sqrt{s_1^2 + s_2^2 + s_3^2}}{s_0} = \overline{\omega} \le 1 \tag{12.55}$$

Therefore, their distance to the centre characterizes the intrinsic degree of polarization. In particular, the centre of the sphere represents a completely depolarized wave.

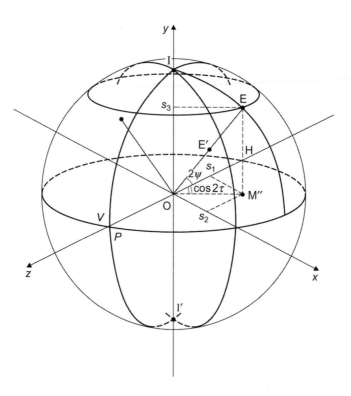

Fig. 12.13 Poincaré sphere. OE′ = $\overline{\omega}$.

12.3.8 Polarization coupling and Stokes vectors

Consider a receiving antenna, whose polarization is defined by the unit vector $\mathbf{h}$ or the column matrix

$$\mathbf{h} = \begin{bmatrix} \alpha \\ \beta \end{bmatrix} \quad ; \quad \overline{|\alpha|^2} + \overline{|\beta|^2} = 1$$

The corresponding Stokes vector $\mathbf{S}$ is

$$\mathbf{S} = \begin{cases} s_0 = 1 \\ s_1 \\ s_2 \\ s_3 \end{cases} \qquad r = \sqrt{s_1^2 + s_2^2 + s_3^2} = 1$$

Let us now consider a partially polarized wave, such that its available power is given the unit vector $\mathbf{E}$, or the column matrix

$$\mathbf{E} = \begin{bmatrix} \alpha'(t) \\ \beta'(t) \end{bmatrix} \quad ; \quad \overline{|\alpha'|^2} + \overline{|\beta'|^2} = 1$$

$$\mathbf{S}' = \begin{cases} s_0' = 1 \\ s_1' \\ s_2' \\ s_3' \end{cases} \qquad r' \leq 1$$

The average power received is given by the expression

$$\overline{w} = \overline{\left| \mathbf{h}'\mathbf{E}^* \right|^2} = \overline{\left| \alpha\alpha'^* + \beta\beta'^* \right|^2}$$

from which we obtain

$$\overline{w} = \frac{1}{2}\left(s_0 s_0' + s_1 s_1' + s_2 s_2' + s_3 s_3' \right) \tag{12.56}$$

which is half the scalar product of the Stokes vectors defined previously

$$\overline{w} = \frac{1}{2}\mathbf{S}.\mathbf{S}' \tag{12.57}$$

Remark

We can also write

$$\overline{w} = \frac{1}{2}(1 + s_1 s_1' + s_2 s_2' + s_3 s_3') \tag{12.58}$$

The scalar product of the vectors $(\mathbf{OE}.\mathbf{OE'})$ that characterize the two polarizations on the Poincaré sphere is the term between brackets (Fig. 12.13)

$$\overline{w} = \frac{1}{2}(1 + \mathbf{OE}.\mathbf{OE'}) \tag{12.59}$$

Typical cases :

Depolarized incoming wave:
$$|\mathbf{OE'}| = 0$$

$$\overline{w} = \frac{1}{2}$$

Half of the available power is received, independently of antenna polarization: this is the same result that was obtained in §12.3.3.

Polarized incident wave:
- Matched polarizations: $\mathbf{OE} = \mathbf{OE'}$ $w = 1$ (all available power is received)

- Cross-polarizations: $\mathbf{OE} = -\mathbf{OE'}$ $w = 0$ (power delivered is zero)

Partially polarized jammer with degree of polarization $\overline{\omega}$.
Minimum power delivered to the antenna is required. The chosen polarization at reception is represented by a unit vector (fully polarized wave)

$$\mathbf{OE} = -\frac{\mathbf{OE'}}{|\mathbf{OE'}|}$$

Then the power received will be

$$\overline{w} = \frac{1}{2}\left(1 - \frac{|\mathbf{OE'}|^2}{|\mathbf{OE'}|}\right) = \frac{1}{2}(1 - \overline{\omega})$$

Thus a completely polarized jamming signal can be suppressed ($\overline{\omega}=1$). On the other hand, if it is fully unpolarized only half of the power can be suppressed.

12.4 POLARIMETRIC REPRESENTATION OF RADAR TARGETS

12.4.1 Introduction

Applications of polarimetry are found in telecommunications, remote sensing and more specifically in radar. The analysis of polarized waves reflected by useful targets and parasitic echoes allows: the reduction of echo fluctuations, the improvement of detection and the improvement of contrast between useful and unwanted echoes. Ultimately, it also permits improved echo identification owing to the symmetrical properties revealed by the polarizations.

Even if we do not intend to develop here these important but specific subjects, some examples will be given in order to illustrate the notions introduced in the preceding sections. In the preceding pages, the antenna polarization and the polarization of the incident wave were considered independently. What characterizes the polarized wave reflected by a radar target is that it depends not only on the target but also on the polarization transmitted by the antenna.

The relationship between polarization transmitted and received will be studied using two new tools that we shall define now:

- the Sinclair matrix for completely polarized waves;

- the Mueller matrix for partially polarized waves.

12.4.2 Sinclair diffraction matrix

Definition

Let us consider an antenna transmitting in direction **w** towards the target and a polarized wave **E** in a Cartesian system (**u**, **v**) whose components (*a*, *b*) are represented by the following column matrix

$$\mathbf{E} = \begin{bmatrix} a \\ b \end{bmatrix}$$

The wave $\mathbf{E}'$ reflected by the target, in the same coordinate system, has the following components (a', b').

$$\mathbf{E}' = \begin{bmatrix} a' \\ b' \end{bmatrix}$$

It can be demonstrated that these components are linearly related to those of $\mathbf{E}$ by means of the $\mathbf{S}$ matrix

$$\mathbf{S} = \begin{bmatrix} s_{11} & s_{12} \\ s_{21} & s_{22} \end{bmatrix} \tag{12.60}$$

Using reciprocity it can be demonstrated that the matrix is symmetrical, therefore $\mathbf{S} = \mathbf{S}^t$ and we have

$$\begin{bmatrix} a' \\ b' \end{bmatrix} = \begin{bmatrix} s_{11} & s_{12} \\ s_{12} & s_{22} \end{bmatrix} \begin{bmatrix} a \\ b \end{bmatrix}$$

We remark that the $\mathbf{S}$ matrix depends on

- the target;
- the target attitude with respect to the radar;
- the frequency;
- the coordinate system used $(\mathbf{u}, \mathbf{v})$.

Characteristic polarizations

The preceding matrix relationship allows us to define some characteristic polarizations of the target with respect to the radar.
   In this manner, we define:

- characteristic polarizations (reflected and transmitted polarizations are identical);
- null polarizations (reflected and transmitted polarizations are orthogonal).

If $\mathbf{P}$ is the transmitted polarized wave, an identical reflected wave will be denoted $\mathbf{P}^*$. In effect, inverting the direction of wave propagation inverts the sign of the phase. Thus, we obtain

$$\mathbf{SP} = \gamma \mathbf{P}^* \tag{12.61}$$

where γ is the associated eigenvalue.
   This is a slightly different problem from the classical problem of the determination of the eigenvalues. Nevertheless, when the target presents certain symmetrical properties the characteristic polarizations become evident, for instance,

if the target has a plane of symmetry containing the radar, a linearly polarized wave parallel to this plane is necessarily an eigenpolarization.

In the general case we multiply the preceding equation by $\mathbf{S}^+ = \mathbf{S}^*$ and we proceed as follows.

$$\mathbf{S}^+\mathbf{SP} = \gamma\,\mathbf{S}^+\mathbf{P}^* = \gamma\,\mathbf{S}^*\mathbf{P}^* = \gamma\,(\mathbf{SP})^*$$

or

$$\left(\mathbf{S}^+\mathbf{S}\right)\mathbf{P} = \gamma\gamma^*\mathbf{P}$$

We are now brought to the classical problem of the eigenvalues of the hermitian and positive Graves matrix

$$\mathbf{G} = \mathbf{S}^+\mathbf{S} \qquad\qquad (12.62)$$

This, then, results in the following theorem

THEOREM: There always exists, at least, a couple (P_1, P_2) whose characteristic polarizations are orthogonal. Referred to the coordinate system $(\mathbf{P}_1, \mathbf{P}_2)$, the Sinclair matrix has the following form

$$\mathbf{S} = \begin{bmatrix} \gamma_1 & 0 \\ 0 & \gamma_2 \end{bmatrix}$$

$$(12.63)$$

where (γ_1, γ_2) are the associated eigenvalues.

Remark

The backscattered power is maximum (a situation very often required) if the transmitted characteristic polarization corresponds to the eigenvalue of maximum magnitude.

Example: cylindrical target (Fig. 12.14): The characteristic polarizations are, by symmetry, P_1 coplanar with the cylinder axis (eigenvalue γ_1) and the orthogonal polarization P_2.

Generally, we have $|\gamma_1| > |\gamma_2|$ particularly when the cylinder diameter is small.

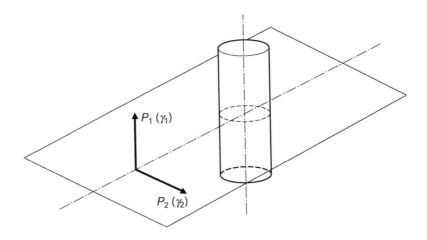

Fig. 12.14 Eigen-polarizations of a cylindrical radar target.

Expression of the Sinclair matrix referred to an orthogonal coordinate system
(u, v)

In the coordinate system of the characteristic polarizations the column matrices of the polarizations, the incident wave **E** and the reflected wave **E′** are related according to the relationship

$$\mathbf{E}' = \mathbf{SE} = \begin{bmatrix} \gamma_1 & 0 \\ 0 & \gamma_2 \end{bmatrix} \mathbf{E}$$

(12.64)

In any coordinate system $(\mathbf{u}, \mathbf{v})$, waves of the same polarization are represented by the column matrices **F** and **F′**, which are related to the preceding ones by means of a matrix **R** such that

$$\mathbf{E} = \mathbf{RF}$$

$$\mathbf{E}' = \mathbf{RF}'$$

Then, we have

$$\mathbf{RF'} = \mathbf{SRF}$$

$$\mathbf{R^{-1}RF'} = \mathbf{R^{-1}SRF}$$

$$\mathbf{F'} = \mathbf{R^{-1}SRF}$$

In the coordinate system $(\mathbf{u}, \mathbf{v})$ the Sinclair matrix is of the form

$$\mathbf{S'} = \mathbf{R^{-1}SR} \tag{12.65}$$

Example: rotation through an angle φ (which corresponds to the case of a symmetrical target whose characteristic polarizations are linear). In this case, we have

$$\mathbf{R} = \begin{bmatrix} \cos\varphi & -\sin\varphi \\ \sin\varphi & \cos\varphi \end{bmatrix} \tag{12.66}$$

Then we find

$$\left.\begin{aligned} s'_{11} &= \frac{\gamma_1 + \gamma_2}{2} + \frac{\gamma_1 - \gamma_2}{2}\cos(2\varphi) \\ s'_{22} &= \frac{\gamma_1 + \gamma_2}{2} - \frac{\gamma_1 - \gamma_2}{2}\sin(2\varphi) \\ s'_{12} &= \frac{\gamma_1 - \gamma_2}{2}\sin(2\varphi) \end{aligned}\right\} \tag{12.67}$$

Example: slanted linear wire

$$\gamma_1 = 1, \quad \gamma_2 = 0 \tag{12.68}$$

$$\mathbf{S'} = \begin{bmatrix} \cos^2\varphi & \sin\varphi\cos\varphi \\ \sin\varphi\cos\varphi & \sin^2\varphi \end{bmatrix}$$

Null polarizations

Which polarizations, denoted by $\mathbf{N}$, should the antenna transmit so that the returning polarization should be orthogonal? This situation arises, for instance, when certain types of echoes or signals are not desired (jammers).

The condition for orthogonality is written as follows

$$\mathbf{N}'\mathbf{N}' = 0$$

and

$$\mathbf{N}' = \mathbf{SN}$$

therefore

$$\mathbf{N}^t\mathbf{SN} = 0 \tag{12.69}$$

Let (α, β) be the unknown components of $\mathbf{N}$ in the eigenvector coordinate system. Then we have

$$[\alpha \quad \beta]\begin{bmatrix} \gamma_1 & 0 \\ 0 & \gamma_2 \end{bmatrix}\begin{bmatrix} \alpha \\ \beta \end{bmatrix} = 0$$

and

$$\gamma_1\alpha^2 + \gamma_2\beta^2 = 0$$

from which we obtain the following solutions.

$$\frac{\beta}{\alpha} = \pm j\sqrt{\frac{\gamma_1}{\gamma_2}} \tag{12.70}$$

For instance

$$\alpha = \sqrt{\gamma_2} \qquad \beta = \pm j\sqrt{\gamma_1} \tag{12.71}$$

Example: rectangular target orthogonal to the line of sight (Fig. 12.15) The characteristic polarizations are parallel to the target axes and the ratio γ_1/γ_2 is real and < 1.

The null polarizations are defined by

$$\mathbf{N} = \begin{bmatrix} \sqrt{\gamma_2} \\ \pm j\sqrt{\gamma_1} \end{bmatrix} \tag{12.72}$$

We obtain two elliptic polarizations whose major axes are parallel to the width of the target.

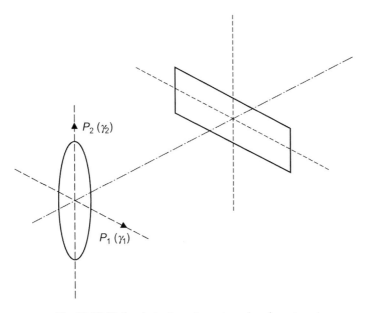

Fig. 12.15 Null polarization of a rectangular planar target.

Particular cases and target classification

Certain particular properties of the eigenvalues and of eigenvectors permit us to characterize and to classify the targets:

Linear targets: They correspond to those for which one of the eigenvalues of the **S** matrix is zero, for instance $\gamma_2 = 0$.
   Null polarizations are, then, identical

$$\mathbf{N} = \pm j\sqrt{\gamma_1}\begin{bmatrix} 0 \\ 1 \end{bmatrix}$$

(12.73)

The reflected polarization is independent of the transmitted polarization, such as is the case with a straight wire, which always radiates a linear polarization parallel to the wire. These targets are said to be 'linear'.
   Another example of a linear target is obtained from a helix oriented toward the radar. Such a target always radiates a circular polarization.
Isotropic targets: The magnitudes of the eigenvalues of isotropic targets are equal, $|\gamma_1| = |\gamma_2|$. In such a case, the total backward power is independent of the transmitted polarization. Therefore, we have

- transmitted polarization $\begin{bmatrix} \alpha \\ \beta \end{bmatrix}$

- reflected polarization $\begin{bmatrix} \alpha' \\ \beta' \end{bmatrix} = \begin{bmatrix} \gamma_1 & 0 \\ 0 & \gamma_2 \end{bmatrix} \begin{bmatrix} \alpha \\ \beta \end{bmatrix}$

The reflected power is

$$w = |\alpha'|^2 + |\beta'|^2 = |\gamma_1|^2 |\alpha|^2 + |\gamma_2|^2 |\beta|^2 = |\gamma_1|^2 \left(|\alpha|^2 + |\beta|^2 \right)$$

(1) $\gamma_1 = \gamma_2$: This corresponds to the case of a target having an axis of revolution aligned with the radar, for instance, a spherical target. This is approximately the case with water droplets comprising a cloud. The form of the **S** matrix is then

$$\mathbf{S} = \gamma_1 \begin{bmatrix} 1 & 0 \\ 0 & 1 \end{bmatrix}$$

(12.74)

We verify that all linear polarizations are eigenpolarizations. Null polarizations are the two circular polarizations.

This result is of great interest. In effect, to prevent wanted radar echoes (aircraft) from being buried in clutter echoes reflected by clouds or rain, circular polarization is used. Generally, the polarization of useful echoes is not circular.

(2) $\gamma_2 = -\gamma_1$: This case may be obtained by means of a rectangular corner reflector.

Such a reflector possesses the very interesting property of reflecting the energy in the direction of the transmitting radar, after double reflection from the two faces of the wedge (Fig. 12.16), and the echo becomes quite perceptible on the radar screen. Many radar passive beacons are of the corner reflector type, and can often be seen at the masthead of ships or on navigation buoys, to enhance their radar cross section.

It can be seen that the corner reflector, owing to its symmetrical particularities, has two linear orthogonal characteristic polarizations, one of which is parallel to the edge line.

It can be shown that the associated eigenvalues are opposite. In fact, when the polarization is parallel to the edge line of the dihedral angle $(\mathbf{P}_1)$ the incident electric field on each face is parallel to that face. The boundary conditions require that the resultant field should be zero, therefore a phase inversion of the reflected field occurs. The double reflection produces a double inversion and the reflected beam recovers its initial sign (its phase shift is due to the path-difference of its trajectory).

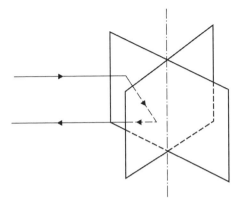

Fig. 12.16 Rectangular corner reflector.

When the polarization is orthogonal to the edge line (P_2), the reflected field remains in the incidence plane. The polarization of the field undergoes a double rotation of 90°, therefore a polarity inversion occurs (Fig. 12.17).

Then the Sinclair matrix may be written

$$S = \gamma_1 \begin{bmatrix} 1 & 0 \\ 0 & -1 \end{bmatrix}$$

(12.75)

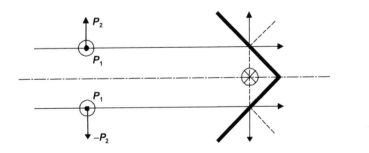

Fig. 12.17 Wedge-shaped dihedral reflector.

Circular incident polarization: Such a polarization is represented by the matrix

$$\begin{bmatrix} 1 \\ j \end{bmatrix}$$

The reflected polarization matrix is

$$\begin{bmatrix} 1 \\ -j \end{bmatrix}$$

In order to take into account the direction reversal we must consider the complex conjugate and we find the incident circular polarization. The circular polarizations are then eigenpolarizations. This property underlines the interest of corner reflectors for the realization of radar beacons. They are particularly easily detectable by circularly polarized radars as well as vertically and horizontally polarized radars.

Null polarization of a corner reflector: The polarization column matrices are written as follows

$$\begin{bmatrix} \sqrt{\gamma_2} \\ \pm j\sqrt{\gamma_1} \end{bmatrix} = \begin{bmatrix} \pm j\sqrt{\gamma_1} \\ \pm j\sqrt{\gamma_1} \end{bmatrix} = j\sqrt{\gamma_1} \begin{bmatrix} 1 \\ \pm 1 \end{bmatrix}$$

These correspond to linear polarizations tilted at 45° with respect to the vertical axis.

Symmetrical targets: We already have presented an example of a symmetrical cylinder, in order to put in concrete form the notions of eigenvectors and eigenvalues. Generally a symmetrical target is characterized by orthogonal linear characteristic polarizations of any eigenvalue. We have deduced that the null polarizations have opposite ellipticities Any target having a symmetrical plane passing through the radar is symmetrical. This is true for all targets which possess an axis of revolution, such as the cylinder. This category includes isotropic targets as well as some particular linear targets (straight wires).

Classification of simple targets: Table 12.1 summarizes the main properties of simple targets. The transmitted and reflected polarizations are denoted, respectively, by **E** and **E′** .

Target type	*Linear*	*Symmetrical*	*Isotropic*
Definition	**E′** fixed independent of **E**	Linear orthogonal eigen-polarizations	Reflected power density independent of **E**
Null polarizations	Superposed, orthogonal to **E**	Inverted elliptical	Orthogonal
Eigen-polarizations	Reflected polarization is fixed	See definition	Orthogonal eigen-polarizations
Examples	Straight wire (polarization fixed in the plane of the wire); helix (circular right or left according to helix sense), fixed polarization	Target possessing a plane of symmetry passing through the radar. Any target having an axis of revolution	Target with axis of revolution coaxial with radar axis; rectangular corner reflector

Table 12.1 Main properties of simple radar targets.

12.5 PARTIALLY POLARIZED WAVES: THE MUELLER MATRIX

12.5.1 The Mueller matrix

In most cases the target is not constant, but changes with respect to the radar and alters its shape, particularly when it is composed of a number of independent elements (multiple targets, ground and atmospheric clutter). Then, the resulting reflected wave fluctuates quite rapidly and can be partially or totally unpolarized. The Sinclair matrix is not well suited to this case. It is better to describe the incident and reflected polarizations by means of the Stokes vectors $\mathbf{S}$ and $\mathbf{S}'$ whose coordinates represent points on the Poincaré sphere (Fig. 12.13). Here again the relationships between the components are linear. The Stokes vectors are interrelated by a 4×4 Mueller matrix. Then, we have

$$\mathbf{S}' = \mathbf{MS}$$

or

$$\begin{bmatrix} s'_0 \\ s'_1 \\ s'_2 \\ s'_3 \end{bmatrix} = \mathbf{M} \begin{bmatrix} s_0 \\ s_1 \\ s_2 \\ s_3 \end{bmatrix}$$

(12.77)

12.5.2 Application example

Chaff cloud

In order to prevent radar detection the target may be buried in an artificial cloud of chaff, which consist of metal dipoles having random distribution orientation. What are the properties of such clouds regarding polarization? In order to answer this question let us consider a single dipole. Consider it as transverse and making an angle φ with the horizontal axis. This angle is a random variable whose probability density is constant

$$p(\varphi) d\varphi = \frac{1}{2\pi} d\varphi$$

We have seen that the elements of the Sinclair matrix related to this dipole are given by equation (12.68)

$$s_{11} = \cos^2 \varphi \qquad s_{22} = \sin^2 \varphi \qquad s_{12} = \sin \varphi \cos \varphi$$

The scattered polarization $\mathbf{E}'$ is related to the incident polarization by the Sinclair matrix

$$\begin{bmatrix} \alpha' \\ \beta' \end{bmatrix} = S \begin{bmatrix} \alpha \\ \beta \end{bmatrix}$$

from which the coherence matrix $\mathbf{J}$ may be deduced

$$\mathbf{J} = \begin{bmatrix} \alpha' \\ \beta' \end{bmatrix} \begin{bmatrix} \alpha'^* & \beta'^* \end{bmatrix} = \begin{bmatrix} |\overline{\alpha'}|^2 & \overline{\alpha'\beta'^*} \\ \overline{\alpha'^*\beta'} & |\overline{\beta'}|^2 \end{bmatrix}$$

Its components may be obtained from the second-order average of the matrix elements (we suppose them to be identical to the statistical average over the entire dipole cloud)

$$\left.\begin{aligned} \overline{|s_{11}|^2} &= \int_0^{2\pi} |s_{11}(\varphi)|^2 \, p(\varphi) d\varphi = \frac{1}{2\pi} \int_0^{2\pi} \cos^4 \varphi \, d\varphi = \frac{3}{8} = \overline{|s_{22}|^2} \\[2mm] \overline{|s_{12}|^2} &= \cdots = \frac{1}{8} = \overline{|s_{11}s_{22}^*|} \end{aligned}\right\} \quad (12.78)$$

from which we can deduce

$$\left.\begin{aligned} \overline{|\alpha'|^2} &= \frac{3\overline{|\alpha|^2} + \overline{|\beta|^2}}{8}, \qquad \overline{|\beta'|^2} = \frac{\overline{|\alpha|^2} + 3\overline{|\beta|^2}}{8} \\[2mm] \overline{\alpha'\beta'^*} &= \frac{1}{4} \mathrm{Re}\left(\alpha\beta^*\right) \end{aligned}\right\} \quad (12.79)$$

Then, we obtain the Stokes parameters of the scattered wave

$$s_0' = \overline{|\alpha'|^2} + \overline{|\beta'|^2} = \ldots \frac{\overline{|\alpha|^2 + |\beta|^2}}{2} = \frac{s_0}{2}$$

$$s_1' = \ldots \frac{\overline{|\alpha|^2 - |\beta|^2}}{4} = \frac{s_1}{4}$$

$$s_2' = \frac{1}{4}\mathrm{Re}\left(\alpha\beta^*\right) = \frac{s_2}{4}$$

$$s_3' = 0$$

(12.81)

from which we obtain the Mueller matrix

$$\mathbf{M} = \begin{bmatrix} 1/2 & 0 & 0 & 0 \\ 0 & 1/4 & 0 & 0 \\ 0 & 0 & 1/4 & 0 \\ 0 & 0 & 0 & 0 \end{bmatrix}$$

(12.82)

12.5.3 Examples of responses to different incident polarizations (Fig. 12.18)

Circular polarization: $\alpha = 1$, $\beta = j$

Diffracted coherence matrix:

$$\mathbf{J} = \frac{1}{2}\begin{bmatrix} 1 & 0 \\ 0 & 1 \end{bmatrix}$$

(12.83)

The wave is completely unpolarized
Let us check the Stokes parameters and their images in the Poincaré sphere:

- incident wave: $s_0 = 2$, $s_1 = s_2 = 0$, $s_3 = -1$ (south pole) - Pole C;

- scattered wave: $s_1'/s_0' = 0$, $s_2'/s_0' = 0$, $s_3'/s_0' = 0$ (centre O).

Linear polarization: $\alpha = 1$, $\beta = 0$ (point L on the equator of the Poincaré sphere)
Coherence matrix:

$$\mathbf{J} = \begin{bmatrix} 3/8 & 0 \\ 0 & 1/8 \end{bmatrix}$$

(12.84)

We observe three more times power in the vertical (incident) component than in the horizontal component. The corresponding signals are uncorrelated.

The Stokes parameters are:

- incident wave: $s_0 = 1$, $s_1 = 1$, $s_2 = 0$, $s_3 = 0$;a

- diffracted wave: $s_1'/s_0' = 1/2$, $s_2'/s_0' = 0$, $s_3'/s_0' = 0$.

We obtain a partially polarized wave whose intrinsic polarization degree is

$$p = \sqrt{\left(\frac{s_1'}{s_0'}\right)^2 + \left(\frac{s_2'}{s_0'}\right)^2 + \left(\frac{s_3'}{s_0'}\right)^2} = \frac{1}{2}$$

(12.85)

It is represented on the Poincaré sphere by the vector **OA** (Fig. 12.18) such that **OA/OL** $= 1/2$.

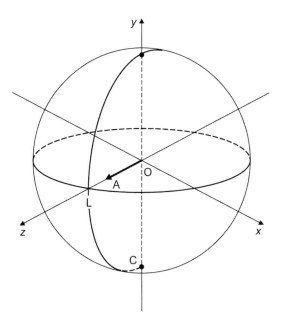

Fig. 12.18 Stokes vector of a chaff cloud.

12.6 POLARIZERS AND POLARIZATION SEPARATORS FOR TELECOMMUNICATIONS ANTENNAS AND POLARIMETRIC RADARS

12.6.1 Introduction

In the introduction to this chapter (§12.1), the application was explained of polarimetric techniques to radars and to telecommunications ('frequency re-use techniques'). In this section we consider the structure and principles of various polarizing elements and polarization separators, used particularly in primary feeds, but also sometimes in phased array elements.

12.6.2 Non-symmetrical polarization separator

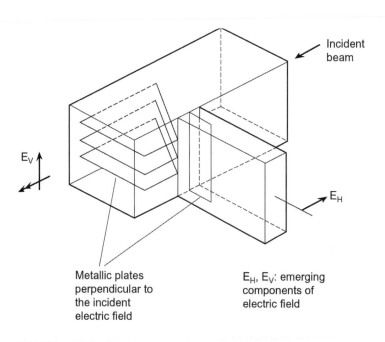

Fig. 12.19 Polarization separator for a polarimetric receiver (non-symmetrical version).

A polarization separator is used on receive to compare the amplitudes and phases of orthogonal polarization components. The parameters of the polarization ellipse can be deduced using the methods of §12.2 (equation (12.22)).

The non-symmetric polarization separator (Fig. 12.19) is simple, and can be used in narrow-band systems. It consists of a square waveguide carrying the incident signal (from a symmetric primary feed: square horn or circular horn, preferably dual-mode or multi-mode). A set of parallel horizontal metal fins filter out the vertical component E_V, but reflect the horizontal component E_H into a lateral waveguide made up of parallel vertical fins filtering out the horizontal component. The impedances presented and the transmission coefficients are different, in both amplitude and phase, for the two channels, and need to be taken into account in calculating the parameters of the polarization ellipse of the incident signal. This is relatively easy for narrow-band systems.

12.6.3 Semi-symmetrical polarization separator (Fig. 12.20)

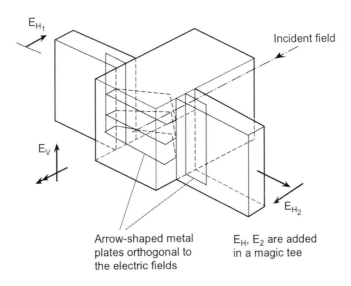

Fig. **12.20** Polarization separator for a polarimetric receiver (semi-symmetrical version).

As with the previous separator, the incident field is fed to a square waveguide with a set of parallel fins which filter the vertical component E_V. However, these fins are symmetrically tapered to separate the horizontal component into two equal parts E_{H1} and E_{H2} into two lateral guides. The horizontal components E_{H1} and E_{H2} are summed in a tee (preferably a magic tee which absorbs the antisymmetric components due to imperfections in fabrication). This tee is not shown in Fig. 12.20.

12.6.4 Symmetrical polarization separator (turnstile)

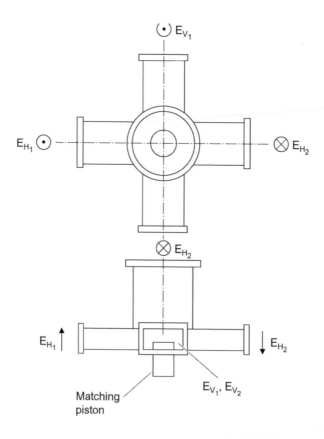

Fig. 12.21 'Turnstile' polarization separator.

This separator is shown in Fig. 12.21 in two projection planes. The incident field is fed to a symmetric waveguide; this is of circular cross-section in the figure, but it is also possible to use a square cross-section. The waveguide is closed by a symmetrical adjustable matching piston. The vertical and horizontal components are both separated and fed to four symmetrically-placed rectangular waveguides. The vertical components (E_{V1}, E_{V2}) on one side, and the horizontal components (E_{H1}, E_{H2}) on the other, are summed in two magic tees (not shown). Here the insertion losses are the same for the two components, and this structure is well suited to operation over a broad bandwidth.

12.6.5 Dielectric vane polarizer (Fig. 12.22)

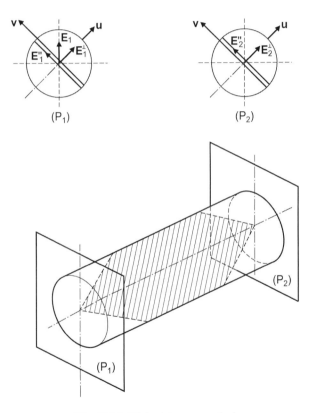

Fig. 12.22 Dielectric vane polarizer.

Suppose here that the antenna is operating on transmit. The vertical transmitted polarization E_1 must be transformed into a circular polarization before being fed to a symmetrical radiating element (a corrugated horn, for example).

For this we use a section of circular waveguide with a dielectric vane inclined at 45° with respect to the incident polarization (Fig. 12.22), whose length ℓ is adjusted so that between a wave of polarization parallel to the plane of the sheet and a wave perpendicular to this plane, the total phase shift between these planes P_1 and P_2 along the sheet is $\pi/2$ (i.e. 90°).

Under these conditions, let us decompose the transmitted signal $\mathbf{E}_1$ in the plane P_1 into two orthogonal components, one ($\mathbf{E}_1^{\|}$) parallel to the vane and the other ($\mathbf{E}_1^{\perp}$) perpendicular, such that

$$\mathbf{E}_1 = \mathbf{E}_1^{\perp} + \mathbf{E}_1^{\|} \tag{12.86}$$

The emerging signals in the plane P_2, $\mathbf{E}_2^{\perp}$ and $\mathbf{E}_2^{\|}$, are equal, and phase shifted, one with respect to the other, by $\pi/2$. Here $\mathbf{u}$ and $\mathbf{v}$ are orthogonal unit vectors, respectively perpendicular and parallel to the vane, so that we can write :

$$\mathbf{E}_2^{\perp} = A\mathbf{u} \ , \qquad \mathbf{E}_2^{\|} = jA\mathbf{v} \tag{12.87}$$

If there are no losses, conservation of energy implies that

$$A = \frac{|\mathbf{E}_1|}{\sqrt{2}} \exp(-j\varphi) \tag{12.88}$$

where φ is the insertion phase of the vane for a wave parallel to $\mathbf{u}$. The total emerging field is

$$\mathbf{E}_1 = \mathbf{E}_2^{\perp} + \mathbf{E}_2^{\|} = A(\mathbf{u} + j\mathbf{v}) = |\mathbf{E}_1| \exp(-j\varphi) \left(\frac{\mathbf{u} + j\mathbf{v}}{\sqrt{2}} \right) \tag{12.89}$$

This is a circularly-polarized wave. We can verify that energy is conserved, since $|\mathbf{E}_2|^2 = |\mathbf{E}_1|^2$.

FURTHER READING

1. Huynen, J.R., 'Phenomenological theory of radar targets', Chapter 11 of *Electromagnetic Scattering*, P.L. Uslenghi (ed.), Academic Press, New York, 1978.
2. Ulaby, F. T. and Elachi, C. (eds), *Radar Polarimetry for Geoscience Applications*, Artech House, Dedham, MA, 1990.
3. Proceedings of the 16th ESA Workshop on Dual Polarization Antennas, 8-9 June 1993, ESTEC, ESA Scientific & Technical Publications Branch, Noordwijk, The Netherlands
4. Note also the series of conferences on radar polarimetry organized at IRESTE, Nantes, France, entitled *Journées Internationales de la Polarimetrie Radar.*

EXERCISES

Q1 Consider the dielectric vane polarizer (§12.6.5). Suppose that the angle between the incident polarization $\mathbf{E}_1$ in the plane P_1 and the dielectric vane is different from $\pi/4$, so that

$$\alpha = \frac{\pi}{4} + \varepsilon$$

What is the polarization of the emerging wave ?

ANSWER

$$\mathbf{E}_2 = |\mathbf{E}_1| \exp(-j\varphi)(\mathbf{u}\cos\alpha + j\mathbf{v}\sin\alpha)$$

This is an ellipse with principal axes $(\mathbf{u}, \mathbf{v})$ and of ellipticity

$$\tan\alpha = \tan\left(\frac{\pi}{4} + \varepsilon\right) = \frac{1 + \tan\varepsilon}{1 - \tan\varepsilon}$$

13
Antennas and signal theory

Claude Shannon (1916–2001)

13.1 INTRODUCTION

The subject of signal theory was developed particularly after the Second World War [Shannon, Wiener]; developments in telecommunications, sonar and radar opened numerous fields of applications. Using this theory, the ability of a system to transmit information may be quantitatively evaluated. About the same time, optical engineers had the idea of treating an optical image as a 2-D spatial message and to evaluate the quantity of information that it contained. In 1947, Duffieux[1] developed on a theoretical and experimental basis the harmonic analysis of images; optical instruments could be evaluated by their ability to reproduce periodic test patterns of shorter and shorter spatial periods, that is to say, of higher and higher *spatial frequency*, until a cut-off frequency is reached beyond which the image is no longer reproduced.

---

[1] Duffieux, internal publication, University of Besançon, France.

The Fourier transform leads to the conclusion that an image provided by an optical instrument, or an antenna, results from a filtering of the object by a low pass filter.

Nevertheless, post-processing allows this image to be refined if we take into account the amount of *a priori* information that we possess about the object; this is the domain of 'signal processing antennas'. Radio astronomers have often played a pioneering role in this domain by devising methods which have subsequently been used in other domains (radar, sonar etc.) [Chapter 14].

In this chapter, we shall see the role of signal theory in antenna design, and the limits and synthesis possibilities for antennas.

13.2 EQUIVALENCE OF AN APERTURE AND A SPATIAL FREQUENCY FILTER

13.2.1 Concept of spatial frequency

We know that the far field radiated by an aperture is expressed in terms of illumination function by means of a Fourier transform. We recall here the results of Chapter 7 (§7.4.3, §7.4.4, Fig. 7.29).

The coordinates (x, y) of a point M of the aperture are written in terms of the wavelength (see equation (7.27))

$$x/\lambda = v, \qquad y/\lambda = \mu \tag{13.1}$$

The radiation in the direction of the unit vector **u** is expressed as a function of the direction cosines (α, β) in the form (see equation (7.28))

$$F_T(\alpha, \beta) = \iint_S f_T(v, \mu) \exp\left[j2\pi(\alpha v + \beta \mu) \right] dv d\mu \tag{13.2}$$

Since the illumination f is zero outside the aperture the integral can formally be extended to infinity. Note that this integral is of the form of a classical Fourier transform.

Consider the case of a rectangular aperture of extent D, D' and whose illumination function is separable. The double integral is decomposed into the product of two single integrals (equation (7.39)) in the form

$$F_1(\alpha) = \int_{-v_0}^{+v_0} f_1(v) \exp(j2\pi v \alpha) dv \tag{13.3}$$

$$F_2(\beta) = \int_{-\mu_0}^{+\mu_0} f_2(\mu) \exp(j2\pi\mu\beta) d\mu \qquad (13.4)$$

with

$$v_0 = \frac{D}{2\lambda}, \quad \mu_0 = \frac{D'}{2\lambda} \qquad (13.5)$$

Equation (13.3) also corresponds to a linear aperture such as a continuous linear array (§11.3.1). In such a case we use, as for arrays, the notation

$$\tau = \sin\theta \qquad (13.6)$$

where θ is the angle between the direction of unit vector **u** and the normal to the array of unit vector **n**.

Equation (13.3) is a Fourier transform *identical to that relating a time-domain signal $F(\tau)$ to its frequency spectrum $f(v)$ whose frequency bounds are $-v_0$ and $+v_0$*. Here the function $F(\tau)$ may be complex. The parameter $v_0 = D/2\lambda$ corresponds to the cut-off frequency of a filter; that is why we term the parameter $v = x/\lambda$ (the abscissa of an aperture point expressed in wavelengths) the 'spatial frequency'. *This analogy allows us to transpose to the antenna and optics domain most of the properties related to filtering and processing of time-domain signals.*

Note that the symmetry of the Fourier transform makes the designation of the parameter playing the role of frequency somewhat arbitrary. The choice made so far is legitimate in the sense that it is the parameter v_0, determined by the antenna dimensions, which limits the capacity of the antenna regarding the transfer of angular information.

We proceed with the study of the filter equivalent to an antenna by using the linear aperture model expressed by equation (13.3). This model allows us to emphasize the essential properties that can easily be generalized subsequently.

13.2.2 Consequences of the limitation of the aperture dimensions on the properties of the radiation characteristic function

Real or visible domain and imaginary or invisible domain

Since any practical aperture is bounded, its radiation is represented by a function $F(\tau)$ whose spectrum is limited to the closed interval $(-v_0, +v_0)$. It follows that the function $F(\tau)$ *is not bounded*. Now, the parameter $\tau = \sin\theta$ only leads to real directions θ of space when $|\tau| \leq 1$. The necessity of taking into consideration other values for calculation purposes leads to the introduction of *imaginary* radiation directions

$$\theta = \pm \frac{\pi}{2} + j\theta' \tag{13.7}$$

In effect, we have

$$|\tau| = \cosh \theta' \geq 1 \tag{13.8}$$

which is why in the representation of the pattern $F(\tau)$, we separate the *visible* domain ($|\tau| \leq 1$) from the *invisible* domain ($|\tau| > 1$). This interpretation allows us to express the properties of waves whose phase varies rapidly along the aperture in terms of surface waves. Also, it allows us to clearly express the features of so-called superdirective antennas (§13.4). There are other applications of this representation in the field of arrays (§11.3).

Beams in the invisible domain (imaginary space) are reactive, and correspond to reactive power stored in the vicinity of the antenna, i.e. as evanescent surface waves.

Application of the sampling theorem

Since the radiation pattern $F(\tau)$ possesses a bounded spectrum $(-v_0, +v_0)$, it follows that it can be sampled with a sampling interval $\delta\tau$ such that

$$\delta\tau = \frac{1}{2v_0} = \frac{\lambda}{D} \tag{13.9}$$

and $F(\tau)$ is obtained from the values in the sampled directions using the classical Whitaker interpolation formula

$$F(\tau) = \sum_{n=-\infty}^{+\infty} F\left(\frac{n}{2v_0}\right) \frac{\sin(2\pi v_0 \tau - n\pi)}{2\pi v_0 \tau - n\pi} \tag{13.10}$$

Fig. 13.1 represents a sampled pattern of this kind.

The functions

$$\frac{\sin(x - n\pi)}{(x - n\pi)}$$

are mutually orthogonal, thus the maximum of one of them coincides with the nulls of all the others. The correct sampling of $F(\tau)$ requires an infinite set of samples extending over the invisible domain.

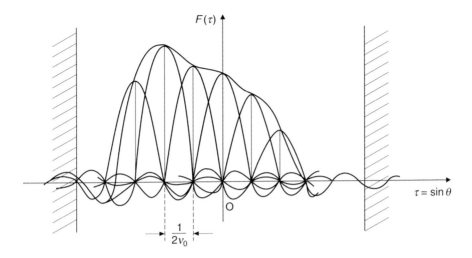

Fig. 13.1 Sampling of a radiation pattern.

Interpretation of the sampling of the illumination function: The linearity of the Fourier transform allows us to write the illumination function $f(v)$, in the domain $(-v_0, +v_0)$, in the form of a Fourier series

$$f(v) = \sum_{n=-\infty}^{+\infty} F\left(\frac{n}{2v_0}\right) \exp\left(j2\pi v \frac{n}{2v_0}\right) \qquad (13.11)$$

Thus, the illumination may be considered as the superposition of a set of distinct plane waves in the directions of sampling, and whose amplitudes are the sampled values of the pattern. We find again a representation of the illumination in the form of a plane wave spectrum (§ 2.1), but here it is a 'ray spectrum' which is only valid inside the aperture $\left(|v| \leq v_0\right)$.

This form of the illumination is the basis of the methods of pattern synthesis (§13.3).

Number of degrees of freedom: The notion of the number of degrees of freedom of an aperture allows us to easily express the limitation imposed on the antenna performance by its finite dimensions. If, on the other hand, the pattern radiated by the aperture in a certain angular domain (τ_1, τ_2), for instance the whole visible domain $(-1, +1)$, we see that the form of the patterns depends particularly on the samples located in the corresponding domain (the others only intervene via their sidelobes). According to this point of view, we can say that the pattern only possesses a finite number of degrees of freedom, that is to say a finite number of samples in the corresponding angular domain

$$N = \frac{|\tau_2 - \tau_1|}{\delta \tau} = 2v_0 |\tau_2 - \tau_1| \qquad (13.12)$$

For the whole visible domain, $N = 4v_0 = 2D/\lambda$. This severely limits the possibilities for the shape of the radiated pattern. Nevertheless, to be rigorous we should consider the total number of samples, as will be done for superdirective antennas (§13.3.2).

Application of Bernstein's theorem

If F_1 is an upper limit of the radiation function $F(\tau)$, it follows from Bernstein's theorem that the relative slope of the pattern is limited by the spatial frequency bandwidth of the antenna

$$\frac{1}{F_1} \frac{dF}{d\tau} \le 2\pi v_0 = \pi \frac{D}{\lambda} \qquad (13.13)$$

This slope limitation plays an important role, such as with the angular sensitivity of tracking antennas (§10.2). This is also the case with radar antennas where it is desired that the gain falls rapidly to zero below the horizon to avoid ground clutter. More generally, Bernstein's theorem implies a limit to all derivatives of bounded signals with bounded spectra

$$\frac{1}{F_1} F^{(n)}(\tau) \le (2\pi v_0)^n \qquad (13.14)$$

We can deduce a relationship between the dimension D of the aperture and the half-power beamwidth $\theta_{3\,\mathrm{dB}}$ of the main lobe of the radiation pattern.

In the vicinity of the main lobe, the pattern may be treated as a parabola (expanded to second order). We have

$$F(\tau) = F(0) - \frac{\tau^2}{2} F''(0)$$

Since $F''(0)$ is bounded according to Bernstein's theorem, we find for the half-power beamwidth $\theta_{3\,\mathrm{dB}}$

$$\theta_{3\,\mathrm{dB}} \ge \frac{1}{2} \frac{\lambda}{D} \ (\text{radians}) \qquad (13.15)$$

The angular width of the main lobe cannot be smaller than a certain value. In fact, due to the desire for low sidelobes, in general we have

$$\theta_{3dB} \geq \frac{\lambda}{D}$$

(13.16)

This leads to a limitation of the antenna *resolving power*. A rule formulated by Lord Rayleigh (1879) proposed that two directions cannot be distinguished if they are separated by an angular width less than λ/D. In fact, we shall see that we cannot assign an absolute limit to the resolution power, which should actually be expressed in terms of probability by taking into account the accuracy of the measurements and the particular context (Chapter 14).

13.2.3 Consequences of the limitation of the aperture dimensions on the 'gain' function of the antenna

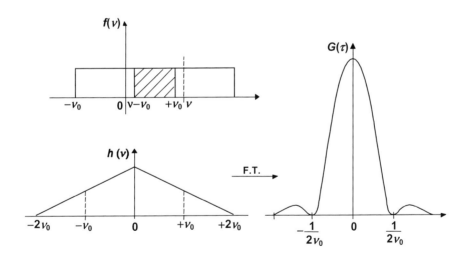

Fig. 13.2 Autocorrelation function of uniform illumination of a linear aperture. The gain is its Fourier Transform.

The gain function of an antenna is proportional to the square of the magnitude of its characteristic function

$$G(\mathbf{u}) = k^2 |F(\mathbf{u})|^2 \tag{13.17}$$

(**u** is the unit vector of a direction in space)

The function G, just like the function F, possesses a spatial frequency spectrum $h(v)$ which is bounded. This spectrum is the autocorrelation function of the aperture illumination function

$$h(v) = \int_{-\infty}^{+\infty} f(\mu) f^*(v - \mu) d\mu \tag{13.18}$$

This results from the properties of the Fourier transform of the square magnitude of a signal

$$G(\tau) = k^2 \int_{-\infty}^{+\infty} h(v) \exp(j2\pi v\tau) dv \tag{13.19}$$

Consequence: spatial frequency spectrum bandwidth of the 'gain' function: Since the illumination function of the aperture is limited to a closed interval $(-v_0, +v_0)$, it follows that the autocorrelation function $h(\tau)$ is limited to twice that interval $(-2\, v_0, +2\, v_0)$.

The example for a uniform illumination is illustrated in Fig. 13.2.

13.3 SYNTHESIS OF AN APERTURE TO RADIATE A GIVEN RADIATION PATTERN

In many practical applications, a design engineer is confronted with the problem of the synthesis of an antenna with a pattern as close as possible to an ideal pattern defined by the operational target. This is the case with surveillance radar and satellite telecommunication antennas etc. In each case, a more or less arbitrary approximation criterion is considered. This leads to an optimal result that takes into account the various constraints imposed.

13.3.1 Statement of problem

Let us start here with the linear aperture model. The dimension of the aperture is D, thus it is equivalent to a filter of finite bandwidth $(-v_0, +v_0)$. An ideal pattern $F_0(\tau)$ is desired. Is it possible to define an illumination $f_0(v)$ leading to the pattern F_0? If not, what is the physically realizable function $f_1(v)$ that gives a pattern $F_1(\tau)$ as close as possible to F_0?

If f_0 is the solution to the problem, f_0 is necessarily the Fourier transform of F_0. In order to be physically realizable f_0 must be bounded on the closed interval $(-v_0, +v_0)$.

Considering the important and particular case where the required pattern F_0 is bounded (pattern equal to zero outside a certain interval) it follows that f_0 is not bounded and therefore is not physically realizable. If we search for a better realizable function f_1, it is necessary to define a deviation criterion, i.e. a permissible error between F_0 and F_1. The criterion of the minimum mean square error leads to a relatively simple solution. This means minimizing the following integral.

$$I = \int_{-\infty}^{+\infty} |F_0 - F|^2 d\tau \qquad (13.20)$$

Thus, according to Parseval's theorem

$$I = \int_{-\infty}^{+\infty} |f_0 - f|^2 dv \qquad (13.21)$$

Since f is zero outside $(-v_0, +v_0)$, the result is obtained if $f = f_1 \equiv f_0$ over this interval. Thus, the best approximation is obtained with an *illumination f_1 which is the Fourier transform of the ideal pattern over the antenna aperture*. The corresponding pattern F_1 is the Fourier transform of this illumination

$$F_1 = \int_{-v_0}^{+v_0} f_1 \exp(j2\pi v\tau) dv = \int_{-\infty}^{+\infty} f_0 \mathrm{rect}\left(\frac{v}{v_0}\right) \exp(j2\pi v\tau) dv \qquad (13.22)$$

which is then the convolution product

$$F_1 = \int_{-\infty}^{+\infty} F_0(\tau') \frac{\sin[2\pi v_0(\tau-\tau')]}{2\pi v_0(\tau-\tau')} d\tau' \qquad (13.23)$$

Example: F_0 is a rectangular pattern: $F_0 = \mathrm{rect}(\tau/\tau_0)$. We obtain

$$f_0 = \frac{\sin(2\pi v\tau_0)}{2\pi v\tau_0}, \quad f_1 = f_0 \mathrm{rect}\left(\frac{v}{v_0}\right) \qquad (13.24)$$

$$F_1 = \int_{-\tau_0}^{+\tau_0} \frac{\sin[2\pi v_0(\tau-\tau')]}{2\pi v_0(\tau-\tau')} d\tau' = \mathrm{si}[2\pi v_0(\tau-\tau_0)] - \mathrm{si}[2\pi v_0(\tau+\tau_0)] \qquad (13.25)$$

where $\mathrm{si}(X) = \int_0^X \frac{\sin x}{c} dx$.

Fig. 13.3 represents F_0 and F_1 for $\nu_0 = 4$, $\tau_0 = 1/2$.

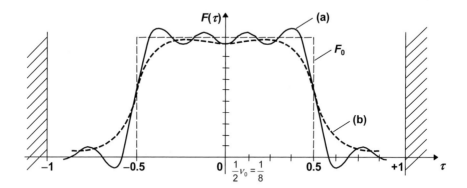

Fig. 13.3 Synthesis of a radiation pattern: (a) least mean square criterion; (b) use of Fejer triangular weighting function.

Disadvantage of the method - Gibbs' phenomenon

This method can locally give very significant deviations between F_0 and F_1 particularly in the vicinity of discontinuities. In the preceding example, it can be shown that if ν_0 tends to infinity, $F_1(\tau_0)/F_0(\tau_0)$ tends to 1.09 (Gibbs' phenomenon).

13.3.2 Generalization of the approximation method

The approximation F_1 of F_0 is the filtering of F_0 by means of a filter of transfer function constant over the bandwidth $(-\nu_0, +\nu_0)$ (i.e. $\mathrm{rect}\,\nu/\nu_0$). We can consider replacing this function by a different filtering function $T(\nu)$ with the same bandwidth. If $\Psi(\tau)$ is its Fourier transform, the approximation obtained is expressed by the convolution

$$F_T(\tau) = \int_{-\infty}^{+\infty} F_0(\tau')\Psi(\tau-\tau')d\tau' \tag{13.26}$$

$T(\nu)$ and $\Psi(\tau)$ respectively play the role of the transfer function and of the filter impulse response. The aperture illumination function to be realized is then

$$f_T(v) = f_0(v)T(v) \tag{13.27}$$

To avoid ripple and sidelobes, we intuitively look for a function Ψ with low sidelobes (taper function) which can be obtained, as we know, by a $T(v)$ function of the 'bell' type. The choice of a triangular function leads to the so-called Fejer approximation

$$T(v) = 1 - \frac{|v|}{v_0} \qquad \Psi(\tau) = \left(\frac{\sin(\pi v_0 \tau)}{\pi v_0 \tau}\right)^2 \tag{13.28}$$

Numerical example:

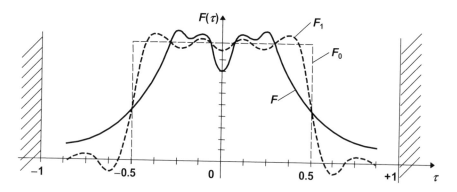

Fig. 13.4 Synthesis of a radiation pattern. F: stationary phase method; F_1: least mean square criterion.

Consider the preceding numerical example. We observe that the 'sides' of F_T are less abrupt than before. On the other hand, ripple and sidelobes are significantly smaller (Fig. 13.3b).

There are a large number of transfer functions $T(v)$ that could be employed. Only the uniform function gives the minimum mean square error. Nevertheless, the choice of the function $T(v)$ allows us to control, to some extent, the angular distribution of the deviations between the ideal and actual patterns.

13.3.3 Use of sampling methods

Once the pattern F_T and the corresponding illumination function defined by equation (13.27) are chosen, it is worth making use of the sampling possibilities of the function F_T in realizing the illumination f_T. The former may be represented by an infinite sum of sampling patterns

$$F_T(\tau) = \sum_{n=-\infty}^{+\infty} F_T\left(\frac{n}{2v_0}\right) \frac{\sin(2\pi v_0 \tau - n\pi)}{2\pi v_0 \tau - n\pi} \tag{13.29}$$

As we saw (§13.2.2), this is equivalent to representing the illumination f_T by a 'plane-wave spectrum' (see equation (13.11))

$$f_T(v) = \sum_{n=-\infty}^{+\infty} F_T\left(\frac{n}{2v_0}\right) \exp\left(j2\pi v \frac{n}{2v_0}\right) \tag{13.30}$$

For obvious practical reasons, it is convenient to use only a finite number N of samples. That is what we do in the process now to be described.

Woodward's method [2]

This consists of directly sampling the *non filtered* function F_0. We obtain then the following approximation.

$$F_W(\tau) = \sum_{-N/2}^{+N/2} F_0\left(\frac{n}{2v_0}\right) \frac{\sin(2\pi v_0 \tau - n\pi)}{2\pi v_0 \tau - n\pi} \tag{13.31}$$

if the pattern F_0 is bounded $(-v_0, +v_0)$. The illumination results from the superposition of N plane waves

$$N = 4 v_0 \tau_0 \tag{13.32}$$

This approximation has the merit of being easy to use, but it deviates from the approximations F_1 and F_T defined above.

Raabe's method [3]

We start by sampling the chosen filtered pattern $T(v)$. Since the angular position of samples is arbitrary, the function F_T may be written

---

[2] Woodward, P.M., 'A method for calculating the field over a plane aperture required to produce a given polar diagram', *J.IEE*, Vol. 93, Pt. IIIA, pp1554-1558, 1946.

[3] Raabe, H.P., 'Antenna pattern synthesis of the most truthful approximation', *IRE Wescon Convention Record*, Part 1, Antennas, August 1958.

$$F_T(\tau) = \sum_{n=-\infty}^{+\infty} F_T\left(\frac{n}{2v_0} + \varepsilon\right) \frac{\sin\left[2\pi v_0\left(\tau - \dfrac{n}{2v_0} - \varepsilon\right)\right]}{2\pi v_0\left(\tau - \dfrac{n}{2v_0} - \varepsilon\right)} \tag{13.33}$$

Then, we search for a value for ε such that most of the samples outside the interval $(-v_0, +v_0)$ of F_0 lie in the vicinity of the nulls between the sidelobes. This result in general is easy to reach since, in these regions, the sample periodicity coincides approximately with that of the sidelobes. These samples are then neglected and we obtain a good approximation F_{T_1} to F_T comprising only a finite number N_1 of samples. The illumination function f_{T_1} we are looking for is the superposition of the N_1 corresponding plane waves.

Representation of the pattern approximated by a sampling of tapered patterns

In practice, we often seek to synthesize the antenna pattern by a weighted sum of tapered patterns shifted in angle with respect to each other (case of multiple element antennas: §13.3.5). It is possible to give to the approximation $F_1(\tau)$ of the pattern a form corresponding to this process. In effect, it is known that over the interval $(-v_0, +v_0)$, the illumination f_0 can be represented by a Fourier series as follows.

$$f_0(v)\,\mathrm{rect}\left(\frac{v}{v_0}\right) = \sum_{n=-\infty}^{+\infty} F_1\left(\frac{n}{2v_0}\right)\exp\left(j2\pi\frac{n}{2v_0}v\right) \tag{13.34}$$

which can be written

$$f_0(v)\,T(v)\,\mathrm{rect}\left(\frac{v}{v_0}\right) = \sum_{n=-\infty}^{+\infty} F_1\left(\frac{n}{2v_0}\right)T(v)\exp\left(j2\pi\frac{n}{2v_0}v\right) \tag{13.35}$$

Taking the Fourier transform of both sides, we obtain the representation that we seek

$$F_T(\tau) = \sum_{n=-\infty}^{+\infty} F_1\left(\frac{n}{2v_0}\right)\Psi\left(\tau - \frac{n}{2v_0}\right) \tag{13.36}$$

The approximation F_T can be obtained from the pattern F_1 resulting from the filtering of F_0 by a uniform filter by sampling with the patterns Ψ corresponding to the illumination T. Note that in general the functions $\Psi[\tau - n/2v_0]$ are not orthogonal.

13.3.4 Role of phase - stationary phase method

Let us set

$$F_0(\tau) = A_0(\tau)\exp\left[j\Phi_0(\tau)\right] \tag{13.37}$$

$$f(v) = a(v)\exp\left[j\varphi(v)\right] \tag{13.38}$$

In general, only the amplitude $A_0(\tau)$ of the desired pattern is constrained, and Φ_0 is important only when the choice may influence the degree to which A_0 is realizable. Thus, we can impose a supplementary constraint. We shall study the case where the illumination amplitude function $a(v)$ is constant, the available parameter being the phase function $\varphi(v)$.

This condition is frequently found in practice. Let us give two examples:

- *Case of a reflector*: the amplitude $a(v)$ of the illumination is principally imposed by the pattern of the primary source. On the other hand, on reflection it is relatively easy to modify the phase $\varphi(v)$ of the illumination by giving an appropriate shape to the reflector (§10.3.2).

- *Case of an electronically scanned array*: the elements of the array are fed by a feed network which imposes the amplitude function $a(v)$; in contrast, the phase function $\varphi(v)$ is defined by electronically controlled phase shifters, thus it may be modified arbitrarily.

According to the stationary phase method, it can be shown that a curved surface wave, with a given illumination, generates a far field whose angular distribution depends both on this illumination and on the *curvature* of the wave surface at the point of stationary phase. We shall here find again this phenomenon in a slightly different form.

Let us start from an illumination in the form (equation (13.38))

$$f(v) = a(v)\exp\left[j\varphi(v)\right]$$

where $a(v)$ is constant and the synthesis is obtained by varying $\varphi(v)$. The corresponding pattern may be written from the equation

$$F(\tau) = \int_{-v_0}^{+v_0} a(v)\exp\left[j\varphi_1(v,\tau)\right]dv \tag{13.39}$$

with

$$\varphi_1(v,\tau) = 2\pi v\tau + \varphi(v) \tag{13.40}$$

Let us suppose that for each direction τ_1 there is one point (and only one point) on the aperture, of coordinate v_1, for which the phase $\varphi_1(v)$ is stationary, such that

$$\left.\frac{\partial\varphi_1(v,\tau_1)}{\partial v}\right|_{v=v_1} = 0 \tag{13.41}$$

or

$$\tau_1 + \frac{1}{2\pi}\varphi'(v_1) = 0 \tag{13.42}$$

where $\varphi'(v_1)$ is the first derivative of $\varphi_1(v)$.

In a similar way, let us assume that in the vicinity of this point, the phase φ_1 varies 'rapidly'. Under these conditions, we can say that the major contribution of the function in the integrand of equation (13.39) is given precisely when v is close to v_1. We can then represent the function φ_1 by its Taylor series expansion limited to second order. Since the first term is zero, according to the stationary phase condition (13.41) we have

$$\varphi_1(v,\tau_1) = \varphi_1(v_1,\tau_1) + \frac{(v-v_1)^2}{2}\varphi''(v_1) \tag{13.43}$$

If in the vicinity of v_1, $a(v)$ varies slowly, the integral (13.39) may be approximately written

$$F(\tau_1) = a(v_1)\exp\left[j\varphi_1(v_1,\tau_1)\right]\int_{-v_0}^{+v_0}\exp\left[j\frac{(v-v_1)^2}{2}\varphi''(v_1)dv\right] \tag{13.44}$$

Let us set

$$\frac{\pi}{2}y^2 = \varphi''(v_1)\frac{(v-v_1)^2}{2}$$

Ignoring the subscripts

$$F(\tau) = a(v)\exp\left[j\varphi_1(v,\tau)\right]\sqrt{\frac{\pi}{\varphi''(v)}}\int_{-v_0}^{+v_0}\exp\left(j\frac{\pi}{2}y^2\right)dy \tag{13.46}$$

The Fresnel-type integral, when v_0 is sufficiently large, is approximately

$$\sqrt{2}\exp\left(j\frac{\pi}{4}\right)$$

we obtain

$$|F(\tau)|^2 \approx 2\pi \frac{[a(v)]^2}{\varphi''(v)} \qquad (13.47a)$$

with

$$\tau = \frac{1}{2\pi}\varphi'(v) \qquad (13.47b)$$

Conclusion: The gain in the direction τ is defined by the illumination at the point of stationary phase corresponding to v given by equation (13.47b)[4]. It is proportional to the illumination intensity and *inversely proportional to the 'curvature' of the phase* at this point. We here find a known result of time-domain signals. If for example, the phase is quadratic as a function of v, the 'spectrum' $a^2(v)$ reproduces approximately the shape of the 'signal' (pattern) $F(\tau)$.

Pattern synthesis: The above calculations provide a relatively simple synthesis method on condition nevertheless that the stationary phase point should be unique for any considered direction (this does not apply to equiphase apertures or linear phase with v). Under these conditions, starting from a pattern $F_0(\tau)$ to be synthesized and an illumination of *fixed* amplitude $a(v)$, the relation (13.47a) gives the phase curvature $\varphi''(v)$ at each point of coordinate v fulfilling the condition (13.47b). We obtain a differential equation defining the phase function $\varphi(v)$. Since the illumination is known, we deduce the pattern actually synthesized $F(\tau)$ by performing a Fourier transform. Of course, the   on obtained is generally not as good as that in which the amplitude function is not constrained.

Numerical example: A 'sectoral' pattern is desired

$$F_0(\tau) = \text{rect}\left(\frac{\tau}{\tau_0}\right)$$

(for instance, in the domain such that $\tau_0 = 1/2$, we have $\pm\theta_0 = \pm 30°$)

We start from a constant illumination $a(v) = 1$.

Equation (13.47a) imposes the condition $\varphi''(v) = \text{const.} = 2m$, from which we have

$$\varphi'(v) = 2mv \quad (m \text{ being a parameter})$$

---

[4] This condition expresses that at this point the direction τ is normal to the emerging wave.

Equation (13.47b) imposes the condition: $\varphi'(v) = 2\pi\tau$. This must be satisfied at the aperture edges (for example $v_0 = 4$, $m\,v = \pi\tau_0$). Then $\varphi(v) = mv^2$, from which we obtain the pattern

$$F(v) = \int_{-v_0}^{+v_0} \exp(jmv^2) \exp(j2\pi v\tau)dv$$

Let us denote the Fresnel integral by $\phi(X)$. We find, to within a constant

$$F(\tau) = \phi\left(\sqrt{\frac{v_0\tau_0}{2}} + \sqrt{\frac{2\tau_0}{\tau_0}}\tau\right) - \phi\left(\sqrt{\frac{v_0\tau_0}{2}} - \sqrt{\frac{2\tau_0}{\tau_0}}\tau\right)$$

In the example chosen, the modulus of $F(t)$ represents the segment joining the points of coordinates $(4\tau + 2)$ and $(4\tau - 2)$ on the Cornu spiral. We obtain the pattern shown in Fig. 13.4. We have also represented the pattern F_1 obtained without constraining the amplitude illumination function (§13.2.1). We observe the distortion due to this constraint.

13.3.5 Pattern synthesis for a focusing system

Since the pattern F_T synthesized by an aperture is represented (equation ((13.26)) by the convolution of the ideal pattern F_0 and a filtering function Ψ, we can, in the case of a focusing system, identify the function with the diffraction pattern of the system. The pattern sought is obtained by creating in the focal plane an illumination function similar to that of the ideal pattern F_0.

The sampling of the pattern F_T (§13.2.3) allows us with a good approximation to sample the illumination F_0 itself. The samples outside the bandwidth of F_1 may be purely and simply suppressed as with Raabe's method (§13.2.3).

Thus, it is useless to control the primary illumination in a continuous but discrete way by means of multiple elements (horns or dielectric rods). The sampling spacing X in the focal plane is defined by multiplication of the angular sampling spacing $(1/2v_0)$ by the equivalent focal length of the system

$$\Delta X = F_e \frac{1}{2v_0} = F_e \frac{\lambda}{D} \tag{13.48}$$

This is approximately the radius of the central spot of the diffraction pattern (equation (9.102)). The total number of elements to be used is equal to the number of degrees of freedom of the pattern in the angular domain considered.

Example: A geostationary satellite antenna (i.e. at h = 36,000 km) is designed to illuminate Australia, that is to say an almost elliptical zone of axes: $a \approx$ 3000 km, $b \approx$ 4000 km. If the focal length of the antenna is F_e = 3 m, we arrange in the focal plane an array of elements limited by a contour similar to that of Australia, with an overall scale equal to the ratio of the focal length to the altitude, i.e. with axes

$$a' = a\,F_e/h = 1/4\,\text{m} \ (=25\ \text{cm}), \quad b' = b\,F_e/h = 1/3\,\text{m} \ (\approx 33\ \text{cm})$$

If the diameter of the antenna is D = 6 m, and the wavelength λ = 4 cm, the interelement spacing of the array will be

$$\Delta X = 2\ \text{cm}$$

This leads to a total number of elements of the array $N \approx 150$.

13.4 SUPERDIRECTIVE ANTENNAS

13.4.1 Introduction

The theoretical possibility for a finite array to reach an arbitrarily high gain has long been recognized. Nevertheless, it is generally admitted that the maximum gain of an aperture is obtained by an equiphase and uniform illumination. It can be shown, as we shall see later, that such an aperture can theoretically not only have an arbitrarily high directivity but also a pattern of any form. The physical origin of this possibility lies in the properties of the illuminations whose phase varies very rapidly along the aperture. The sampling method proposed by Woodward gives a relatively simple explanation of these phenomena[5].

13.4.2 Role of the 'invisible' domain of radiation

Let us recall the sampled expression for a pattern $F(\tau)$ of finite aperture

$$F(\tau) = \sum_{k=-\infty}^{+\infty} F(\tau_k) \frac{\sin\left[2\pi v_0\left(\tau - \tau_k\right)\right]}{2\pi v_0\left(\tau - \tau_k\right)}, \quad \tau_k = \frac{k}{2v_0} \qquad (13.49)$$

---

[5] Woodward, P.M. and Lawson, J.D., 'The theoretical precision with which an arbitrary radiation pattern may be obtained from a source of a finite size', *J.IEE*, Vol. 95, Pt III, No. 37, pp363-370, September 1948.

Even in the 'visible' domain, the pattern depends, strictly, on an *infinite* set of parameters $F_k = F(\tau_k)$. In particular, coefficients F_k corresponding to values $|\tau_k| > 1$ (i.e. imaginary directions) are involved in the 'visible' domain by means of the sidelobes of sampling functions whose main lobes are 'invisible'. Clearly, these samples intervene much less than those of the visible ones, but we can impose the pattern F, in the visible domain, to pass through a large arbitrary number N of finite points. It is sufficient for that to utilize the N samples closest to the visible domain. We obtain then a set of N linear relations in N unknowns, which are the sampling coefficients of F_k.

Calculation of an illumination function whose pattern is defined

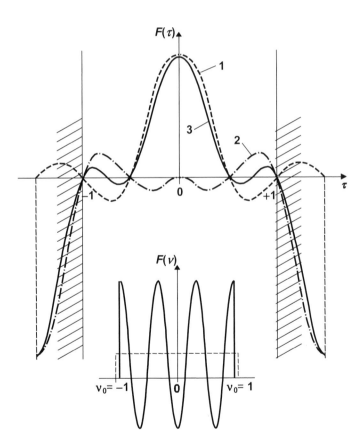

Fig. 13.5 Superdirective radiation pattern of a linear aperture, and its illumination function.

Let $F_n = F(\tau_n)$ ($n = 1, 2 \ldots N$) be the values imposed on the pattern F in the N directions τ_n, and $X_k = F(\tau_k)$ the coefficients of the samples.

Let us set

$$a_{n,k} = \frac{\sin\left[2\pi v_0\left(\tau_n - \dfrac{k}{2v_0}\right)\right]}{2\pi v_0\left(\tau_n - \dfrac{k}{2v_0}\right)} \tag{13.50}$$

The sampling formula can be written

$$F_n = \sum_{k=-N/2}^{k=+N/2} X_k a_{n,k} \tag{13.51}$$

The coefficients X_k are obtained by solving this linear system.

Numerical example: Let us set: $v_0 = 1$, $F(0) = 1$, $F(\pm 1/2) = F(\pm 3/4) = F(\pm 1) = 0$ (see Fig. 13.5)

We find

$$F(\tau) = \frac{\sin(2\pi\tau)}{2\pi\tau} - \frac{3}{2}\left[\frac{\sin(2\pi\tau - 3\pi)}{2\pi\tau - 3\pi} + \frac{\sin(2\pi\tau + 3\pi)}{2\pi\tau + 3\pi}\right]$$

with the illumination $F(v) = 1 - 3\cos(3\pi v)$.

The pattern obtained has a narrower main lobe and lower sidelobes than that resulting from the uniform function, due to the interference in the visible domain of the 'invisible' lobes. We observe the rapid and significant oscillations of the corresponding illumination function. This type of illumination function is evidently very difficult to realize, even over a very narrow bandwidth. It is associated with the presence of relatively significant *reactive power* in the vicinity of the aperture. A more realistic approach is considered in the next section.

13.5 THE ANTENNA AS A FILTER OF ANGULAR SIGNALS[6,7]

13.5.1 Introduction

Given an unknown 'object' distribution x, the image s that an antenna can give results from a filtering operation.

In the time domain, to characterize a filter we use an impulse input signal. At a time t' such an impulse is represented by the Dirac function $\delta(t-t')$ which is zero except at $t=t'$.

The impulse response ψ of the filter has a form which is independent of time t'. It is denoted $\psi(t-t')$. The filter is said to be *stationary*.

By decomposing the *input* signal of the filter $x(t)$ in the form of a sum of elementary impulses, if the filter is *linear*, the output signal $s(t)$ is the *sum* of the corresponding impulse responses, weighted by the input signal. Thus, it is expressed in the form of a convolution

$$s(t) = \int x(t')\psi(t-t')\,dt' = x(t)*\psi(t) \tag{13.52}$$

If we take the Fourier transform of both sides, we obtain the following relationship between the 'frequency spectra'

$$S(f) = X(f)T(f) \tag{13.53}$$

where $T(f)$, the Fourier transform of the impulse response, is the transfer function of the filter

$$T(f) = \int_{-\infty}^{+\infty} \psi(t)\exp(-j2\pi ft)\,dt \tag{13.54}$$

Any sinusoidal input signal gives a sinusoidal output signal except when the transfer function is zero beyond the cut-off frequency $f = f_0$.

---

[6] Arsac, J., *Transformation de Fourier et Théorie des Distributions*, Dunod, Paris, 1966.

[7] Ksienski, A., 'A survey of signal processing arrays', AGARD Conference on Signal Processing Arrays, Düsseldorf, 1966.

13.5.2 Optical or microwave imaging and linear filters

In the case of an antenna observing a distributed object (for instance a stellar constellation in radio astronomy), this distribution plays the role of the input signal.

For simplicity, take the case of a one-dimensional distribution $X(\theta)$ on a plane, for instance horizontal radar echoes observed by the rotating antenna of a scanning radar. Here the complex numbers $X(\theta)$ express the amplitudes and phases (relative to an implicit carrier signal) coming from directions θ with respect to a reference direction (geographical north, for instance). A point source in the direction θ' is represented by the Dirac function δ weighted by its amplitude: $\delta(\theta-\theta')X(\theta)$.

If the source is observed by means of a receiving antenna pointed towards the direction θ, the characteristic function of the antenna may be denoted $F(\theta-\theta')$; its form does not depend on the pointing direction, we find again the *stationary* property. The signal received by the antenna is weighted by this function and takes the form: $X(\theta)F(\theta-\theta')$. We see that the characteristic function plays the role of the impulse response.

Consider now a distribution of point-target objects in the directions

$$\theta' = \theta_1, \theta' = \theta_2, \ \dots \ \theta' = \theta_n$$

Due to the linearity of the process, the antenna output signal is the sum of the individual responses

$$s(\theta) = \sum_n X(\theta_n) F(\theta_n - \theta)$$

Extending this result to a continuous distribution, we obtain the antenna signal as the *convolution* of the unknown object distribution X with the characteristic function F

$$s(\theta) = \int X(\theta') F(\theta-\theta') d\theta' = X(\theta) \otimes F(\theta) \qquad (13.55)$$

Comment 1: In the usual case of a rotating radar antenna (N revolutions per second) the pointing direction takes the form $\theta = 2\pi Nt$ and the image signal $s(\theta)$ becomes a time-domain signal.

Comment 2: In the case of a transmitting-receiving radar the amplitude of the echo $X(\theta')$ illuminated by the antenna is itself proportional to the amplitude of the characteristic function: $X(\theta') = X'(\theta')F(\theta-\theta')$. Thus, this is the square of the characteristic function which appears in the convolution in the form

$$s(\theta) = \int X'(\theta') F^2(\theta-\theta') d\theta' \qquad (13.56)$$

13.5.3 False echoes and resolving power

The image $s(\theta)$ is a complex function. In general, only the magnitude (or its square) is considered. However, by demodulating s coherently (in I and Q form), we can obtain the real and imaginary components. Fig. 13.6 gives the form of the functions $|X(\theta)|$, assumed to be discontinuous, and the corresponding function $|s(\theta)|$. We observe that the *sidelobes* of the pattern $G(\theta)$ give rise to some relative maxima (1) which may be interpreted as echoes (*false echoes*). On the other hand, two close angular echoes may give only one maximum which can be interpreted as a single echo (2). The Rayleigh criterion defines the resolution limit (or resolving power) as the minimum angular interval between the maxima of two echoes (3). This interval $\Delta\theta$ is approximately equal to the -3 dB width of the main lobe of the antenna. If the antenna is an aperture of dimension D, we have

$$\Delta\theta \approx \theta_{3dB} \approx \frac{1}{2v_0} = \frac{\lambda}{D} \tag{13.57}$$

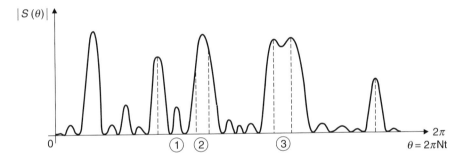

Fig. 13.6 Output signal from a rotating radar antenna: (1) false echo; (2) and (3) resolution limit of the antenna.

13.5.4 Case where the antenna is treated as an aperture

Let us take the case of a linear aperture of dimension D, operating in a space plane. Directions are referenced to $\tau = \sin\theta$ (θ is the angle between the directions and the aperture). The characteristic function $F(\tau)$ is the Fourier transform of the illumination $f(v)$ over the aperture (v is the abscissa x of a point on the aperture divided by the wavelength: $v = x/\lambda$)

$$F(\tau) = \int_{-\infty}^{+\infty} f(v)\exp(j2\pi v\tau)d\tau' \qquad (13.58)$$

The convolution relation between the object distribution and the antenna signal $s(\tau)$ takes the form

$$s(\tau) = \int_{-\infty}^{+\infty} F(\tau')X(\tau-\tau')d\tau' \qquad (13.59)$$

If we take the Fourier transform of both sides, we obtain the spectral relation

$$S(v) = f(v)x(v) \qquad (13.60)$$

We see that equivalent filter *transfer function is identified with the illumination function of the aperture*; this justifies the definition previously given of the cut-off spatial frequency

$$v_0 = \frac{D}{2\lambda} \qquad (13.61)$$

Remark: imaging problem: The determination of the unknown object distribution $X(\tau)$ is a 'deconvolution' operation; it may be obtained, in principle, from equation (13.60). The existence of a cut-off frequency v_0 makes impossible the determination of the spectrum values $x(v_0)$ for $|v| > v_0$ Certain processing methods (maximum entropy method) are based on the estimation of these values from the extrapolated values determined for $|v| \le v_0$. Other methods use *a priori* information to improve the estimation of the object distribution (Chapter 14).

13.5.5 Spectrum of fixed echoes of a rotating radar

Consider a radar with a rotating antenna of width D, radiating a continuous wave. This radar gives an angular image of the echoes. It also can give their radial velocity v by measuring the frequency shift (Doppler effect) Δf, and applying the equation

$$\Delta f = \frac{2v}{\lambda} \qquad (13.62)$$

Now the scanning of the radiation pattern produces a modulation of the echoes. Their spectrum is no longer a single impulse but has a certain bandwidth. The duration of an echo corresponds in practice to the time during which it is illuminated by the 3 dB beamwidth of the pattern

$$\theta_{3dB} = \frac{\lambda}{D} \tag{13.63}$$

If T is the rotation period of the antenna, this duration is

$$\Delta T = \frac{\theta_{3dB}}{2\pi} T = \frac{\lambda}{2\pi D} T$$

It is known that the associated spectrum bandwidth is approximately measured by the inverse of the illumination duration

$$\Delta F = \frac{1}{\Delta T} = \frac{2}{\lambda} \left(\frac{\pi D}{T} \right) = \frac{2v_D}{\lambda} \tag{13.64}$$

This is the expression of a Doppler shift where the ratio $\pi D/T$ represents the linear velocity v_D of an extremity of the antenna of length D.

Conclusion: The frequency spectrum of echoes (fixed or mobile) possesses a minimum width represented by the Doppler shift related to the *linear* velocity of the sides of the antenna: thus, this *velocity represents the accuracy with which it is possible to measure radial velocities.*

Example: $D = 15$ m, $T = 3$ s, $v_D = 15$ m/s.

Remark: A more rigorous study of the echo spectrum can be carried out from the convolution function, showing that the spectrum is identified with that of the square of the pattern. If we assume it to be equiphase (real), it is then proportional to the 'gain' function. We then find, to a good approximation, a spectrum whose shape is that of the autocorrelation function of the illumination function (§13.2.3).

FURTHER READING

1. Hansen, R.C., 'Array pattern control and synthesis', *Proc. IEEE*, Vol. 80, No. 1, pp141-151, January 1992.
2. Milne, K., 'Synthesis of power radiation patterns for linear array antennas', *IEE Proc.*, Vol. 134, Pt. H, No. 3, pp285-296, June 1987.
3. Schell, A.C., 'The antenna as a spatial filter', Chapter 26 in *Antenna Theory* (Part 2), Collin, R.E. and Zucker, F.J. (eds), *Antenna Theory*, Parts 1 and 2, McGraw-Hill, New York, 1969.
4. Woodward, P.M., *Probability and Information Theory, with Applications to Radar*, Pergamon Press, 1953; republished by Artech House, 1980.

EXERCISES

13.1 Woodward synthesis

An antenna with the far-field voltage pattern shown in Fig. 13.7 is required

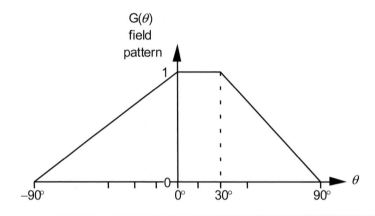

Fig. 13.7 Required radiation pattern.

Derive the value of the Fourier coefficients required to synthesize this pattern, using Woodward's technique, for a linear aperture 4 wavelengths wide. Sketch the resulting pattern.

ANSWER

The closest spacing for orthogonal $\sin x/x$ beams is $a\Delta u = 1$, where $u = (\sin\theta)/\lambda$. Thus $\Delta(\sin\theta) = \lambda/a = 1/4$ in this example, so the sampling points for Woodward synthesis are $\sin\theta = 0, \pm 0.25, \pm 0.5, \pm 0.75$ and ± 1, i.e. $\theta = 0, \pm 14.48°, \pm 30°, \pm 48.59°$ and $\pm 90°$.

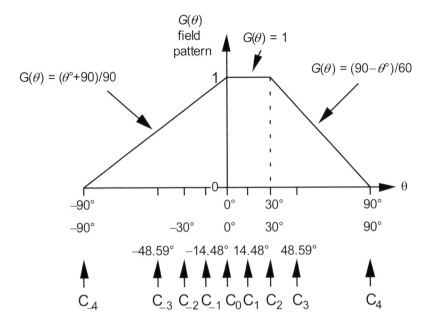

Fig. 13.8. Woodward synthesis technique.

From the diagram it is clear that the required Fourier coefficients are
$C_0 = C_1 = C_2 = 1$, $C_4 = C_{-4} = 0$, $C_{-1} = 0.84$, $C_{-2} = 0.67$, $C_{-3} = 0.46$ and
$C_3 = 0.69$.

Note: in fact the first sample need not be at $0°$; one *could* sample at $\sin\theta = \pm 0.125$, ± 0.375, ± 0.625, ± 0.875, thus giving 8 non-zero samples rather than 7.

13.2 Optimum 'difference' illumination
Using Schwartz's inequality (§7.4.8, eqn. 7.75), show that the amplitude
illumination $f_D(x,y)$ of a rectangular aperture which gives the *maximum slope* of a

'difference' pattern $F_D(\alpha,\beta)$ close to the normal axis, in the plane $\beta = 0$ is of the form:

$$f_D(x,y) = ax$$

where a is an arbitrary coefficient.

Show that the *gain* difference, close to the axis is given by eqn. 10.101.

14

Signal processing antennas

The Reverend Thomas Bayes (1702–1761)

14.1 INTRODUCTION

As a general rule, in a conventional complex system such as a radar or sonar, each constituent element - antenna, receiver etc. - performs its own specific function. Thus the directional properties of the system are due to the antenna, the estimation of target distance and speed is due to the receiver, and so on.

However, in certain modern systems it can be argued that the angular directivity of a radar is due both to the antenna *and* the particular processing of the signal in the receiver. This is the case with monopulse and conical scanning tracking antennas, where the angular precision is not only limited by the antenna performance but also by the choice of the signal processing algorithm (scanning or monopulse) and finally by the signal-to-noise ratio in the receiver.

The domain of 'signal processing antennas' results from a generalization of this example. In a piece of equipment designed on this basis, the set of directional properties (precision, angular resolution, spatial filtering) results from the symbiosis of the properties of the antenna and the processing of the signals received or transmitted.

In this chapter we consider, in succession, a number of different types of signal processing antennas. We begin by considering *synthetic antennas*, particularly aircraft-borne or satellite-borne Synthetic Aperture Radar (SAR) systems which are able to provide high resolution microwave images of a target scene. We then consider the problem of imaging of objects consisting of *coherent sources*, i.e. where the amplitudes and relative phases are constant, and then *incoherent sources*. The properties of incoherent sources are governed by the Van Cittert–Zernike theorem, which emphasizes the fundamental role of the covariance matrix of the observed field.

Methods of *high-resolution imaging* are then considered, in particular the method due to Burg, known as the Maximum Entropy Method (MEM). (The concept of entropy in information theory is covered in Appendix 14A). Other methods ('MUSIC') are also described.

The chapter concludes with a study of *adaptive antennas*, which perform a form of spatial filtering to present maximum gain towards wanted signals whilst directing nulls towards unwanted signals such as jammers.

14.2 SYNTHETIC ANTENNAS IN RADAR AND SONAR

14.2.1 Principles of synthetic antennas

This type of antenna is most usually known in the form of Synthetic Aperture Radar (SAR) systems, where a small antenna carried by an aircraft or satellite allows, by appropriate signal processing, high resolution images of a target scene to be obtained.

The concept of a synthetic antenna is, however, more general. It is linked with active radar or sonar systems making use of a transmitting antenna in order to form an image of the environment. While a classical system transmits the same form of signal in all directions of observation, a 'synthetic' system is characterized by the fact that the transmitted signal is coded with different modulation in different directions; this is sometimes spoken of as a 'coloured' transmission. The transmitted signals therefore form a spatio-temporal coding. On reception, a set of filters or correlators matched to a set of directions allow the echoes to be identified at the same time as performing the pulse compression.

The notation used is essentially the same as those of Chapter 13. We consider signals of sufficiently narrow bandwidth that we can use Rice's notation of the analytic signal as a complex number. The antennas under consideration will generally be linear arrays, continuous or otherwise, radiating into a planar region.

The conclusions obtained can readily be generalized to the case of three-dimensional space.

14.2.2 Synthetic receive array with non-directional beam

Consider a transmit array radiating in each direction $\tau = \sin\theta$ a different temporal signal $F(\tau, t)$. If there is a target in the direction τ it will give rise to an echo of the corresponding temporal signal. On receive, a filter matched to this signal will therefore allow the target to be identified (Fig. 14.1).

More specifically, if the transmit array is of finite length D $(v_0 = D/2\lambda)$, the transmitted signal can be sampled in the form

$$F(\tau_1,t) = \sum_{(n)} f_n(t)\,\mathrm{sinc}\,(2\pi v_0 - n\pi) \tag{14.1}$$

where $\mathrm{sinc}\,X = \sin X/X$ and where the functions $f_n(t)$ represent orthogonal signals of duration T, bandwidth f and the same energy

$$W = \int_0^T |f_n|^2\,dt$$

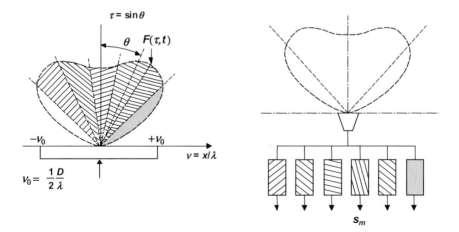

Fig. 14.1 Synthetic receive array with coded transmission and non-directional beam.

Suppose there is an echo in a direction τ, close to τ_m. If the receive antenna (of wide beamwidth) is provided with a set of filters (or correlators) matched to the signals, the signal at the output of the mth correlator will be a composite signal, of effective duration $T' \approx 1/\Delta f$ and of peak amplitude

$$s_m = \int_0^T F(\tau,t) f_m^*(t)\,dt = W \operatorname{sinc} 2\pi v_0 (\tau - \tau_m)$$

$$(\tau_m = m/2v_0) \tag{14.2}$$

Conclusion: With a simple receive antenna it is possible to achieve the equivalent of a multibeam antenna at the same time as a pulse compression effect with processing gain T/T'.

14.2.3 Synthetic receive array with multiple beams

The directional effect can be augmented by using a multiple-beam antenna with a filter or correlator at each beam output matched to the corresponding direction (Fig. 14.2).

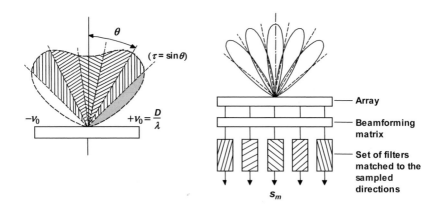

Fig. 14.2 Synthetic receive array with multiple beams.

In effect, the signal s_m (equation (14.2)) is weighted by the directivity of the associated beam, and takes the form

$$s_m' = s_m \operatorname{sinc} 2\pi v_0 (\tau - \tau_m) = W \operatorname{sinc}^2 2\pi v_0 (\tau - \tau_m) \tag{14.3}$$

Conclusions:
- we achieve the same properties as the previous scheme in respect of target identification and pulse compression;
- the directivity and gain are doubled (in dB): a sidelobe at −20 dB in the first scheme becomes −40 dB in this scheme.

14.2.4 Examples of spatio-temporal coding

Binary phase coding

The array radiates via 1-bit phase shifters (0 or π, which we represent by + or −), so that it transmits binary sequences which are (if possible) orthogonal. The structure of the system on receive follows the previous descriptions. The spatio-temporal coding of the array can be symbolized as follows

$$
\begin{array}{cccc}
\rightarrow & \text{space (array)} & & \\
+ & + & + & + \\
+ & + & - & - \\
+ & - & - & + \\
+ & - & + & - \\
\end{array}
$$

time ↓

Each one of these spatial codes 'radiates' - as can be verified by taking its Fourier transform - a pattern consisting of two (generally distinct) main lobes, in phase or in antiphase according to the parity of the code, and whose direction in space depends on the spatial period of the transmitted code.

Binary amplitude coding

The array radiates by means of switches, which give a 1-bit amplitude code (0, 1). We note that the example of orthogonal sequences shown below is equivalent to a discrete linear displacement of a single source.

$$
\begin{array}{ccccc}
\rightarrow & \text{space (array)} & & & \\
1 & 0 & 0 & 0 & 0 \\
0 & 1 & 0 & 0 & 0 \\
0 & 0 & 1 & 0 & 0 \\
0 & 0 & 0 & 1 & 0 \\
0 & 0 & 0 & 0 & 1 \\
\end{array}
$$

time ↓

Here, at each instant the radiated pattern has the same wide beam, but the phase centre moves as a quasi-linear phase modulation, proportional to the sine of the pointing angle. This provides an introduction to the following important application.

Sideways-looking Synthetic Aperture Radar (SAR)[1,2]

Consider a moving platform M (aircraft, satellite, ...) moving in a straight line from A to B through a distance D with a velocity V, so that $D = VT$ (Fig. 14.3). We make the assumption that during the duration T the directions of the echoes considered do not change.

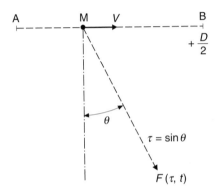

Fig. 14.3 Synthetic array with linear displacement.

The spatio-temporal transmitted signal is characterized by a linear phase modulation due to the Doppler effect

$$F(\tau,t) = \exp\left(j2\pi \frac{V\tau}{\lambda} t \right) \mathrm{rect}\left(\frac{2t}{T} \right) \qquad (14.4)$$

The echo from a target in direction τ is the square of (14.4)

[1] Wiley, C.A., 'Pulse Doppler radar methods and apparatus', US Patent No. 3196436, 1954.

[2] Wiley, C.A., 'Synthetic Aperture Radars - a paradigm for technology evolution', *IEEE Trans. Aerospace and Electronics Systems*, Vol. AES-21, pp440-443, 1985.

$$G(\tau,t) = \exp\left(j4\pi \frac{V\tau}{\lambda} t \right) \text{rect}\left(\frac{2t}{T} \right) \qquad (14.5)$$

This represents a signal of duration T with a linear Doppler rate $2V\tau/\lambda$ with respect to the signal transmitted by the radar.

The signal may be sampled in frequency. A bank of contiguous filters each of bandwidth $\Delta F = 1/T$ allows the directions τ to be discriminated with an angular precision such that

$$\Delta F = \frac{1}{T} = \frac{2V\Delta\tau}{\lambda} \qquad (14.6)$$

i.e.

$$\Delta\tau = \frac{\lambda}{2VT} = \frac{\lambda}{2D} \qquad (14.7)$$

This is the angular precision of a radar using an aperture $D = VT$. This is also the limit of angular resolution, in the classical sense, of two targets, and in the sense that the radar is actually a synthetic aperture. These results allow us to specify the condition for stationarity: the range R of the target should be such that its angular variation during the time T should be small compared to the resolution $\Delta\tau$ of equation (14.7)

$$\frac{D}{R} < \frac{\lambda}{2D}$$

i.e.

$$R > \frac{2D^2}{\lambda} \qquad (14.8)$$

The *linear* resolution in this case is therefore proportional to the range R and inversely proportional to the length of the synthetic aperture

$$\Delta X = R\Delta\tau = R\frac{\lambda}{2D} \qquad (14.9)$$

Note: If the target is moving with a radial velocity V_R, there will be an ambiguity between velocity and angle, giving an angular error $\delta\tau$ such that

$$V\delta\tau = V_R \qquad (14.10)$$

Focused synthetic aperture (Fig. 14.4)

The preceding limitations can be overcome by using a 'focused' synthetic array. The platform M carries an antenna of physical dimension L, illuminating the target scene with a beam whose beamwidth is approximately

$$\theta_1 = \lambda/L \tag{14.11}$$

The height of the radar platform above the surface is H_1, so the length of its footprint is approximately

$$X_1 = \theta_1 H = \lambda H/L \tag{14.12}$$

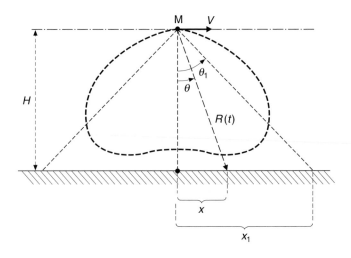

Fig. 14.4 Focused synthetic aperture.

A target C on the surface is illuminated for a duration

$$T = \frac{X_1}{V} = \frac{\lambda H}{VL} \tag{14.13}$$

At an instant t the (one-way) range from the radar to the target is

$$R(t) = \sqrt{H^2 + V^2 T^2} \approx H + \frac{V^2 t^2}{2H} \tag{14.14}$$

Since the two-way range from the radar to the target and back is $2R$, the effect of this is to cause a modulation of the signal in phase - and hence in also in frequency (as rate of change of phase) via the Doppler effect, in the form

$$S(t) = \exp\left[j\Phi(t) \right] \mathrm{rect}\left(\frac{2t}{T} \right) \tag{14.15}$$

with

$$\Phi(t) = \frac{4\pi}{\lambda} R(t)$$

The Doppler frequency

$$f_D = \frac{1}{2\pi} \frac{d\Phi}{dt} \tag{14.15}$$

varies between extreme values $\left(+f_D, -f_D \right)$, which for $t = T/2$ are

$$f_D = \frac{2 \left| \dfrac{dR}{dt} \right|}{\lambda} = \frac{V^2 T}{R} \approx \frac{V}{L} \tag{14.16}$$

This represents the approximate limits of the Doppler spectrum (14.15). If this Doppler variation is matched-filtered with a correlator, the useful duration of the output signal will be

$$\Delta T = \frac{1}{2 f_D} = \frac{L}{2V} \tag{14.17}$$

which corresponds to a spatial precision ΔX for the target position of

$$\Delta X = V \Delta T = L/2 \tag{14.18}$$

Conclusion: With a focused synthetic aperture, the maximum linear resolution is in principle equal to half the physical dimension of the antenna carried by the platform. This is independent of range (altitude), platform velocity, and frequency.

At first sight this may seem strange, since it implies that the smaller the antenna carried by the radar, the better the resolution. This is explained by the fact that a smaller antenna has a wider radiation pattern, so that a target is illuminated for a longer time and the Doppler bandwidth $2B$ of its echoes is correspondingly greater (equation (14.16)). In practice there is no benefit from an angular aperture of greater than 120°, which corresponds to a half-wave dipole.

The preceding analysis has given the resolution in the direction of motion of the platform (i.e. *azimuth resolution*). The equivalent resolution in the orthogonal direction is the *range resolution*.

Remark: the term 'focused processing' is justified in the following way. Consider an antenna of dimension D equal to the *synthetic* aperture, i.e. from equation (14.13)

$$D = VT = \frac{\lambda H}{L} \qquad (14.19)$$

Suppose that this aperture is focused at a distance H (the focal length), where it creates a diffraction spot of effective aperture $\delta/2$, given by equation (12.97), i.e. approximately

$$\frac{\delta}{2} = \frac{\lambda H}{D} = L \qquad (14.20)$$

Because the operation of the radar involves the square of the radiation pattern, this doubles the directivity. This gives the same result as equation (14.18).

Frequency-scanning synthetic antenna

We have already considered the scanning of the beam of an array by varying the frequency (§11.6). Suppose that a very short pulse (i.e. large bandwidth) is transmitted from such an array; a bank of filters allows the direction of arrival of echoes to be distinguished, just as with a synthetic antenna. The two types of antenna work essentially on the same principle: the short pulse propagates down the feed network in the same way as the moving platform with a speed equal to the *group velocity* V_g in the guide. It is easy to see that the frequencies associated with the different directions are just the Doppler frequencies corresponding to V_g.

14.3 IMAGING OF COHERENT SOURCES

14.3.1 Introduction

We have seen in Chapter 13 (§13.5) that an antenna behaves as a spatial filter. With an object that consists of a number of scatterers (in the horizontal plane, for example), a rotating antenna provides a filtered image, the convolution of the object angular distribution with the radiation pattern of the antenna (equation (13.56)). The quality of the image is therefore limited by the 'limit of resolution' of the antenna, i.e. the 3 dB beamwidth of the main lobe.

   The question is therefore whether it is possible to improve upon this limit. Can we improve the resolution of an antenna of a given aperture size? This question, already touched on in Chapter 13 in the consideration of 'superdirective' antennas, can be thought of in another way, making use of additional *a priori* information about the target configuration.

There are certain operational situations where one knows at the outset that the target is of finite extent and consists of N point sources such that the distribution and the amplitudes and relative phases are unknown.

If these quantities remain constant during the measurement period, we say that the sources are coherent. Under these conditions, if the receiving array consists of a number of elements (or radiation patterns), it is possible to determine the position of the targets with a precision which depends only on the signal-to-noise ratio.

More precisely, if we know *a priori* the probability function governing the spatial distribution of the object (for example, the objects are at an elevation angle that is positive and below a certain limit), and if the measurement errors (noise) obey a known probability function (for example, Gaussian probability density function), then the probability density function of the object distribution can be deduced *a posteriori*. Then the most probable distribution of the objects can be determined, according to some decision criterion (for example, the object distribution which results in the maximum *a posteriori* probability). We now illustrate these considerations by a specific example.

14.3.2 Two-source distribution

The case of a single point target has already been considered under the heading of tracking systems in Chapter 10. The techniques of conical scanning or monopulse allow a single point target to be located with a precision measured with respect to the 3 dB beamwidth of the aperture and proportional to the square root of the signal-to-noise ratio. However, in certain situations this technique will not work, for example a twin-engined aircraft at short range does not behave as a point target. As another example, when tracking an aircraft at constant elevation over the sea, as the direction of the signal reflected from the sea surface approaches the 3 dB beamwidth in elevation, the servo loops that position the antenna in the vertical plane will become unstable, so that tracking is no longer possible. This is because the direct echo and the image reflected in the sea surface constitute a two-source target, so that the tracking algorithms which assume a point target do not work. A different tracking algorithm is therefore necessary.

14.3.3 Estimation of the elevation angle of a low-altitude target above a reflecting plane

Suppose that the radar antenna is located at a height h above the plane, and a target O_1 makes an elevation angle θ_1 at the radar. Its image O_2 makes an elevation angle θ_2 at the radar. O_1 and O_2 form a two-source target. Suppose that the angular separation $\theta_1 - \theta_2$ of the signals is less than the antenna beamwidth (Fig. 14.5).

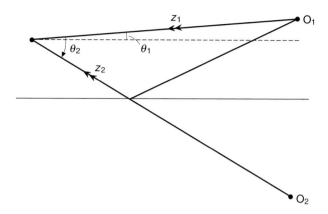

Fig. 14.5 Low-altitude target above a reflecting plane.

If the radar antenna takes the form of a vertical array fed by a beamforming matrix (§11.7), we can make use of three adjacent orthogonal beams, for example, of the form (Fig. 14.6)

$$f_1(\tau) = \mathrm{sinc}(2\pi v_0 - \pi)$$
$$f_2(\tau) = \mathrm{sinc}(2\pi v_0)$$
$$f_3(\tau) = \mathrm{sinc}(2\pi v_0 + \pi) \tag{14.21}$$

where $v_0 = D/2\lambda$ and $\tau = \sin\theta$, according to the usual notation.

The two-source configuration (O_1, O_2) is characterized by two real parameters (the directions $\tau_1 = \sin\theta_1$, $\tau_2 = \sin\theta_2$) and two complex parameters (their complex amplitudes Z_1 and Z_2), making a total of six real quantities. A measurement of the signals s_1, s_2, s_3 received by the three beams provides three complex quantities, i.e. six real quantities, so the problem is in principle soluble.

In fact, in the absence of noise (or with a sufficiently large signal-to-noise ratio) the signals received by the beams are the convolutions of the distributions with the radiation patterns, i.e.

$$\left.\begin{array}{l} s_1 = Z_1 f_1(\tau_1) + Z_2 f_1(\tau_2) \\ s_2 = Z_1 f_2(\tau_1) + Z_2 f_2(\tau_2) \\ s_3 = Z_1 f_3(\tau_1) + Z_2 f_3(\tau_2) \end{array}\right\} \tag{14.22}$$

The solution of this system of linear equations requires that the determinant should be zero

$$\Delta = \begin{vmatrix} s_1 & s_2 & s_3 \\ f_1(\tau_1) & f_2(\tau_1) & f_3(\tau_1) \\ f_1(\tau_2) & f_2(\tau_2) & f_3(\tau_2) \end{vmatrix} = 0 \qquad (14.23)$$

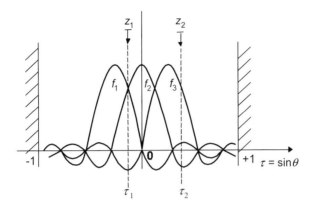

Fig. 14.6 Set of three orthogonal beams.

This complex equation leads to two real equations, which may be solved to give τ_1 and τ_2. Note that the complex amplitudes Z_1, Z_2 are eliminated, so that the coherence condition (stability of these amplitudes) is not actually a constraint here.

Particular examples:
(1) *Processing channel*: Instead of using the directional beams as in the example above, it can be more convenient to use the following linear combinations, which lead to the following 'sum' and 'difference' patterns, just as with the monopulse technique, plus a third pattern which is a quadratic function of τ, named *voie ecart* (processing channel) by its originator. Restricting ourselves to the principal parts gives

$$\left. \begin{aligned} F_1(\tau) &= f_1 + f_2 + f_3 \approx 1 \\ F_2(\tau) &= f_3 - f_1 \approx \alpha\tau \\ F_3(\tau) &= f_3 + f_1 \approx \beta\,\tau^2 \end{aligned} \right\} \qquad (14.24)$$

Here α and β characterize respectively the *slope* of F_2 and the *curvature* of F_3 in the neighbourhood of the axis ($\tau = 0$).

It is convenient to introduce the normalized hermitian products of the signals by means of correlators. Separating out the real and imaginary parts gives

$$\frac{s_1^* s_2}{|s_1|^2} = a + ja' \; ; \quad \frac{s_1^* s_3}{|s_1|^2} = b + jb'$$

Equating the determinant to zero gives the sum and the product of the unknown quantities, which leads to a second-order equation

$$\alpha'\beta\tau^2 - b'\alpha\tau + ab' - ba' = 0$$

(2) It is also possible to use three receiving elements separated by *D*/2. Their patterns are of the form

$$F_1(\tau) = \exp(j2\pi v_0\tau), F_2(\tau) = 1, F_3(\tau) = \exp(-j2\pi v_0\tau)$$

14.3.4 Effect of noise: *a posteriori* probabilities and decision theory

The preceding analysis has ignored the effect of noise. To explain the role of noise and the principle of the following method, we can simplify the problem and suppose that the *measured* elevation angle τ' includes a noise signal $n(t)$ which masks the *true* elevation angle τ, so that

$$\tau' = \tau + n(t) \tag{14.25}$$

We assume that the noise $n(t)$ is represented by a Gaussian random function with probability density function $q(n)$, with standard deviation σ. Also, we can make use of certain *a priori* information concerning the possible values of τ, namely the elevation angle cannot be negative, and cannot exceed a certain maximum value $\tau_{1_{\max}}$, from which the target is no longer considered as point-like. We define an *a priori* probability density function $p_1(\tau)$ (in particular, $p_1(\tau) = 0$ for $\tau < 0$). Under these conditions, the 'conditional' probability density function $r(\tau'|\tau)$ of the measured angle τ' when τ is supposed known, is derived from the probability density function $q(n)$ by a simple 'translation'

$$r(\tau'|\tau) = q(\tau' - \tau)$$

The crucial question is now: what is the *a posteriori* probability density function of τ once τ' has been measured ? This is the conditional probability density function $p_2(\tau|\tau')$. This leads to the following result, using Bayes' formula[3]

$$p_2(\tau|\tau') = \frac{r(\tau'|\tau) p_2(\tau)}{\int\limits_{-\infty}^{+\infty} r(\tau'|\tau) p_1(\tau) d\tau} \tag{14.26}$$

If p_1 is constant for $\tau > 0$, we find

$$p_2(\tau|\tau') = \frac{2}{1+\Theta\left(\dfrac{\tau'}{\sigma\sqrt{2}}\right)} \frac{1}{\sigma\sqrt{2\pi}} \exp\left(-\frac{(\tau'-\tau)^2}{2\sigma^2}\right) \tag{14.27}$$

where Θ is the error function

$$\Theta(x) = \frac{2}{\sqrt{\pi}} \int\limits_0^x \exp\left(-t^2\right) dt \tag{14.28}$$

The *a posteriori* distribution function gives the probability that the angle τ is less than a value τ_0 given by

$$p_2(\tau < \tau_0|\tau') = \int\limits_0^{\tau_0} p_2(\tau|\tau') d\tau \tag{14.29}$$

$$p_2(\tau < \tau_0|\tau') = \frac{\Theta\left(\dfrac{\tau_0-\tau'}{\sigma\sqrt{2}}\right) + \Theta\left(\dfrac{\tau'}{\sigma\sqrt{2}}\right)}{1+\Theta\left(\dfrac{\tau'}{\sigma\sqrt{2}}\right)} \tag{14.30}$$

Choice of decision criterion

Once the measurement has been made, what is the most likely value for the angle τ? Several criteria could be used, according to what is known as Decision Theory. We could, for example, choose the angle τ which maximizes the *a posteriori* probability $p_2(\tau|\tau')$. Another criterion, perhaps more practical, is that of the 'median value', that is, the value which is not exceeded with a probability of 50%. The function found gives the estimated value τ_0 corresponding to a measurement τ' writing the distribution function as a constant equal to 1/2, as follows

---

[3] Papoulis, A., *Probability, Random Variables and Stochastic Processes*, p84, McGraw-Hill, New York, 1965.

$$p_2\left(\tau<\tau_0\middle|\tau'\right)=1/2$$

We then obtain the function shown in Fig. 14.7. Note that the angles chosen are entirely compatible with the assumptions that have been made: if a noise spike gives a negative measurement, the resulting value is nevertheless positive.

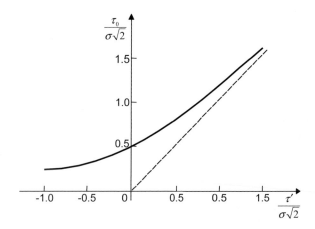

Fig. 14.7 Estimated elevation angle.

14.4 IMAGING OF INCOHERENT SOURCES

14.4.1 Introduction

Radiating sources are said to be incoherent if their relative phases (and hence their amplitudes) vary rapidly over the duration of a measurement. This situation is encountered frequently in nature whenever there is a set of independent sources. Such is the case in radioastronomy, with the radiation from stars or from radio sources in general. Such is also the case in radiometry, where it is the thermal nature of the radiation which causes the incoherence. It is also the case with passive sonars. With active sonars, the decorrelation of the echoes is often caused by the propagation conditions.

In the domain of radar the situation is less clear cut. At microwave frequencies it is only necessary to have a relative displacement of the echoes by a few centimetres during the measurement to cause their decorrelation, but this is not always the case. In contrast, with natural or man-made noise sources, the signals will always be decorrelated.

The condition for incoherence of sources is important, because it forms part of the *a priori* information which can be made use of, and which, in combination with the results of the measurements, allows improved image resolution (so-called 'high-resolution' methods) or to perform spatial filtering to eliminate noise sources or jammers (the 'optimum array', adaptive arrays).

14.4.2 Conditions for incoherence

Consider two sources at large distance in the directions of the unit vectors u_1 and u_2, and of complex amplitudes which vary as a function of time according to the stationary random functions

$$z_1 = z(u_1, t), \qquad z_2 = z(u_2, t) \tag{14.31}$$

These sources are said to be incoherent if for $u_2 \neq u_1$ the mean value (in the sense of probabilities) of the hermitian product is zero

$$\overline{z_1 z_2^*} = 0 \tag{14.32}$$

For $u_2 = u_1$ we define the intensities in terms of the same average products

$$T(u_1) = \overline{|z_1|^2}; \ T(u_2) = \overline{|z_2|^2} \tag{14.33}$$

The symbol $T(u)$ is used by analogy with the 'black body' temperature.

If the random functions z_1 and z_2 are stationary and ergodic, the mean values can be estimated in terms of *temporal* averages

$$\overline{z_1 z_2^*} = \lim_{T \to \infty} \frac{1}{T} \int_0^T z(u_1, t) \, z^*(u_2, t) \, dt \tag{14.34}$$

In practice it is adequate that the integration time T is 'sufficiently long' with respect to the autocorrelation times of the functions z. This gives

$$T(u) = \lim_{T \to \infty} \frac{1}{T} \int_0^T |z(u,t)|^2 dt \qquad (14.35)$$

14.4.3 Multiplicative arrays

The *a priori* knowledge of the incoherent nature of the sources allows us to improve the image obtained by an antenna. The example of a multiplicative array provides a first demonstration of this.

Consider a linear array of length L, with a uniform illumination function. Its radiation pattern is of the form

$$f_1(\tau) = \operatorname{sinc} 2\pi v_0 \tau \qquad (14.36)$$

where

$$\tau = \sin \theta, \quad v_0 = L/2\lambda$$

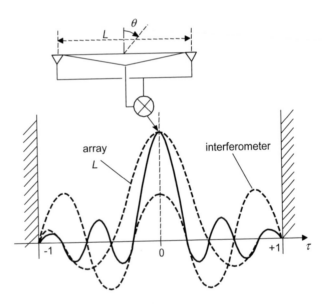

Fig. 14.8 Multiplicative array. The resulting pattern (continuous line) has twice the directivity of the original pattern.

Suppose now that a simple interferometer is added to the array, consisting of two small auxiliary elements at the extremities of the array (i.e. with a separation approximately equal to L), and fed in phase (Fig. 14.8). The radiation pattern of the interferometer is then of the form

$$f_2(\tau) = \cos 2\pi v_0 \tau \tag{14.37}$$

If, on receive, we form the hermitian product of the signals received by the two antennas we obtain a product signal of the form

$$f_1(\tau) f_2^*(\tau) = \text{sinc} \, 4\pi v_0 \tau \tag{14.38}$$

This gives the same radiation pattern as an aperture of $2L$, i.e. *twice that of the original array*.

Suppose now that this system is placed in an environment of N incoherent point sources in directions τ_n $(n \in N)$ and complex amplitudes z_n. The signals received by the two antennas are the convolutions

$$s_1(\tau) = \sum_n z_n \cos 2\pi v_0 \tau_n$$

$$s_2(\tau) = \sum_m z_m \cos 2\pi v_0 \tau_m \tag{14.39}$$

If the average product of the signals is formed by means of a coherent demodulator (or a phase-amplitude demodulator), we obtain the signal S in the expressions where the terms for which $n = m$ and those for which $n \neq m$ are separated

$$S = \overline{s_1 s_2^*} = \sum_n \overline{(z_n)^2} \, \text{sinc}(4\pi v_0 \tau_n) + \sum_{n \neq m} \overline{z_n z_m^*} (\text{sinc} \, 2\pi v_0 \tau_n) \cos(2\pi v_0 \tau_m) \tag{14.40}$$

Because of the condition for incoherence the second series reduces to zero and the product signal appears as the convolution of the *intensities* of the sources with the radiation pattern defined above, so that the directivity of the antenna has been doubled

$$S = \sum_n T(\tau_n) \, \text{sinc}(4\pi v_0 \tau_n) \tag{14.41}$$

If electronic scanning is used, by means of phase shifters, the signal S gives an image of the distribution T in the form of the convolution

$$S(\tau) = \sum_n T(\tau_n) \, \text{sinc}(4\pi v_0 (\tau_n - \tau)) \tag{14.42}$$

Note: Effect of noise: a more detailed study implies an analysis of the effect of noise. This system is effective only if the signal-to-noise ratio is high. It is used in harbour surveillance radars where the ranges are short[4].

Another example: the Mills Cross

Here, in contrast, the ranges are quite literally astronomical !

Two linear arrays are arranged at right angles on the ground, one aligned on an east–west axis $x'Ox$ and the other on a north–south axis $y'Oy$. A given direction in space is specified by the unit vector $\hat{u}$ or by the direction cosines α, β, γ. The arrays, of length L, have a uniform illumination function and hence have characteristic radiation surfaces which are cylindrical (centred on the axis of each array) and with characteristic functions

$$F_1(\alpha) = \mathrm{sinc}(2\pi v_0 \alpha)$$
$$F_2(\beta) = \mathrm{sinc}(2\pi v_0 \beta) \tag{14.43}$$

The product of F_1 and F_2 is equivalent to a radiation pattern

$$F(\alpha, \beta) = \mathrm{sinc}(2\pi v_0 \alpha)\,\mathrm{sinc}(2\pi v_0 \beta) \tag{14.44}$$

To obtain such a pattern directly it is necessary to use a two-dimensional square array of side L. The system due to Mills allows an important economy in amount of antenna hardware; if each array uses M elements, a fully-populated array would need M^2 elements, so the Mills Cross involves an economy of $M^2/2M = M/2$.

Considering a number of incoherent sources, it can be shown that the product pattern does indeed provide the convolution of the angular distribution of the sources (in intensity) with the 'product' pattern derived above.

14.4.4 Relationship between an angular distribution of incoherent sources and the observed field: the Van Cittert-Zernicke Theorem

Consider firstly a plane space defined by the axes $x'Ox$ (horizontal) and $z'Oz$ (vertical) - Fig. 14.9. Suppose there is an arrangement of incoherent sources in directions τ_n ($n = 1, 2, ... N$), and with variable complex amplitudes $z(\tau_n, t)$, all at a large distance. The signals that they radiate are therefore effectively plane at the axis $x'Ox$. The resulting field at this axis is the sum of these plane waves, and takes the form

---

[4] Ksienski, A., 'Multiplicative processing antenna systems for radar applications', *The Radio and Electronic Engineer*, January 1965.

$$E(v,t) = \sum_n z(\tau_n,t)\exp(j2\pi v\tau_n) \qquad (v=x/\lambda) \qquad (14.45)$$

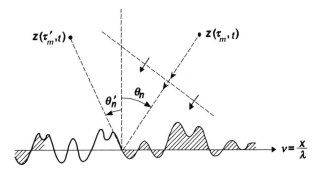

Fig. 14.9 Instantaneous hologram created by a set of distributed point sources.

Note that this field constitutes a 'hologram' of the sources. It contains all the information on the sources, and it is well known that a hologram allows an image of a set of sources to be reconstructed by illumination with coherent light[5]. Here, the observed field $E(v, t)$ is fluctuating, so it cannot be used directly. Rather, we can define a second-order average: the *spatial autocorrelation function* of the field, or the 'coherence function'. If the field E is observed at two points x and x', their average hermitian product is defined as

$$C(v,v') = \overline{E(v,t)E^*(v',t)}; \quad (v=x/\lambda, \ v'=x'/\lambda) \qquad (14.46)$$

Calculation of the coherence function:

We have

$$E^*(v',t) = \sum_m z^*(\tau_m,t)\exp(-j2\pi v'\tau_m) \qquad (14.47)$$

In the expression for the hermitian product we can separate the terms for which $n = m$ from those for which $n \neq m$

[5] Gabor, D., 'A new microscope principle', *Nature*, Vol. 161, pp777-778, May 1948.

$$C(v,v') = \sum_n \overline{|z(\tau_n,t)|^2} \exp\left[j2\pi(v-v')t \right]$$
$$+ \sum_{n \neq m} \sum \overline{z(\tau_n,t)z^*(\tau_m,t)} \exp\left[j2\pi(v\tau_n - v'\tau_m) \right] \quad (14.48)$$

The second term is zero by hypothesis.

In the first term the intensities of the radiating sources can be recognized

$$T(\tau_n) = \overline{|z(\tau_n,t)|^2}$$

We therefore have

$$C(v,v') = \sum_n T(\tau_n) \exp\left[j2\pi(v-v')t \right] \quad (14.49)$$

Note that the coherence function does not depend on the absolute position of the observation points, but only on their algebraic position $\Delta v = v - v'$. The coherence function is therefore *stationary* along the axis xOx', and as such it is a *test* for the incoherence of the input signals.

The result can be formally extended to a continuous distribution of incoherent sources in the classical integral form

$$C(\Delta v) = \int_{-\infty}^{\infty} T(\tau) \exp(j2\pi\Delta v\tau) d\tau \quad (14.50)$$

From which we have the Van Cittert-Zernicke Theorem

'The Coherence Function is the Fourier Transform of the angular distribution of the intensity of the sources'

Generalization to three dimensions

In an *xyz* system of axes, the directions of the sources are specified in terms of their direction cosines (α, β, γ), and in the horizontal plane by their coordinates expressed in terms of wavelengths

$$v = \frac{x}{\lambda}, \quad \mu = \frac{y}{\lambda} \quad (14.51)$$

When the field is observed at two points $M(v, \mu)$ and $M'(v', \mu')$, the algebraic distances are defined as

$$\Delta\nu = \nu - \nu', \quad \Delta\mu = \mu - \mu' \tag{14.52}$$

The Van Cittert-Zernicke theorem is then expressed in terms of the Fourier Transform

$$C(\Delta\nu, \Delta\mu) = \int_{-\infty}^{\infty}\int_{-\infty}^{\infty} T(\alpha, \beta)\exp\left[j2\pi(\Delta\nu\alpha + \Delta\nu\beta)\right]d\Omega \tag{14.53}$$

Remarks

- The limits of the integrals are formally extended to infinity. In fact the extent of the function T is limited; the sources are supposed to be above the horizon: $\alpha^2 + \beta^2 \leq 1$.

- The relations (14.50) and (14.53) can be inverted to give (in principle) the angular distribution of the sources to be determined from their spatial coherence function. The problem of imagery, as we shall see, is that such a measurement can only give a partial knowledge of the coherence function.

14.4.5 Sampling of the coherence function

Since the angular distribution $T(\tau)$ is limited in extent to the domain $(-1, +1)$, its Fourier Transform - the coherence function - may be sampled with a sample interval a such that $a/\lambda = 1/2$.

If we consider two points with coordinates ν and ν', which are respectively multiples n and n' of this sample interval, then

$$\nu = n\frac{a}{\lambda} = \frac{n}{2}, \quad \nu' = n'\frac{a}{\lambda} = \frac{n'}{2}$$

The sampled coherence function is then written

$$C(n - n') = \int_{-1}^{+1} T(\tau)\exp\left[j\pi(n - n')\tau\right]d\tau \tag{14.54}$$

Setting $m = n - n'$, this becomes

$$C(m) = \int_{-1}^{+1} T(\tau)\exp(j\pi m\tau)d\tau \tag{14.55}$$

We recognize this expression as the coefficient in a Fourier series expansion of $T(\tau)$, which, over the interval $(-1, +1)$ is expressed in the form

$$T(\tau) = \sum_{-\infty}^{+\infty} C(m) \exp(-j\pi m \tau) \qquad (14.56)$$

14.4.6 Measurement of the coefficients of correlation or covariance $C(n\text{-}n')$

Consider an array of elements at $\lambda/2$ spacing (Fig. 14.10). The signals $E_n(t)$ and $E_{n'}(t)$ received by the elements n and n' allow the covariances to be evaluated by integration with respect to time of their hermitian product. Stationarity supposes that the incident signals allow the mean with respect to time to be assimilated by the probabilistic mean

$$C(n,n') = C(n-n') = \lim_{T \to \infty} \frac{1}{T} \int_{t}^{t+T} E_n(t) E_{n'}^* dt \qquad (14.57)$$

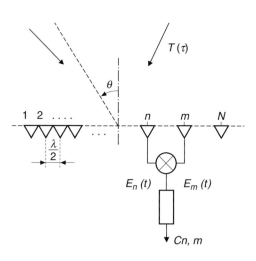

Fig. 14.10 Coherence function: sampling of the field received by means of an array, evaluation of the covariances C_{nm}.

In fact, since the integration time is necessarily finite, it is only possible to make an *estimate* $\hat{C}_{n,n'}$ of the covariances.

To separate the real and imaginary parts of the covariances, one of the elements is chosen as a reference ($n = 0$, for example). The signal $E_0(t)$ that it receives is split into two parts, of which one is shifted by $\pi/2$ with respect to the other. Each element signal $E_n(t)$ is multiplied by this reference by means of a complex multiplier, giving components in phase and in quadrature with $E_0(t)$

$$a_{0,n}(t) = \mathrm{Re}\{E_0(t)E_n^*(t)\}$$
$$b_{0,n}(t) = \mathrm{Im}\{E_0(t)E_n^*(t)\} \qquad (14.58)$$

Then, the required estimate of the covariances are provided by integrators

$$\hat{C}_{0,n} = \overline{a_{0,n}(t)} + j\overline{b_{0,n}(t)} \qquad (14.59)$$

Of course, this processing can equally be done digitally rather than in analogue form (after frequency downconversion).

In radioastronomy, two antenna elements are often used: one fixed, and the other mounted on a carriage which moves along a rail track, so that the measurement of the covariances is only limited by the length of the track.

14.4.7 The covariance matrix

The covariances $C(n,n')$ defined in the preceding way can be arranged in matrix form, to give the 'covariance matrix' associated with the distribution $T(\tau)$. If, for example, we have an array of N elements denoted from 0 to $N-1$, the covariance matrix is hermitian, of order N. Thus

$$\left.\begin{array}{c} C_{n'-n} = C_{n-n'}^* \\[2mm] 0 \leq n - n' \leq N - 1 \end{array}\right\} \qquad (14.60)$$

$$
\mathbf{C} =
\begin{bmatrix}
C_0 & C_1 & C_2 & \cdots & & C_{n-1} \\
C_1^* & C_0 & C_1 & \cdots & & C_{n-2} \\
C_2^* & C_1^* & C_0 & \cdots & & \\
\vdots & & & & & \vdots \\
\vdots & & & & & \\
\vdots & & & & & \\
C_{n-1}^* & & & & C_1^* & C_0
\end{bmatrix}
$$

$$(14.61)$$

If the sources are truly incoherent, the elements on any diagonal parallel to the principal diagonal are equal, which is the characteristic of a Toeplitz matrix. This provides an experimental way of verifying whether a set of sources is actually incoherent.

14.5 HIGH RESOLUTION IMAGERY AND THE MAXIMUM ENTROPY METHOD

14.5.1 Introduction

An unknown spatial distribution of incoherent sources $T(\tau)$ (Fig. 14.11) can be considered as the spectral density of a random function represented by the field $E(v, t)$ observed at an array located on the $x'x$ axis ($v = x/\lambda$). We have seen that $T(\tau)$ can be expressed in terms of the covariance samples C_m by a Fourier series (§3.4, equation (14.56)). Since the number of array elements is necessarily finite, so is the number N of covariances (from C_0 to C_{N-1}). We know also that $C_{-m} = C_m^*$. The problem under consideration is *to find an estimate* $\hat{T}(\tau)$ *of T(τ) given only these N samples*. In 1967, the concept of *entropy* (considered more fully in Appendix 14A) allowed Burg to find a solution to this problem. Other methods are described in §14.6.

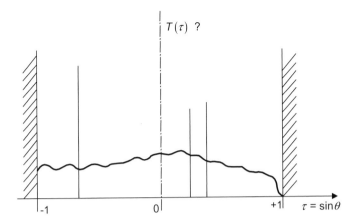

Fig. 14.11 Unknown spatial distribution $T(\tau)$ of incoherent sources.

14.5.2 Classical method of 'correlogram'

The classical method consists of performing a filtering operation on $T(\tau)$, limiting the trigonometric series (equation (14.56)) to the covariance samples. This gives the estimate

$$\hat{T}_1(\tau) = \sum_{n=-(N-1)}^{N-1} C_n \exp\left(-j\pi n\tau\right) \tag{14.62}$$

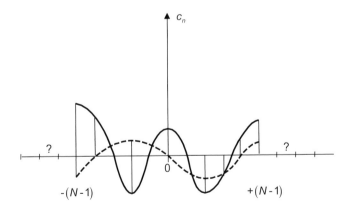

Fig. 14.12 Measured covariance samples.

It is important to realize that this procedure *arbitrarily assigns the value zero to all samples not actually measured* (Fig. 14.12)

$$C_N = 0, \; C_{N+1} = 0, \quad \text{etc.} \tag{14.63}$$

This produces an estimate which, although closest to $T(\tau)$ in the least-squares sense, can give *negative* values which are non-physical (see Chapter 13). To avoid this difficulty, the usual method is to distort the measured values C_n by weighting them with a function W_n (this is equivalent to multiplying the distribution of C_n by a 'window' function). This gives

$$\hat{T}_2(\tau) = \sum_{n=-(N-1)}^{N-1} W_n C_n \exp(-j\pi n\tau) \tag{14.64}$$

This usually avoids negative values, but sometimes leads to soft maxima which degrade the angular resolution.

14.5.3 Method of Maximum Entropy[6]

In 1967, at a conference on geophysics, Burg presented a method which overcomes the arbitrary choice of C_n for $n > N$.

The idea is the following: among the ensemble of estimates $\hat{T}(\tau)$ of $T(\tau)$ which give the same first N values of the covariances C_n, choose for the non-measured covariances values such that the entropy density *remains stationary* for $n \geq N$ (see Appendix 14A).

The entropy S is a measure of the *uncertainty* once the first N samples have been taken. The choice of values assigned to the samples for $n \geq N$ *should not reduce this uncertainty* (which is not the case with the correlogram method).

14.5.4 Estimation of *T* under conditions of Maximum Entropy

It can be shown that the entropy density of a random series whose covariances C_n are known can be simply expressed in the form (Appendix 14A):

$$H(C_n) = \int_{-1}^{+1} \log T(\tau)\, d\tau \tag{14.65}$$

---

[6] Burg, J.P., *Maximum Entropy Spectral Analysis*, Stanford University, 1975.

The stationarity of H with respect to the values assigned to the covariances for $n > N$ is expressed by

$$\frac{\partial H}{\partial C_n} = 0, \quad |n| \geq N \tag{14.66}$$

Making use of the expression (14.56) for $T(\tau)$ in terms of the covariances by means of a Fourier series, this equation can be written by differentiating under the summation sign

$$\int_{-1}^{+1} \frac{1}{\hat{T}(\tau)} \frac{\partial}{\partial C_n} \exp(-j\pi n\tau) d\tau = 0 \tag{14.67}$$

or

$$\int_{-1}^{+1} \frac{1}{\hat{T}(\tau)} \exp(-j\pi n\tau) d\tau = 0 \quad (|n| \geq N) \tag{14.68}$$

For $|n| < N$ we can write

$$\int_{-1}^{+1} \frac{1}{\hat{T}(\tau)} \exp(-j\pi n\tau) d\tau = \alpha_n \tag{14.69}$$

We recognize in this expression the coefficients of a Fourier series expansion of $1/\hat{T}(\tau)$, limited to $N-1$ terms. This therefore gives

$$\hat{T}(\tau) = \frac{1}{\displaystyle\sum_{-(N-1)}^{N-1} \alpha_n \exp(jn\pi\tau)} \tag{14.70}$$

14.5.5 Factorization of $T(\tau)$ - properties

The positive nature of $T(\tau)$ allows us to factorize equation (14.70) in the following forms

$$T(\tau) = \frac{b_0^2}{\left| \displaystyle\sum_0^{N-1} a_n \exp(jn\pi\tau) \right|^2} \tag{14.71}$$

$$T(\tau) = \frac{b_0^2}{\displaystyle\prod_{k=1}^{N} \left[1 - p_k \exp(-j\pi\tau) \right]\left[1 - p_k^* \exp(+j\pi\tau) \right]} \tag{14.72}$$

This latter form justifies the use of the term 'all-pole model' sometimes used for the form of $\hat{T}(\tau)$. We obtain a discrete spectrum. This type of spectrum is very appropriate to represent distributions of point sources, and demonstrates much better angular resolution than classical methods (Fig. 14.13).

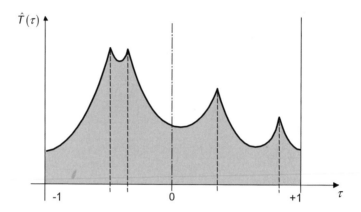

Fig. 14.13 Type of angular spectral density $T(\tau)$ obtained with Burg's method.

14.5.6 Determination of the coefficients a_n in equation (14.71)

From the known samples of the covariances we can identify the expressions (14.56) and (14.71) for the distribution $T(\tau)$. We end up with the Yule-Walker equations[7] in the form of a set of linear relations

$$[\mathbf{C}]\begin{bmatrix} 1 \\ a_1 \\ a_2 \\ \vdots \\ a_{N-1} \end{bmatrix} = \begin{bmatrix} b_0 \\ 0 \\ 0 \\ \vdots \\ 0 \end{bmatrix} \qquad (14.73)$$

Inversion of the covariance matrix $\mathbf{C}$ therefore allows us to evaluate the coefficients a_n. Taking into account the Toeplitz nature of the matrix, a number of interesting

[7] Papoulis, A., *Probability, Random Variables and Stochastic Processes*, p410 *et seq*, McGraw-Hill, New York, 1965.

algorithms have been developed for this, of which the best-known is that due to Levinson and Durbin[8].

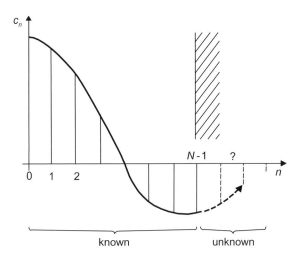

Fig. 14.14 Prediction of unmeasured covariances. AR method.

The relations in (14.73) can equally be interpreted as an extrapolation formula which allows the unmeasured covariances to be evaluated (under the hypothesis of maximum entropy) from the measured ones (Fig. 14.14). We have, in effect, for $n \neq 0$

$$C_n = -\sum_{k=1}^{N-1} C_{n-k} a_k \qquad (14.74)$$

This is a type of *linear prediction* whose form is equivalent to that of a process known in time-domain signal processing as 'auto-regressive'. This is why this technique is often known as the AR method.

[8] Levinson, N., 'The Wiener RMS criterion in filter design and prediction' *Journal of Mathematics and Physics*, Vol. 25, 1947.

Durbin, J., 'The fitting of time series models', *Revue de l'Institut Internationale de Statistique*, Vol. 28, 1960.

14.5.7 Generalization: ARMA model

Since the Maximum Entropy method - or the all-pole method - represents the angular energy density $T(\tau)$ in the form of the inverse of a polynomial in the variable $Z = \exp(j\pi\tau)$, the correlogram method gives a polynomial representation of $T(\tau)$ in the same variable, sometimes called the 'all-zero model', equivalent to a spectral representation in the time domain called MA ('Moving Average'). We can combine the two methods in the form of a representation of a rational function of Z, known as the ARMA model. In principle, the interest in such a generalized representation is to represent the angular or spectral densities to include both zeros as well as sharp maxima.

14.5.8 Numerical example

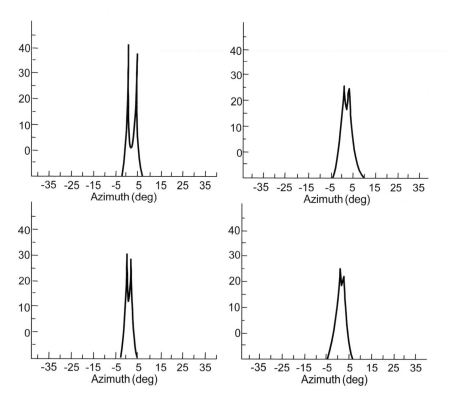

Fig. 14.15 AR model of order 9. Response of an array to two point sources. Angular separation (from left to right and top to bottom): 4°, 3°, 2° and 1.5°. In the conventional sense the array has a 3 dB beamwidth of 5°.

The example considered here (Fig. 14.15) considers a linear array whose resolution (beamwidth) in the conventional sense is 5 degrees. There are two point sources, of the same strength (producing a signal-to-noise ratio at each element of +30 dB). The estimation of the covariances is based on the average of 100 independent samples.

The AR model used is of order 9. The angular separations between the sources are respectively 4°, 3°, 2° and 1.5°. It is clear that the sources are clearly resolved, even though the separations are less than the classical resolution of the array.

14.5.9 Minimum redundance arrays

Consider a conventional array of N elements. It can be seen that the covariance C_0 can be measured N times, the covariance C_1 $(N–1)$ times, the covariance C_2 $(N–2)$ times, and so on. The covariance of maximum rank is $C_{N–1}$. Radioastronomers would say that this is a *redundant* array.

The problem posed in the introduction (§14.1) leads here to the following question: 'is it possible to configure the N elements to obtain a non-redundant array and, at the same time, to obtain a greater number of samples of the covariance C?'

It appears that it was Arsac[9] who first proposed the use of a non-redundant array, to analyse microwave radiation from the sun.

We give here some examples of non-redundant arrays for $N = 1, 2, 3, 4$.

The set of measurable covariance samples with such arrays is called the co-array (Fig. 14.16).

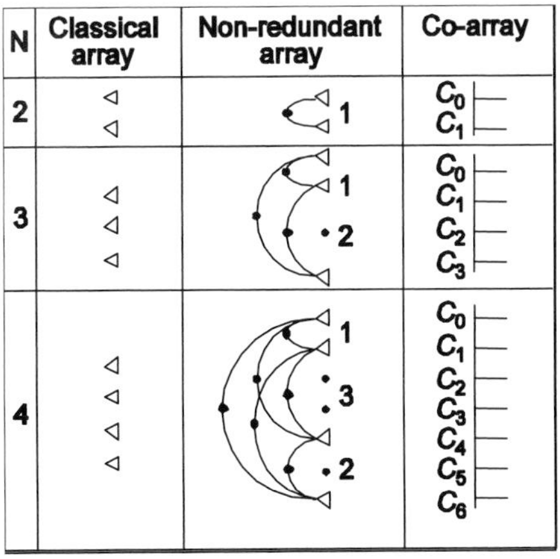

Fig. 14.16 Non-redundant arrays and co-arrays for $N = 2, 3, 4$.

---

[9] Arsac, J., 'Nouveau réseau pour l'observation radio-astronomique de la brillance du soleil à 9350 Mc/s', *C.R. de l'Académie des Sciences de Paris*, Vol. 240, pp942-945, February 1955.

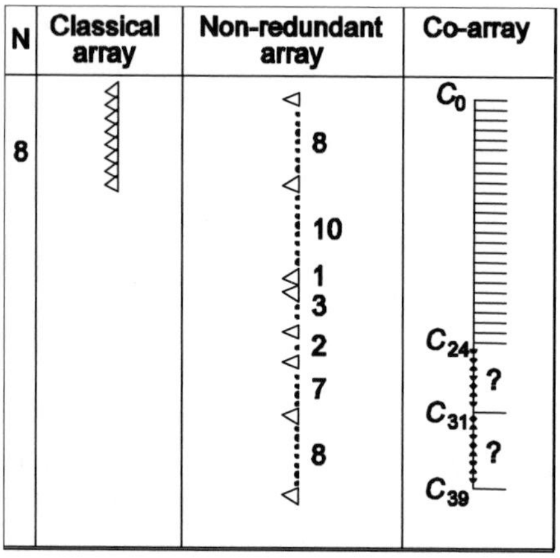

Fig. 14.17 Non-redundant arrays and co-arrays for $N = 8$.

It can be seen already that for $N = 4$, the use of a non-redundant array allows us to *double* the number of samples C_n, and hence the resolution of the system. Bracewell[10] has shown that arrays of four elements at most can give zero redundancy. For greater numbers of elements there does not seem to be, at present, any general method to determine the array configuration for minimum redundancy. This problem has been examined by Leech[11] in the context of number theory. This has given solutions up to $N = 11$ and gives approximately $R = 4/3$ for the degree of redundancy for large values of N.

For a perfectly non-redundant array the number of combinations of pairs of elements, which is $N(N-1)/2$, is also the number of covariance samples *without gaps*. In the general case this number is given by

$$N_{max} = \frac{N(N-1)}{2R} \tag{14.75}$$

Above N_{max} the covariance samples can be found, but with certain 'errors' (Fig. 14.17).

We are therefore forced to use spectral estimation methods with an incomplete set of samples; the interpolation of the missing covariance elements is the same type of problem as their extrapolation.

---

[10] Bracewell, R.N., *The Fourier Transform and its Applications*, McGraw-Hill, 1978.
[11] Leech, J., 'On the representation of 1, 2, ... n by differences', *J. London Math. Soc.*, Vol. 31, pp. 160-169, 1956.

14.6 OTHER METHODS OF SPECTRAL ESTIMATION

14.6.1 Introduction

Besides the condition for incoherence (§14.4.20, the maximum entropy and ARMA methods no longer need *a priori* hypotheses about the unknown object distribution. In the following methods a discrete object distribution of radiating point sources is assumed.

The maximum entropy method described in §14.5.3 is just one of a class of so-called superresolution methods, all of which attempt to resolve signals of closer angular separation than the classical Rayleigh λ/D limit (where λ is the signal wavelength and D the array length). A more intuitive demonstration that this can be done is provided by considering the 'tree' structure of phase shifters and combiners shown in Fig. 14.18, and known as the 'Davies Beamformer'[12].

Suppose that there are a set of incident signals S_1, S_2, ... incident on a uniformly-spaced linear array at angles θ_1, θ_2, ... to the array normal. The first signal can be nulled at the outputs of the first set of combiners, by setting all the phase shifters in the first layer to

$$\phi_1 = \frac{2\pi d}{\lambda}\sin\theta_1 \qquad (14.76)$$

where d is the interelement spacing.

The signals at the outputs of the first layer therefore contain all the signals except S_1, and the phase shifter settings ϕ_1 allow the direction of arrival of S_1 to be determined, via equation (14.76). In an exactly similar way, the second signal can be nulled at the outputs of the second set of combiners by setting all the phase shifters in the second layer to

$$\phi_2 = \frac{2\pi d}{\lambda}\sin\theta_2 \qquad (14.77)$$

and the phase shifter settings ϕ_2 allow the direction of arrival of S_2 to be determined.

This procedure can be repeated, nulling each signal at the appropriate layer of the tree. Clearly, then, the directions of arrivals of the signals can be determined, to a

[12] Davies, D.E.N., 'Independent angular steering of each zero of the directional pattern of a linear array', *IEEE Trans. Antennas & Propagation*, Vol. AP-15, March 1967, pp296-298.

precision which can in principle be significantly better than the λ/D Rayleigh limit, and depends in practice on their signal-to-noise ratios.

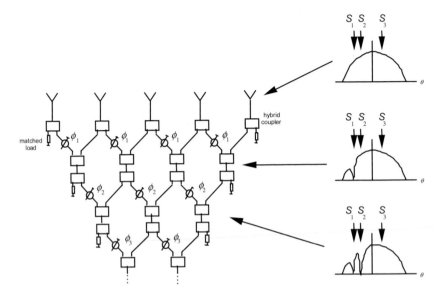

Fig. 14.18 The Davies Beamformer.

14.6.2 The MUSIC algorithm

In 1979, Schmidt presented one of the earliest, and certainly one of the best-known of the superresolution algorithms, which he named MUltiple SIgnal Classification, or MUSIC[13],[14].

[13] Schmidt, R.O., 'Multiple emitter location and signal parameter estimation', *Proc. RADC Spectrum Estimation Workshop*, RADC-TR-79-63, Rome Air Development Center, Rome, NY, USA, Oct 1979, pp243-258; reprinted in *IEEE Trans. Antennas & Propagation*, Vol. AP-34, March 1986, pp276-280.

[14] A similar formulation was published earlier the same year by Bienvenu: Bienvenu, G., 'Influence of the spatial coherence of the background noise on high resolution passive methods', *Proc. IEEE International Conference on Acoustics, Speech and Signal Processing*, Washington DC, IEEE Publication No. 79CH1379-7 ASSP, 2-4 April 1979, pp306-309.

Consider an *M*-element array of arbitrary geometry, with *n* incident signals F_1, F_2, ... ,F_n ($n \leq M$). The element signals X_1, X_2, ... ,X_M can be written as

$$
\begin{aligned}
X_1 &= a_{11}F_1 + a_{21}F_2 + \cdots + a_{D1}F_n + W_1 \\
X_2 &= a_{12}F_1 + a_{22}F_2 + \cdots + a_{D2}F_n + W_2 \\
\vdots &= \vdots + \vdots + \cdots + \vdots + \vdots \\
X_M &= a_{1M}F_1 + a_{2M}F_2 + \cdots + a_{DM}F_n + W_M
\end{aligned}
$$

or in matrix notation

$$\mathbf{X} = \mathbf{AF} + \mathbf{W} \tag{14.78}$$

Here the terms W_1, W_2, ... W_M represent the noise at the array elements. This noise may be either internally or externally-generated.

The matrix **A** is known as the *array manifold*, and its coefficients depend on the element positions and directional responses. The *j*th column of **A** is a mode vector **a**(θ) of responses of the array to the direction of arrival θ_j The mode vectors and **X** can each be visualized as vectors in *M*-dimensional space, and **X** is a linear combination of mode vectors, where the coefficients are the elements of **F**.

The covariance matrix **C** is formed by averaging a number of 'snapshots' of the element signals

$$\mathbf{C} = \overline{\mathbf{X}\mathbf{X}^*}$$

$$= \overline{\mathbf{AFF}^*\mathbf{A}^*} + \sigma^2\mathbf{I} \tag{14.79}$$

under the assumption that the signals and noise are uncorrelated, and where the elements of the noise vector **W** are zero mean and of variance σ^2.

The covariance matrix can be broken down into its eigenvectors and eigenvalues

$$\mathbf{C} = \sum_{i=1}^{N} \lambda_i \mathbf{e}_i \mathbf{e}_i^* = \mathbf{ELE}^* \tag{14.80}$$

The eigenvalues λ_i of the covariance matrix will be

$$\lambda_i > \sigma^2 \text{ for } i = 1, \dots, n, \text{ and } \lambda_i = \sigma^2 \text{ for } i = n+1, \dots, M \tag{14.81}$$

The number of incident signals can therefore be determined by inspection of the relative magnitudes of the eigenvalues, or in some situations this information may be known *a priori*.

Consequently, the covariance matrix can be partitioned into an *n*-dimensional subspace spanned by the incident signal mode vectors and an *M–n* dimensional subspace spanned by the *M–n* noise eigenvectors

$$\mathbf{C} = \mathbf{E}_S \mathbf{L}_S \mathbf{E}_S^* + \mathbf{E}_N \mathbf{L}_N \mathbf{E}_N^* \qquad (14.82)$$

where $\mathbf{E}_S$ is the M by n signal subspace and $\mathbf{E}_N$ is the M by $M-n$ noise subspace.

The MUSIC algorithm then estimates the angular spectrum $P(\theta)$ of the incident signals according to

$$P(\theta) = \frac{1}{\mathbf{a}^*(\theta)\mathbf{E}_N \mathbf{E}_N^* \mathbf{a}(\theta)} \qquad (14.83)$$

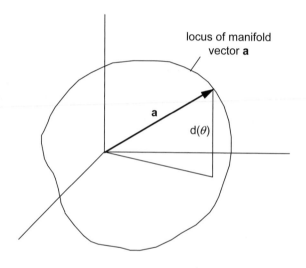

Fig. 14.19 Geometric interpretation of MUSIC algorithm. When the projection d(θ) of **a** is equal to zero, the vector lies in the signal subspace.

Figs. 14.20 and 14.21 show a simulation of the operation of the MUSIC algorithm with an 8-element linear array, with three signals incident at angles at $-30°$, $+30°$ and $+50°$, at signal-to-noise ratios of 25 dB, 15 dB and 20 dB respectively. Fig. 14.20 shows the eigenvalue spectrum, from which it can be seen that there are three large eigenvalues corresponding to the three signals, and five smaller ones corresponding to noise level. Fig. 14.21 plots the MUSIC function (equation (14.83)). Also shown, for comparison is the response obtained from the conventional beamforming algorithm, using an amplitude taper to lower the sidelobes. The MUSIC algorithm is able to resolve the three signals comfortably, whilst the conventional beamformer algorithm is unable to resolve the two closely spaced signals at all.

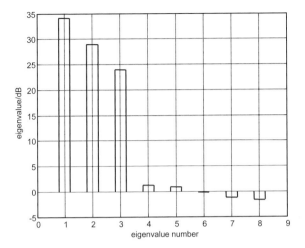

Fig. 14.20 Eigenvalue spectrum. There are three large eigenvalues, corresponding to the three signals, and five smaller ones, corresponding to noise. (figure courtesy of Dr David Brandwood).

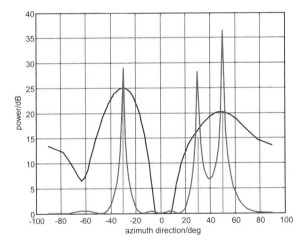

Fig. 14.21 Performance of the MUSIC algorithm with an 8-element linear array. Signal incident at −30°, +30° and +50°, at signal-to-noise ratios of 25 dB, 15 dB and 20 dB. Also shown, for comparison is the response obtained from the conventional beamforming algorithm, using an amplitude taper to lower the sidelobes. The MUSIC algorithm is able to resolve the three signals comfortably, whilst the conventional beamformer algorithm is unable to resolve the two closely spaced signals at all (figure courtesy of Dr David Brandwood).

14.6.3 Illustration of the MUSIC algorithm

Consider an array of $M = 3$ isotropic and aligned elements, at $\lambda/2$ spacing, and with $n = 2$ incoherent incident sources.

The dimension of the 'noise' subspace is $M - n = 1$. It has a single eigenvector whose minimum eigenvalue is λ_0.

Letting

$$\mathbf{e} = \begin{bmatrix} a \\ b \\ c \end{bmatrix}$$

(14.84)

The steering vector is

$$\mathbf{F}(\tau) = \begin{bmatrix} 1 \\ \exp j\pi\tau \\ \exp 2 j\pi\tau \end{bmatrix}$$

(14.85)

Let τ_1, τ_2 be the wanted directions, and let $\alpha = \exp j\pi\tau_1$, $\beta = \exp j\pi\tau_2$. The components of $\mathbf{e}$ are obtained from their *orthogonality to the signal subspace*:

$$\mathbf{F}^+(\tau_1)\mathbf{e} = 0; \quad \mathbf{F}^+(\tau_2)\mathbf{e} = 0$$

(14.86)

or

$$\begin{bmatrix} 1 & \alpha & \alpha^2 \end{bmatrix}\begin{bmatrix} a \\ b \\ c \end{bmatrix} = 0; \quad \begin{bmatrix} 1 & \beta & \beta^2 \end{bmatrix}\begin{bmatrix} a \\ b \\ c \end{bmatrix} = 0$$

We find that:

$$a = c\alpha\beta; \quad b = -c(\alpha + \beta)$$

(14.87)

In practice the eigenvalue is determined from the estimation of the covariance matrix [**C**]. The wanted directions are then given by:

$$\mathbf{F}^+(\tau)\mathbf{e} = 0$$

(14.88)

or:

$$a + b\exp j\pi\tau + c\exp 2 j\pi\tau = 0$$

(14.89)

In fact, because of estimation errors the roots of this equation are not all of unit modulus. The MUSIC spectrum is given by:

$$S(\tau) = \frac{1}{\left| a + b\exp j\pi\tau + c2 j\pi\tau \right|^2}$$

(14.90)

As τ varies, $S(\tau)$ has sharp maxima in the neighborhood of τ_1 and τ_2. Also, by replacing a, b, c by their values from (14.87), we obtain, to within a multiplicative factor:

$$S(\tau) = \frac{1}{16\sin^2\frac{\pi}{2}(\tau - \tau_1)\sin^2\frac{\pi}{2}(\tau - \tau_2)}$$

(14.92)

14.6.4 Other superresolution algorithms

A large number of other superresolution algorithms have been formulated and evaluated, giving improved performance over the basic MUSIC algorithm. These include search-free methods such as ESPRIT (Estimation of Signal Parameters by Rotational Invariance Technique) and TAM, one-dimensional parameter search methods such as MUSIC and Capon's MVDR, and multidimensional search schemes such as IMP (Incremental MultiParameter), stochastic and deterministic max-likelihood, and WSF. These algorithms are described in detail in reference [8] in the list at the end of this Chapter, which the interested reader should consult for full details.

	Arrays of arbitrary geometry	Equi-spaced linear arrays
Translational invariance	ESPRIT [LS, TLS] †	TAM
One-dimensional parameter search	MVDR (Capon) MUSIC (Schmidt)	Maximum Entropy (Burg) minimum norm (KT)
Multi-dimensional parameter search	IMP (Clarke) WSF Stochastic max. likelihood Deterministic max. likelihood	† array must possess at least one translational invariance

Table. 14.1 Classification of superresolution algorithms.

Table 14.1 shows an attempt to classify these algorithms, into those which work with arrays of arbitrary geometry and those which are formulated for uniformly-spaced linear arrays. The algorithms are also divided into those based on translational invariance, one-dimensional parameter searches, and multi-dimensional parameter searches.

A particular problem occurs when the incident signals are correlated - such as would be the case with multipath, for example. In this case the MUSIC algorithm does not perform well, and a 'pre-whitening' process is necessary to decorrelate the input signals applied to the algorithm. This may be done by taking successive subaperture samples of the complete array, and using these as the inputs to the algorithm[15].

14.6.5 Superresolution with circular arrays

Superresolution estimators are usually directly applied to the signals received by the array elements, but it can often be beneficial to preprocess the array outputs by means of a beamforming network, transforming the superresolution scheme from 'element space' to 'beam space'. Similarly, with circular arrays excited in terms of phase modes (see §11.8.3), it is possible to operate on the phase mode signals themselves, thereby giving superresolution in 'phase mode space'[16]. This works because the phase modes can be treated in exactly the same way as the elements of an equispaced linear array, and so any processing technique developed for a linear array can equally be applied to the phase modes of a circular array. This technique has the advantages that the phase modes are omnidirectional, and hence that the same performance is obtained over the full 360° of azimuth, and also that the performance is broadband (over potentially an octave or more).

[15] Shan, T.J. and Kailath, T., 'Adaptive beamforming for coherent signals and interference', *IEEE Trans. Acoustics, Speech & Signal Processing*, Vol. ASSP-33, No. 3, pp527-536, June 1985.

[16] Eiges, R. and Griffiths, H.D., 'Mode-space spatial spectral estimation for circular arrays'; *IEE Proc. Radar, Sonar and Navigation*, Vol. 141, No. 6, pp300-306, December 1994.

14.7 SPATIAL FILTERING

14.7.1 Introduction

The goal of all types of imaging technique, whether they are traditional (convolution) or high resolution, is to provide the most faithful image possible of the electromagnetic (or acoustic) environment. But among the signals present in the environment, some are useful and desired, and some are not useful, or even undesired, since they interfere with the wanted signals (e.g. jammers).

The concept of spatial filtering is to modify the antenna pattern on receive according to some criterion which optimizes reception of the wanted signals. The antenna is no longer a passive subsystem acting just as a transparent transducer, but an active device which controls its behaviour intelligently.

We describe here two conceptual approaches. The first is to regard the antenna as responding dynamically to the environment as it changes, i.e. an adaptive antenna. The second approach is to directly derive the optimal array parameters according to some particular optimization criterion, i.e. an optimum array. This latter approach is effectively an extension to the spatial domain of the concept of the optimal Wiener filter.

14.7.2 What is an adaptive array ?

An adaptive array is characterized by the ability to adapt its radiation pattern (i.e. its characteristic radiation function) as a function of the electromagnetic environment, according to some prescribed optimization criterion, by means of a processor which controls the antenna excitation.

For example, a phased array forms the sum of the signals from each array element, with each element signal weighted by a phase shifter and an attenuator. These are controlled by a processor, operating on the received signals according to a particular optimization criterion (Fig. 14.22).

There are several different optimization criteria that can be used. Two examples are:

- in the absence of the wanted signal, to minimize the total received noise;
- in the presence of the wanted signal, to maximize the ratio of wanted signal to total noise. This criterion leads to the concept of the optimum array.

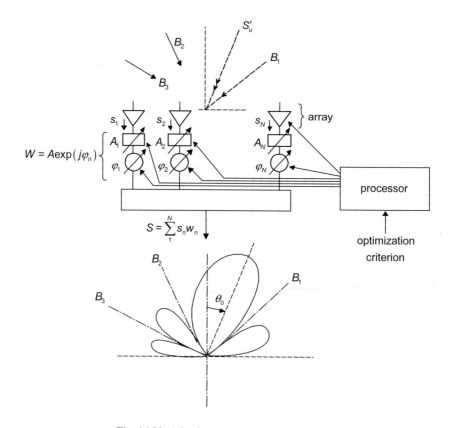

Fig. 14.22 Adaptive array and resulting pattern.

14.7.3 Simple example: two-element array

Consider an array of two omnidirectional elements (Fig. 14.23), of separation a. A plane wave is incident from a jammer B in the direction **u** making an angle θ with the normal to the array. The analysis which follows is in terms of narrowband signals.

If s_1 is the signal received by the first element, the signal at the second element is phase shifted with respect to the first, and is of the form

$$s_2 = s_1 \exp\left(j \frac{2\pi a}{\lambda} \sin \theta_B \right) \qquad (14.93)$$

This signal is weighted by a coefficient W, which consists of a phase shift φ and an attenuation A

$$W = A\exp(j\varphi) \tag{14.94}$$

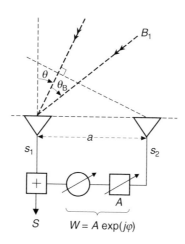

Fig. 14.23 Two-element adaptive array and jammer B.

The resultant signal is thus of the form

$$s_B = s_1 + Ws_2 \tag{14.95}$$

The problem is therefore to choose W so that the signal s_B from the jammer is cancelled. The solution is obtained immediately from

$$s_1\left[1 + A\exp(j\varphi)\exp\left(j\frac{2\pi a}{\lambda}\sin\theta_B\right)\right] = 0 \tag{14.96}$$

from which

$$\hat{A} = 1, \quad \hat{\varphi} = \left(\pi - \frac{2\pi a}{\lambda}\sin\theta_B\right) \tag{14.97}$$

so

$$\hat{W} = -\exp\left(-j\frac{2\pi a}{\lambda}\sin\theta_B\right) \tag{14.98}$$

What is the radiation pattern of the array ? With a signal incident from an arbitrary direction θ, the signal received is

$$s(\theta) = s_1 \left[1 + \hat{W} \exp\left(j\frac{2\pi a}{\lambda} \sin \theta \right) \right] \tag{14.99}$$

The received signal is therefore proportional to the characteristic function

$$F(\theta) = 1 + \hat{W} \exp\left(j\frac{2\pi a}{\lambda} \sin \theta \right)$$

or

$$F(\theta) = 1 - \exp\left[j\frac{2\pi a}{\lambda} (\sin \theta - \sin \theta_B) \right] = 2j\sin\left[\frac{\pi a}{\lambda} (\sin \theta - \sin \theta_B) \right] \tag{14.100}$$

The radiation pattern has a zero in the direction of the jammer, as expected (Fig. 14.24).

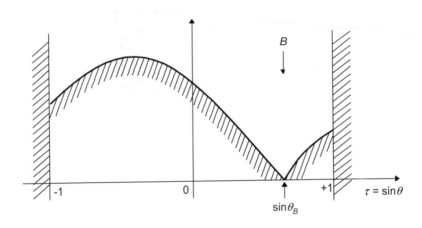

Fig. 14.24 Optimized pattern.

Note: the values A and φ must be set very accurately, since errors result in imperfect suppression of the jammer (i.e. finite null depth). Fig. 14.25 shows contours of constant null depth as a function of errors in A and φ.

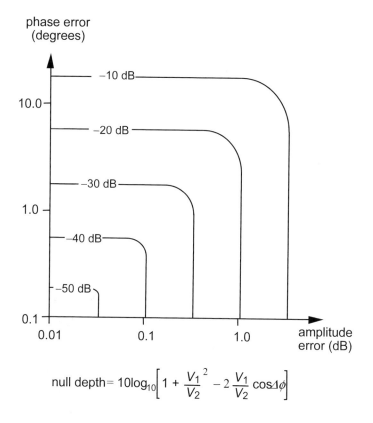

$$\text{null depth} = 10\log_{10}\left[1 + \frac{V_1}{V_2}^2 - 2\frac{V_1}{V_2}\cos\Delta\phi\right]$$

Fig. 14.25 Effect of amplitude and phase errors on null depth.

14.7.4 Howells-Applebaum correlation loop[17]

In the preceding example, the direction of the jammer was supposed known. The major question, on which rests all of the theory that follows, is: 'Is it possible to suppress a jammer s_B *without* knowing its direction ?' The answer turns out to be 'yes'. The method uses the correlation loop due to Howells and Applebaum, which

---

[17] Howells, P., 'Intermediate frequency side-lobe canceller', US Patent 3202990, 24 August 1965.

Applebaum, S.P., 'Adaptive arrays', Syracuse Univ. Res. Corp. Tech. Rep. SPL TR 66-1, August 1966.

works on the criterion that: 'the correlation between the resultant signal s and the signal received by one of the array elements should be zero'.

This criterion is justified by the fact that the signal, s_2 for example, from an isotropic element necessarily includes the jammer. If the signal S also includes the jammer then the correlation between them cannot be zero. A null therefore implies the absence of the jammer in S.

This correlation is performed by means of a coherent demodulator, or more precisely by two mixers in quadrature (Fig. 14.26). The mixers are followed by low-pass filters or integrators, and the filtered signals modify the phase φ and the attenuation A so as to null the jammer.

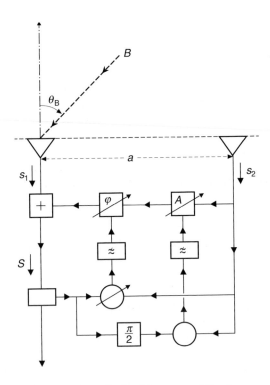

Fig. 14.26 Howells-Applebaum correlation loop.

Calculation of the weights

The criterion leads to

$$\overline{\left(s_1 + Ws_2\right)s_2^*} = 0 \qquad (14.101)$$

so that

$$\hat{W} = -\frac{\overline{s_1 s_2^*}}{|s_2|^2} \tag{14.102}$$

Example: an incident noise signal $f(t)$ comes from an unknown direction θ_B. We can write, where K is a constant

$$s_1 = Kf(t) , \qquad s_2 = Kf(t)\exp\left(j\frac{2\pi a}{\lambda}\sin\theta_B \right)$$

$$\hat{W} = -\exp\left(-j\frac{2\pi a}{\lambda}\sin\theta_B \right)$$

which gives the same result (14.89) as before, without knowing θ_B.

Important note: with this criterion, how do we avoid nulling a wanted signal rather than a jammer ? To avoid this problem it is necessary to use some additional method to distinguish one signal from the other.

(1) *Difference in spectrum*: the spectrum of the jammer will generally be broader than that of the signal. It is therefore possible to use a filter within the correlation loop so that the nulling process is controlled by the jammer and not the wanted signal. More generally, whatever the spectrum of the wanted signal, the filter should have a transfer function which is orthogonal to it.

(2) *Difference in time*: in the case of a radar, the loop is only enabled at the end of each pulse repetition interval, i.e. after all the echoes from a given pulse have been received.
    Example: if a radar has a maximum range $r_0 = 250$ km, the echoes from the most distant targets have a propagation delay of $2r_0/c \approx 1.67$ ms. If the pulse repetition frequency of the radar is 500 Hz there is, at the end of each pulse repetition interval an interval of about 0.33 ms during which the loop can be enabled.

(3) *Difference in polarization*: if the jammer is unpolarized the adaptive loop can be orthogonally-polarized to the wanted signal. If the jammer is polarized, the wanted signal should be transmitted using a polarization orthogonal to the jammer.

(4) *Difference in direction*: in this case the criterion is different, and is considered in the section entitled 'optimum array'.

14.7.5 Minimum noise criterion

We have seen that a two-element array allows, in principle, a single jammer to be nulled. More generally, an N-element array allows $N-1$ jammers to be nulled (we say it has $N-1$ degrees of freedom). However, even if the number of jammers is greater than the number of degrees of freedom, the criterion adopted minimizes the received noise level.

Consider again the previous case, with two or more jammers. The total noise is, as we have seen, of the form

$$s(t) = s_1 + W s_2 \qquad (14.103)$$

The average noise power received is therefore

$$B = \overline{s(t)s^*(t)} = \overline{|s_1|^2} + WW^* \overline{|s_2^2|} + W^* \overline{s_1 s_2^*} + W \overline{s_1^* s_2} \qquad (14.104)$$

When B is minimum, the differential is zero:

$$dB = \left(W \overline{|s_2^2|} + \overline{s_1 s_2^*} \right) dW^* + \left(W^* \overline{|s_2^2|} + \overline{s_1^* s_2} \right) dW = 0$$

Finally, the optimum weight is

$$\hat{W} = -\frac{\overline{s_1 s_2^*}}{\overline{|s_2|^2}} \qquad (14.105)$$

This is the same result as (14.93), independent of the number of sources. The derivation can be generalized to an N-element array and gives the same results.

From (14.95) we obtain the minimum of noise

$$B_{\min} = \overline{|s_1^2|} - \frac{\overline{s_1 s_2^*}}{\overline{|s_2^2|}}$$

14.7.6 Effect of internal receiver noise

The internal receiver noise contributions, which are of course uncorrelated, may be considered as equivalent to an external jammer which is omnidirectional rather than localized. This additional contribution therefore means that the number of jammers is greater than the number of elements, so that the total received noise cannot be completely nulled; this is a particular application of the preceding case.

Consider again the case of a two-element array with internal element noise contributions $b_1(t)$ and $b_2(t)$. These are uncorrelated, with the same variance

$$\overline{b_1 b_2^*} = 0 \ , \quad \overline{\left|b_1\right|^2} = \overline{\left|b_2\right|^2} = N \tag{14.106}$$

The ratio of external noise to internal noise is then

$$R = \frac{N}{\overline{\left|s_1\right|^2}} \tag{14.107}$$

This is zero in the case considered in §14.7.4.

Here the received signals are corrupted by the internal noise contributions. They take the form

$$s_1' = s_1 + b_1(t)$$

$$s_2' = s_2 + b_2(t)$$

The resulting signal is formed

$$s' = s_1' + W s_2'$$

Using the criterion of decorrelation - or of minimum noise - leads to the optimum weight

$$\hat{W}' = \frac{\hat{W}}{1+R} = \frac{1}{1+R}\exp\left(-\frac{j2\pi a}{\lambda}\tau_B\right) \quad \text{(where } \tau_B = \sin\theta_B)$$

The modulus of the characteristic function takes the form

$$\left|F'(\theta)\right| = \left|2\sin\frac{\pi a}{\lambda}(\tau - \tau_B) - jR\cos\frac{\pi a}{\lambda}(\tau - \tau_B)\right| \tag{14.108}$$

The radiation pattern has a minimum of finite depth R, rather than a null, in the direction of the jammer.

14.7.7 Multiple correlation loops: the coherent sidelobe canceller (CSLC)

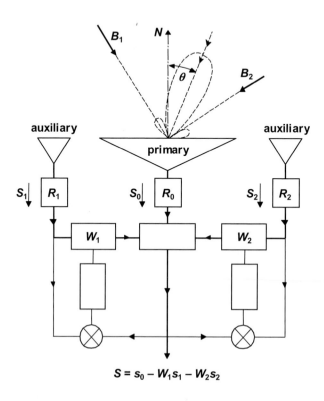

$$S = s_0 - W_1 s_1 - W_2 s_2$$

Fig. 14.27 Adaptive array: coherent sidelobe canceller.

Consider a large directional antenna, such as a parabolic reflector or an electronically-scanned array antenna. This is denoted the 'primary (or principal) antenna'. Associated with the primary antenna are N auxiliary elements, where in principle N is greater than the number of jammers (Fig. 14.27). The signals $s_1, \dots s_N$ which they receive are added to the signal s_0 received by the primary antenna, after multiplication by weight coefficients $W_1, \dots W_N$. The resultant signal is therefore

$$S = s_1 - \sum_{n=1}^{N} s_n W_n \qquad (14.109)$$

These weights are determined from the measured correlations between S and the auxiliary element signals s_n; they become stable when the correlations become zero, and are determined by a set of N linear relations of the form

$$
[\mathbf{C}] \begin{bmatrix} W_1 \\ \vdots \\ W_n \\ \vdots \\ W_N \end{bmatrix} = \begin{bmatrix} \overline{s_0 s_1^*} \\ \vdots \\ \overline{s_0 s_n^*} \\ \vdots \\ \overline{s_0 s_N^*} \end{bmatrix}
\tag{14.110}
$$

This relation contains the covariance matrix $\mathbf{C}$ between the auxiliary element signals

$$
\mathbf{C} = \begin{bmatrix} \overline{|s_1|^2} & \overline{s_1 s_2^*} & \cdots & \overline{s_1 s_N^*} \\ \overline{s_2 s_1^*} & \overline{|s_2|^2} & & \\ \vdots & & \ddots & \\ \vdots & & & \\ \overline{s_N s_1^*} & & & \overline{|s_N|^2} \end{bmatrix}
\tag{14.111}
$$

This is a hermitian matrix. If the jammers are incoherent and the auxiliary elements regularly spaced, then the matrix is Toeplitz in form.

The resulting radiation pattern has nulls in the directions of the jammers. This is explained by the fact that the weight coefficients W_n are optimized in amplitude and phase such that the sidelobes of the primary antenna, combined with the patterns of the auxiliary elements, give a minimum resultant signals in the directions of the jammers (Fig. 14.28).

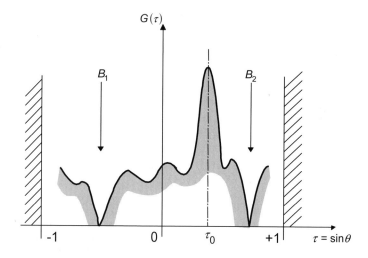

Fig. 14.28 Form of radiation pattern after convergence of the weight coefficients W.

Of course, this system does not work if the jammer lies in the main beam of the primary antenna. But the criterion used here does not need to distinguish between the jammers and the wanted signal in the direction of the main beam.

14.7.8 The optimum array

Consider an N-element array, of $\lambda/2$ spacing, in an environment of jammers of which the angular distribution, in intensity, is fixed by a function $T(\tau)$ (Fig. 14.29). There is a wanted signal in the direction τ_0.

The optimization criterion chosen here is to maximize the ratio between the array gain in this direction, $G(\tau_0)$, and the total noise power T_A received by the antenna. This ratio is sometimes called the 'factor of merit' (§7.2.2, p176 and exercise 7.1, p229)

$$M = \frac{G(\tau_0)}{T_A} \qquad (14.112)$$

The gain $G(\tau_0)$ is proportional to the square modulus of the characteristic function, which in turn is the Fourier Transform of the set of weight coefficients. We can therefore write

$$G(\tau_0) = \left| \sum_1^N W_n \exp\left(j2\pi n\tau_0 \right) \right|^2 \qquad (14.113)$$

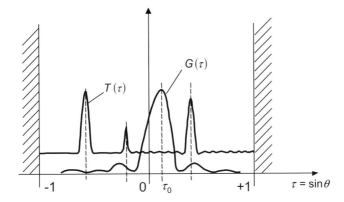

Fig. 14.29 Angular distribution of jammers $T(\tau)$ and array gain $G(\tau)$.

In the following, it is convenient to use matrix notation. We define the 'weight matrix' or 'weight vector' by the column matrix

$$\mathbf{W} = \begin{bmatrix} W_1 \\ \vdots \\ W_N \end{bmatrix} \qquad (14.114)$$

The steering vector defining the pointing direction is written in the same way

$$\mathbf{D}_0 = \begin{bmatrix} 1 \\ \exp(j\pi\tau_0) \\ \vdots \\ \exp(j\pi N\tau_0) \end{bmatrix} \qquad (14.115)$$

The gain $G(\tau_0)$ is therefore written

$$G(\tau_0) = \left| \mathbf{W}^t \mathbf{D}_0 \right|^2 \qquad (14.116)$$

(The symbol $\mathbf{W}^t$ denotes the transpose of $\mathbf{W}$).

In the same way as the gain, the noise power received can be expressed in terms of the weight vector $\mathbf{W}$

$$S = \sum_{1}^{N} s_n W_n \tag{14.117}$$

Using vector notation for the set of element signals

$$\sigma = \begin{bmatrix} s_1 \\ \vdots \\ s_N \end{bmatrix} \tag{14.118}$$

we can write

$$S = \sigma^t W = W^t \sigma \tag{14.119}$$

The noise power is therefore

$$T_A = \overline{|S|^2} = \overline{S^* S} = \overline{(W^t \sigma) \sigma^t W} = W^\dagger \overline{\sigma^* \sigma^t} W \tag{14.120}$$

where $W^\dagger$ is the adjoint matrix of W.

The covariance matrix C can be recognized in the product $\overline{\sigma^* \sigma^t}$, thus

$$\overline{\sigma^* \sigma^t} = \overline{\begin{bmatrix} s_1^* \\ s_2^* \\ \vdots \\ s_N^* \end{bmatrix} [s_1 \ldots s_N]} = \begin{bmatrix} \overline{|s_1|^2} & \cdots & \overline{s_1^* s_N} \\ \vdots & & \vdots \\ \overline{s_N^* s_1} & \cdots & \overline{|s_N|^2} \end{bmatrix} = C \tag{14.121}$$

The factor of merit is therefore written

$$M = \frac{G(\tau_0)}{T_A} = \frac{|W^t D_0|^2}{W^\dagger C W} \tag{14.122}$$

The problem of the optimum array is now to find a weight vector W which maximizes the factor of merit M. This reduces to finding the vector W such that the factor of merit is stationary with respect to a perturbation dW. This is the solution to the matrix equation

$$dM = 0 \tag{14.123}$$

The calculation does not present any particular difficulty, but requires a certain amount of care. We find that

$$\mathbf{C\hat{W}} = k\mathbf{D}_0 \qquad\qquad (14.124)$$

where k is an arbitrary scalar constant.

If the covariance matrix can be inverted, we obtain

$$\boxed{\mathbf{\hat{W}} = k\mathbf{C}^{-1}\mathbf{D}_0} \qquad\qquad (14.125)$$

14.7.9 Interpretation

The result obtained is very similar to that of the transfer function of the 'optimal Wiener filter', namely the ratio of the spectral density of the signal to the spectral density of the noise. We show elsewhere that if the angular distribution of jammers has sharp maxima in certain directions, the optimum radiation pattern has sharp minima in these directions.

14.7.10 Digital implementation

Each of the N array elements is followed by a receiver channel R_n and an analogue-to-digital converter (ADC) (Fig. 14.30) which extracts the in-phase and quadrature components (x_n, y_n) of the signal $(s_n = x_n + jy_n)$, using a common local oscillator reference. This allows an estimate $\mathbf{C}$ to be made of the covariance matrix, with a precision proportional to the number of samples integrated. Reed, Mallett and Brennan showed that the number of samples K should be such that $K \geq 2N-3$ where N is the number of elements; this result is sometimes known as 'Brennan's law'[18]. The inversion of the covariance matrix often uses an approximate method such as that of Gauss. Knowledge of the desired pointing direction then allows the optimum weight coefficients W_n to be calculated.

The disadvantage of digital implementation is the large computational load, which can be proportional to the square or even the cube of the number of receive channels, as well as to the data rate (proportional to the signal bandwidth) which can be high. The main advantage lies in the flexibility of the scheme, for example several different patterns can be formed according to different optimization criteria. Also, in contrast to analogue adaptive array methods, there are no problems of convergence.

An array realized in a completely digital form may on the other hand be prohibitively expensive. The best compromise may lie in a combination of analogue and digital techniques. The use of analogue beamforming matrices, lenses or

---

[18] Reed, I.S., Mallett, J.D. and Brennan, L.E., 'Rapid convergence rate in adaptive arrays', *IEEE Trans. Aerospace & Electronic Systems*, Vol.AES-10, November 1974, pp853-863.

reflectors would allow a restricted number of adjacent, steerable directional beams to be formed. These in turn may be processed digitally to achieve the optimum result.

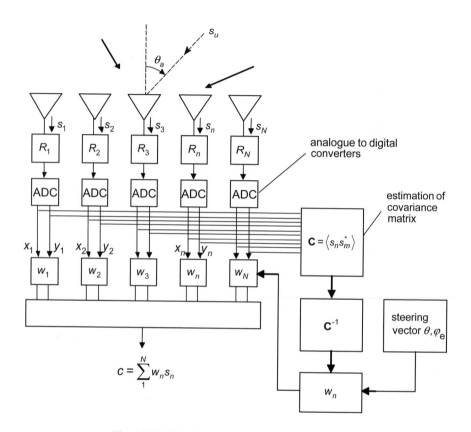

Fig. 14.30 Digital implementation of optimum array.

14.7.11 Smart antennas

The term 'smart antenna' has been coined to describe the combination of adaptive arrays with signal processing, in the context of modern radiocommunications, radar and even acoustic (underwater or medical imaging) systems. Particular impetus has been given to the subject by the explosion of interest in the last decade in mobile communications. Smart antennas may use adaptivity and/or diversity in several domains: the spatial (angular) domain, time domain, frequency domain, coding domain and even the polarization domain. The following section attempts to introduce some of these applications in more detail.

Applications in radiocommunications

The requirement in modern radiocommunications systems is essentially one of separating wanted signals from interference and multipath, and of making optimum use of the available spectrum by multiplexing signals in time (Time Division Multiplexing), frequency (Frequency Division Multiplexing), and coding (Code Division Multiplexing).

In mobile communications channels the received signal is the sum of direct and multipath components arriving from various directions, so the net signal fluctuates randomly with distance (and hence, for a moving transmitter or receiver, with time). The fluctuations are known as *fading*, and the net signal may vary by as much as 20 dB within a distance of one wavelength.

Two types of fading may be distinguished: *large-scale fading* is due to shadowing caused by hills and large buildings, and determines the local mean signal power; *small-scale fading* is due to local multipath, and describes the rapid fluctuation of the signal around the slowly-varying local mean. Small-scale fading may be represented by the Rayleigh model or by the Nagamaki model[19].

A wide range of signal processing techniques have been devised to provide optimal communications performance in the presence of multipath and co-channel interference. Some of these are based on the use of antennas, and all of the array signal processing techniques described in this chapter are relevant. It can readily be appreciated that it should be possible to combine several signals with independent fading statistics in an optimal manner to reduce the effect of fading, and this process is known as *diversity combining*. There are a number of different approaches:

- *temporal diversity*, in which copies of the signal are transmitted in different time slots, taking advantage of the fact that the fading characteristics will be different for each version of the signal;
- *frequency diversity*, in which copies of the signal are transmitted simultaneously with different carrier frequencies;
- *spatial diversity*, using versions of the received signal from antenna elements separated in space.
- *polarization diversity*, in which orthogonally-polarized versions of the signal are used. This may allow a more compact antenna system that for spatial diversity.

---

[19] Proakis, J.G. *Digital Communications* (third edition), McGraw-Hill, 1995.

The techniques used to combine the diverse signals include[20]:

- *switch diversity*, in which the strongest signal is selected;
- *equal gain*, in which the versions of the signal are adjusted to be cophasal before adding;
- *maximum ratio combining* (MRC), co-phasing the signals and weighting them according to their signal-to-noise ratio before addition.

The choice between these is a compromise between simplicity (switch diversity is the simplest) and performance (maximum ratio combining has the highest performance).

A further topic of great current interest is that of *Multiple-Input, Multiple-Output* (MIMO) arrays. Here, a transmitter may radiate several signals from different antenna elements or beams. The receiver will utilize an array and form several patterns, each receiving one of the transmitted signals and rejecting the other signal streams as well as any interference (Figure 14.31). Such a scheme can greatly increase the capacity of the link.

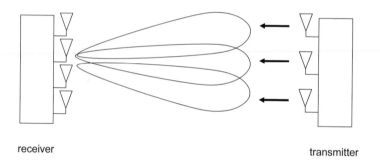

receiver   transmitter

Fig. 14.31 Multiple-Input, Multiple-Output (MIMO) array system.

Applications in radar: Space-Time Adaptive Processing (STAP)

The digital adaptive array algorithm described in §14.7.10 can equally be applied to adaptive Doppler filtering of a sequence of radar echoes at a given range from successive pulses, thereby suppressing clutter with a time-varying Doppler spectrum (due to, for example, sea clutter or wind-blown vegetation) and allowing weak

---

[20] Murch, R.D. and Ben Letaief, K., 'Antenna systems for broadband wireless access', *IEEE Communications Magazine*, April 2002, pp76-83.

moving target echoes to be detected[21]. The term *Space-Time Adaptive Processing* (STAP) describes a set of techniques used with aircraft-borne or spaceborne radar (usually sideways-looking) to filter adaptively in both the time domain and the frequency domain, to suppress clutter and interference[22].

The Doppler frequency of clutter echoes from a sideways-looking radar is related to azimuth angle (Figure 14.32) by

$$f_D = \frac{2vf_0}{c}\sin\theta \qquad (14.126)$$

The spectrum of clutter echoes experienced by the radar can therefore be plotted as a two-dimensional function of azimuth angle (extending over $-1 \leq \sin\theta \leq +1$) and Doppler (extending over $-PRF/2 \leq f_D \leq +PRF/2$, where PRF is the pulse repetition frequency of the radar), and including one or more jamming or interference signals (Figure 14.33). The radiation pattern of the antenna traces out a characteristic diagonal ridge of clutter.

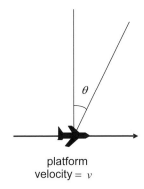

platform
velocity $= v$

Fig. 14.32 Relationship between clutter Doppler shift and azimuth angle θ for a sideways-looking radar.

The process of suppressing the clutter and the jamming, hence allowing the detection of weak targets, will involve adaptive filtering in both angle and Doppler domains. The processing is essentially a two-dimensional version of the optimum array derived in §14.7.8, and involves forming an estimate of the covariance matrix from the STAP data cube (an arrangement of radar data by pulse number, array element

---

[21] Brennan, L.E. and Reed, I.S., 'Theory of adaptive radar', *IEEE Trans. Aerospace & Electronic Systems*, Vol.AES-9, No.2, March 1973, pp237-252.

[22] Klemm, R., *Space-Time Adaptive Processing*, Peter Peregrinus, Stevenage, 1999.

number and range), and inverting it to yield the weight vectors. The process becomes somewhat more complicated if the clutter is nonhomogeneous, both spatially and temporally, and algorithms which exploit prior knowledge of the clutter have been developed to give improved performance.

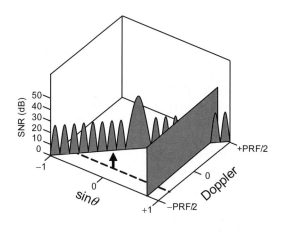

Fig. 14.33 Clutter and jamming plotted as a function of azimuth angle and Doppler shift in a sideways-looking airborne radar. There is a weak target at $\theta = 0$ and a jamming signal at $\sin \theta \cong 0.8$.

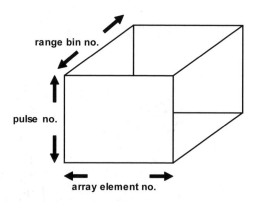

Fig. 14.34 The STAP data cube.

Waveform diversity

The multifunction radar depicted at the beginning of Chapter 11 is able to exploit the flexibility of an electronically-steered phased array to perform surveillance, tracking and several other functions, prioritizing and scheduling the various tasks dynamically according to their instantaneous importance. Thus tracking of threat targets at close range would be a high priority; surveillance of targets at long range a lower priority.

The concept of waveform diversity extends this by performing several functions (which might include radar surveillance, radar tracking, communications, …) all at the same time, so the array is forming different patterns, with differently-coded signals, on transmit and on receive, in different directions for different purposes simultaneously. It may exploit adaptivity in several domains. The radar waveforms may be chosen adaptively to optimally detect particular targets[23]. The MIMO array in the previous section is an example of this. In another example, a radar may radiate in addition to the waveform used for radar purposes a second 'masking waveform' which is orthogonal to the radar waveform both in a spatial sense and a coding sense, to prevent an adversary using the radar waveform as an illumination for a bistatic radar, by denying the adversary a coherent reference of the radar signal via the radar antenna's sidelobes[24].

---

[23] Grieve, P.G. and Guerci, J.R., 'Optimum matched illumination-reception radar', US Patent S517522, 1992.

[24] Griffiths, H.D., Wicks, M.C., Weiner, D., Adve, R., Antonik, P.A. and Fotinopoulos, I., 'Denial of bistatic hosting by spatial-temporal waveform design', *IEE Proc. Radar, Sonar and Navigation*, Vol.152, No.2, pp81–88, April 2005.

APPENDIX 14A: ENTROPY AND PROBABILITY

(following a treatment due to Professor E. Roubine)

14A.1 Uncertainty of an event A of probability $p(A)$

A first evaluation of the relative uncertainty of a possible event A is simply the inverse of the probability $p(A)$. It is convenient to use the logarithm to base 2, so that we have

$$I(A) = \log_2 \frac{1}{p(A)} = -\log_2 p(A) \qquad (14A.1)$$

The uncertainty is thus expressed in the form due to Shannon, or explicitly in bits. The use of the logarithm means that when we are dealing with a set of independent events, the uncertainties can be summed.

14A.2 Information gained by the knowledge of an event

The observation of an event A constitutes information measured by the *a priori* uncertainty of A, and is thus expressed in the form due to Shannon. As an example, a child is born, and its sex is announced. What is the quantity of information contained in this announcement ?

14A.3 Uncertainty relative to an alternative

Consider an event A_1 of probability p, and a complementary event 'non-A_1', denoted A_2, whose probability is evidently

$$q = 1 - p$$

The notion of entropy is a second way of evaluating the relative uncertainty of this situation. It is defined as the average of the relative uncertainties of the events A_1 and A_2, weighted by their respective probabilities of occurrence. The entropy is therefore

$$H = \frac{p \log_2 \dfrac{1}{p} + q \log_2 \dfrac{1}{q}}{p+q} = -p \log_2 p - q \log_2 q \qquad (14A.2)$$

If we draw a graph of the variation of H for values of p from 0 to 1, we obtain the curve shown in Fig. 14.35.

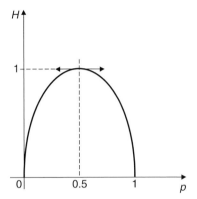

Fig. 14.35 Entropy relative to an alternative.

We see that the entropy (the uncertainty) is zero if one or other of the occurrences is certain. It is maximum if the probabilities are 50%. The approach of Shannon is linked closely with that of uncertainty.

Example: tossing a coin is a situation of maximum entropy.

14A.4 First generalization: entropy of a set of events

Consider a set of possible events: A_1, A_2, ..., A_i, ..., A_N, of respective probabilities $p(A_1)$, ... , $p(A_i)$, ... , $p(A_N)$, so that the individual uncertainties (in the first sense) are

$$I(A_i) = \log_2 \frac{1}{p(A_i)} \tag{14A.3}$$

The relative entropy of this situation is still defined as *the average of the partial uncertainties*

$$H = \sum_1^N p(A_i)I(A_i) = \sum_1^N p(A_i)\log_2 \frac{1}{p(A_i)}$$

or in abbreviated notation

$$H = -\sum_{1}^{N} p_i \log_2 p_i \qquad (14\text{A}.4)$$

14A.5 Second generalization: random variable

Consider a random variable x of probability density $f(x)$. By simple generalization of the preceding definition, we define its entropy as the average of the uncertainties:

$$H_x = -\int_{-\infty}^{\infty} f(x) \log_2 f(x) dx \qquad (14\text{A}.5)$$

We define the average (or expectation) and the variance of the variable x (Fig.14.36) by

$$m = \bar{x} \ \left(= E(x)\right)$$

$$\sigma^2 = \overline{(x-m)^2} \qquad (14\text{A}.6)$$

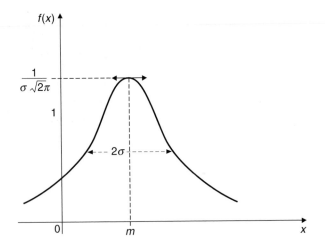

Fig. 14.36 Entropy of a random variable: probability density.

The following theorem can be demonstrated:

The entropy H_x is maximum if the probability density $f(x)$ is Gaussian, i.e.

$$\hat{f}(x) = \frac{1}{\sigma\sqrt{2\pi}} \exp\left[-\frac{(x-m)^2}{2\sigma^2} \right] \tag{14A.7}$$

In this case, the value of the entropy is easily calculated:

$$H = \log_2\left(\sigma\sqrt{2\pi e}\right) \tag{14A.8}$$

We see that the entropy of the random variable increases as the logarithm of its variance σ^2. This result serves to justify the relation (14A.9) between entropy and spectral density.

14A.6 Decision theory: Maximum Entropy

In order to take a decision, it is necessary to evaluate the relative probabilities of the various outcomes. For example, I am going out today and I do not know whether it will rain. Should I take an umbrella ? My decision is partly subjective: (is the inconvenience of having to carry an umbrella if it does not rain greater or less than getting wet if it does rain ?) and partly objective: (the likelihood of it raining). In the absence of any reliable information (are the weather forecasts often wrong ?), common sense suggests assuming a probability of 50%. This is, as we saw in paragraph 14A3, the choice of *maximum entropy*[25].

14A.7 Entropy and spectral density

The method of spectral estimation known as 'maximum entropy' relies on the relationship between the spectral density $T(\tau)$ that we seek and the entropy H of the process. The stationarity of the entropy is equivalent in effect to the condition that the spectrum can in turn be expressed just in terms of the measured covariances. From this, the following relationship can be established

$$H = \int_{-1}^{1} \log_2 T(\tau)\, d\tau \tag{14A.9}$$

---

[25] The historic origin of decision theory can be found in the works of the French mathematician and philosopher Blaise Pascal (*Pensées* VII, 2).

14A.8 Justification of this relationship

Justification for this can be found in several books. Its origin can be intuitively appreciated in the relation between the entropy of a Gaussian process and its variance.

Suppose we have a white noise signal $a(t)$ passed through a narrow band filter of bandwidth Δf centred on a frequency f_0, and with transfer function $T(f)$. The energy spectral density of the signal is, by hypothesis, constant

$$A(f) = A_0$$

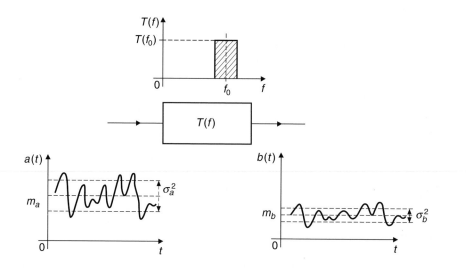

Fig. 14.37 Entropy variation of white noise passing through a narrow band filter.

According to the Wiener-Khintchine theorem, the autocorrelation function is the Fourier transform

$$p(\theta) = \overline{a(t)a^*(t-\theta)} = \int_{-\infty}^{+\infty} A(f)\exp(j2\pi f\theta)\,dfa \qquad (14A.10)$$

The variance of the incident signal is deduced by setting $\theta = 0$

$$\sigma_a^2 = \overline{\left|a(t)\right|^2} = p_a(0) = \int_{-\infty}^{+\infty} A(f)\,df = A_0\Delta f \qquad (14A.11)$$

The output signal $b(t)$ has an energy spectral density

$$B(f) = A(f)T(f) = A_0 T(f) \tag{14A.12}$$

The variance is obtained from the autocorrelation function

$$\sigma_b^2 = p_b(0) = \int B(f)\, df = A_0 \int T(f)\, df \tag{14A.13}$$

Under the 'narrowband' approximation, this becomes

$$\sigma_b^2 \approx A_0 T(f_0)\Delta f \tag{14A.14}$$

We can define a variation in entropy ΔH between the input signal $a(t)$ and the output signal $b(t)$, both assumed to be Gaussian, from their variances

$$\Delta H = \log \sigma_b \sqrt{2\pi e} \; - \; \log \sigma_a \sqrt{2\pi e} \; = \; \frac{1}{2}\log_2 \frac{\sigma_b^2}{\sigma_a^2} \tag{14A.15}$$

from which

$$\Delta H = \frac{1}{2}\log_2 T(f_0) \tag{14A.16}$$

The general case of a wide bandwidth F can be considered as the resultant of a set of N narrowband filters, each of bandwidth ΔF and centred on frequencies f_1, $f_2 \ldots$ f_N, and in parallel (Fig. 14.38). Here, again, we can define a global variation of entropy as the average of the partial entropies

$$H = \frac{1}{2N}\sum_{n=1}^{N}\log_2 T(f_n) = \frac{1}{2N\Delta f}\sum_{1}^{N}\log_2 T(f_n)\Delta f = \frac{1}{2F}\sum_{1}^{N}\log_2 T(f_n)\Delta f \tag{14A.17}$$

We can therefore form the expression

$$H = \frac{1}{2F}\int_F \log_2 T(f)\, df \tag{14A.18}$$

In the case of imaging of incoherent sources, it is the *angular distribution of intensities*, $T(\tau)$ which plays the role of the energy spectrum. Its extent is limited to the domain $(-1, +1)$, so that we can write

$$H = \int_{-1}^{+1} \log_2 T(\tau) \, d\tau \qquad\qquad (14A.19)$$

This result is used (§14.5.4) for an estimation of $T(\tau)$ under the condition of maximum entropy.

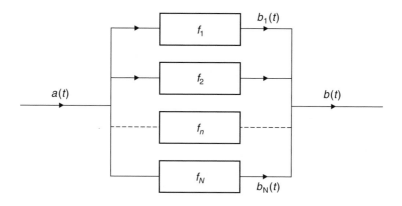

Fig. 14.38 Wide bandwidth case.

FURTHER READING

1. Amuso, V. (ed), *Proc. First International Conference on Waveform Diversity and Design*, Edinburgh, November 2004.
2. Benjamin, R., 'Optimum use of fully populated, over populated and sparsely filled antenna apertures', *IEE Proc.*, Pt. H., No. 3, pp117-120, June 1980.
3. Benjamin, R. and Griffiths, H.D., 'Aperture-domain signal processing', *Electronics & Communication Engineering Journal*, Vol. 1, No. 2, pp71-80, March/April 1989.
4. Compton, R.T. Jr., *Adaptive Antennas - Concepts and Performance*, Prentice Hall, 1988.
5. Curlander, J.C. and McDonough, R.N., *Synthetic Aperture Radar*, Wiley, New York, 1991.
6. Davies, D.E.N., Corless, K.G., Hicks, D.S. and Milne, K., 'Array signal processing'; Chapter 13 in *The Handbook of Antenna Design*, Vol.II, A.W. Rudge, K. Milne, A.D. Olver and P. Knight (eds), Peter Peregrinus, Stevenage, 1983.
7. El Zooghby, A. (ed.), *Smart Antenna Engineering*, Artech House, 2005.
8. Farina, A., *Antenna-Based Signal Processing Techniques for Radar Systems*, Artech House, Dedham, MA, 1992.
9. Gething, P.J.D., *Radio Direction Finding and Superresolution*, Peter Peregrinus, Stevenage, 1991.
10. Gjessing, D.T., *Target Adaptive Matched Illumination Radar*, Peter Peregrinus, Stevenage, 1986.
11. Godara, L.C. (ed), *Handbook of Antennas in Wireless Communications*, CRC Press, 2002.
12. Godara, L.C., *Smart Antennas*, CRC Press, 2004.
13. Guerci, J., *Space-Time Adaptive Processing for Radar*, Artech House, 2003.
14. Haykin, S. (ed) *Advances in Spectrum Analysis and Array Processing* (Vols. I and II), Prentice-Hall, 1991.
15. Hudson, J.E., *Adaptive Array Principles*; Peter Peregrinus, Stevenage, 1989 (second edition).
16. Klemm, R., *Space-Time Adaptive Processing*, Peter Peregrinus, Stevenage, 1999.
17. Klemm, R. (ed.) *Applications of Space-Time Adaptive Processing*, Peter Peregrinus, Stevenage, 2004.
18. Krim, H. and Viberg, M., 'Two decades of array signal processing research', *IEEE Signal Processing Magazine*, pp67-94, July 1996.
19. Monzingo, R.A. and Miller, T.W., *Introduction to Adaptive Arrays*, Wiley-Interscience, 1980.
20. Sarkar, T.K., Wicks, M.C., Salazar-Palma, M. and Bonneau, R.J., *Smart Antennas*, Wiley, 2003.
21. Scott, K.K., 'Transversal filter techniques for adaptive array applications', *IEE Proc. Pt.F. and H.*, Vol.130, pp29-35, 1983.
22. Thompson, A.R., Moran, J.M. and Swenson, G.W. Jr., *Interferometry and Synthesis in Radio Astronomy*, Wiley-Interscience, 2001 (second edition).
23. Vaughan, R.G. and Andersen, J.B., 'Antenna diversity in mobile communications', *IEEE Trans. Vehicular Technology*, Vol.36, No.4, pp149-172, 1987.
24. Vaughan, R.G., 'Polarization diversity in mobile communications', *IEEE Trans. Vehicular Technology.*, Vol. 39, No. 3, pp177–186, 1990.

25. Vaughan, R.G. and Andersen, J.B., *Channels, Propagation and Antennas for Mobile Communications*, Peter Peregrinus, Stevenage, 2003.

26. Winters, J.H., 'Smart antennas for wireless systems', *IEEE Pers. Communications Magazine*, Vol.5, No.1, pp23-27, February 1998.

27. Wirth, W-D., *Radar Techniques Using Array Antennas*, Peter Peregrinus, Stevenage, 2001.

28. Special Issue of *IEEE Trans. Antennas & Propagation* on Adaptive Antennas, Vol. 24, No. 5, September 1976.

29. Special Issue of *Proc. IEEE* on Spectral Estimation, September 1982.

EXERCISES

14.1 Coherence function of the Sun

We wish to find the *apparent diameter* and the *direction* of the Sun at a wavelength of 21 cm, from the coherence function measured on the ground.

In this example we consider the problem in a plane; the direction of the Sun is given by an angle θ_0, measured with respect to the local zenith. The Sun is considered as a segment, radiating incoherently, with apparent diameter $\Delta\theta$. Its temperature T_0 is constant within this segment.

The coherence function $C(v)$ is measured on the ground, along a horizontal axis Ox where the *x*-coordinate is normalized by the wavelength ($v = x/\lambda$).

Q1 Calculate the coherence function $C(v)$ in amplitude and in phase.

Q2 It is found that the minimum spacing between two receivers such that the coherence function is zero is $\Delta x_0 = 25$ metres. Deduce from this the apparent diameter of the Sun at this frequency.

Q3 It is found that the minimum spacing between two receivers such that the phase of the coherence function is 2π radians is $\Delta x_1 = 42$ centimetres. Deduce from this the direction θ_0 with respect to the zenith.

Note: In these calculations use the angular parameter $\tau = \sin\theta$. We can also assume that the apparent diameter $\Delta\theta$ is sufficiently small that the corresponding variation of τ is given by $\Delta\tau = \Delta\theta\cos\theta_0$.

ANSWERS

Q1   *Coherence function of the Sun*: The angular distribution $T(\tau)$ of the noise temperature of the Sun is uniform over the interval $\Delta\tau$ around the direction $\tau_0 = \sin\theta_0$ of the Sun

$$T(\tau) = T_0 \, \mathrm{rect}\left(2\,\frac{\tau - \tau_0}{\Delta\tau}\right)$$

The coherence function is the Fourier transform of this

$$C(v) = \left(T_0\Delta\tau \frac{\sin\pi v\Delta\tau}{\pi v\Delta\tau}\right)\exp\left(j2\pi v\tau_0\right)$$

Q2   The interval between the first zeroes of the function is given by $v\Delta\tau = 1$.

i.e. $\dfrac{\Delta x_0}{\lambda}\Delta\tau = 1$

We have therefore $\Delta\tau = \dfrac{\lambda}{\Delta x_0} = \dfrac{0.21}{25} \approx \dfrac{1}{120}$

The value of the apparent diameter $\Delta\theta$ depends on the direction θ_0.

Q3   The interval for 2π phase shift is given by $v\tau_0 = 1$

i.e. $\dfrac{\Delta x_1}{\lambda}\Delta\tau = 1$

It follows that $\sin\theta_0 = \tau_0 = \dfrac{\lambda}{\Delta x_1} = \dfrac{0.21}{0.42} = \dfrac{1}{2}$ i.e. $\theta_0 = 30°$.

The Sun is therefore $30°$ from the zenith. From this we can deduce the apparent diameter

$$\Delta\theta_{radians} = \frac{\Delta\tau}{\cos\theta_0} = \frac{\sqrt{3}}{180}$$

i.e. $\Delta\theta_{degrees} = \dfrac{180}{\pi}\Delta\theta_{radians} = \dfrac{\sqrt{3}}{\pi} \approx 0.6°$

15

Antenna measurements

Planar near-field scanner at EADS Astrium, Portsmouth, England. This is used to
measure antennas on remote sensing and communications satellites, and has a scan
area of 22 m x 8 m. It is believed to be the largest such facility in the world. (photo:
EADS Astrium)

15.1 INTRODUCTION

Antenna measurements are important both to the antenna designer and the antenna
manufacturer. Measurement is an integral part of the design process, and the results
of measurement of a prototype may be fed back to iteratively refine the design. In
manufacture, it will be necessary to check that the performance of a product meets
certain specifications.

There are a wide range of antenna parameters that can be measured. The key ones
are the gain and the radiation pattern, but input impedance (and hence bandwidth)
can also be important. For the radiation pattern, usually a 1-D cut as a function of
azimuth angle, at a single polarization, is adequate, though the full 2-D co-polar and

cross-polar patterns may be required for some applications. The purpose of this chapter is therefore to describe the principles of the various measurement techniques that are used and explain their advantages and disadvantages. A brief and selective description of some current research topics is also given in each case.

15.2 GAIN MEASUREMENTS

15.2.1 Comparison with a standard-gain horn

The signal obtained from the antenna under test is compared by substitution with that from an antenna of known gain, such as a standard-gain horn, with constant power transmitted from the source antenna, under controlled conditions such as an anechoic chamber or test range. The gain G of the antenna under test is simply given by

$$G = P_R \cdot \frac{G_S}{P_S}$$

(15.1)

where P_R is the power accepted by the antenna under test, G_S is the gain of the standard-gain horn and P_S is the power accepted by the standard gain horn, allowing for any mismatches. The advantage of this method is that it does not rely on knowledge of the path loss due to the distance between the source antenna and the antenna under test.

15.2.2 Two-antenna measurement

The Friis transmission equation (see §4.2.1) establishes the power received by the antenna under test as

$$P_R = \frac{P_S G_S G \lambda^2}{(4\pi)^2 r^2}$$

(15.2)

So if P_S, G_S, λ and r are known, a measurement of P_R allows G to be determined. If G_S is not known, then identical test antennas may be used at both the transmit and receive ends of the path to establish their gain.

Alternatively, the return signal from a single test antenna facing a large conducting sheet may be used to measure its gain (the effective chamber length being twice the separation of the antenna and the sheet). This is known as *Purcell's method*. It is important that the inherent mismatch of the test antenna is taken into account.

15.2.3 Three-antenna measurement

If three antennas of unknown gains G_1, G_2 and G_3 are used as source and test antennas in all three possible combinations, denoting the source power by P_S, the received powers are

$$P_{12} = \frac{P_S G_1 G_2 \lambda^2}{(4\pi)^2 r^2} \qquad P_{23} = \frac{P_S G_2 G_3 \lambda^2}{(4\pi)^2 r^2} \qquad P_{13} = \frac{P_S G_1 G_3 \lambda^2}{(4\pi)^2 r^2} \qquad (15.3)$$

These can be solved for G_1, G_2 and G_3, giving

$$G_1 = \frac{4\pi r}{\lambda} \sqrt{\frac{P_{13} P_{12}}{P_S P_{23}}} \qquad G_2 = \frac{4\pi r}{\lambda} \sqrt{\frac{P_{23} P_{12}}{P_S P_{13}}} \qquad G_3 = \frac{4\pi r}{\lambda} \sqrt{\frac{P_{13} P_{23}}{P_S P_{12}}} \qquad (15.4)$$

This method, again, does not rely on the use of a standard gain antenna, but does require knowledge of the path length between the source and test ends of the chamber, and of the source power P_S.

15.2.4 Extrapolation

The most accurate method currently in use is that developed by Newell *et al.*[1] at the National Institute of Standards and Technology (NIST) in the USA, and refined by workers at the National Physical Laboratory (NPL), Teddington, UK. These laboratories (amongst others) provide a calibration service for antenna gains.

The method is a development of the three-antenna technique, but eliminates the uncertainty in knowledge of the path loss. This may arise because of difficulties in locating the exact phase centre of each antenna (is it at the aperture of a horn or at its 'throat' ?). To overcome this, measurements are made at a number of separations of source and test antennas and a polynomial fitted to the measurements using a least-squares technique, and then extrapolated to infinite range. In the NPL method, the polynomial also includes sine and cosine terms to model the oscillations due to multiple reflections.

If the measurements are to be made at a number of frequencies over a band (as would usually be the case), then the procedure has to be automated under computer control. An overall accuracy of better than ±0.04 dB is claimed. Fig. 15.1 shows this technique being used for the measurement of the gain of a log spiral antenna.

---

[1]   Newell, A.C., Baird, R.C. and Wacker, P.F., 'Accurate measurement of antenna gain and polarisation at reduced distances by an extrapolation technique', *IEEE Trans. Antennas & Propagation*, Vol. AP-21, No. 4, April 1973.

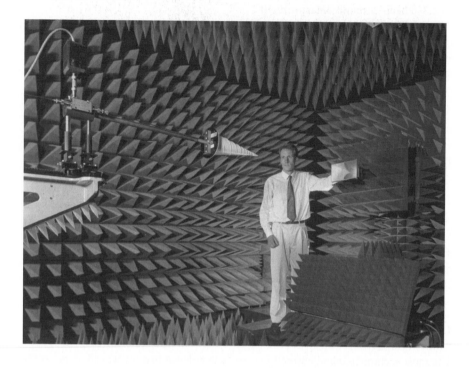

Fig. 15.1 Gain calibration of a log-spiral antenna, being carried out in an anechoic chamber at the National Physical Laboratory, Teddington, UK (Crown copyright).

15.3 RADIATION PATTERN MEASUREMENTS

Measurements of radiation pattern and of gain are best performed in an unobstructed medium without the presence of any objects that might give rise to reflections. Either a large outdoor test range or (more commonly) an anechoic chamber may be used. Outdoor test ranges allow a greater distance between the source antenna and the antenna under test (and hence a better approximation to a plane wave), but may be undesirable if one does not want to reveal the antenna under test for reasons of commercial or military security. Fig. 15.2 shows an example of a typical outdoor test range.

Fig. 15.2 The outdoor test range at Funtington in Southern England (Crown copyright).

15.3.1 Anechoic chambers and far-field ranges

An anechoic chamber (Fig. 15.3) consists of an enclosed room with a source antenna at one end which is used to excite the antenna under test at the other end. The antenna under test is mounted on a turntable or multi-axis positioner which is rotated to obtain its radiation pattern. The chamber is lined with Radio Absorbing Material (RAM) which minimizes reflections from the walls, floor and ceiling. Chambers are sometimes tapered to prevent the formation of standing waves.

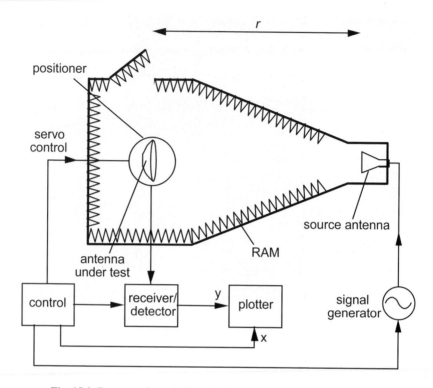

Fig. 15.3 Features of a typical anechoic chamber and its instrumentation.

RAM [1] is made from carbon-impregnated polyurethane foam (or similar), and usually has a tapered pyramidal shape, which in effect provides a gradual transition between the impedance of free space and that of the metal sheet that forms the walls of the chamber. Figure 15.4 shows typical RAM of this kind. Reflection coefficients of typically −20 to −40 dB are obtainable over a range of incidence angles and frequencies, for absorber thicknesses of the order of a few wavelengths. More sophisticated absorbers can be made from multi-layer structures, but these are only used in specialized applications.

As well as being rotatable in azimuth, the antenna under test may be tilted in elevation. Also, the source antenna may be rotated about its axis for polarization measurements (see §15.2.6). The chamber is usually arranged to transmit from the source antenna and receive on the antenna under test, although this arrangement may be reversed if it is more convenient. The turntable positioner, plotter, receiver and transmit source are usually all under computer control, and purpose-built suites of instrumentation are available from manufacturers such as Orbit, Agilent and Scientific Atlanta.

Fig. 15.4 Typical pyramidal RAM in an anechoic chamber (photo: H. Griffiths).

These types of antenna measurements are performed in the far field - that is to say, the antenna under test is illuminated by a plane wave. To obtain a perfect plane wave the range would have to be infinitely long, but for practical purposes it is usual to adopt a criterion such that the phase error at the edge of the curved wavefront should not exceed $\pi/8$ radians ($22.5°$).

From Fig. 15.5 it can be seen that the spherical wavefront from the source antenna gives rise to a path difference at the edge of the antenna under test of

$$\Delta x = \left(r^2 + D^2/4\right)^{1/2} - r$$

The bracket can be expanded, and taking just the first two terms of the expansion

$$\Delta x = \left(r + \frac{D^2}{8r} + \ldots\right) - r \tag{15.5}$$

which corresponds to a phase difference of

$$\Delta\phi = \frac{2\pi}{\lambda}.\Delta x = \frac{\pi D^2}{4r\lambda}$$

Setting this $= \pi/8$ gives

$$r = \frac{2D^2}{\lambda}$$ (15.6)

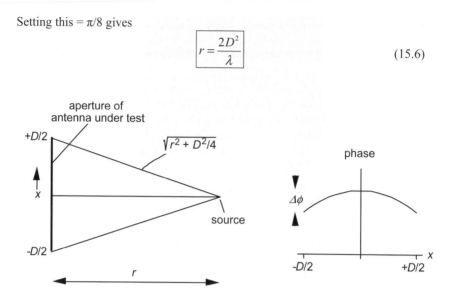

Fig. 15.5 Derivation of the Rayleigh distance criterion.

This is known as the *Rayleigh distance*. In practice one would like to arrange that the distance between the source antenna and the antenna under test is at least one Rayleigh distance, and preferably more. Clearly this constraint becomes more of a problem the larger the antenna under test, and the higher the frequency.

There are two other minimum chamber length criteria which may need to be considered, particularly at low frequencies. Firstly, the reactive near field will be significant within about one wavelength of the antenna, so we require that $r > \lambda$. Secondly, when the antenna is rotated, the distance from one end to the source varies from $r + D/2$ to $r - D/2$. The $1/r$ field dependence can give rise to pattern errors. If, for example, ± 0.5 dB amplitude error is acceptable, then $r > 10D$ is required.

The volume within which the phase and amplitude errors of the wavefront are sufficiently low is known as the *quiet zone*. A criterion such that the errors should be less than ± 0.5 dB (amplitude) and $\pm 5°$ (phase) is usual (this question is considered further in §15.2.3).

At low frequencies (< 1 GHz, say) an open test range would normally be used. This is because any chamber would need to be very long to operate in the far field, and also a large volume of RAM would be needed, since the RAM depth needs to be two wavelengths or more to minimize the reflectivity.

The usable dynamic range of an antenna range is limited by two main factors:

(1) The source transmit power and receiver sensitivity. This is not a serious limitation provided that the signal is detected in a narrow bandwidth and a wide dynamic range low-noise amplifier is used at the test antenna.

(2) The finite reflectivity of the chamber walls, or other objects within the chamber. These reflections degrade the quality of the plane wave with which the antenna under test is illuminated. Multiple reflections 'fill in' nulls in the radiation pattern and create 'phantom' sidelobes. Naturally, these effects will be most troublesome in the measurement of low-sidelobe antennas. This question is considered further in §15.2.3.

The peak power received by the antenna under test (when it is pointing directly at the source antenna) is given by equation (15.2). If the required dynamic range of the measurement is X dB, then the minimum detectable signal must be at least X dB below this figure, and preferably slightly more if measurements of nulls between sidelobes are not to be corrupted by noise.

15.3.2 Compact ranges

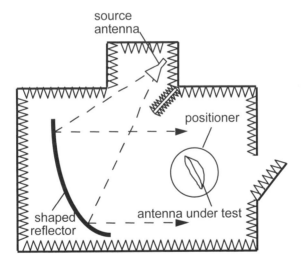

Fig. 15.6 A single-offset-fed compact antenna test range (CATR).

Evidently, for large antennas and high frequencies the amount of 'real estate' needed to get far enough away to achieve a good approximation to a plane wave can become prohibitive, and this has caused engineers to consider other techniques. Chief among these is the *compact range*, or *compact antenna test range* (CATR), originated by R.C. Johnson[2] at Georgia Tech, where much of the innovative work on antenna measurements has been done during the past few decades. The compact range uses an offset-fed parabolic reflector to synthesize a plane wave in a considerably more compact configuration than that of a conventional range. The reflectors and their feeds are designed using the techniques established in §10.3.

Fig. 15.6 shows the typical features of a compact range. Note the shielding between the source antenna and the turntable, to minimize leakage of the direct signal. The key component in a compact range is the reflector, and this is designed in exactly the same way as for an offset-fed reflector antenna. It is usually constructed from a set of shaped panels, whose positions are individually set up during alignment (often using laser-ranging techniques) by means of lead screws. To minimize diffraction from the edges of the reflector, the edges may either be serrated or rolled.

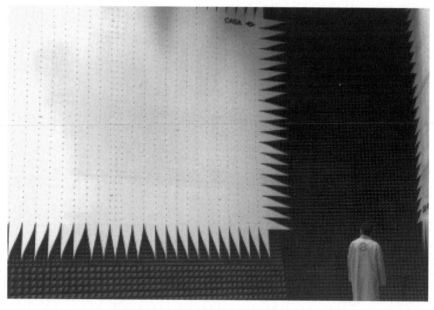

Fig. 15.7 The enormous compact range at the European Space Agency's ESTEC facility at Noordwijk in the Netherlands. This is used for pre-launch testing of antennas on satellites, so the positioner has to be large enough to accommodate complete satellites. The range uses a dual-reflector configuration to give high polarization purity for measurements on dual-polarized antennas. Note also the serrated edges of the reflectors. (photo: H. Griffiths).

---

[2] Johnson, R.C., 'Antenna range for providing a plane wave for antenna measurements', US patent 3302205, 31 January 1967.

In a variant of the scheme two reflectors are used in a reflex configuration. This gives better control over generation of cross-polarization, so the dual-reflector compact range is often preferred for measurements on antennas where high cross-polar performance is important, such as satellite communications antennas. Two interesting examples of compact ranges are shown in Figs. 15.7 and 15.8.

Fig. 15.8 The mm-wave compact range at Queen Mary College, University of London. This is capable of working up to 200 GHz. The reflector is a single offset antenna of 5.4 m focal length, made up of 18 panels, with an rms accuracy of approximately 60 microns. The quiet zone is 1 m × 1 m × 1 m. The antenna under test is a symmetric-pair array developed for direction-finding applications (§11.8.6). (photo: Dr W. Titze).

15.3.3 Wavefront quality

We have already remarked that the accuracy of radiation pattern measurements depends on the quality of the plane wave with which the antenna is illuminated, and both the finite distance of the source antenna and reflections from the walls of the anechoic chamber will cause the illumination to depart from a perfect plane wave.

These effects can be quantified in terms of plane wave spectra[3] [4], as introduced in §2.1.

A perfect plane wave travelling normal to the aperture corresponds to uniform phase and uniform amplitude over that aperture. If we assume for the moment that this aperture is infinite in extent, then the plane wave spectrum is just a single Dirac delta function at the origin. On the other hand, a plane wave incident from a different direction will give uniform amplitude, but phase varying linearly with direction over the aperture, and the plane wave spectrum will be a delta function whose position corresponds to the direction of incidence.

By taking a set of samples over a plane in the quiet zone and computing the 2-D Fourier Transform, we can therefore determine the directions and levels of the wavefront components. Clearly, if we want to measure sidelobe levels (for example) < −40 dB, we need to be sure that no other component of the plane wave spectrum exceeds this level. A more recent idea is that if the plane wave spectrum of the illumination can be measured, it should in principle be possible to deconvolve it from the measurement of the antenna radiation pattern, and hence remove the effects of imperfections in the illumination.

15.3.4 Near-field techniques

An alternative to illuminating the antenna under test with a plane wave, and rotating it to obtain one cut of the radiation pattern, is to transmit a signal from the antenna under test and sample the radiated signal (in both amplitude and phase) in the near field. From a knowledge of the near-field signal, the far-field radiation pattern can be computed. This permits a very compact measurement system, and the whole 3-D radiation pattern can be derived rather than just one cut; however, the full sampling of a large aperture may take some time, and also the probe antenna must be positioned very accurately. Furthermore, the flexible cable that connects the probe to the measurement receiver must have very stable phase characteristics, and it is necessary to ensure that the probe does not itself perturb the field that it is trying to measure.

There are three geometries that are used. In *planar near-field scanning* the antenna under test is fixed and the probe is scanned over the antenna aperture plane by means of an *x-y* positioner. In *cylindrical near-field scanning* the probe is scanned vertically, while the antenna under test is rotated. The calculation of the far-field pattern is carried out in cylindrical polar co-ordinates. The cylindrical geometry

---

[3] Booker, H.G. and Clemmow, P.C., 'The concept of an angular distribution of plane waves and its relation to that of a polar diagram and aperture distribution', *JIEE*, Vol. 97, 1950, pp11-17.

[4] Bennett, J.C. and Farhat, K.S., 'Waveform quality in antenna pattern measurement: the use of residuals', *Proc. IEE*, Vol. 134, Pt. H, No. 1, pp30-34, February 1987.

avoids the complication of an *x-y* positioner, since the probe is only scanned in one dimension, though a turntable is required to rotate the antenna under test, as well as a phase-stable rotating joint. In *spherical near-field scanning* the probe is scanned over a spherical surface (or the probe held fixed and the antenna under test rotated).

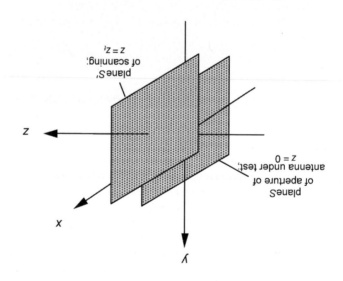

Fig. 15.9 Geometry of planar near-field scanning.

Planar scanning is probably the most frequently-used, because most antennas of practical interest are directive, and have a well-defined planar region over which to make the measurement. Following the treatment of Wang [4], and noting also the results on plane wave spectra obtained in §2.1, suppose that the antenna aperture lies in plane S defined by the *xy* plane at $z = 0$. The field is sampled in the plane S', at $z = z_i$ (Fig. 15.9). The plane wave spectrum $\mathbf{A}\left(k_x, k_y\right)$ of the radiation is related to the field $\mathbf{E}\left(x, y, z_i\right)$ sampled at S' by

$$E_x\left(x, y, z_i\right) = \frac{1}{2\pi} \int_{-\infty}^{\infty} \int_{-\infty}^{\infty} A_x\left(k_x, k_y\right) . \exp\left(-jk_z z_i\right) \exp\left[-j\left(k_x x + k_y y\right)\right] dk_x dk_y$$

(15.7)

$$E_y\left(x, y, z_i\right) = \frac{1}{2\pi} \int_{-\infty}^{\infty} \int_{-\infty}^{\infty} A_y\left(k_x, k_y\right) \exp\left(-jk_z z_i\right) \exp\left[-j\left(k_x x + k_y y\right)\right] dk_x dk_y$$

(15.8)

From these, the aperture field $\mathbf{E}(x,y,0)$ can be obtained

$$E_x(x,y,0) = \frac{1}{2\pi} \int_{-\infty}^{\infty} \int_{-\infty}^{\infty} A_x(k_x,k_y) \exp\left[-j(k_x x + k_y y)\right] dk_x dk_y$$

(15.9)

$$E_y(x,y,0) = \frac{1}{2\pi} \int_{-\infty}^{\infty} \int_{-\infty}^{\infty} A_y(k_x,k_y) \exp\left[-j(k_x x + k_y y)\right] dk_x dk_y$$

(15.10)

and hence the far-field radiation pattern, by Fourier transformation of $\mathbf{E}(x,y,0)$.

The required sample spacing in the plane S', according to the sampling theorem, should be no greater than $\lambda/2$. In practice a sample spacing of ~$\lambda/3$ would be appropriate.

Fig. 15.10 Spherical near-field scanning system at University College London (installed by MI Technologies, Atlanta, USA, May 2005).

The simple theory assumes that the probe acts as a perfect point receiver. In practice this will not be the case, so the technique of *probe compensation*[5] was developed, characterizing and compensating for the angular response of the probe antenna.

Near-field techniques are therefore quite attractive, and a few years ago an entire Special Issue of *IEEE Transactions on Antennas and Propagation* was devoted to advances in the subject [8].

15.3.5 Other techniques

Rather than measuring the aperture field directly with a probe, it is also possible to scatter it in some way back into the antenna under test and measure the change in reflection coefficient at the antenna feedpoint due to the scattered signal. If the scattering is modulated, then the scattered signal can be distinguished from the static component of the reflection coefficient by synchronous detection. This technique was originally developed at University College London back in the 1950s by Cullen and Parr, using so-called 'spinning dipoles'[6] to modulate the scattered field[6], and also independently by Richmond in the USA[7]. The technique has been used more recently by Bolomey and co-workers at *Supelec* near Paris and in their spin-out company *SATIMO*, using an array of dipoles modulated electronically by PIN diodes at their feedpoints. In one implementation[8] the antenna under test is rotated slowly through 360° while the dipoles are modulated sequentially (Fig. 15.11). The use of an array of modulated scatterers has advantages in respect of the time taken to complete the measurement. The scattered signal may be received either monostatically by the antenna under test, or bistatically by a second, fixed test antenna, which gives advantages in respect of dynamic range. This gives a rapid measurement of the full 3-D radiation pattern in a single scan.

5 Kerns, D.M., 'Correction of near-field antenna measurements made with an arbitrary but known measuring antenna', *Electronics Letters*, Vol. 6, pp346-347, May 1970.

6 Cullen, A.L. and Parr, J.C., 'A new perturbation technique for measuring microwave fields in free space', *Proc. IEE*, Vol. 102, Part B, No. 6, pp836-844, November 1955.

7 Richmond, J.H., 'A modulated scattering technique for the measurement of field distributions', *IRE Trans. Microwave Theory and Techniques*, Vol.3, pp13-15, 1955.

8 Bolomey, J.-Ch. *et al.*, 'Rapid near-field antenna testing via arrays of modulated scattering probes', *IEEE Trans. Antennas & Propagation*, Vol. AP-36, No. 6, pp804-814, June 1988.

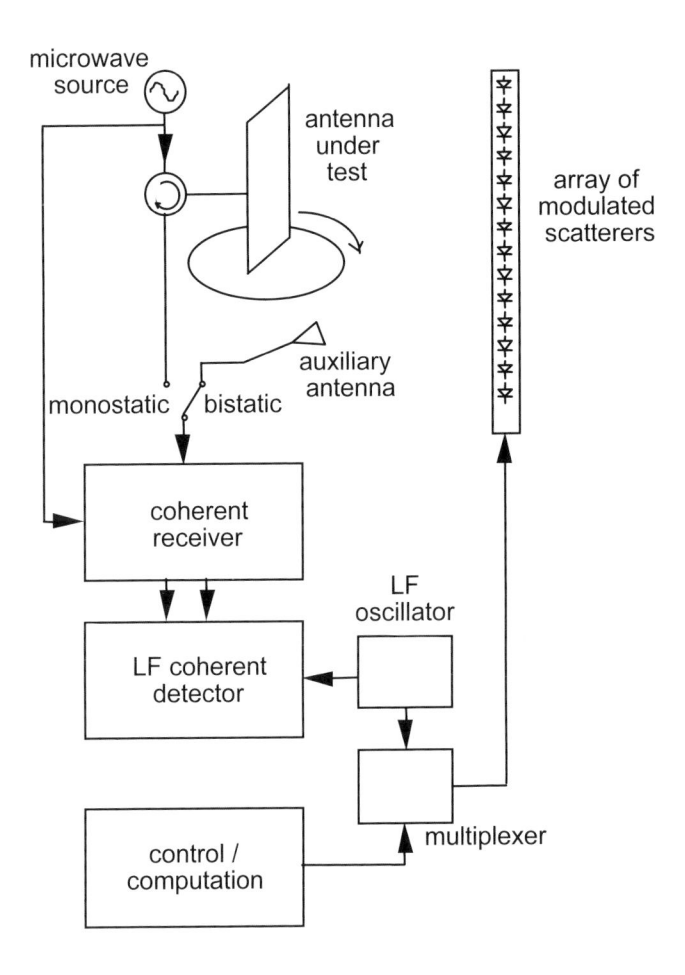

Fig. 15.11 Experimental arrangement of modulated scatterer radiation pattern measurement technique employed by Bolomey *et al.* (adapted from reference 8).

Another implementation of this type of technique uses the scattering from a wire, which is moved across the aperture of the antenna under test[9]. It can be shown that the change in reflection coefficient measured at the antenna feed is proportional to the square of the aperture field at the position of the wire, integrated along the wire.

[9] Calazans, E.T., Griffiths, H.D., Cullen, A.L., Davies, D.E.N. and Benjamin, R., 'Antenna radiation pattern measurement using a near-field wire scattering technique'; *IEE Proc. Microwaves, Antennas and Propagation*, Vol.145, No.3, pp263–267, June 1998.

The change in reflection coefficient (amplitude and phase) is therefore measured as a function of wire position, and the square root taken, giving a quantity directly proportional to the aperture field. The Fourier Transform of the aperture field yields the far-field radiation pattern directly. The theory is only rigorously valid for antennas whose aperture field (and hence radiation pattern) is separable - that is, expressible as the product of an x-variation and a y-variation, but this is true for many practical types of antenna. Fig. 15.12 shows a comparison of the radiation pattern of a 2×2 array of open-ended waveguides measured by this technique, and on a conventional far-field range.

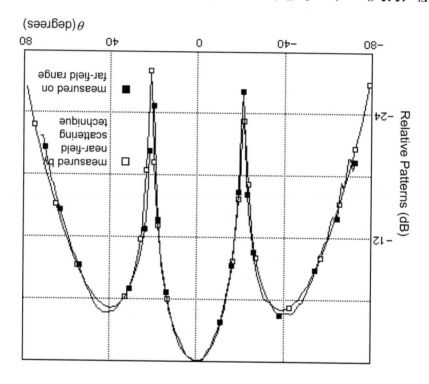

Fig. 15.12 Comparison of radiation patterns of open-ended waveguide array, computed using the wire-scattering technique (clear squares) and measured on a conventional far-field range (dark squares).

As a final comment, there are some very simple techniques that can be used with scanning radar antennas. If a probe antenna is used in the far-field, connected to a spectrum analyser in 'zero-scan' mode, tuned to the radar carrier frequency (and with IF bandwidth matched to the radar pulse length), then the time-domain sweep on the spectrum analyser screen will trace out the antenna radiation pattern (cf

Figure 13.6). The sweep speed should be adjusted to correspond to the antenna rotation rate.

15.3.6 Polarization

The polarization properties of an antenna may be measured by rotation of the source antenna. For a linearly polarized antenna, it is the co-polar and cross-polar patterns that are of interest, and these may be obtained by measuring the pattern with the source antenna aligned with the intended polarization of the antenna under test, and then repeating the measurement with the orthogonal source antenna polarization. With an elliptically-polarized antenna, the axial ratio can be measured by rotating the source antenna at a rapid rate compared with the rotation rate of the turntable. This modulates the pattern, and the peak-to-peak value of the modulation determines the axial ratio of the antenna. A pattern produced in this manner is shown in Fig. 15.13.

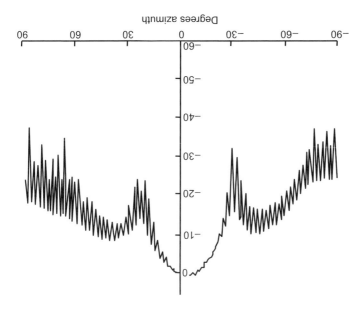

Fig. 15.13 Measurement of an elliptically-polarized antenna using a rotating linearly-polarized feed antenna. The peak-to-peak value of the modulation shows the axial ratio of the antenna under test (courtesy of Dr P. Brennan).

It should be remembered that, when measuring the gain of a circularly-polarized antenna using a linearly-polarized source antenna, the 3 dB polarization loss has to be taken into account.

15.4 TIME-DOMAIN GATING

15.4.1 Principles

A technique developed in recent years is to measure the antenna pattern with the equivalent of a short pulse, so that unwanted reflections from the chamber walls, floor and ceiling can easily be distinguished and 'gated out', thereby removing their effect[10]. In practice the time-domain response is obtained by making measurements sequentially at several frequencies and then Fourier transforming from the frequency domain into the time domain. The gating operation is then performed, and the gated signal transformed back into the frequency domain, giving 'cleaned-up' versions of the responses at the original frequencies (Fig. 15.14). The computation is made easy by routines such as those available on the Agilent HP8510 network analyser.

Denoting the measurements of transmission coefficient at a set of regularly-spaced frequencies $f_0, f_1, \cdots f_{N-1}$ by $a_0, a_1, \cdots a_{N-1}$, the samples A_k of the time-domain response are obtained by Discrete Fourier Transform of the samples a_n

$$A_k = \frac{1}{N} \sum_{n=0}^{n=N-1} a_n \exp\left[j(2\pi/N)nk\right] \qquad (15.11)$$

The resolution in the time domain is approximately equal to the reciprocal of the total bandwidth over which the measurements are made. The wanted response can then be gated out, using a suitably-shaped window function $W(k)$, and then the inverse DFT computed

$$a'_n = \frac{1}{N} \sum_{n=0}^{n=N-1} A_k W(k) \exp\left[-j(2\pi/N)nk\right] \qquad (15.12)$$

10 Chaloupka, H., Galka, M. and Schlendermann, A., "Determination of antenna radiation pattern from frequency-domain measurements in reflecting environment", *Electronics Letters*, Vol. 15, No. 17, pp512-513, August 1979.

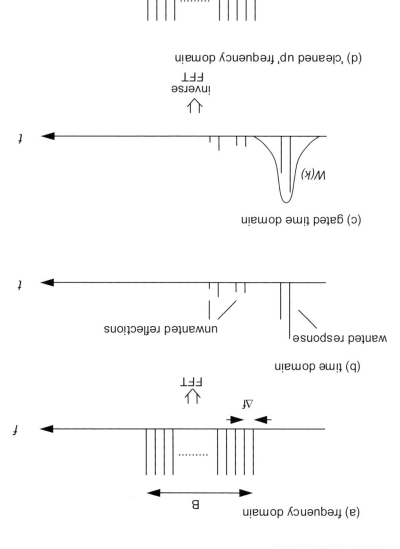

Fig. 15.14 Principle of time-domain gating.

(a) frequency domain

B

Δf

FFT

(b) time domain

wanted response

unwanted reflections

(c) gated time domain

W(k)

inverse
FFT

(d) 'cleaned up' frequency domain

15.4.2 Limitations

The technique has to be used with a certain amount of care. Because it makes use of measurements at other frequencies to 'clean up' the response at a single frequency, it should be used with caution with antennas whose properties are a strong function of frequency over the band implied by the range resolution. On the other hand, if the antenna is actually to be used over an instantaneously-wide band of frequencies (such as a radar using a short pulse), then arguably it is the frequency-smoothed radiation pattern that is relevant. The effect can be thought of in two (equivalent) ways. Firstly, a radiation pattern will, in general, be a function of frequency so that as the frequency is increased the main beam narrows and the sidelobe structure 'telescopes' inward. The radiation pattern measured using time-domain gating is thus an average of the patterns over a bandwidth of $1/\Delta t$, where Δt is the gate width. The averaging effect will be greatest at angles close to $\pm 90°$, tending to 'smear out' the sidelobe structure (Fig. 15.15a). Alternatively, the measurement can be considered as being made by a short pulse. Once the projected aperture of the antenna is less than the spatial extent of the pulse, then the pulse cannot excite the whole antenna simultaneously, and the sidelobe structure 'smears out' (Fig. 15.15b). Again, this effect will be greatest at angles close to $\pm 90°$.

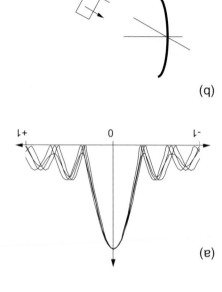

Fig. 15.15 Time-domain gating is equivalent to measurement of antenna radiation patterns over a bandwidth of $1/\Delta t$: (a) the effective pattern is an average over the bandwidth; (b) the effect can also be considered in terms of a short pulse incident at an off-boresight angle.

15.5 ANTENNA NOISE TEMPERATURE AND G/T

15.5.1 Measurement of antenna noise temperature

The noise temperature T_A of an antenna was defined in §4.3 as the average, weighted by the gain, of the temperatures surrounding the antenna

$$T_A = \frac{1}{4\pi} \int^{(4\pi)} T(\mathbf{n}) \, G(\mathbf{n}) \, d\Omega \qquad (15.13)$$

Clearly this depends both on the orientation of the antenna and the environment in which it is placed. For this reason, a measurement of antenna temperature must be carried out *in situ*, if the radiation pattern $G(\mathbf{n})$ is accurately known, and the function $T(\mathbf{u})$ is known or can accurately be approximated (for example, 290 K below the horizon and an appropriate value of sky temperature, according to the frequency, above the horizon), then the integration of equation (15.13) can be performed and T_A determined.

Alternatively, if the receiver noise temperature T_R is known (which will usually be the case), the receiver noise output with the antenna connected can be compared with that when a 50 Ω load (at 290 K) is connected to the receiver input. Denoting this ratio by Y

$$Y = \frac{k(T_A + T_R)B}{k(290 + T_R)B} \qquad (15.14)$$

where k is Boltzmann's constant and B is the receiver noise bandwidth.

From this

$$T_A = 290Y + T_R(Y - 1) \qquad (15.15)$$

15.5.2 Direct measurement of G/T using solar noise

Probably the simplest way of deriving G/T is to make separate measurements of G and T and to combine them. There is, however, an elegant way of measuring the ratio of gain to system noise temperature directly, using the Sun as a noise source.

Suppose firstly that the antenna is pointed at 'cold' sky. The receiver noise power, referred to the antenna terminal, is given by

$$P_n = kT_{sys}B \qquad (15.16)$$

T_{sys} is the system noise temperature, and includes sky noise, ground noise via the antenna sidelobes and backlobes, and loss prior to the LNA (both as a loss and as a source of noise itself), the LNA noise temperature, and the noise temperature of any subsequent stages, weighted in each case by the inverse of the gain preceding them. P_n is therefore the noise power against which any signal must compete for detection.

Suppose now that the antenna is pointed at the Sun. The receiver noise power, referred to the antenna terminal, is now

$$P'_n = \frac{G\lambda^2}{4\pi} \cdot \Phi B + kT_{sys}B \qquad (15.17)$$

where G is the antenna gain at boresight and Φ is the solar flux, in $W/m^2/Hz$. The ratio of these noise powers may be measured (both subject to the same receiver gain), giving

$$\frac{P'_n}{P_n} = \frac{G\lambda^2\Phi B/4\pi + kT_{sys}B}{kT_{sys}B} = \frac{G\lambda^2\Phi B/4\pi}{kT_{sys}B} + 1 \qquad (15.18)$$

from which the G/T ratio is given directly by

$$\boxed{\frac{G}{T_{sys}} = \frac{4\pi k}{\lambda^2\Phi}\left[\frac{P'_n}{P_n} - 1\right]} \qquad (15.19)$$

The solar flux Φ is not actually constant, and three components can be distinguished: (i) a constant component (the so-called 'quiet Sun'); (ii) a slowly-varying component, varying on timescales of the order of a week; and (iii) sporadic radioemission (bursts), which last from a few seconds to a few minutes. Φ is expressed in 'solar flux units', where one solar flux unit is equivalent to 10^{-22} $W/m^2/Hz$, and values range from below 50 to over 300, a total variability of the order of 8 dB. It is also a fairly strong function of frequency, though an interpolation formula has been derived[11]:

---

[11] Harris, J.M., 'Measurement of phased array gain-to-temperature ratio using the solar method', *Proc. XI Antenna Measurement Techniques Association Symposium, Monterey*, pp.4.26-4.31, 9-13 October 1989.

$$\Phi = \Phi_2 + (\Phi_2 - \Phi_1)\left[\frac{\sqrt{\frac{f}{f_2}} - 1}{1 - \sqrt{\frac{f_1}{f_2}}}\right]$$

(15.20)

where Φ is the desired solar flux density at frequency f, and Φ_1 and Φ_2 are the solar flux densities at frequencies f_1 and f_2 respectively.

Measurements of Φ are made at a frequency of 2.8 GHz each day at approximately 1700 UT by the Penticton Radio Observatory, British Columbia, Canada, and the results are available from various sources[12].

The solar flux will be subject to any atmospheric attenuation due to water vapour and precipitation, which under extreme conditions might be as much as 10 dB at microwave frequencies. Obviously it is best to perform the measurement on a clear, dry day when such attenuation can be ignored.

With most types of antenna likely to be measured by this method, the angular extent of the Sun (~ 0.5°) will be significant compared with the antenna beamwidth, so the solar flux is not received with equal gain across the whole of the antenna beam. The value used for the solar flux must be corrected to account for this, and the correction factor is given by

$$L_b = \frac{1}{1 - 0.163\left[\left(\frac{\phi_s}{\phi_a}\right)^2 + \left(\frac{\theta_s}{\theta_a}\right)^2\right]}$$

(15.21)

where $\phi_s = \theta_s$ = angular extent of the Sun, and ϕ_a and θ_a are the −3 dB azimuth and elevation beamwidths (respectively) of the antenna under consideration.

It is also possible to use this technique with 'radio stars', such as Cassiopeia-A or Cygnus-A. These are much weaker but more constant, and are a good approximation to a point source. An accuracy of ±0.3 dB has been claimed for a measurement using Cassiopeia-A of the G/T_{sys} of a 10-metre earth station.

15.6 IMPEDANCE AND BANDWIDTH

A vector network analyser may be used to measure the complex reflection coefficient ρ at the antenna feed, from which the complex antenna impedance may be derived. The antenna must be measured in a situation that is representative of its

---

12 See, for example, http://www.drao.nrc.ca/icarus

operational use for this, as with any other measurement, since any structure close to the antenna could modify its impedance. In many cases only the magnitude of the mismatch is of interest, in which case the return loss, $1/|\rho|^2$, is measured. The voltage standing wave ratio (*VSWR*) and the reflection coefficient are related by

$$VSWR = \frac{1 + |\rho|}{1 - |\rho|} \qquad (15.22)$$

from which the return loss is given by

$$\frac{1}{|\rho|^2} = \left(\frac{VSWR + 1}{VSWR - 1}\right)^2 \qquad (15.23)$$

The transmission coefficient is $|\tau|^2 = 1 - |\rho|^2$, so that the VSWR loss is

$$\boxed{|\tau|^2 = \frac{4\ VSWR}{(VSWR + 1)^2}} \qquad (15.24)$$

The variation of return loss and VSWR loss as a function of VSWR is shown in Table 15.1. Evidently, a high value of return loss is desirable for good matching and low VSWR loss. A device might be considered to be well matched if it has a VSWR of 2 or less, corresponding to a return loss of at least 9.5 dB and a VSWR loss of no more than 0.51 dB.

VSWR	$|\rho|^2$	$|\tau|^2$	Return loss (dB)	VSWR loss (dB)
1	0	1	∞	0
1.5	0.04	0.96	14	0.18
2	0.11	0.89	9.5	0.51
3	0.25	0.75	6	1.25
5	0.44	0.56	3.5	2.6

Table 15.1 Return loss and VSWR loss as a function of VSWR.

The bandwidth of an antenna is often limited by its input impedance. The 3 dB points are the frequencies for which the return loss (and VSWR loss) are 3 dB, corresponding to a VSWR of 5.83.

15.7 MEASUREMENTS OF CELLULAR RADIO HANDSET ANTENNAS

In recent years it has become important to be able to make reliable measurements of the radiation from mobile phone handset antennas, both to quantify the effect of any potential radiation hazard to the user, and to measure the directional properties. The radiation is significantly affected by the presence of the user's head – indeed, in some present models, more than half the radiated power may be absorbed in the user's head. This is most usually taken into account by using a 'phantom', which is a physical model of the human head with equivalent dielectric properties. Fig. 15.16 shows a typical phantom of this kind, which is filled with liquid.

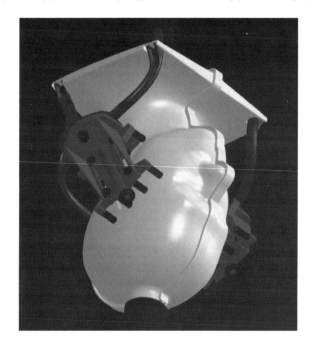

Fig. 15.16 A phantom used for measurements of handset antennas. (photograph: Schmidt and Partner Engineering AG).

15.7.1 Specific Absorption Rate

The power absorbed in materials exposed to radiation is expressed in Watts per kilogram by means of the specific absorption rate (SAR), defined as follows:

$$\text{SAR} = \frac{dP_{dep}}{dm} = \frac{\sigma |E|^2}{2\rho} \qquad (15.25)$$

where ρ (kg/m^3) is the tissue density, σ (S/m) is the tissue conductivity and E is the electric field within the tissue. The recommended safe SAR values are internationally agreed and depend on a variety of factors[13] relating to how the phone is used; as a guide, a SAR of about 4 W/kg produces a temperature rise of approximately 1°C in human tissue. For mobile phones the European standard is a maximum of 2 W/kg averaged over 10g of tissue, while in the USA the value is 1.6 W/kg averaged over 1g of tissue.

Fig. 15.17 The DASY equipment (photograph: Schmidt and Partner Engineering AG).

13 Fujimoto, K. and James, J.R., *Mobile Antenna Systems Handbook*, Artech House, 2nd edition, pp334-360, 2001.

Extensive simulation work has been carried out globally using anatomical head models to investigate SAR together with the power level contours in the brain, but most confidence is placed in field probe measurements within a tissue phantom using an actual handset. One such instrument for SAR measurement is the DASY (Dosimetric Assessment SYstem) equipment[14], manufactured by Schmidt and Partner Engineering AG in Switzerland, and shown in Fig. 15.17. This uses a probe to measure the electric field within the phantom, and hence with a knowledge of the dielectric properties the SAR can be computed from eqn. (15.25).

15.7.2 Reverberation chambers

The problem of measurement of handsets in a realistic propagation environment is also difficult. In contrast to an anechoic chamber, the requirement is to set up a realistic fading environment (see §14.7.11) in respect of the fading statistics both in amplitude and polarization.

For measurement purposes the wave polarization can be resolved into vertical polarization and horizontal polarization so in moving the mobile handset along a typical urban route the mean powers P_v and P_h respectively associated with the two polarized components can be measured. The handset antenna will not be able to capture all the polarized power and $P_r < (P_v + P_h)$, where P_r is the mean handset power that would have been received over the same route, in the same time period. The antenna performance is defined[15] in terms of its Mean Effective Gain (MEG):

$$\text{MEG} = \frac{P_r}{(P_v + P_h)}$$

$$= \int_0^{2\pi} \int_0^\pi \left[\frac{P_v}{(P_v + P_h)} G_\theta(\theta,\phi) P_\theta(\theta,\phi) + \frac{P_h}{(P_v + P_h)} G_\phi(\theta,\phi) P_\phi(\theta,\phi) \right] \sin\theta \, d\theta \, d\phi \quad (15.26)$$

where (θ,ϕ) are spherical angular coordinates, $G_\theta(\theta,\phi)$ and $G_\phi(\theta,\phi)$ are the θ and ϕ components of the power patterns respectively, and $P_\theta(\theta,\phi)$ and $P_\phi(\theta,\phi)$ are the θ and ϕ components of the angular density functions of the incoming plane waves, respectively.

The MEG measurement process has proved useful for designing mobile antennas that give optimized performance in realistic propagation scenarios. However, the recent trend towards extreme handset compactness and internal integrated antennas (see §6.5) gives little scope for optimizing the antenna in isolation from the handset

---

[14] http://www.dasy4.com

[15] Fujimoto, K. and James, J.R., *Mobile Antenna Systems Handbook*, Artech House, 2nd edition, pp63-68, 2001.

itself. The use of reverberation chambers to test antennas in a Rayleigh fading environment is well established, but its recent application to mobile handset testing appears to offer manufacturers an efficient, low-cost method of assessing handset performance. The reverberation chamber consists of a large metal-walled cavity capable of supporting many modes which are perturbed by rotating reflectors within the excited chamber. The mode stirring is sufficiently random to create a Rayleigh distributed fading characteristic between the transmit and receive antennas within the chamber.

A demonstration chamber of this kind has recently been described[16], which has many attractive features. For example, a phantom tissue head model can be placed in close proximity to the handset under test to enable the real operating conditions to be accurately reproduced. The chamber can also be used to test the diversity functions and MIMO systems (§14.7.11). The installation costs and operating times are low, and compatible with handset manufacturing requirements.

---

[16] Kildal, P.S., 'Characterisation of small antennas and active mobile terminals in Rayleigh fading using reverberation chamber', *Proc. Loughborough Antennas and Propagation Conference*, pp234-239, April 2005.

FURTHER READING

1. Arai, H., *Measurement of Mobile Antenna Systems*, Artech House, 2005.

2. Bolomey, J.-Ch. and Gardiol, F.E. *Engineering Applications of the Modulated Scatterer Technique*, Artech House, 2002.

3. Evans, G.E., *Antenna Measurement Techniques*, Artech House, Dedham, MA, 1990.

4. Hansen, J.E. (ed), *Spherical Near-field Antenna Measurements*, Peter Peregrinus, Stevenage, 1988.

5. Olver, A.D., 'Compact antenna test ranges', *Proc. 7th IEE Intl. Conference on Antennas & Propagation*, York, IEE Conf. Publ. No. 333 Part 1, pp99-108, April 1991.

6. Tuley, M.T., 'Radar absorbing materials', Chapter 8 in *Radar Cross Section* (second edition), E.F. Knott, J.F. Shaeffer and M.T. Tuley, Artech House, Dedham, MA, 1992.

7. Wang, J.J.H., 'An examination of the theory and practice of planar near-field measurement', *IEEE Trans. Antennas & Propagation*, vol. AP-36, No. 6, pp746-753, June 1988.

8. Special Issue of *IEEE Trans. Antennas & Propagation* on Near-field Scanning Techniques, Vol. AP-36, No. 6, June 1988.

9. IEEE Standard Test Procedures for Antennas: IEEE Std 149-1979, IEEE/Wiley Interscience, New York, 1979.

10. There is a series of conferences on antenna measurements, organized by the Antenna Measurement Techniques Association (*AMTA*), and held each year in the USA. See http://www.amta.org.

EXERCISES

15.1 *Purcell's method*

An antenna, operating at 10 GHz, faces a large conducting sheet at a distance of 2 m. Under these conditions the antenna has a reflection coefficient of magnitude −15 dB. Assuming the antenna is perfectly matched, what is its gain?

ANSWER 24.7 dBi

15.2 *Three antenna measurement*

Three antennas, of identical and aligned polarizations, are to be measured in an anechoic chamber of length 4 m at a frequency of 10.6 GHz. The power supplied at the source end is +10 dBm and the power measured at the test end is −20 dBm using antennas 1 and 2, −31 dBm using antennas 2 and 3 and −24 dBm using antennas 1 and 3. What is the gain of each antenna? Given that the antennas are parabolic reflectors with a radiation efficiency of 60%, check that the Rayleigh criterion is satisfied for this measurement.

ANSWER G_1 = 21 dBi; G_2 = 14 dBi; G_3 = 10 dBi

15.3 *Rayleigh distance*

The antenna on the Synthetic Aperture Radar carried by the European Space Agency's ERS-1 satellite is approximately 10 m in length. The radar operates at a frequency of 5.3 GHz. If the radiation pattern of this antenna is to be measured on a far-field range, how long must the range be? What do you conclude?

ANSWER 3.53 km

15.4 *Solar method of G/T measurement*

The G/T of a satellite Earth Station receiving system is to be estimated using the solar flux method (§15.5.2). The solar flux is quoted from the website as 111 solar flux units. The beamwidth of the antenna, in both azimuth and elevation planes, is

0.25°, and the difference between the receiver noise output when the system points at the Sun, compared to cold sky is 24.5 dB. What is the G/T of the system?

ANSWER 43.1 dB K^{-1}.